AF595003

Synthetic Peptides and Peptidomimetics: From Basic Science to Biomedical Applications

Synthetic Peptides and Peptidomimetics: From Basic Science to Biomedical Applications

Editors

Nunzianna Doti
Menotti Ruvo

MDPI • Basel • Beijing • Wuhan • Barcelona • Belgrade • Manchester • Tokyo • Cluj • Tianjin

Editors

Nunzianna Doti
Institute of Biostructures and Bioimaging (IBB)
CNR
Naples
Italy

Menotti Ruvo
Institute of Biostructures and Bioimaging (IBB)
CNR
Naples
Italy

Editorial Office
MDPI
St. Alban-Anlage 66
4052 Basel, Switzerland

This is a reprint of articles from the Special Issue published online in the open access journal *International Journal of Molecular Sciences* (ISSN 1422-0067) (available at: www.mdpi.com/journal/ijms/special_issues/synthetic_peptides).

For citation purposes, cite each article independently as indicated on the article page online and as indicated below:

LastName, A.A.; LastName, B.B.; LastName, C.C. Article Title. *Journal Name* **Year**, *Volume Number*, Page Range.

ISBN 978-3-0365-4396-3 (Hbk)
ISBN 978-3-0365-4395-6 (PDF)

Contents

Preface to "Synthetic Peptides and Peptidomimetics: From Basic Science to Biomedical Applications"

Synthetic peptides are a very appealing class of compounds as both basic and applied science research tools. Furthermore, the high affinity and specificity of peptides towards biological targets, together with their advantageous pharmacological properties, such as poor immunogenicity and low toxicity, has greatly encouraged their development as therapeutic and diagnostic agents in the clinical setting. More than 60 peptide drugs have reached the market and several hundred new therapeutic peptides are in preclinical and clinical development. However, to date, this class of molecules represents only 2% of the world drug market due to their low in vivo stability, short half-lives, poor cell permeability and low oral bioavailability. Many approaches have been developed and are continually being explored to improve peptide stability and pharmacological properties, while maintaining biological potency and selectivity and avoiding toxicity. For example, cyclization, the protection of N- and C-termini, side chain or backbone modifications of natural structures, replacement with D or non-natural amino acids, and the addition of membrane permeability elements are commonly used strategies to improve the in vivo stability and cellular permeability of this class of molecules. Furthermore, the conjugation of peptides with large polymers, fusion with long-lived plasma proteins and lipidation have also been exploited to prolong plasma half-life and increase oral bioavailability, significantly expanding the applicability of peptides as effective drugs.

Trying to cover all the progresses made in this field, this Special Issue, entitled "Synthetic Peptides and Peptidomimetics: From Basic Science to Biomedical Applications", has included both reviews and original research contributions focused on the chemical design and biomedical applications of structurally modified bioactive peptides. In this framework, Doti et al. provide a wide overview on the applications of retro-inverso-modified peptides in anticancer therapies, in immunology, in neurodegenerative diseases, and as antimicrobials, highlighting the benefits and limits of this interesting subclass of molecules as bioactive compounds. Tarvirdipour et al. review the properties of peptides that promote the site-specific localization of nucleic acids and peptide-based nano-assemblies. Romanowska et al. show the chemical design of a set of DNA-binding peptide-based polymers composed of *N*-substituted L-2,3-diaminopropionic acid (DAPEG) residues, and investigate their relative cellular permeability and localization, cytotoxicity and DNA-binding capacity. The experimental results of this study pave the way for the further development of peptide-based nanocarriers. Pandey et al. summarize the current status on the use of peptides as probes both in non-imaging and imaging diagnostic platforms. Moreover, they discuss the applicability of peptide-based diagnostics in deadly diseases, mainly COVID-19 and cancers. In the same research field, Krajcovicova et al. report on the preparation of a new RGD-based radiolabeled peptide targeting $\alpha v\beta 3$ integrin. The binding properties of [68Ga]Ga-DFO-c(RGDyK) towards the $\alpha v\beta 3$ integrin were studied in vitro and in vivo with various techniques, including PET/CT imaging in a mouse tumor model. In addition, Siepe et al. describe the characterization of a set of bioactive peptides conjugated with environment-sensitive labels, such as luciferin and aminoluciferin, which are used to study their interactions with model membranes, SDS micelles, lipopolysaccharide micelles and bacterial cells. The results demonstrate that luciferin and aminoluciferin are environment-sensitive labels with widespread potential applications in the study of peptides interacting with membranes.

The contributions included in this book also provide several examples of peptides employed as protein–protein interaction (PPI) modulators, both to elucidate the molecular mechanisms underlying diseases and to use them as a starting template for developing new potential therapeutics. In this framework, Dvorak et al. focus their attention on the interaction between a voltage-gated Na+ channel (Nav1.6) and fibroblast growth factor 14 (FGF14), which plays a role in the regulation of neuron excitability in the central nervous system. Notably, using an FGF14-derived synthetic peptide, they demonstrate that pharmacologically targeting the FGF14 interaction site on the C-terminal domain (CDT) of Nav1.6 results in a powerful strategy to achieve the selective modulation of the isoform activity. The data also show that, more generally, the interaction of the CTDs of Nav channels with auxiliary proteins are a target candidate for developing new therapies. Conte et al. focus their attention on the interaction between the Apoptosis-Inducing Factor and the Cyclophilic A (AIF/CypA), which mediates neuronal cell death in vivo and in vitro. Using AIF(370-394) as a prototypical inhibitor, they elucidate the role of the complex in SH-SY5Y cells treated with high concentrations of staurosporine, which is a well-known cell model to study Parkinson's Disease (PD). The results obtained highlight the role of the AIF/CypA complex in the pathophysiological mechanisms leading to PD, suggesting the complex as a promising target for developing first-in-class therapeutics to treat this currently incurable disease. Levi et al. propose the targeting of the integrin $\alpha v\beta 3$, which is involved in different stages of cancer progression, metastasis, invasion, and angiogenesis, with a cyclic non-RGD synthetic peptide (ALOS4). This peptide, nine residues, was tested in a subcutaneous xenograft model of A375 human melanoma to evaluate tumor growth, tumor tissue development and the expression of downstream targets of $\alpha v\beta 3$. The stability and toxicity of ALOS4 in mice together with the blood cell profile in healthy mice were also evaluated. The results suggest that ALOS4 is stable in the proposed formulations, presents no overt toxicity risks and is effective in melanoma tumor shrinkage by a mechanism related to $\alpha v\beta 3$ and possibly other mechanisms. Di Micco et al. focus on the main proteases (Mpro) of SARS-CoV2, a protein essential for viral replication, modifying and repurposing the active peptide AT1001 (Larazotide acetate). AT1001 and five derivatives were designed and assayed in vitro for their ability to interfere with Mpro catalytic activity. The data provide useful information for the development of new generations of antiviral agents for treating SARS-CoV-2, which so far lacks selective therapeutic treatments. The review by Vanzolini et al. provides an overview of the peptides that are currently used as antimicrobials and their mechanism of action.

Highly ordered secondary structure motifs, α-helix, β-sheets and turns, are scaffolds for key amino acid residues in protein–protein hot-spots. The development of peptides that adopt conformations suitable for their biological activity is one of the most important goals of protein chemists, not only in relation to the design of PPI inhibitors. In the article of Makura et al., this is addressed by stapling peptides at i,i+1 positions using hydrocarbon linkers introduced by ring-closing metathesis reactions and analyzing their structures through X-ray crystallography. The authors show how their approach is valid for short oligopeptides where stapling is achieved using two adjacent residues. In this field, Kovačević et al. investigated the conformational behavior and antiproliferative activity of peptidomimetics obtained by the conjugation of methyl-1′-aminoferrocene-1-carboxylate with homo- and heterochiral Pro-Ala dipeptides. The results show a promising outcome which could serve for further research and the development of compounds with antitumor activity.

The papers on peptides collected in this Special Issue, which have been proposed and published by groups operating in various parts of the world, show how successful this class of molecules still is, both as model molecules for studying the structure of proteins, and as potential therapeutics and diagnostics, and also as laboratory tools for advanced basic and applied studies. The large scientific community working in this field is very active and productive, and is making the most of the potential and versatility of these molecules to generate increasingly interesting and innovative molecules of therapeutic interest and to understand the fundamental molecular mechanisms of life.

Nunzianna Doti and Menotti Ruvo
Editors

International Journal of
Molecular Sciences

Article

Novel Cell Permeable Polymers of *N*-Substituted L-2,3-Diaminopropionic Acid (DAPEGs) and Cellular Consequences of Their Interactions with Nucleic Acids

Anita Romanowska [1], Katarzyna Węgrzyn [2], Katarzyna Bury [2], Emilia Sikorska [1], Aleksandra Gnatek [1], Agnieszka Piwkowska [1,3], Igor Konieczny [2], Adam Lesner [1] and Magdalena Wysocka [1,*]

1 Faculty of Chemistry, University of Gdansk, Wita Stwosza 63, 80-308 Gdansk, Poland; anita_romanowska@o2.pl (A.R.); emilia.sikorska@ug.edu.pl (E.S.); ognatek@gmail.com (A.G.); apiwkowska@imdik.pan.pl (A.P.); adam.lesner@ug.edu.pl (A.L.)
2 Intercollegiate Faculty of Biotechnology, University of Gdansk Abrahama 58, 80-308 Gdansk, Poland; katarzyna.wegrzyn@biotech.ug.edu.pl (K.W.); katarzyna.bury@biotech.ug.edu.pl (K.B.); igor.konieczny@ug.edu.pl (I.K.)
3 Mossakowski Medical Research Institute, Polish Academy of Sciences, Wita Stwosza 63, 80-308 Gdansk, Poland
* Correspondence: magdalena.wysocka@ug.edu.pl

Abstract: The present study aimed to synthesize novel polycationic polymers composed of *N*-substituted L-2,3-diaminopropionic acid residues (DAPEGs) and investigate their cell permeability, cytotoxicity, and DNA-binding ability. The most efficient cell membrane-penetrating compounds (O2Oc-Dap(GO2)$_n$-O2Oc-NH$_2$, where n = 4, 6, and 8) showed dsDNA binding with a binding constant in the micromolar range (0.3, 3.4, and 0.19 μM, respectively) and were not cytotoxic to HB2 and MDA-MB-231 cells. Selected compounds used in the transfection of a GFP plasmid showed high transfection efficacy and minimal cytotoxicity. Their interaction with plasmid DNA and the increasing length of the main chain of tested compounds strongly influenced the organization and shape of the flower-like nanostructures formed, which were unique for 5/6-FAM-O2Oc-[Dap(GO2)]$_8$-O2Oc-NH$_2$ and typical for large proteins.

Keywords: polymers; peptidomimetics; AFM; transfection; molecular modelling

Citation: Romanowska, A.; Węgrzyn, K.; Bury, K.; Sikorska, E.; Gnatek, A.; Piwkowska, A.; Konieczny, I.; Lesner, A.; Wysocka, M. Novel Cell Permeable Polymers of *N*-Substituted L-2,3-Diaminopropionic Acid (DAPEGs) and Cellular Consequences of Their Interactions with Nucleic Acids. *Int. J. Mol. Sci.* **2021**, 22, 2571. https://doi.org/10.3390/ijms22052571

Academic Editor: Nunzianna Doti

Received: 25 January 2021
Accepted: 27 February 2021
Published: 4 March 2021

1. Introduction

Gene transfection can be defined as the transmission of DNA to regulate or induce specific gene expression in the target cells or organs. This mechanism is very important in biosciences, pharmaceutics, and clinical applications. In the past decades, several transfection systems including viral and non-viral vectors have been developed. DNA chains are negatively charged polymers, repelling each other owing to the intrachain and interchain electrostatic repulsion among fragments, a major drawback of which is low transfection efficiency [1–4]. To enhance the effective transfer of a plasmid or short linear molecule of nucleic acid through the cell membrane, two major barriers need to be overcome. The first problem is the negative charge of DNA molecules, which should be neutralized or masked. The second issue, in some cases, is the size of DNA molecules, especially large vectors or plasmids. There are several reports describing the methodology to cause DNA condensation, which results in the concurrent reduction in size and charge. This strategy mimics the natural processes of genome organization by positively charged proteins or small cationic compounds such as bivalent or multivalent metal cations and polyamines [5,6].

In the late 1980s, a group of positively charged molecules referred to as cell-penetrating peptides (CPPs) was developed by many research groups [7,8]. Owing to their high content of basic amino acid residues (Arg and Lys) and positive charge at physiological pH, CPPs passively diffuse across the lipid bilayer, chiefly owing to endocytosis. Moreover,

CPPs can cross the plasma membrane at low micromolar concentrations in vivo and in vitro without using any receptors and without causing any significant membrane damage [9,10]. Another advantage of using CPPs for the therapeutic delivery of numerous molecules, including DNA, is the lack of toxicity in comparison with other cytoplasmic delivery systems, such as liposomes and polymers [11]. Additionally, the strong positive charge of most CPPs neutralizes the DNA, resulting in its condensation, which is beneficial for cellular delivery [12,13].

Recently, our group reported the synthesis of a novel class of peptidomimetics [14]. Such molecules are synthesized using two building blocks; the beta amino group of diaminopropionic acid was decorated by a functionalized oxa acid (Figure 1). We were able to manipulate the length and properties of the side chain and functional groups, producing novel amino acid mimetics. The deconvolution of a 400-member library composed of *N*-substituted L-2,3-diaminopropionic acid residue (DAPEG) building blocks facilitated the selection of an efficient and selective fluorogenic probe or substrate of neutrophil serine protease **4**. The obtained substrate is cleaved by neutrophil serine protease **4** in an efficient and selective manner [14]. More recently, a fluorogenic probe for a trypsin-like subunit of 20S proteasome, composed of DAPEG building blocks, was synthesized and found to be useful in the diagnostics of bladder cancer [15].

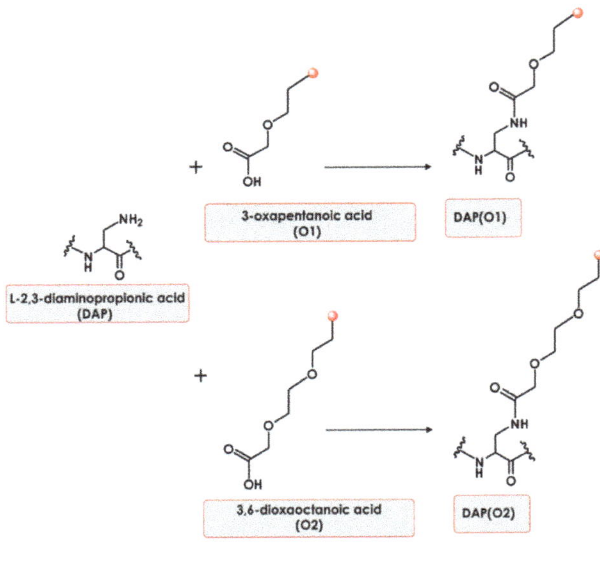

Figure 1. Examples of *N*-substituted L-2,3-diaminopropionic acid residues (DAPEGs).

In the present study, we synthesized homopolymers containing a different number of L-2,3-diaminopropionic acid residues (**2–8**) that were decorated by functionalized oxa acids (see Figure 2) with a variety of side chain groups. Moreover, three compounds containing arginine (Arg) and its analogs (D-arg and homoarginine (Har)) in their structure were synthesized and treated as controls. Thus, nine different molecules were obtained; their N-terminal amino groups were labeled with a 5/6-carboxyfluorescein succinimidyl ester (5/6-FAM) fluorophore, and the molecules were selected using 5/6-carboxytetramethylrhodamine succinimidyl ester (5/6-TAMRA) derivatives. The aim of this study was to synthesise the panel of DNA binding polymers that, in a complex with target DNA, are able to efficiently penetrate the cell membrane allowing gene deliery. To do so, cytotoxicicty studies and DNA binding efficacy, along with cell penetration assay were performed. Based on a combination of the above, the selected compounds were subjected to transfection experiments using a model GFP vector, and the most effective was selected for further structural studies.

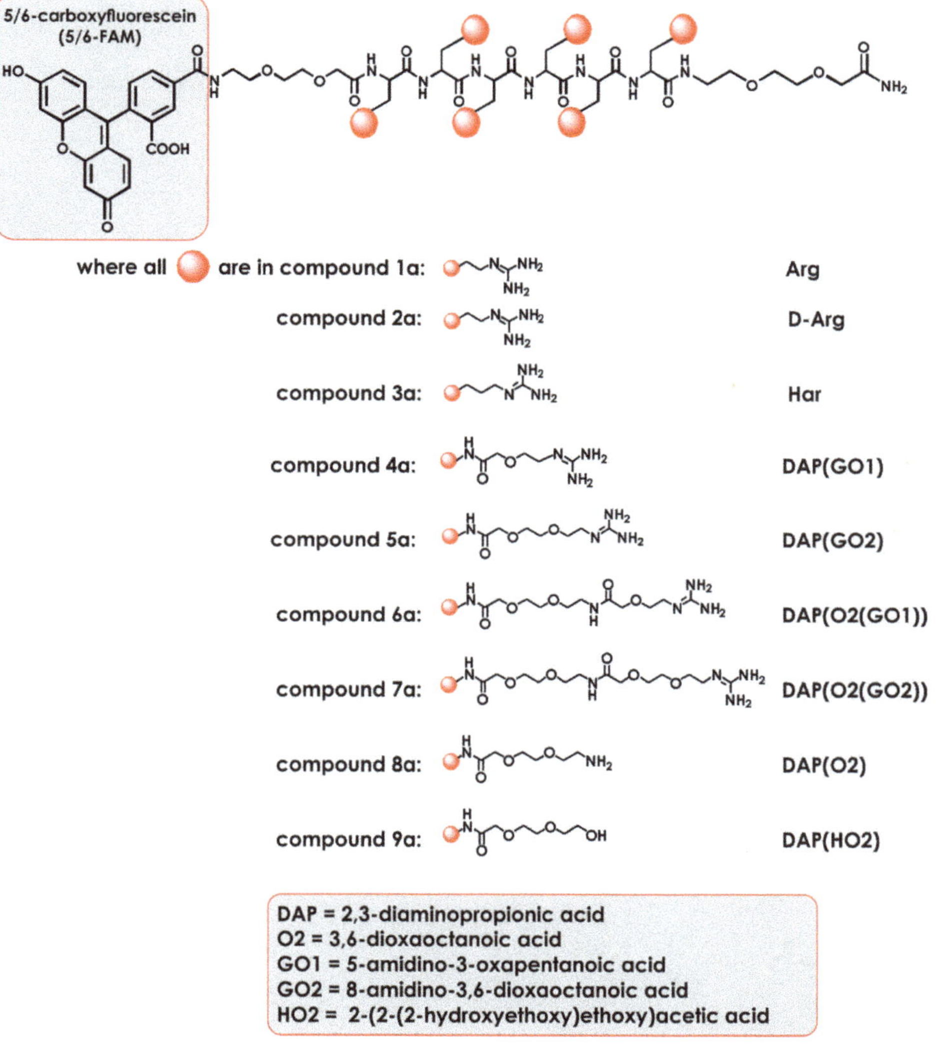

Figure 2. Chemical formulae of compounds from series **1**.

2. Results and Discussion

2.1. Synthesis

In total, nine compounds were synthesized using a synthetic method (see Figure S1) [14]. All synthesized compounds were labeled with a 5/6-FAM fluorophore and two with an N-terminal 5/6-TAMRA group. The physicochemical characteristics of all compounds are listed in Tables 1 and 2.

Table 1. Physicochemical characteristics of series **1** compounds along with their DNA binding constant, cell permeability, and cytotoxicity.

No	Sequence	Retention Time * [min]	Molecular Weight Calculated/ Determined **	Cell Permeablility (Target) ***	Binding Constants Double [μM] # / Single-Stranded DNA [μM] #	Cytotoxicity ##
1	O2Oc(Arg)$_6$-O2Oc-NH$_2$	2.11	1244.5/1245.4	ND	Moderate	10%
1a	5′,6-FAM-O2Oc-(Arg)$_6$-O2Oc-NH$_2$	9.12/9.26	1603.8/1604.4	++ (cytoplasm)	7.4 ± 0.1 / 0.02 ± 0.04 &	10%
2	O2Oc(D-arg)$_6$-O2Oc-NH$_2$	2.12	1244.5/1245.5	ND	Moderate	15%
2a	5′,6-FAM -O2Oc-(D-arg)$_6$-O2Oc-NH$_2$	8.46/8.71	1603.8/1604.8	++ (cytoplasm)	–	15%
3	O2Oc-(Har)$_6$-O2Oc-NH$_2$	2.45	1329.6/1330.4	ND	Weak	10%
3a	5′,6-FAM -O2Oc-(Har)$_6$-O2Oc-NH$_2$	8.55/8.83	1687.9/1689.0	++ (cytoplasm)	–	10%
4	O2Oc-[Dap(GO1)]$_6$-O2Oc-NH$_2$	3.15	1682.8/1683.6	ND	8.8 ± 2.1/ / 1.8 ± 1.1 &&	–
4a	5′,6-FAM -O2Oc-[Dap(GO1)]$_6$-O2Oc-NH$_2$	7.89/8.13	2042.1/2043.0	++ (nucleus)	3.3 ± 2.6/ / 1.7 ± 2.0 &	–
4b	5′,6-TAMRA-O2Oc-[Dap(GO1)]$_6$-O2Oc-NH$_2$	8.17/8.59	2096.2/2097.0	++ (nucleus)	ND	–
5	O2Oc-Dap(GO2)$_6$-O2Oc-NH$_2$	3.40	1947.1/1947.9	ND	1.8 ± 0.9/ / 2.8 ± 1.2 &&	–
5a	5′,6-FAM -O2Oc-[Dap(GO2)]$_6$-O2Oc-NH$_2$	9.10/9.83	2306.4/2307.6	+++ (nucleus)	3.4 ± 3.0/ / 2.3 ± 4.6 &	–
5b	5′,6-TAMRA-O2Oc-[Dap(GO2)]$_6$-O2Oc-NH$_2$	9.82/10.15	2360.6/2361.4	+++ (nucleus)	ND	–
6	O2Oc-[Dap(O2(GO1))]$_6$-O2Oc-NH$_2$	3.55	2554.7/2555.3	ND	ND	–
6a	5′,6-FAM-O2Oc-[Dap(O2(GO1))]$_6$-O2Oc-NH$_2$	12.99/13.43	2913.0/2913.8	–	ND	–
7	O2Oc-[Dap(O2(GO2))]$_6$-O2Oc-NH$_2$	4.09	2819.1/2819.9	ND	ND	–
7a	5′,6-FAM-O2Oc-[Dap(O2(GO2))]$_6$-O2Oc-NH$_2$	14.35/15.03	3177.3/3178.3	–	ND	–
8	O2Oc-[Dap(O2)]$_6$-O2Oc-NH$_2$	3.45	1694.9/1696.0	ND	Weak	–
8a	5′,6-FAM-O2Oc-[Dap(O2)]$_6$-O2Oc-NH$_2$	9.12/9.38	2053.2/2054.1	–	ND	–
9	O2Oc-[Dap(HO2)]$_6$-O2Oc-NH$_2$	3.71	1700.8/1701.5	ND	Weak	–
9a	5′,6-FAM-O2Oc-[Dap(HO2)]$_6$-O2Oc-NH$_2$	9.45/9.78	2059.1/2060.4	–	ND	–

Table 1. *Cont.*

No	Sequence	Retention Time * [min]	Molecular Weight Calculated/ Determined **	Cell Permeablility (Target) ***	Binding Constants Double [μM] # / Single-Stranded DNA [μM] #	Cytotoxicity ##
10	5′,6-FAM	11.21	376.3	–	639 ± 703/ 399 ± 497 &	–
11	5′,6-TAMRA	12.17	431.5	–	ND	–

* Ultra performance liquid chromatography UPLC analysis (Nexera X2 LC-30AD (Shimadzu, Japan)) equipped with a Phenomenex column (150 × 2.1 mm), with a grain size of 1.7 μm (peptide XB-C18) equipped with a UV-Vis detector and a fluorescence detector. A linear gradient from **2** to 80% B within 15 min was applied (A: 0.1% trifluoroacetic acid; B: 80% acetonitrile in A); for fluorescent-labeled compounds, two retention times correspond to two diasereoisomers being provided; ** HR MALDI analysis with 2,5-dihydroxybenzoic acid as a matrix; *** confocal fluorescence microscope Olympus I51 (Olympus, Japan); # binding constant determined with single- or double-stranded DNA using microscale thermophoresis (MST) (labeled compounds) or surface plasmon resonance (SPR) (unlabeled compounds); ## MTT assay for MDA-MB-231 or HB-2 cell lines (numbers indicate the percentage of dead cells) performed at the greatest concentration of 50 μg/mL; & MST binding constants; && SPR binding constants; ND: not determined; "-" means no cytotoxic effect was observed; "moderate" or "weak" DNA binding was aribitrary set based on the SPR experiment (Figure S3); weak: compounds below 10 RU, moderate: up to 60 RFU.

Table 2. Physicochemical characteristics of series **2** compounds, including their DNA binding constant, cell permeability, and cytotoxicity.

No	Sequence	Retention Time * [min]	Molecular Weight Calculated/ Determined **	Cell Permeablility (Target) ***	Binding Constant Double DNA [μM] # / Single-Stranded DNA [μM] #	Cytotoxicity ##
12	O2Oc-[Dap(GO2)]$_2$-O2Oc-NH$_2$	1.94	853.9/854.6	ND	weak	–
12a	5′,6-FAM-O2Oc-[Dap(GO2)]$_2$-O2Oc-NH$_2$	10.11/10.43	1213.2/1213.5	–	–	–
13	O2Oc-[Dap(GO2)]$_4$-O2Oc-NH$_2$	2.09	1400.5/1401.2	ND	–	–
13a	5′,6-FAM-O2Oc-[Dap(GO2)]$_4$-O2Oc-NH$_2$	9.78/10.01	1759.8/1759.8	++ (cytoplasm)	0.3 ± 0.1/ 0.5 ± 0.2 &	–
5	O2Oc-[Dap(GO2)]$_6$-O2Oc-NH$_2$	3.40	1948.1/1947.1	ND	1.8 ± 0.9/ 2.8 ± 1.2 &&	–
5a	5′,6-FAM-O2Oc-[Dap(GO2)]$_6$-O2Oc-NH$_2$	9.10/9.83	2306.4/2307.6	+++ (nucleus)	3.4 ± 3.0/ 2.3 ± 4.6 &	–
5b	5,6-TAMRA-O2Oc-[Dap(GO2)]$_6$-O2Oc-NH$_2$	9.82/10.15	2360.6/2361.4	+++ (nucleus)	ND	–
14	O2Oc-[Dap(GO2)]$_8$-O2Oc-NH$_2$	2.97	2493.7/2493.5	ND	ND	10%
14a	5′,6-FAM-O2Oc-[Dap(GO2)]$_8$-O2Oc-NH$_2$	8.2/8.8	2852.0/2853.7	+++ (nucleus)	0.2 ± 0.1/ 3.3 ± 2.34 &	15%
14b	5′,6-TAMRA-O2Oc-[Dap(GO2)]$_8$-O2Oc-NH$_2$	8.2/8.5	2907.1/2908.5	+++ (nucleus)	ND	15%

* UPLC analysis (Nexera X2 LC-30AD (Schimadzu, Japan)) equipped with a Phenomenex column (150 × 2.1 mm), with a grain size of 1.7 μm (peptide XB-C18) equipped with a UV-Vis detector and a fluorescence detector. A linear gradient from 2 to 80% B within 15 min was applied (A: 0.1% trifluoroacetic acid; B: 80% acetonitrile in A); for fluorescent-labeled compounds, two retention times correspond to two diasereoisomers being provided; ** HR MALDI analysis with 2,5-dihydroxybenzoic acid as a matrix; *** confocal fluorescent microscope Olympus IX51 fluorescence microscope (Olympus, Japan); # binding constant determined with single strand ss or double strand dsDNA using MST (labeled compounds) or SPR (unlabeled compounds); ## MTT assay for MDA-MB-231 or HB-2 cell lines; & MST binding constants; && SPR binding constants; ND: not determined; "-" means no cytotoxic effect was observed; "weak" DNA binding was aribitrarily set based on the SPR experiment (Figure S3).

2.2. Cell Permeability and Localization

Preliminary cell permeability experiments facilitated the identification of three groups of compounds. The first group was visible in the nuclei of HB2 or MDA-MB-231 cells incubated with the compounds at micromolar level (10 μM) (Figure 3). The first group includes compound **4** and its derivatives (**4a** and **4b**) and the family of compound **5** (**5a** and **5b**). Compounds **4** and **5** are similar in structure and are composed of six residues of diaminopropionic acid (Dap) decorated with side chains of different lengths, terminating with the same highly positive guanidine moiety. However, compounds **5a** and **5b** showed a significantly higher intensity of nuclear penetration/accumulation than compounds **4a** and **4b**. Both compounds display unique cellular localization. The second group of compounds (**1a**, **2a**, and **3a**) are able to cross the cell membrane, and their presence is partially visible in the cytoplasm in the form of granules (see Figure 4). The compounds in the third group (**6a**, **7a**, **8a** and **9a**) were not visible by fluorescence microscopy indicating that they did not cross the cell membrane. We believe that such effective nuclear localisation of the first group of tested compounds is due to the effective mimicking of the nuclear localisation sequence (NLS). Such sequence consists of several positively charged lysines or arginines present in the protein structure. Additionally, side chain length of the polymer molecules seems to be a crucial factor that is optimal for compounds belonging to first group of polymers.

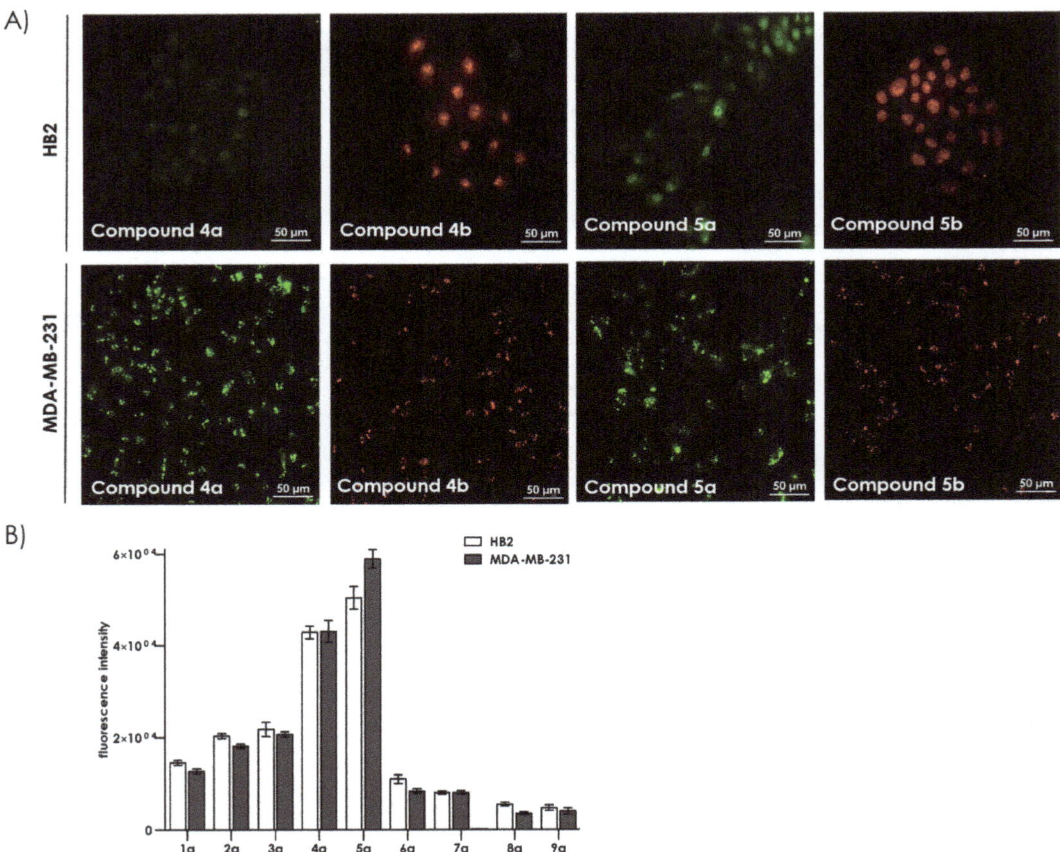

Figure 3. (**A**) Cell membrane permeability in HB2 and MDA-MB-231 cells incubated with 10 μM of compounds **4a**, **4b**, **5a**, and **5b** for 24 h. Magnification 20×. (**B**) Fluorescence intensity in HB2 and MDA-MB-231 cells incubated with 10 μM of compounds **1a**–**9a**.

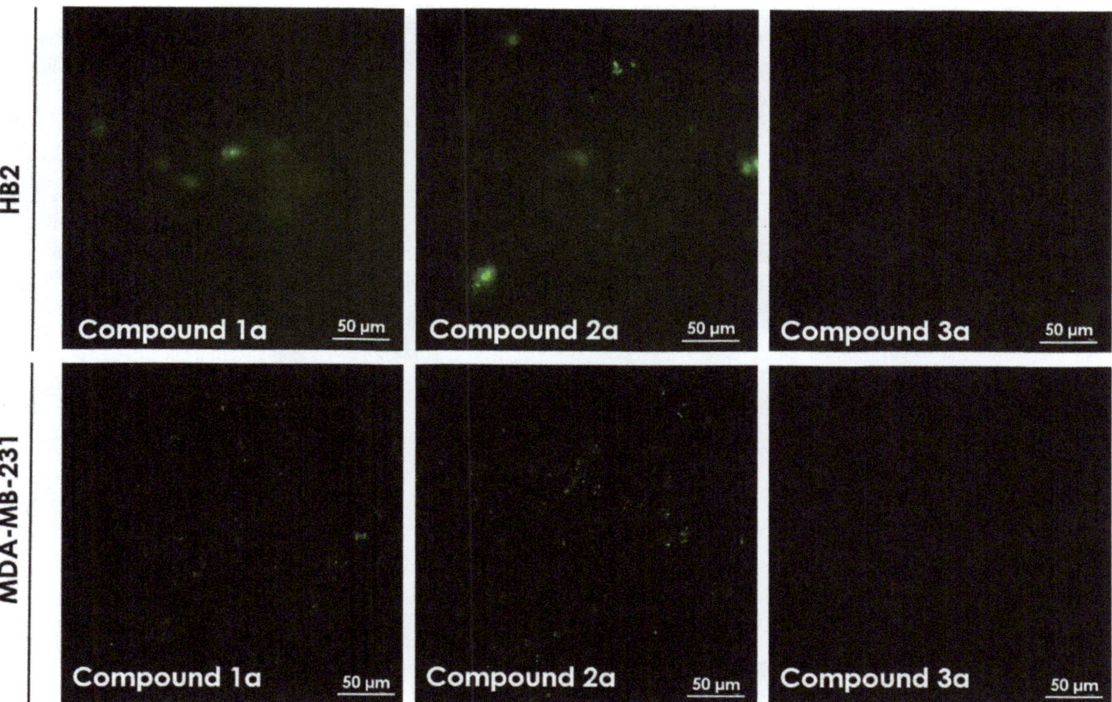

Figure 4. Cell membrane permeability in HB2 and MDA-MB-231 cells incubated with 10 μM of compounds **1a**, **2a**, and **3a** for 24 h. Magnification 20×.

Notably, most of these compounds did not show cytotoxicity at the concentration used in the cell experiments (10 μM). However, compounds **1a**, **2a**, and **3a** were slightly cytotoxic; the number of cells incubated with these compounds was slightly more reduced than those in control (Figure S2).

Next, we aimed to explore the effect of active processes (such as endocytosis) and passive, energy-independent mechanisms (such as pore formation) on the cellular internalization of compounds **4b** and **5b**. Incubation of HB2 or MDA-MB-231 cells cultured with compounds **4b** and **5b** at 4 °C, at which all energy-dependent uptake processes are significantly reduced, resulted in an almost complete inhibition of **4b** and **5b** uptake into the cells. These findings indicate that the cellular internalization of the peptide largely occurs through an active, energy-dependent uptake mechanism. To examine whether this energy-dependent uptake process includes a specific endocytic route, the cells were preincubated with a set of endocytosis inhibitors (methyl-β-cyclodextrin, cytochalasin D, and chlorpromazine) [16–18]. As seen in Figure 5, cytochalasin D and chlorpromazine significantly reduced the fluorescence intensity of compound **4b** within the cells; however, methyl-β-cyclodextrin did not reduce itsfluorescence intensity, indicating that it did not affect the uptake process. These findings suggest the presence of a mixed mechanism of uptake of compound **4b**. For compound **5b** (data not shown), the same observation was made, indidating that both compounds follow the clathrin-dependent and actin-dependent endocytosis.

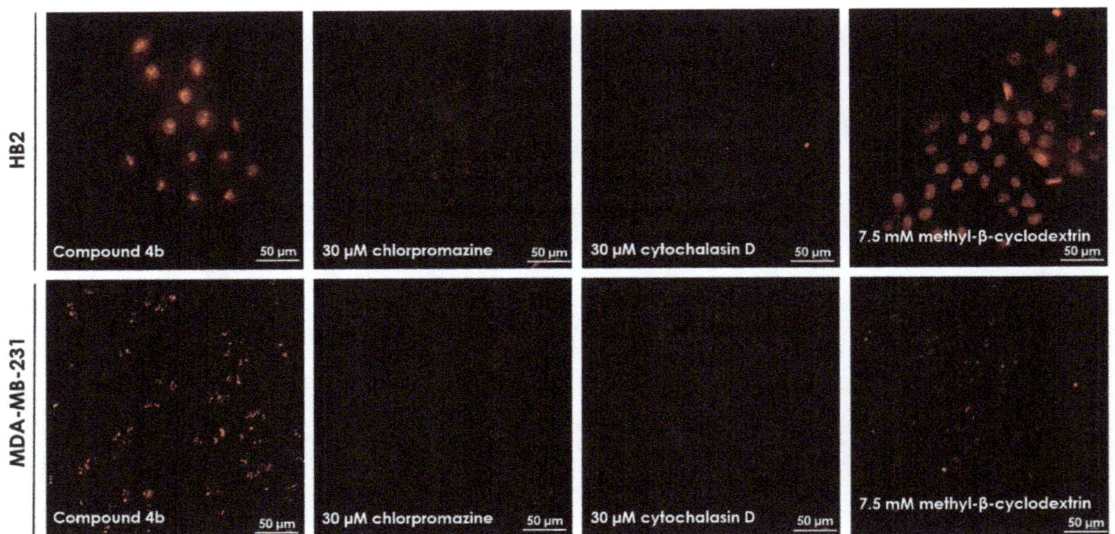

Figure 5. Cellular uptake study of compound **4b** in HB2 and MDA-MB-231 cells (incubation time 24 h, 37 °C). All other details are provided in individual images. Magnification 20×.

Nucleic acid binding ability of compounds **1–9** by electrophoretic mobility shift assay (EMSA), surface plasmon resonance (SPR), and microscale thermophoresis (MST).

Owing to the nuclear localization of the compounds **4a**, **4b**, **5a**, and **5b** and their highly positive charge, we decided to investigate the nucleic acid-binding ability of compounds **1–9**. Initially, we used a short linear model 76 bp dsDNA fragment containing the sequence of beta-actin (*Homo sapiens*). Polyacrylamide gel separation of the compounds incubated with the model DNA resulted in significant retardation, indicating the interaction of the synthesized compounds (**1–9**) with dsDNA (Figure 6). This is visible for compounds **4** (lines 13, 14) and **5** (lines 15, 16) and slightly visible for compounds **6** (21, 22), **7** (23, 24), and **8** (25, 26). The minimal C/P ratio resulting in forming complex of each compound is provided in Table S1. Preliminary surface plasmon resonance (SPR) analysis performed using two different concentrations (50 and 100 μM) of the compounds with dsDNA as a ligand indicated that at the lower concentration, compound **5** showed the highest DNA-binding ability, followed by compound **4** and compounds **1–3**, which are guanidine-rich polymers (Figure S3). At the concentration of 100 μM, compound **4** showed stronger DNA binding, and it is followed by compounds **1–3**. Substitution of guanidine groups by amino groups (compound **8**) significantly reduced the strength of interaction. Moreover, the analog of compound **5** with six hydroxyl moieties on its side chain (compound **9**) did not show any DNA binding (Figure S4). For compounds with high DNA-binding ability, the binding constant was measured using the SPR technique (Table 1). Both compounds **4** and **5** displayed similar binding constants (see Table 1) in the micromolar range (8.8 $\pm$ 2.1 μM for compound **4** and 1.8 $\pm$ 0.9 μM for compound **5**). To confirm these findings in an alternative system, we performed similar experiments using microscale thermophoresis (MST); the results are presented in Table 1. The binding constant values were in the micromolar range and increased in the following order: **4a** < **5a** < **1a** (Figure 7).

Considering two factors: cell permeability (in which the leading compound was **5a**) and DNA binding (in which compound **5a** was one of the most potent nucleic acid binders), we decided to synthesize a second set of the analogs with varying numbers of monomers in its structure (Figure 8). Thus, compounds **11**, **13**, and **14** (two, four, and eight residues of Dap(GO2), respectively) and their fluorescent derivatives (**11a**, **13a**, **14a**, and **14b**) were obtained. The cell permeability assay indicated that hexamers (**5a**) and octamers (**14a** and **14b**) efficiently penetrate the membrane and localized in the nucleus. Moreover, the DNA binding constants of the tested compounds according to the MST assay were in the micromolar and submicromolar range (Figure 9). Tested compounds displayed no or low cytotoxicity (compound **14**; Figure S5). The cytostatic effect was evaluated using 5-ethynyl-2′-deoxyuridine incorporation assay.

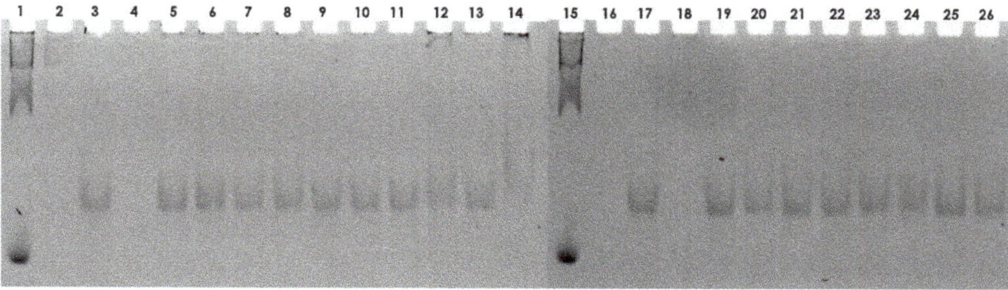

Figure 6. DNA-binding activity of compounds **1–9** was evaluated using a polyacrylamide electrophoretic gel mobility shift assay. (lanes **1, 15**) DNA marker; (lanes **3, 17**) dsDNA 76 bp; (lanes **5, 6**) compound **1** in two different N/P ratios—0.2:1 and 1.5:1 (charge peptidomimetic/charge dsDNA), respectively; (lanes **7, 8**) compound **2** in various N/P ratios—0.2:1 and 1.5:1, respectively; (lanes **9, 10**) compound **3** in various N/P ratios—0.2:1 and 1.5:1, respectively; (lanes **11, 12**) compound **4** in various N/P ratios—0.2:1 and 1.5, respectively; (lanes **13, 14**) compound **5** in various N/P ratios—0.2:1 and 1.5:1, respectively; (lanes **19, 20**) compound **6** in various N/P ratios—0.2:1 and 1.5:1, respectively; (lanes **21, 22**) compound **7** in various N/P ratios—0.2:1 and 1.5, respectively (charge peptidomimetic/charge dsDNA); (lanes **23, 24**) compound **8** in various N/P ratios—0.2:1 and 1.5:1, respectively; (lanes **25, 26**) compound **9** in various N/P ratios—0.2:1 and 1.5:1, respectively; (lanes **2, 4, 16, 18**) intentionally empty lanes. N/P is defined as ratio of positively chargeable polymer amine (N = nitrogen) groups to negatively charged nucleic acid phosphate (P) groups.

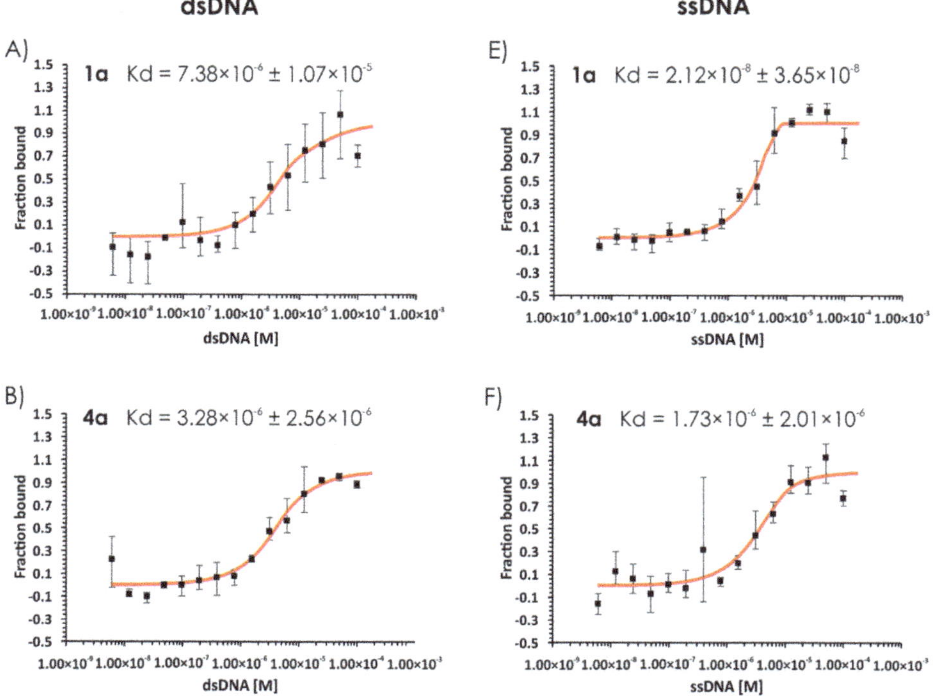

Figure 7. *Cont.*

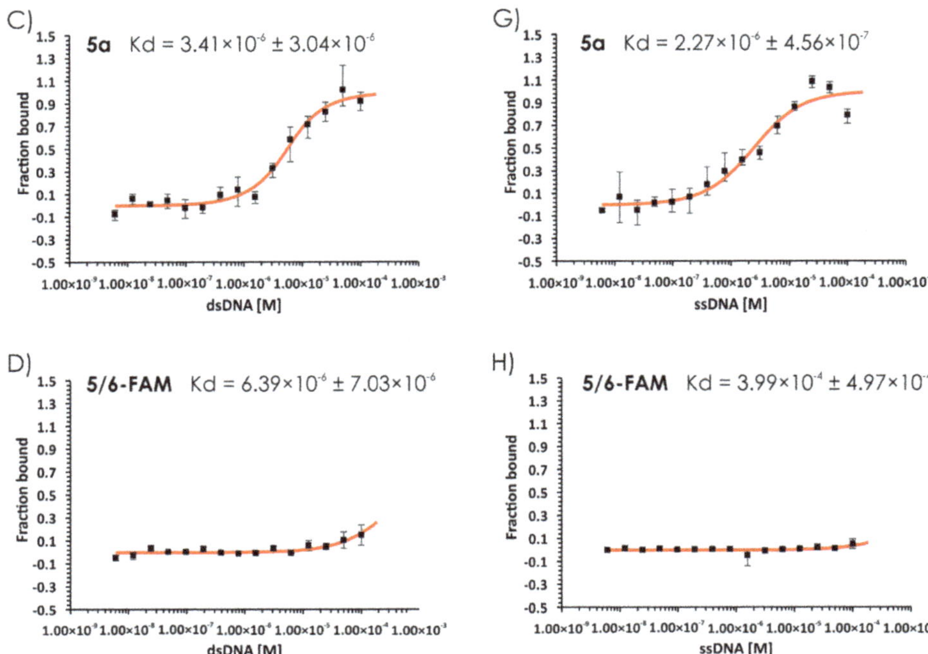

Figure 7. Microscale thermophoresis (MST) analysis of dsDNA binding by 5/6-FAM-labeled compounds. The MST analysis. FAM-labeled compounds **1a** (**A**), **4a** (**B**), **5a** (**C**), and ssDNA binding by 5/6-FAM-labeled compounds **1a** (**E**), **4a** (**F**), **5a** (**G**) was performed using the Monolith NT.115 instrument (NanoTemper). Binding was measured between increasing concentrations (6.1 nM–200 μM) of dsDNA (76 bp fragment) and 0.437 μM of the indicated compounds labeled with 5/6-FAM. Control experiments were performed with DNA and 5/6-FAM dye (**D**,**H**).

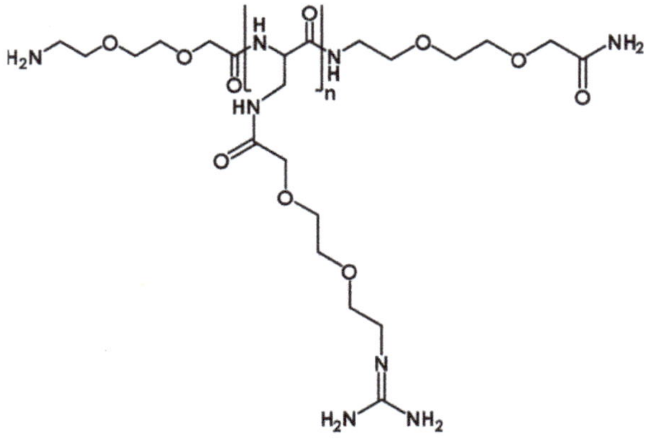

H_2N … NH_2

where n = 2, 4, 6 or 8

compound 11:	O2Oc-[Dap(GO2)]$_2$-O2Oc-NH$_2$
compound 13:	O2Oc-[Dap(GO2)]$_4$-O2Oc-NH$_2$
compound 5:	O2Oc-[Dap(GO2)]$_6$-O2Oc-NH$_2$
compound 14:	O2Oc-[Dap(GO2)]$_8$-O2Oc-NH$_2$

Figure 8. Chemical formulae of second generation compounds with varying lengths of the main chain.

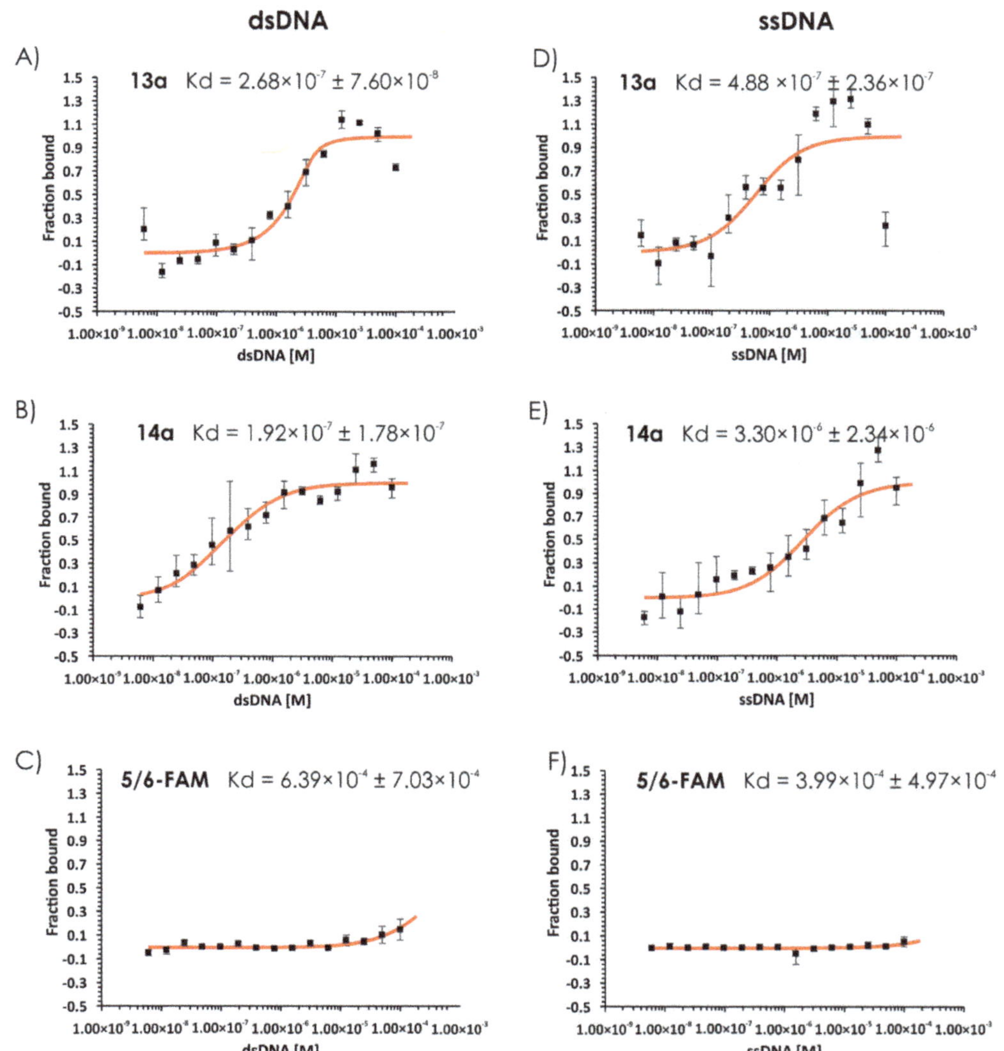

Figure 9. Microscale thermophoresis (MST) analysis of dsDNA binding by 5/6-FAM-labeled compounds. The MST analysis of dsDNA binding by 5/6-FAM-labeled compounds **13a** (**A**) and **14a** (**B**), and ssDNA binding by 5/6-FAM-labeled compounds **13a** (**D**) and **14a** (**E**) was performed using the Monolith NT.115 instrument (NanoTemper). Binding was measured between increasing concentration (6.1 nM–200 µM) of dsDNA (76 bp fragment) and 0.437 µM of the indicated compounds labeled with 5/6-FAM. A control experiment was performed with DNA and 5/6-FAM dye (**C**,**F**).

2.3. Transfection

Next, we used the novel compounds from the second generation series (**5**, **13**, **14**) as transfection agents (Figure 10A). The model plasmid encoding green fluorescent protein (PmaxGFP), size 3486 bzp) was used for transfection. Most cells in the system with these compounds and with the control (a commercially available transfection reagent the commercially available transfection reagent ViaFect) showed green fluorescence that indicates sucesfull transfection. The highest fluorescence intensity was observed for compound **5**, followed by that for compounds **14** and **13**. The post-transfection survival rate was comparable for compounds **13** and **5**, whereas compound **14** showed 30% cytotoxicity.

The control reagent showed the highest cytotoxicity; 50% of cells transfected with the control did not survive, but the fluorescent intensity of the system was comparable with that observed with compound **5** at its optimal concentration (Figure 10B).

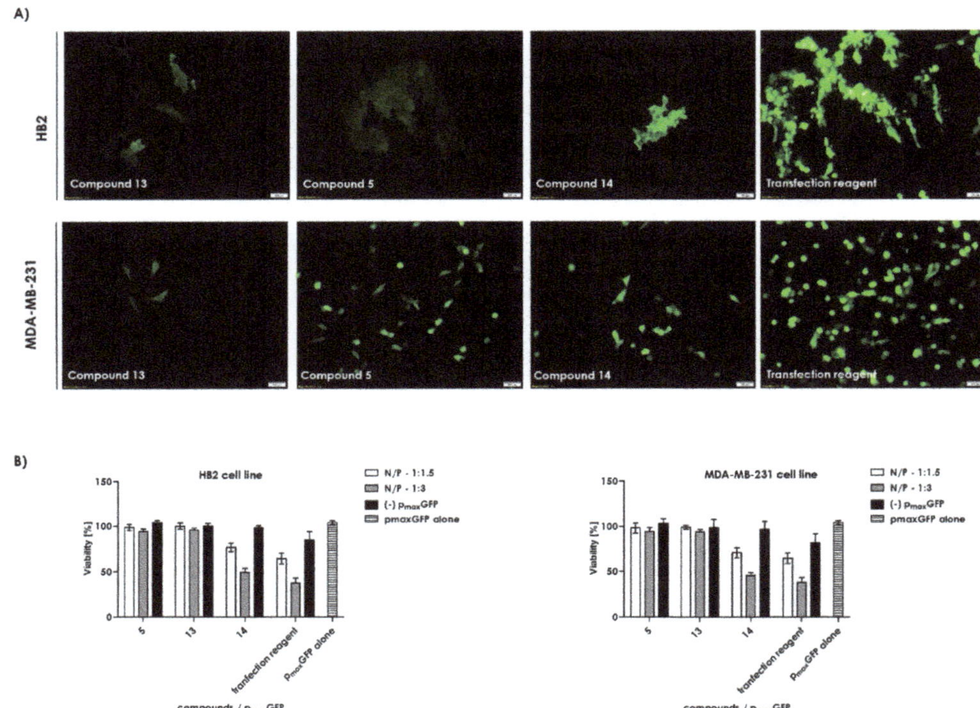

Figure 10. (**A**) Fluorescence imaging of HB-2 and MDA-MB-231 cells transfected with pmaxGFP plasmid mixed with compounds **5**, **13**, **14**, or the commercially available transfection reagent (control). (**B**) Cytotoxicity of plasmid pmaxGFP:compound complex in HB2 and MDA-MB-231 cells at the concentration of the complex equal to 4.62×10^{-10} M.

In the above experiment, we expected a strong correlation between the strength of interaction with DNA and efficient membrane translocation of the compounds tested. However, the relatively low transfection efficacy of compound **13** did not correlate with its high DNA binding constant and rapid cellular internalization. This discrepancy may be explained by the structure and size of the DNA-compound complex. Positively charged compounds interact with DNA by forming electrostatic interactions with the phosphate groups of the DNA backbone. We analyzed the shape and organization of such complexes in the form used for transfection using atomic force microscopy (AFM). As seen in Figure 11B,E,I, the compound itself is observed as a small dot despite its number of building blocks, and plasmid p_{max}GFP is observed to form several conformations (Figure 11A). Figure 11C,D show compound **13a** complexed with DNA in an N/P ratio of 0.2:1 (charge peptidomimetic/charge p_{max}GFP); the structure formed seems to be typical for small cationic molecules such as polyamines and metal ions (e.g., Ca^{2+} and Mg^{2+}) or poliArg [19,20]. Compound **13a** binds to the DNA sequence causing plasmid condensation, leading to a size reduction.

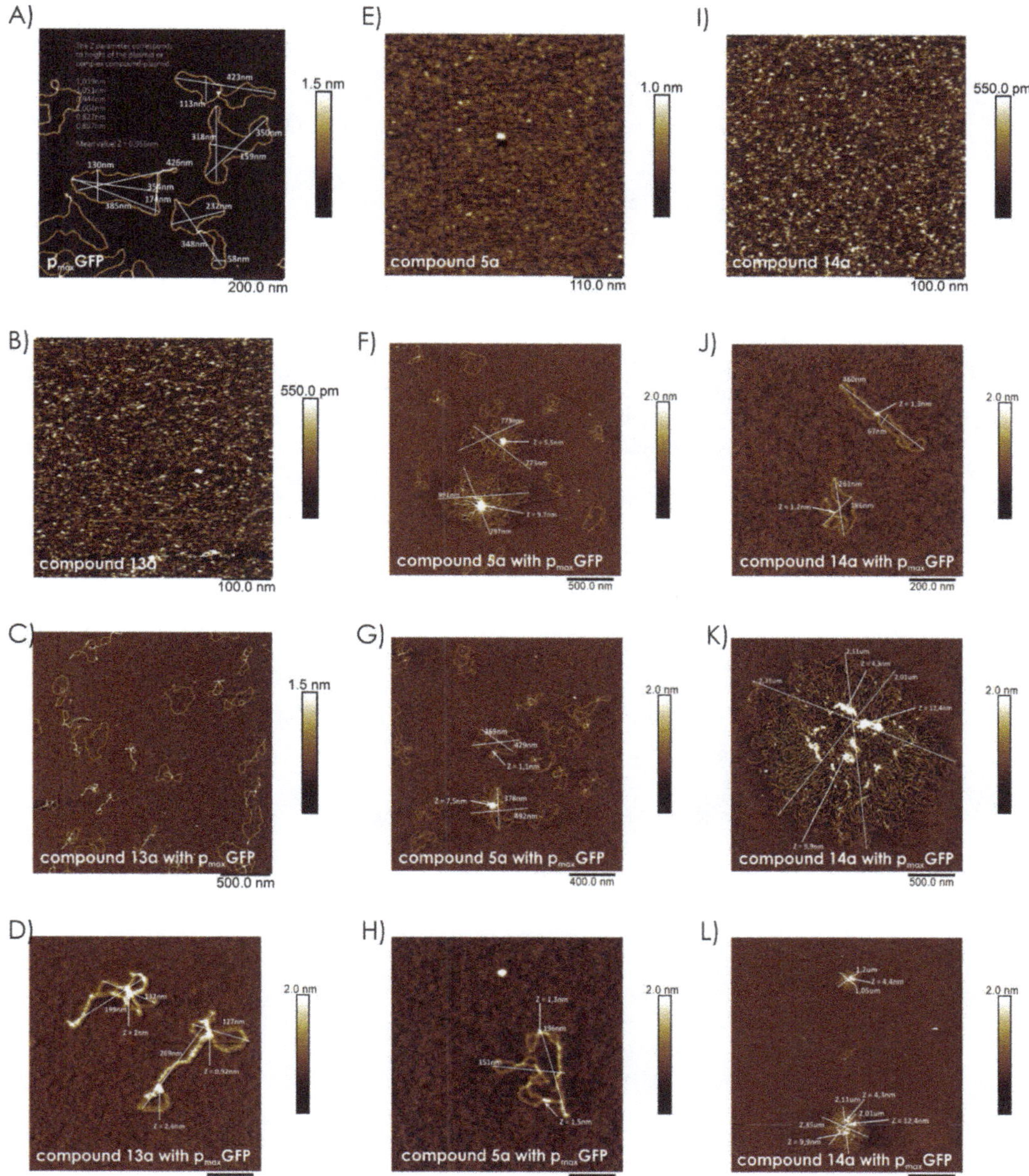

Figure 11. Atomic force microscopy (AFM) findings. (**A**) plasmid p_{max}GFP; (**B**) compound **13a**; (**C**) compound **13a** with plasmid p_{max}GFP–N/P 0.2:1, 2.5 μm; (**D**) compound **13a** with plasmid p_{max}GFP–N/P 0.2:1, 500 nm; (**E**) compound **5a**; (**F**) compound **5a** with plasmid p_{max}GFP–N/P 0.2:1, 2.5 μm; (**G**) compound **5a** with plasmid p_{max}GFP–N/P 0.2:1, 2.5 μm; (**H**) compound **5a** with plasmid p_{max}GFP–N/P 0.2:1, 500 nm; (**I**) compound **14a**; (**J**) compound **14a** with plasmid p_{max}GFP–N/P 0.2:1, 1 μm; (**K**) compound **14a** with plasmid p_{max}GFP–N/P 0.2:1, 2.5 μm; (**L**) compound **14a** with plasmid p_{max}GFP–N/P 0.2:1, 10 μm.

Compound **5a** complexed with DNA, creating flower-like structures [21], with an average size reaching 900 nm (Figure 11F–H). A limited number of single-plasmid compound **5a** structures are also observed in the investigated system. The analysis of the system in which compound **14a** was complexed with the plasmid showed a shift towards larger multiplasmid particles (referred to as connected coils) [22], with sizes up to 2350 nm (Figure 11J–L). However, single-plasmid complexes are also present in the system.

Compound **14a** forms large compact structures or aggregates with a defined high-density core, where several compound **14a** molecules are bound to several plasmid molecules; the structure resembles bacterial chromosome organization, with its average size reaching 2.35 μm. These structures are observed with a charge ratio of 0.2:1, in which the charge of DNA dominates over the charge of compound **14a**. Such a structure seems to act as a DNA scavenger able to bind all DNA molecules present in the system tested.

In general, the observed flower-like structures are highly looped with multiple crossover points. The complexity and compactness of the structure tends to increase with increasing number of monomers in the analyzed compounds (e.g., see compounds **13** versus **14**).

The above findings were confirmed by agarose gel electrophoretic separation of the formed complexes. As seen in Figure 12, the titration of the plasmid using increasing concentrations of compound **13** (lines 3–6) results in relatively minor changes in the electrophoretic mobility of the complexes formed. However, the findings for plasmid complexes with compound **5** were significantly different; the charge ratio of 1.5:1 and greater resulted in the formation of large complexes that were unable to penetrate the agarose gel. When the charge ratio reached the highest value (3:1), the complex was unable to penetrate the gel and remained in the well. This may be explained by either overall charge reduction and/or mass increase in the formed complex. This finding is consistent with the AFM images indicating that compound **13** causes intramolecular condensation of the plasmid, whereas compounds **14** and **5** are able to link together several molecules of the plasmid. Such large complexes were formed at a lower concentration of compound **14** (above charge ratio 1), which confirms the presence of the large structures observed in the AFM images. This is the possible reason for the low transfection efficacy mediated by compound **14**.

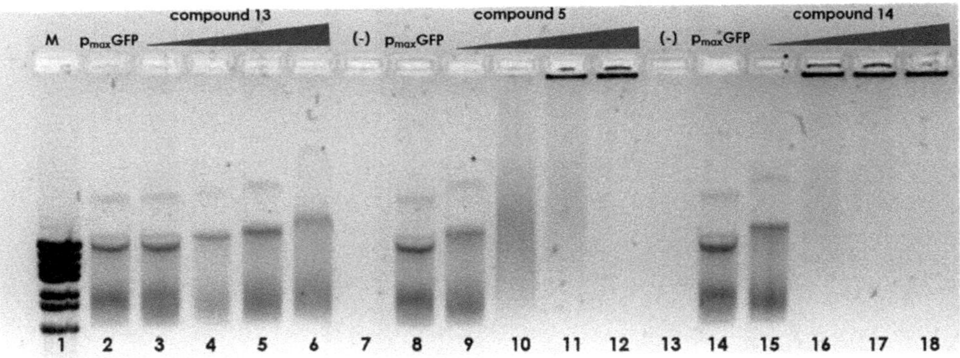

Figure 12. DNA-binding activity of compounds **13**, **5**, and **14** was evaluated using an electrophoretic gel mobility shift assay. (lane **1**) DNA marker; (lanes **2**, **8**, **14**) pmaxGFP plasmid; (lanes **3–6**) compound **13** in various N/P ratios—0.2:1, 1:1, 1.5:1, and 3:1 (charge peptidomimetic/charge plasmid), respectively; (lanes **9–12**) compound **5** in various N/P ratios—0.2:1, 1:1, 1.5:1, and 3:1, respectively; (lanes **15–18**) compound **14** in various N/P ratios—0.2:1, 1:1, 1.5:1, 3:1; (lanes **7**, **13**) intentionally empty lanes.

To further understand the role of the peptidomimetics in DNA condensation, theoretical models representing interactions of compounds **13** and **14** with DNA were created. Due to the complexity of such calculations, we decided to analyze only for two distinct systems and exclude compound **5** that behaved in a mixed mode. The peptidomimetics selected for molecular dynamics (MD) simulations were composed of the same amino acid-like units but differed from each other in sequence length, which is expected to affect the way the dsDNA is condensed.

The results showed that both peptidomimetics induced DNA condensation, and the process began in the early steps of molecular dynamics MD simulations. The peptidomimetics–DNA complexes were stabilized by a hydrogen bond network, including water-bridged hydrogen bonds and salt bridges. In contrast, the control simulation with no peptidomimetics showed no DNA aggregation (Figure 13).

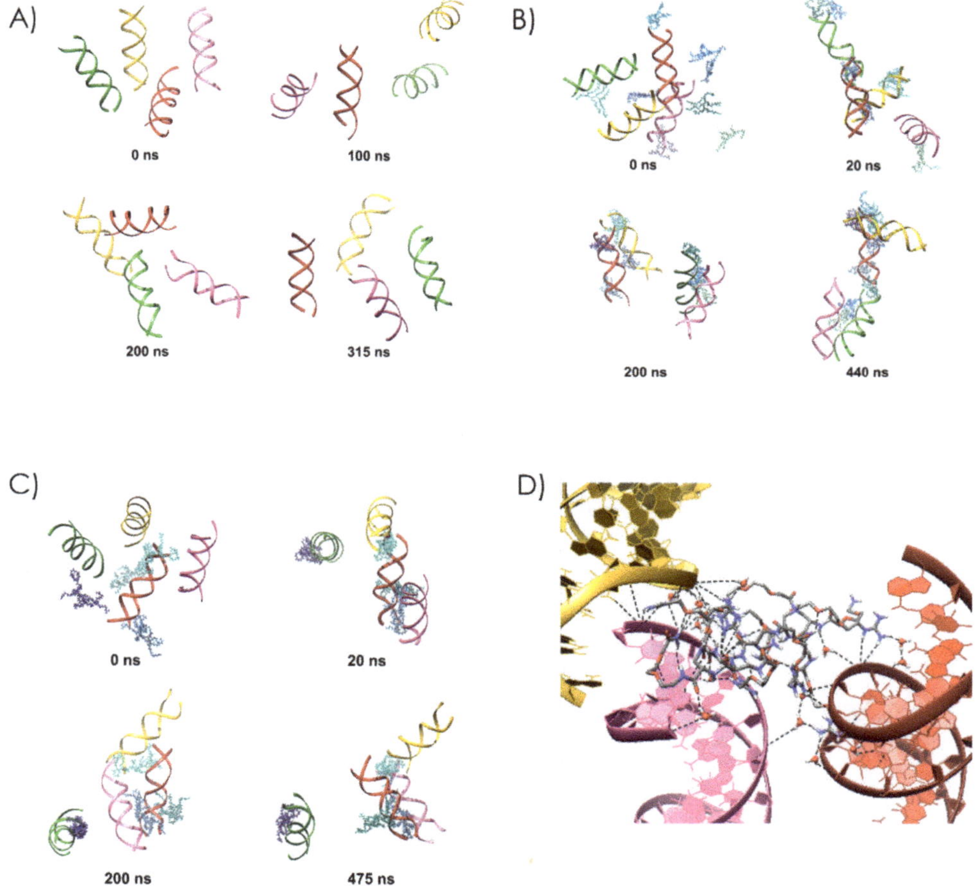

Figure 13. Snapshots from molecular dynamics (MD) simulations examining DNA condensation in a system (**A**) without a peptidomimetic, (**B**) in the presence of compound **13**, and (**C**) in the presence of compound **14**. The four DNA helices are colored gold, green, pink, and red. The peptidomimetic molecules are colored in various shades of blue. (**D**) Interactions between a single molecule of compound **14** and three DNA double helices at the final step of the simulation.

Molecular dynamic simulation of polymer–DNA interaction.

Both polymer-containing systems had a 0.2:1 compound: DNA charge ratio, indicating that in the MD simulation, twice as many compound molecules are needed to achieve the same condition for compound **13** as for compound **14**. This leads to a greater dispersion of the positive charges in the simulation system and affects the subsequent steps of DNA condensation. With compound **13**, the DNA condensation occurs in a clear stepwise manner. In the beginning, the formation of the dsDNA-compound complexes is frequently observed. This complexation shields the charges on the DNA phosphate groups and diminishes the inter-strand phosphate–phosphate repulsion, which facilitates binding of other dsDNA units. In detail, the first 20 ns of the simulation lead to a peptide-mediated condensation of two DNA double helices, whereas two others remain unaggregated until ~80 ns of the MD simulation. For the next 250 ns, two compound-bridged dsDNA assemblies are already observed. Finally, after about 330 ns of the MD simulation, all dsDNA molecules condense together to form a stable four-DNA bundle. In contrast, the longer side chain of compound **14** favors faster DNA condensation. The extended arms of compound **14** can quickly capture neighboring DNA helices during the initial steps of the MD simulation, simultaneously initiating dsDNA compound binding and dsDNA condensation. As seen in Figure 13C, three molecules of compound **14** mediate the aggregation of three dsDNA helices within the first 20 ns of the MD simulation, whereas the fourth one with one the associated peptide molecule stays away in an unaggregated form. This configuration remains unchanged for the rest of the trajectory. The fourth DNA double helix forms occasional contacts with the three-DNA bundle, but does not bind to it permanently, which is well reflected in the number of clusters forming within the whole trajectory (Figure 14). Remarkably, the length of compound **14** enables interactions of a single peptide molecule with as many as three DNA helices, which is not observed for its shorter counterpart (Figure 13D).

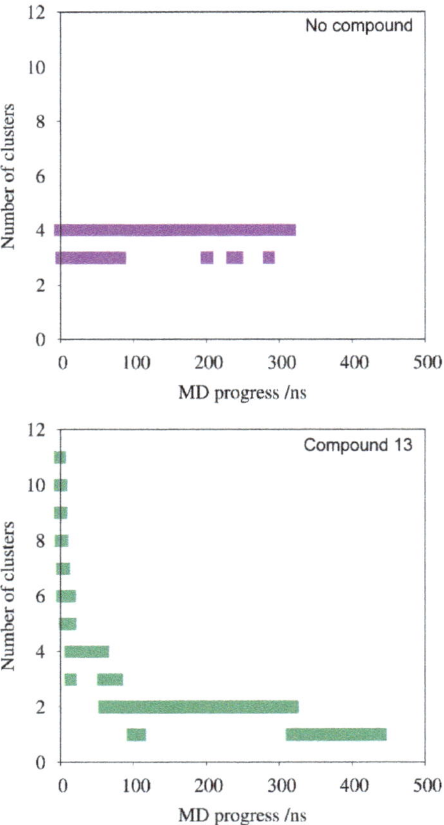

Figure 14. *Cont.*

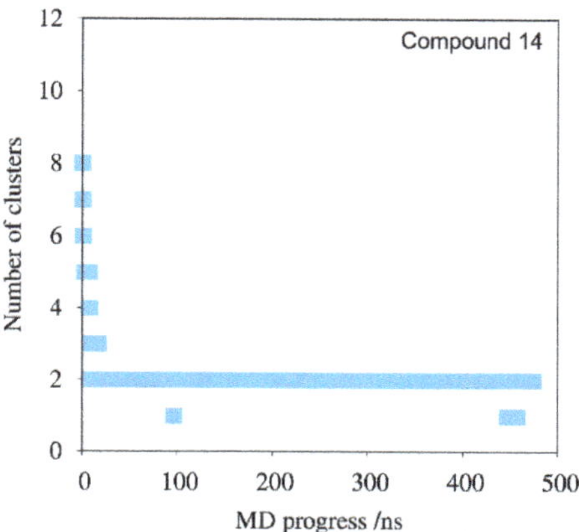

Figure 14. Progression of DNA condensation shown as aggregate numbers versus molecular dynamic (MD) progress. The peptidomimetic and dsDNA molecules at a distance ≤3.5 Å were considered to be in a cluster. The number of clusters at the beginning of a simulation depends on the initial number of DNA and peptide molecules in the system.

The findings of the above experiments suggest that despite the superior properties of compound **14**, it forms complexes with a size exceeding that required for effective cell penetration. The compound **13** complexes are relatively small, and cell penetration of the compound alone is low. Thus, the highest transfection efficacy observed for compound **5** is expected because the size of the DNA; compound **5** complex seems to be optimal for cell membrane crossing. To the best of our knowledge, the large complexes recorded for compound **14** have been reported until now only for large nuclear proteins with dedicated functions, such as viral proteins [22] and bacterial proteins [23]. Such large and dense DNA aggregates may resemble or even mimic a simplified neutrophil extracellular traps (NETs) network [24] or organization of bacterial DNA [25].

Synthesized compounds prove to be good DNA binders comparable to hexaArg with binding constants in the micromolar range. An exceptional feature of such compounds is their ability to rapidly translocation into the nucleus of the cells tested, which is not reported for reference compounds. Additionally, polymer length-dependent DNA structures are unique for such types of molecules. Finally, the polymer transfection efficacy is strongly influenced by length of the molecule used, and it is optimal for hexamer molecules.

The obtained new molecules being polymers of *N*-substituted L-2,3-diaminopropionic acid belong to the class of cell-penetrating inert polymers. Due to DNA binding properties, few of them facilitate gene delivery to the cells studied. In our opinion, the utilization of the above compounds is not limited to non-cytotoxic transfection, but they could also be employed as DNA binding agents. In a broader perspective, they could be DNA sensors or scavengers.

3. Materials and Methods

3.1. Chemistry

Compound Synthesis

All compounds were synthesized on the amide TentaGel S RAM resin (Rapp Polymer, Tubingen, Germany) by Fmoc solid-phase methods. Compounds **1**, **2**, and **3** were synthesized using an automated microwave peptide synthesizer (Liberty Blue, CEM, Matthews, NC, USA); the following reagents were used: Fmoc-O2Oc-OH (where O2Oc is 8-amino-3,6-dioxaoctanoic acid) and Fmoc-L-Arg(Pbf)-OH, or Fmoc-D-arg(Pbf)-OH, or Fmoc-Har(Pbf)-OH. Compounds **1a**, **2a**, and **3a** were synthesized by coupling 5/6-FAM to peptidomimetics **1**, **2**, and **3**, respectively, as described below. To obtain compounds **4**, **5**, **6**, **7**, **8**, **9**, **12**, **13**, and **14**, initially, Fmoc-O2Oc-OH and Fmoc-L-Dap(Mtt)-OH (where

L-Dap-OH is L-2,3-diaminopropionic acid) were used for the main chain synthesis (Fmoc-O2Oc-[Dap(Mtt)]$_n$-O2Oc-R) on an automated microwave peptide synthesizer (Liberty Blue). The scale of each synthesis was 0.1 mmol; amino acid derivatives were dissolved in 0.2 M *N,N*-dimethylformamide (DMF). Piperidine (20%) in DMF was used for deprotection; 0.5 M *N,N′*-diisopropylcarbodiimide in DMF was used as activator; 1.0 M OXYMA in DMF was used as activator base; and DMF was used as main wash. A protocol without final Fmoc deprotection was used. After automatic microwave synthesis, 4-methyltrityl (Mtt) protection groups were removed using the procedure described previously [26] (1% trifluoroacetic acid (TFA) in dichloromethane (DCM) with addition of 2% 1,2-ethanedithiol [EDT]). The mixture was added to compounds **4**, **5**, **6**, **7**, **8**, **9**, **12**, **13**, and **14** and stirred for 15 min. The resin was washed with DCM between each wash. This procedure was repeated until no increase in absorbance at 410 nm was noted. Then, 1% *N,N*-diisopropylethylamine (DIPEA) in DMF solution was added to each portion of the peptidyl resin three times for 10 min. Next, GO2 (compounds **5**, **12**, **13**, **14**), GO1 (compound **4**), O2 (compound **8**), HO2 (compound **9**), and Mtt-O2Oc-NH_2 (compounds **6** and **7**) were coupled to each -NH_2 moiety of Fmoc-O2Oc-(Dap)$_n$-O2Oc-R using equimolar amounts of Amino acid/TBTU/OXYMA/DIPEA in DMF/DCM/2-*N*-methyl-2-pyrrolidone (NMP; 1:1:1, *v*/*v*/*v*) solution. Completeness of the coupling was controlled using Kaiser and chloranil tests. After adding the Mtt-O2Oc moiety to compounds **6** and **7**, Mtt protection groups were removed, and GO1 and GO2 were coupled to resulting free amino groups of compound **6** and **7**, respectively. After confirming the coupling, Fmoc deprotection from *N*-terminal peptidomimetics were performed by using 2 g piperidine in 96 mL NMP with the addition of 2 g 1,8-Diazabicyclo [5.4.0] undec-7-ene for 15 min; during deprotection, the peptidyl resin was washed with DMF, and the whole procedure was repeated six times. Next, fluorophores such as 5/6-FAM or 5/6-TAMRA were attached to the *N*-terminal amino group; a mixture of DIPEA in molar excess of the fluorophore (1:3) in DMF was used. After completing the synthesis, the peptidomimetics were cleaved from the resin, using a TFA/phenol/H_2O/thioanisole/EDT mixture (82.5:5:5:5:2.5, *v*/*m*/*v*/*v*/*v*). The purity of the synthesized compounds and the accuracy of synthesis were confirmed using ultra-performance liquid chromatography using the Nexera X2 LC-30AD system (Schimadzu, Tokyo Japan) equipped with a Phenomenex column (150 × 2.1 mm), with a grain size of 1.7 μm (peptide XB-C18) equipped with a UV-Vis detector and a fluorescence detector. A linear gradient from 2% to 80% B within 15 min was applied (A: 0.1% TFA; B: 80% acetonitrile in A). The peptidomimetics were monitored at 216 nm. The molecular weights of the synthesized compounds were confirmed by analysis of the mass spectra which were recorded on a Biflex III MALDI-TOF mass spectrometer (Bruker Daltonics, Bremen, Germany) using 2,5-dihydroxybenzoic acid as a matrix.

3.2. Biology

3.2.1. Cell Culture

Healthy cell line HB2 (human breast epithelial cells) was obtained from Merck (Hamburg, Germany), and cancer cell line MDA-MB-231 (human breast cancer epithelial cells) was obtained from ATCC (Lomianki, Poland). The MDA-MB-231 and HB2 cells were cultured at 37 °C in 5% CO_2 in Dulbecco's Modified Eagle Medium (DMEM; high glucose) supplemented with 10% fetal bovine serum (FBS) and 1% penicillin–streptomycin solution containing 100 units of penicillin and 100 μg/mL of streptomycin. The HB2 cell line requires the addition of 5 μg/mL insulin and 5 μg/mL alcoholic hydrocortisone solution.

3.2.2. Fluorescence Microscopy

HB2 and MDA-MB-231 cells were seeded on 24-well plates at a density 1.5×10^4/well and 4×10^4/well, respectively, and incubated in 0.5 mL complete medium for 48 h. Subsequently, cells were washed with phosphate-buffered saline (PBS), and fresh medium with a fluorescently labeled peptide was added to each well at a concentration of 10 μM. For nucleus staining a DAPI solution (ThermoFisher Scientific, Waltham, MA, USA, country R37606) in PBS was employed in conditions recommended by the supplier. After the incubation (2 h or 24 h), cells were washed thoroughly with PBS, and then a phenol red-free culture medium (FluoroBrite, DMEM) was added. Subsequently, the cells were examined using an Olympus IX51 fluorescence microscope (Olympus, Tokyo, Japan) using appropriate filters: blue for DAPI, green for FITC, and red TAMRA. Next, the images were merged and colocalization of the dyes' emission were analyzed.

3.2.3. Cytotoxicity Assay: MTT

Cell viability against cell-penetrating compounds was detected by the MTT assay. The HB2 and MDA-MB-231 cells were seeded into 96-well plates at a density of 5×10^3/well and 7×10^3/well, respectively, and incubated in complete medium (100 µL/well) for 48 h. Then, the medium was replaced with fresh medium (100 µL/well), and solutions of test compounds at various concentrations (1, 10, and 50 µM) were added; the cells were incubated for 24 h. Cells incubated in media without any test compounds were used as control. After incubation, medium containing test compounds was removed, and 225 µL fresh medium with 25 µL 3-(4,5-dimethylthiazol-2-yl)-2,5-diphenyltetrazolium bromide (MTT) was added (0.5 mg/mL per well). After incubation at 37 °C for 4 h, supernatants were removed, and the formazan crystals were dissolved overnight with dimethyl sulfoxide (DMSO; 150 µL/well). Results were analyzed using a microplate reader (SPECTROstar Nano, BMG LABTECH, Ortenberg, Germany) at 570 nm and 690 nm.

3.2.4. Cytotoxicity Assay–CCK-8

Cell viability against cell-penetrating peptidomimetics was detected using the cell counting kit-8 (CCK-8) assay (Sigma Aldrich, Poznan, Poland). The HB2 and MDA-MB-231 cells were seeded into 96-well plates at a density of 5×10^3/well and 7×10^3/well, respectively, and incubated in complete medium (100 µL/well) for 48 h. Then, the medium was replaced with fresh medium (100 µL/well), and solutions of compounds at various concentrations (1, 5, 10, 20, 50, and 100 µM) were added; the cells were incubated for 2 h or 24 h. Cells incubated in media without any additives were used as control. After incubation, 10 µL/well of 2-(2-methoxy-4-nitrophenyl)-3-(4-nitrophenyl)-5-(2,4-disulfophenyl)-2H-tetrazolium, monosodium salt (WST-8) reagent was added according to the manufacturers' instructions, and the mixture was incubated at 37 °C for 4 h. Cell viability was analyzed on the basis of formazan absorbance using a microplate reader (SPECTROstar Nano, BMG LABTECH, Ortenberg, Germany) at 450 nm.

3.2.5. Cytotoxicity of Peptidomimetic–Plasmid p_{max}GFP Complexes

In this experiment, second-generation compounds (compounds **5**, **13**, and **14**) were used as transfection reagents. See transfection procedure in methodology below. After gene expression, the medium was removed and cells were lysed with 0.5 M NaOH (100 µL/well), and the fluorescence intensity was analyzed using a microplate reader at excitation and emission wavelengths of 488 and 510 nm, respectively.

3.2.6. Endocytosis Inhibitors

HB2 and MDA-MB-231 cells were seeded into 24-well plates at a density of 1.5×10^4/well and 4×10^4/well and incubated overnight at 37 °C. The medium was then replaced with FBS-free medium and cultured for 24 h more. Afterwards, the cells were treated with endocytosis inhibitors cytochalasin D (final concentration: 1, 2, 5, 10, 20, and 30 µM), chlorpromazine (final concentration: 1, 2, 5, 10, 20, and 30 µM), and methyl-β-cyclodextrin (final concentration: 1, 2.5, 5, and 7.5 mM) for 30 min before adding 10 µM a compound **4b** or **5b**. The cells were incubated with the compounds for 2 h and later washed three times with PBS; the cells were observed under an Olympus IX51 fluorescence microscope (Olympus, Tokyo, Japan). After observation, cells were lysed with 0.5 M NaOH (500 µL/well), and the fluorescence intensity was analyzed using a microplate reader at excitation and emission wavelengths of 558 and 575 nm, respectively.

3.2.7. Transfection

In this experiment, second-generation $(GO2)_n$ compounds (compounds **5**, **13**, and **14**) were used as transfection reagents. HB2 and MDA-MB-231 were seeded in 96-well plates at a density of 5×10^3/well and 7×10^3/well in complete medium and grown until 60–70% confluency. Afterwards, a transfection complex solution was prepared: 200 ng of p_{max}GFP plasmid from Lonza (Basel, Switzerland) was mixed with $(GO2)_n$ peptidomimetics in various N/P ratios (1.5:1 and 3:1; charge peptidomimetic/charge plasmid) (where N/P is defined as the ratio of positively chargeable polymer amine (N = nitrogen) groups to negatively charged nucleic acid phosphate (P) groups) and $CaCl_2$ (4 mM per well)) and incubated for 30 min at room temperature until complex formation. The DNA concentration was determined using UV readout at 260/280 nm. The concentration of polymers using UV signal was at 216 nm. As control, ViaFect (Promega, Walldorf, Germany), a commercially available transfection reagent, was used according to the manufacturers' instructions. Then, the medium in the plate was replaced with serum- and antibiotic-free medium (90 μL/well), and the transfection complex solution was added to each well (10 μL/well). After 5 h incubation, the medium was replaced with 100 μL/well of fresh medium supplemented with 10% FBS, and the plate was incubated for another 48 h to allow for gene expression. Results were observed by fluorescence microscopy.

3.2.8. Electrophoretic Mobility Shift Assay

- DNA polyacrylamide gel electrophoresis. The DNA-binding activity of compounds **1a–9a** was examined using electrophoretic mobility shift assay. The dsDNA; (76 bp) model fragment) was mixed with the peptidomimetics in various N/P ratios (0.2:1 and 1.5:1; charge peptidomimetic/charge dsDNA); after incubating for 30 min, 4 μL of loading buffer was added to the samples. The DNA–peptidomimetic complexes were resolved by 8% polyacrylamide gel electrophoresis, and the migrated DNA was visualized under UV light using the fluorescent dye Midori Green.
- Agarose gel electrophoresis. To test the interactions of compounds **5**, **13**, and **14** with DNA, electrophoretic mobility shift assay was performed. The p_{max}GFP plasmid was mixed with the peptidomimetics in various N/P ratios (0.2:1, 1:1, 1.5:1, and 3:1; charge peptidomimetic/charge plasmid); after incubating for 30 min, 4 μL of loading buffer was added to the samples. The plasmid–peptidomimetic complexes were resolved by 0.7% agarose gel electrophoresis, and the migrated DNA was visualized under UV light using the fluorescent dye Midori Green.

3.2.9. Molecular Dynamics

All-Atom Self-Assembly Simulations

The interaction of selected compounds (compounds **13** and **14**) with DNA was studied by performing all-atom MD simulations using the GPU/CUDA-accelerated implementation of PMEMD in AMBER 16 [27]. Non-standard residues were modelled with the XLEAP module. The point charges were optimized by fitting them to the ab initio molecular electrostatic potential (6–31G* basis set, GAMESS 2013-ab initio molecular electronic structure program) [28] for two different conformations, followed by consecutive averaging of the charges over all conformations, as recommended by the RESP protocol [29]. The initial system for simulations consisted of four double helical DNA molecules with a sequence 5′-ATTGGCAATGAGCGGTTCCG-3′, modeled in an ideal B-form, without and with added selected compounds. In systems with the peptides, a peptidomimetic: DNA charge ratio of 1:5 was maintained to replicate the experimental conditions. Therefore, four molecules of compound **14** and eight molecules of compound **13** were added to the simulation boxes containing four double helical DNA 20-mers in a random position. Each system was solvated and neutralized by adding sodium and chloride ions. The concentration of free salt ions was approximately 100 mM. The 315–475 ns simulations at 300 K with isotropic pressure coupling and 2 fs time step were conducted under periodic boundary conditions with long-range electrostatic interactions evaluated by the particle Mesh Ewald (PME) summation, and a cut-off of 10 Å was used for van der Waals interactions. The SHAKE algorithm was used to constrain bonds involving hydrogen. The temperature was maintained using the Langevin coupling scheme with a friction coefficient of 1 ps^{-1}, whereas a Berendsen barostat maintained the reference pressure set to 1.0 bar. The analyses were performed with the CPPTRAJ module of AmberTool v16 (San Francisco, CA, USA). The aggregation process was investigated with the GROMACS 2019.4 suite (Groningen, The Netherlands) [30]. The visualizations were created using UCSF Chimera v1.15 (San Francisco, CA, USA) [31].

3.2.10. Surface Plasmon Resonance Analysis

Standard surface plasmon resonance (SPR) analyses using a Biacore T200 (GE Healthcare, Warsaw, Poland) were performed essentially as described in the manufacturers' manual. DNA binding by all tested compounds was studied using a 5′-biotinylated 76 nt ssDNA or 76 bp dsDNA fragment containing the sequence of β-actin (*Homo sapiens*), immobilized on a streptavidin matrix-coated sensor chip SA (GE Healthcare, Warsaw, Poland). All oligonucleotides were commercially synthesized (oligo.pl, Poland; Figures S3 and S4). The dsDNA was immobilized on the sensor surface to yield a final value of ~50 RU for dsDNA or ~100 RU. Experiments were performed at 25 °C, and the running buffer was HBS-EP (150 mM NaCl, 10 mM HEPES (pH = 7.4), 3 mM EDTA, and 0.05% Surfactant P20). In binding experiments, the buffer flow rate was set to 15 μL/min, and in kinetic experiments, the buffer flow rate was 30 μL/min. The data were analyzed using Biacore T200 evaluation software (GE Healthcare, Warsaw, Poland). The results are presented as sensorgrams obtained after subtracting the background response signal from a reference flow cell and from a control experiment with buffer injection.

3.2.11. Microscale Thermophoresis

Microscale thermophoresis was performed using the Monolith NT.115 instrument (NanoTemper Technologies GmbH, Munich, Germany). Binding between dsDNA (76 bp) or ssDNA (76 nt) fragments and test compounds labeled with 5/6-FAM was measured. A control experiment was performed with DNA and 5/6-FAM dye. A 16-step dilution series of DNA (400 μM) was prepared in EDBS buffer (25 mM Tris-HCl (pH = 8), 4% (*w*/*v*) sucrose, 4 mM DTT, and 80 μg/mL BSA). Next, 10 μL of 5/6-FAM-labeled compounds diluted in EDBS buffer were added to 10 μL of DNA solution (1:1 dilution series) to reach a final concentration of 0.437 μM. The samples were incubated at 32 °C for 1 h and centrifugated before being transferred to Standard Monolith NT™ Capillaries. The capillaries were scanned at 25 °C using the MST instrument (20% LED, medium MST power). For each compound, at least two independent experiments were performed. All data were analyzed using MO Affinity Analysis software (NanoTemper, Munich, Germany).

3.2.12. Atomic Force Microscopy

The complexes obtained in the reaction between 8 μM peptidomimetics **5a**, **13a**, and **14a** and 2 nM p_{max}GFP plasmid DNA (Lonza, Switzerland) were examined using AFM in 8 mM $MgCl_2$ at room temperature in the PeakForce Tapping mode, using BioScope Resolve AFM (Bruker, Bremen, Germany). The ScanAsyst-Fluid+ probe (Bruker) was used for DNA-peptidomimetic complex imaging (resonant frequency f_0 = 150 kHz; spring constant k = 0.7 N/m). Images were taken at 512 × 512 pixels with a PeakForce Tapping frequency of 1 kHz and an amplitude of 150 nm. Height sensor signal was used to display the protein image using NanoScope Analysis v1.9 (Bruker, Bremen, Germany).

Supplementary Materials: The following are available online at https://www.mdpi.com/1422-0067/22/5/2571/s1.

Author Contributions: Conceptualization, M.W. and A.L.; methodology, M.W., A.P., E.S., K.W., K.B.; investigation, A.R., K.W., K.B., A.G., A.P., E.S., M.W.; data curation A.R., M.W.; writing—original draft preparation, M.W., A.R.; writing—review and editing M.W., A.L., I.K.; A.P., E.S., K.W., visualization, A.R., K.B., K.W., E.S.; supervision, M.W.; project administration, M.W.; funding acquisition M.W. All authors have read and agreed to the published version of the manuscript.

Funding: This work was supported by National Science Center Poland under grant UMO-2017/27/B/ST5/02061 (MW).

Institutional Review Board Statement: Not applicable.

Informed Consent Statement: Not applicable.

Data Availability Statement: Not applicable.

Acknowledgments: Not applicable.

Conflicts of Interest: The authors declare no conflict of interest.

References

1. Li, Q.; Hao, X.; Wang, H.; Guo, J.; Ren, X.K.; Xia, S.; Zhang, W.; Feng, Y. Multifunctional REDV-G-TAT-G-NLS-Cys peptide sequence conjugated gene carriers to enhance gene transfection efficien-cy in endothelial cells. *Colloids Surf. B Biointerfaces* **2019**, *184*, 110510. [CrossRef]
2. Li, L.; Wei, Y.; Gong, C. Polymeric Nanocarriers for Non-Viral Gene Delivery. *J. Biomed. Nanotechnol.* **2015**, *11*, 739–770. [CrossRef]
3. Sun, Y.; Yang, Z.; Wang, C.; Yang, T.; Cai, C.; Zhao, X.; Yang, L.; Ding, P. Exploring the role of peptides in polymer-based gene delivery. *Acta Biomater.* **2017**, *60*, 23–37. [CrossRef] [PubMed]
4. Ghaffari, M.; Dehghan, G.; Abedi-Gaballu, F.; Kashanian, S.; Baradaran, B.; Dolatabadi, J.E.N.; Losic, D. Surface functionalized dendrimers as controlled-release delivery nanosystems for tumor targeting. *Eur. J. Pharm. Sci.* **2018**, *122*, 311–330. [CrossRef]
5. Budker, V.; Trubetskoy, V.; Wolff, J.A. Condensation of nonstochiometric DNA/polycation complexes by divalent cations. *Biopolymers* **2006**, *83*, 646–657. [CrossRef]
6. Thomas, T. Collapse of DNA in packaging and cellular transport. *Int. J. Biol. Macromol.* **2018**, *109*, 36–48. [CrossRef]
7. Oba, M. Cell-penetrating peptide foldamers: Drug-delivery tools. *ChemBioChem* **2019**, *20*, 2041–2045. [CrossRef]
8. He, Y.; Li, F.; Huang, Y. Smart cell-penetrating peptide-based techniques for intracellular delivery of thera-peutic macromolecules. *Adv. Protein Chem. Struct. Biol.* **2018**, *112*, 183–220. [PubMed]
9. Lindberg, S.; Copolovici, D.M.; Langel, U. Therapeutic delivery opportunities, obstacles and applications for cell-penetrating peptides. *Ther. Deliv.* **2011**, *2*, 71–82. [CrossRef] [PubMed]
10. Ye, J.; Liu, E.; Yu, Z.; Pei, X.; Chen, S.; Zhang, P.; Shin, M.C.; Gong, J.; He, H.; Yang, V.C. CPP-assisted intra-cellular drug delivery, what is next? *Int. J. Mol. Sci.* **2016**, *17*, 1892. [CrossRef] [PubMed]
11. Kargaard, A.; Sluijter, J.P.; Klumperman, B. Polymeric siRNA gene delivery—Transfection efficiency versus cytotoxicity. *J. Control. Release* **2019**, *316*, 263–291. [CrossRef]
12. Soltani, F.; Parhiz, H.; Mokhtarzadeh, A.; Ramezani, M. Synthetic and biological vesicular nano-carriers designed for gene delivery. *Curr. Pharm. Des.* **2015**, *21*, 6214–6235. [CrossRef]
13. Rehman, Z.U.; Zuhorn, I.S.; Hoekstra, D. How cationic lipids transfer nucleic acids into cells and across cellular membranes: Recent advances. *J. Control. Release* **2013**, *166*, 46–56. [CrossRef] [PubMed]
14. Wysocka, M.; Gruba, N.; Grzywa, R.; Giełdoń, A.; Bąchor, R.; Brzozowski, K.; Sieńczyk, M.; Dieter, J.; Szewczuk, Z.; Rolka, K.; et al. PEGylated substrates of NSP4 protease: A tool to study protease specificity. *Sci. Rep.* **2016**, *6*, 22856. [CrossRef]
15. Wysocka, M.; Romanowska, A.; Gruba, N.; Michalska, M.; Giełdoń, A.; Lesner, A. A peptidomimetic fluo-rescent probe to detect the trypsin β2 subunit of the human 20S proteasome. *Int. J. Mol. Sci.* **2020**, *21*, 2396. [CrossRef]
16. He, C.; Hu, Y.; Yin, L.; Tang, C.; Yin, C. Effects of particle size and surface charge on cellular uptake and biodistribution of polymeric nanoparticles. *Biomaterials* **2010**, *31*, 3657–3666. [CrossRef]
17. Iversen, T.-G.; Skotland, T.; Sandvig, K. Endocytosis and intracellular transport of nanoparticles: Present knowledge and need for future studies. *Nano Today* **2011**, *6*, 176–185. [CrossRef]
18. Jiang, Z.; He, H.; Liu, H.; Thayumanavan, S. Cellular Uptake Evaluation of Amphiphilic Polymer Assemblies: Importance of Interplay between Pharmacological and Genetic Approaches. *Biomacromolecules* **2019**, *20*, 4407–4418. [CrossRef] [PubMed]
19. Gao, T.; Zhang, W.; Wang, Y.; Guangcan, Y. DNA compaction and charge neutralization regulated by divalent ions in very low pH solution. *Polymers* **2019**, *11*, 337. [CrossRef] [PubMed]
20. Iacomino, G.; Picariello, G.; D'Agostino, L. DNA and nuclear aggregates of polyamines. *BBA Mol. Cell Res.* **2012**, *1823*, 1745–1755. [CrossRef]
21. Kaufman, B.A.; Durisic, N.; Mativetsky, J.M.; Costantino, S.; Hancock, M.A.; Grutter, P.; Shoubridge, E.A. The mitochondrial transcription factor TFAM coordinates the assembly of multiple DNA molecules into nu-cleoid-like structures. *Mol. Biol. Cell* **2007**, *18*, 3225–3236. [CrossRef]

22. Zabolotnaya, E.; Mela, I.; Williamson, M.J.; Bray, S.M.; Yau, S.K.; Papatziamou, D.; Edwardson, J.M.; Robinson, N.P.; Henderson, R.M. Modes of action of the archaeal Mre11/Rad50 DNA-repair complex re-vealed by fast-scan atomic force microscopy. *Proc. Natl. Acad. Sci. USA* **2020**, *117*, 14936–14947. [CrossRef] [PubMed]
23. Speir, J.A.; Johnson, J.E. Nucleic acid packaging in viruses. *Curr. Opin. Struct. Biol.* **2012**, *22*, 65–71. [CrossRef]
24. Pires, R.H.; Felix, S.B.; Delcea, M. The architecture of neutrophil extracellular traps investigated by atomic force microscopy. *Nanoscale* **2016**, *8*, 14193–14202. [CrossRef] [PubMed]
25. Qian, Z.; Macvanin, M.; Dimitriadis, E.K.; He, X.; Zhurkin, V.; Adhya, S. A new noncoding RNA arranges bacterial chromosome organization. *mBio* **2015**, *6*, e00998-15. [CrossRef]
26. Aletras, A.; Barlos, K.; Gatos, D.; Koutsogianni, S.; Mamos, P. Preparation of the very acid-sensitive Fmoc-Lys(Mtt)-OH Application in the synthesis of side-chain to side-chain cyclic peptides and oligolysine cores suitable for the solid-phase assembly of MAPs and TASPs. *Int. J. Pept. Protein Res.* **2009**, *45*, 488–496. [CrossRef]
27. Case, D.A.; Betz, R.M.; Cerutti, D.S.; Cheatham, T.E.; Darden, T.A.; Duke, R.E.; Giese, T.J.; Gohlke, H.; Goetz, A.W.; Homeyer, N.; et al. *AMBER 2016 Reference Manual*; University of California: San Francisco, CA, USA, 2016; pp. 1–923.
28. Schmidt, M.W.; Baldridge, K.K.; Boatz, J.A.; Elbert, S.T.; Gordon, M.S.; Jensen, J.H.; Koseki, S.; Matsunaga, N.; Nguyen, K.A.; Su, S.; et al. General atomic and molecular electronic structure system. *J. Comput. Chem.* **1993**, *14*, 1347–1363. [CrossRef]
29. Bayly, C.I.; Cieplak, P.; Cornell, W.; Kollman, P.A. A well-behaved electrostatic potential based method using charge restraints for deriving atomic charges: The RESP model. *J. Phys. Chem.* **1993**, *97*, 10269–10280. [CrossRef]
30. Van der Spoel, D.; Lindahl, E.; Hess, B.; Groenhof, G.; Mark, A.E.; Berendsen, H.J.C. GROMACS: Fast, flexible and free. *J. Comput. Chem.* **2005**, *26*, 1701–1718. [CrossRef] [PubMed]
31. Pettersen, E.F.; Goddard, T.D.; Huang, C.C.; Couch, G.S.; Greenblatt, D.M.; Meng, E.C.; Ferrin, T.E. UCSF Chimera-a visualization system for exploratory research and analysis. *J. Comput. Chem.* **2004**, *25*, 1605–1612. [CrossRef]

International Journal of
Molecular Sciences

MDPI

Article

X-ray Crystallographic Structure of α-Helical Peptide Stabilized by Hydrocarbon Stapling at *i*,*i* + 1 Positions

Yui Makura [1], Atsushi Ueda [1,*], Takuma Kato [2], Akihiro Iyoshi [1], Mei Higuchi [1], Mitsunobu Doi [2] and Masakazu Tanaka [1,*]

1 Graduate School of Biomedical Sciences, Nagasaki University, 1–14 Bunkyo-machi, Nagasaki 852-8521, Japan; bb55720006@ms.nagasaki-u.ac.jp (Y.M.); bb55621002@ms.nagasaki-u.ac.jp (A.I.); m.higuchi005@gmail.com (M.H.)

2 Faculty of Pharmacy, Osaka Medical and Pharmaceutical University, Osaka 569-1094, Japan; t.kato@gly.oups.ac.jp (T.K.); mitsunobu.doi@ompu.ac.jp (M.D.)

* Correspondence: aueda@nagasaki-u.ac.jp (A.U.); matanaka@nagasaki-u.ac.jp (M.T.); Tel.: +81-95-819-2425 (A.U.); +81-95-819-2423 (M.T.)

Abstract: Hydrocarbon stapling is a useful tool for stabilizing the secondary structure of peptides. Among several methods, hydrocarbon stapling at *i*,*i* + 1 positions was not extensively studied, and their secondary structures are not clarified. In this study, we investigate *i*,*i* + 1 hydrocarbon stapling between *cis*-4-allyloxy-L-proline and various olefin-tethered amino acids. Depending on the ring size of the stapled side chains and structure of the olefin-tethered amino acids, *E*- or *Z*-selectivities were observed during the ring-closing metathesis reaction (*E*/*Z* was up to 8.5:1 for 17–14-membered rings and up to 1:20 for 13-membered rings). We performed X-ray crystallographic analysis of hydrocarbon stapled peptide at *i*,*i* + 1 positions. The X-ray crystallographic structure suggested that the *i*,*i* + 1 staple stabilizes the peptide secondary structure to the right-handed α-helix. These findings are especially important for short oligopeptides because the employed stapling method uses two minimal amino acid residues adjacent to each other.

Keywords: peptide; α-helix; hydrocarbon stapling; ring-closing metathesis; *i*,*i* + 1 staple; X-ray structure

Citation: Makura, Y.; Ueda, A.; Kato, T.; Iyoshi, A.; Higuchi, M.; Doi, M.; Tanaka, M. X-ray Crystallographic Structure of α-Helical Peptide Stabilized by Hydrocarbon Stapling at *i*,*i* + 1 Positions. *Int. J. Mol. Sci.* **2021**, *22*, 5364. https://doi.org/10.3390/ijms22105364

Academic Editors: Menotti Ruvo and Nunzianna Doti

Received: 30 April 2021
Accepted: 17 May 2021
Published: 19 May 2021

1. Introduction

Introducing hydrocarbon stapling on the side chains of peptides is a promising technique for stabilizing the secondary structure of peptides and enhancing their functionalities [1–5]. Hydrocarbon stapling can be easily obtained by ring-closing metathesis reactions between olefin-bearing amino acid residues using Ru catalysts [6,7]. After the report on α-helicity-inducing all-hydrocarbon stapled peptides at *i*,*i* + 4 and *i*,*i* + 7 positions by Verdine et al. [8], several studies focused on the approach (as illustrated in Figure 1a) [9–11]. Currently, all-hydrocarbon stapled peptides are very important in drug development targeting protein–protein interactions because the pharmacophores interact via α-helical motifs [12]. Hydrocarbon stapling at *i*,*i* + 3 positions are reported in the literature [13–15]. For example, O'Leary et al. reported *E*-selective ring-closing metathesis between *O*-allyl-tethered L-serines at *i*,*i* + 3 positions to produce 3_{10}-helical peptides [13]. Other hydrocarbon staples, such as *i*,*i* + 1 and *i*,*i* + 2, were not well researched, and their 3D structures are unknown (as illustrated in Figure 1b) [16–19]. In general, hydrocarbon stapling sacrifices two amino acid residues for the crosslinking motif, and those residues should not include essential residues for their biological activities. Based on this, the development of a large variety of hydrocarbon stapling at different positions can be achieved. Herein, we report hydrocarbon stapling of peptides at *i*,*i* + 1 positions by ring-closing metathesis reactions and the X-ray crystallographic structure of the right-handed α-helical octapeptide stabilized by *i*,*i* + 1 stapling.

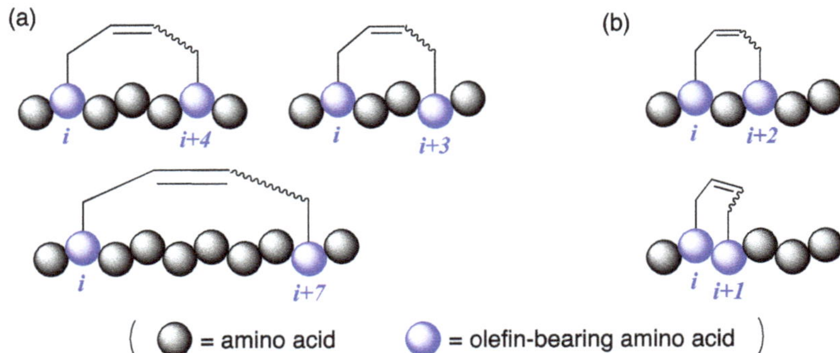

Figure 1. Peptides with hydrocarbon stapling at different positions. (**a**) Commonly used hydrocarbon stapling (at *i,i* + 4, *i,i* + 3, and *i,i* + 7), and (**b**) rarely investigated hydrocarbon stapling (at *i,i* + 2 and *i,i* + 1).

2. Results and Discussion

Our previous report suggests the usefulness of cis-4-hydroxy-L-proline as an olefin-bearing amino acid for peptide stapling [19]. Thus, in this study, we started by optimizing the reaction conditions for *i,i* + 1 peptide stapling using cis-4-hydroxy-L-proline. We screened the ring-closing metathesis reaction at *i,i* + 1 positions using dipeptide **1** as the cyclization precursor (as illustrated in Table 1). The reaction catalyzed by 20 mol% of second-generation Grubbs catalyst in CH_2Cl_2 (20 mM) produced the desired **1′** in 55% yield as a mixture of *E*/*Z*-isomers (*E*/*Z* = 1.0:5.6; Entry 1). A comparable result was obtained using the first-generation Grubbs catalyst (Entry 2). Replacing the reaction solvents, such as toluene, 1,2-dichloroethane (DCE) and tetrahydrofuran (THF), decreased the yields and *Z*-selectivities (Entries 3–5). The reaction under diluted condition (5 mM in CH_2Cl_2) afforded the best yield at 76% (Entry 6). The reactions in refluxing CH_2Cl_2 resulted in insufficient yields due to the degradation of the desired product (Entries 8 and 9).

Table 1. Screening of reaction conditions for ring-closing metathesis of dipeptide **1**.

Grubbs cat. (2nd or 1st, 20 mol %)
solvent, temp., time
1 → **1'**

Entry [1]	Catalyst (mol %)	Solvent (mM)	Temp. (°C)	Time (h)	Yield (%)	*E*/*Z* Ratio [2]
1	Grubbs 2nd (20)	CH_2Cl_2 (20)	rt	2	55	1.0:5.6
2	Grubbs 1st (20)	CH_2Cl_2 (20)	rt	2	53	1.0:4.8
3	Grubbs 2nd (20)	toluene (20)	rt	2	37	1.0:3.0
4	Grubbs 2nd (20)	DCE (20)	rt	2	39	1.0:4.8
5	Grubbs 2nd (20)	THF (20)	rt	2	27	1.0:4.4
6	Grubbs 2nd (20)	CH_2Cl_2 (5)	rt	2	76	1.0:5.0
7	Grubbs 1st (20)	CH_2Cl_2 (5)	rt	2	69	1.0:4.7
8	Grubbs 2nd (20)	CH_2Cl_2 (5)	reflux	2	28	1.0:4.9
9	Grubbs 2nd (20)	CH_2Cl_2 (5)	reflux	0.5	28	1.0:5.3

[1] Condition: 0.05 mmol of **1**. [2] Determined by ^1H NMR.

Further, we investigated the substrate scope for the ring-closing metathesis of peptides at *i,i* + 1 positions using the optimized reaction conditions (as illustrated in Scheme 1). As the ring size of the stapled peptides increased from 13- to 15-membered rings, the yields and *E*-selectivities increased (Entries 1–3). L-Tyrosine and D-serine-derived unstapled peptides **4** and **5** produced the desired stapled peptides **4′** and **5′** in 23% and 21% yields, respectively, with large amounts of unreacted starting material (Entries 4 and 5). Surprisingly, high *Z*-selectivities were observed for the reaction of dipeptides **6** and **7**, which were composed of either *O*-allyl-tethered L-threonine or (*S*)-α-(4-pentenyl)alanine (Entries 6 and 7; *E*/*Z* = 1: >20 for **6′** and 1:14 for **7′**). These results suggest that α-methyl or β-methyl groups of *i* + 1 residue strongly affect the transition state of the ring-closing metathesis to yield *Z*-isomers.

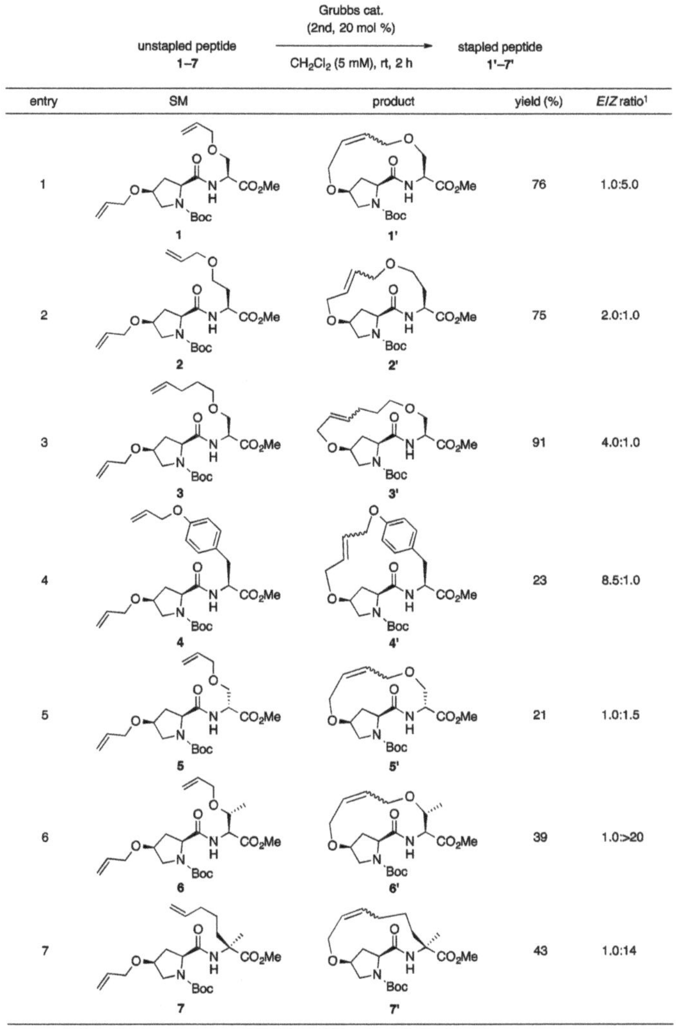

unstapled peptide **1–7** → Grubbs cat. (2nd, 20 mol %), CH_2Cl_2 (5 mM), rt, 2 h → stapled peptide **1′–7′**

entry	SM	product	yield (%)	*E*/*Z* ratio [1]
1	**1**	**1'**	76	1.0:5.0
2	**2**	**2'**	75	2.0:1.0
3	**3**	**3'**	91	4.0:1.0
4	**4**	**4'**	23	8.5:1.0
5	**5**	**5'**	21	1.0:1.5
6	**6**	**6'**	39	1.0:>20
7	**7**	**7'**	43	1.0:14

Scheme 1. Substrate scope for ring-closing metathesis of peptides **1–7** at *i*,*i* + 1 positions. [1] Determined by ^{1}H NMR.

The *i*,*i* + 1 hydrocarbon-stapling reaction of octapeptide **8**, in possession of 1-aminocycl oalkane-1-carboxylic acid [20–33], was investigated under the optimized reaction conditions for the ring-closing metathesis (Scheme 2). In contrast with the moderate *Z*-selectivity of **1** (E/Z = 1:5), much higher Z-selectivity was observed for the ring-closing metathesis reaction of **8** (E/Z = 1: >20). The Z-selectivity could be influenced by their secondary structure. Hydrogenation of **9** afforded saturated stapled peptide **10** in high yield. The high Z-selectivities (E/Z was up to 1: >20) of the *i,i* + 1 hydrocarbon stapling is advantageous for peptide staples compared to those reported for *i,i* + 4 and *i,i* + 7 hydrocarbon stapling (E/Z was up to 1: >9) [15].

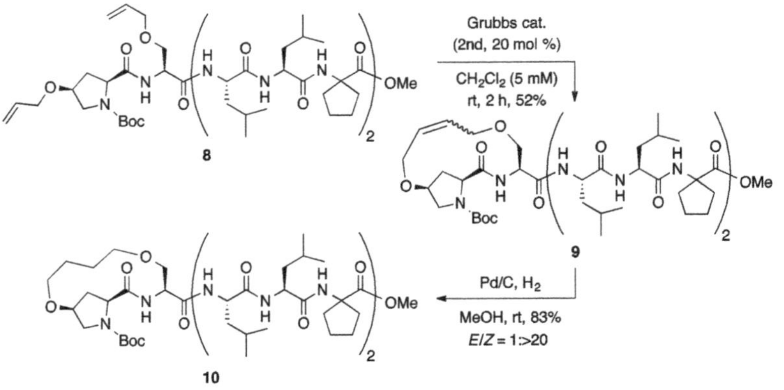

Grubbs cat. (2nd, 20 mol %)

CH_2Cl_2 (5 mM) rt, 2 h, 52%

Pd/C, H_2

MeOH, rt, 83%

E/Z = 1:>20

Scheme 2. Ring-closing metathesis of octapeptide **8** at *i,i* + 1 positions.

Crystals suitable for X-ray crystallographic analyses were successfully obtained by slow evaporation of the solution of **10** in *N,N*-dimethylformamide (DMF)/water at room temperature (20–30 °C) [34]. The structure was solved in the orthorhombic $P2_12_12_1$ space group to give an α-helical structure with a DMF molecule in the asymmetric unit (as illustrated in Figure 2 and Figure S1 and Tables 2 and 3, and Table S1). To the best of our knowledge, this is the first X-ray crystallographic structure of α-helical stapled peptides at *i* and *i* + 1 positions. In the crystal state of the (*i,i* + 1)-stapled peptide **10**, four consecutive intramolecular hydrogen bonds of the *i*←*i* + 4 type, N(4)H···O = C(0) (N···O, 3.09 Å; N–H···O, 163.6°), N(5)H···O = C(1) (N···O, 2.98 Å; N–H···O, 168.6°), N(6)H···O = C(2) (N···O, 2.91 Å; N–H···O, 157.2°), and N(7)H···O = C(3) (N···O, 3.14 Å; N–H···O, 139.7°) were observed. These hydrogen bonds indicate the existence of the α-helical secondary structure in **10**. The average torsion angles of **10** at the N-terminus [avg.(ϕ1–ϕ5) = −62.4° and avg.(Ψ1–Ψ5) = −46.5°] were much closer to the ideal values of a right-handed α-helix [ϕ = −57° and Ψ = −47°] [35]. Therefore, the crosslinkage of the *i,i* + 1 staples at the N-terminus could affect the stabilization of the α-helical structure of **10**. On the C-terminus, weak intramolecular hydrogen bonds of the *i*←*i* + 3 type were observed, N(7)H···O = C(4) (N···O, 3.37 Å; N–H···O, 136.7°) and N(8)H···O = C(5) (N···O, 3.40 Å; N–H···O, 162.7°), while the N(8)–H···O(4) angle of *i*←*i* + 4 type was too small for a hydrogen bond. These bifurcated hydrogen bonds suggest that the conformation of the C-terminus exists as a mixture of α- and 3_{10}-helix. Another intramolecular hydrogen bond between the N(2)–H of the main chain and ethereal oxygen of cis-4-hydroxyproline, N(2)H···O = C(Hyp^4) (N···O, 2.93 Å; N–H···O, 137.6°), was observed. Such hydrogen bond stabilizes the secondary structures of peptides [30,36,37]. On the other hand, no intermolecular hydrogen bonds between peptides were observed in the packing mode (Figure S2). These results suggest that packing contacts have a small or no influence on the secondary structure of right-handed α-helix in this case. Thus, introducing hydrocarbon stapling at *i,i* + 1 positions using cis-4-hydroxyproline could be used for the stabilization of α-helical peptides likewise *i,i* + 4 and *i,i* + 7 staples. In our previous study, we reported asymmetric Michael addition of 1-methylindole to α,β-unsaturated aldehydes catalyzed by Boc-deprotected **10** [19]. We hypothesized that the reactive iminium ion intermediate between cis-4-hydroxy-L-proline and α,β-unsaturated aldehyde was formed inside the helical pipe with a rigid conformation caused by *i,i* + 1 staple. The X-ray crystallographic structure of **10** supports this observed conformation of the intermediate.

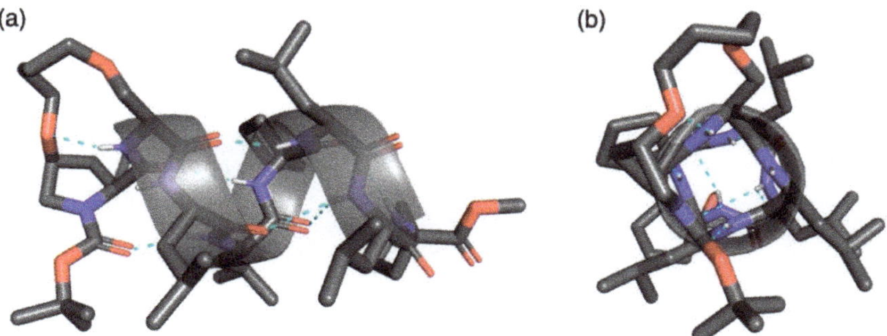

Figure 2. X-ray crystallographic structure of (*i,i* + 1)-stapled peptide **10**: a view (**a**) perpendicular to α-helical axis and (**b**) along helical axis from N-terminus.

Table 2. Crystal and diffraction parameters of peptide **10**.

Empirical Formula	$C_{54} H_{92} N_8 O_{13}$, $C_3 H_7 N O$
Formula weight	1134.45
Crystal dimensions (mm)	0.403 × 0.275 × 0.250
Data collection temp. (K)	93
Crystal system	orthorhombic
Lattice parameters	
a, b, c (Å)	11.357, 19.104, 29.332
α, β, γ (°)	90, 90, 90
V (Å^3)	6363.84
Space group	$P\,2_1\,2_1\,2_1$
Z value	4
D calc (g/cm^3)	1.184
μ (MoKα) (cm^{-1})	0.692
No. of variable	765
No. of observations	11273 ($I > 2\delta$(I))
R_1 ($I > 2\delta$(I))	0.0275
wR_2	0.0707
Crystallizing solvent	DMF/H_2O

Table 3. Intra and intermolecular H-bond parameters for peptide **10**.

Donor D–H	Acceptor A	Distance [Å] D···A	Angle [°] D–H···A	Symmetry Operations
N_4-H	O_0	3.09	163.6	*x,y,z*
N_5-H	O_1	2.98	168.6	*x,y,z*
N_6-H	O_2	2.91	157.2	*x,y,z*
N_7-H	O_3	3.14	139.7	*x,y,z*
N_7-H	O_4	3.37 [1]	136.7	*x,y,z*
N_8-H	O_4	3.23	102.9 [2]	*x,y,z*
N_8-H	O_5	3.40 [1]	162.7	*x,y,z*
N_2-H	O_{Hyp4}	2.93	137.6	*x,y,z*
N_3-H	O_{DMF}	2.90	159.8	$1/2 - x, 1 - y, -1/2 + z$

[1] Distance is a little long for an intramolecular hydrogen bond. [2] N–H···O angle is too small for a hydrogen bond.

In summary, we developed *i,i* + 1 peptide stapling between cis-4-allyloxy-L-proline and various olefin-tethered amino acids. Depending on the ring size of the stapled peptides, *E*- or *Z*-selectivities were observed. The *E*-configured stapled product was preferred when the product was greater than a 14-membered ring, whereas the Z-configured isomer was preferred when the product was a 13-membered ring. The α-or β-methyl substituent of the $i + 1$ residue improved the Z-selectivities of the ring-closing metathesis (E:Z = 1: >20). X-ray crystallographic analysis of the octapeptide **10** revealed a stabilized α-helical structure. These results are useful for developing peptide-based organocatalysts [38–40] (i.e., considering mechanistic insights and structural modification of peptide catalysts based on the X-ray crystal structure), fluorinated peptides [41] (e.g., stabilization effects of using intramolecular hydrogen bonds beside main chain hydrogen bonds), and peptide-based drug delivery systems [42–46] (e.g., introducing $i,i + 1$ hydrocarbon stapling with essential residues for their biological activities remained intact).

3. Materials and Methods

3.1. General Procedure and Method

Melting points were taken on an AS ONE melting point apparatus ATM-01 (AS ONE Corporation, Osaka, Japan) and were uncorrected. Optical rotations were measured on a JASCO DIP-370 polarimeter (JASCO Corporation, Tokyo, Japan) using $CHCl_3$ as a solvent. ^{1}H NMR and ^{13}C NMR spectra were recorded on the JEOL JNM-AL-400 (400 MHz), a Varian NMR System 500PS SN (500 MHz and 125 MHz) spectrometer (Agilent Inc., Santa Clara, CA, USA). Chemical shifts (δ) are reported in parts per million (ppm). For the ^{1}H NMR spectra ($CDCl_3$), tetramethylsilane was used as the internal reference (0.00 ppm), while the central solvent peak was used as the reference (77.0 ppm in $CDCl_3$) for the ^{13}C NMR spectra. The IR spectra were recorded on a Shimadzu IRAffinity-1 FT-IR spectrophotometer (Shimadzu Corporation, Kyoto, Japan). High-resolution mass spectra (HRMS) were obtained on a JEOL JMS-T100TD using electrospray ionization (ESI) (JEOL Ltd., Tokyo, Japan) or direct analysis in the realtime (DART) ionization in time-of-flight TOF mode. Analytical and semipreparative thin layer chromatography (TLC) was performed with Merck Millipore precoated TLC plates (MilliporeSigma, Burlington, VT, USA), silica gel 60 F_{254}, and layer thicknesses of 0.25 and 0.50 mm, respectively. Compounds were observed in UV light at 254 nm and then visualized by staining with iodine, p-anisaldehyde, or phosphomolybdic acid stain. Flash and gravity column chromatography separations were performed on Kanto Chemical silica gel 60N, spherical neutral, with particle sizes of 63–210 μm and 40–50 μm, respectively. All moisture-sensitive reactions were conducted under an inert atmosphere. Reagents and solvents were of commercial grade and were used as supplied, unless otherwise noted. Compounds **1**, **8** [19], **S-1** [47,48], **S-2** [49,50], **S-3** [51], and **S-5** [52] were prepared according to the reported procedures. Copies of NMR Spectra are given in the Supplementary Materials.

3.2. Synthesis of Unstapled Dipeptides **2–7**

H_2N CO_2Me **S-2** → (**S-1**, CO_2H, Boc; EDCI, HOBt) → CO_2Me, Boc **2**

Boc-L-Hyp^{OAll}-L-Hse^{OAll}-OMe (**2**): to a solution of N-*tert*-butoxycarbonyl 4-O-allyl-*cis*-4-hydroxy-L-proline (Boc-L-Hyp^{OAll}-OH, **S-1** [47,48]; 88.1 mg, 0.325 mmol) in CH_2Cl_2 (2 mL) were added N-(3-dimethylaminopropyl)-N'-ethylcarbodiimide hydrochloride (EDCI·HCl, 67.9 mg, 0.354 mmol) and 1-hydroxybenzotriazole hydrate (HOBt·H_2O; 54.2 mg, 0.354 mmol) at 0 °C, and the solution was stirred for 30 min at 0 °C. Then, a solution of O-allyl-L-homoserine methyl ester (H-L-Hse^{OAll}-OMe, **S-2** [49,50], 51.1 mg, 0.295 mmol) in CH_2Cl_2 (1 mL) was added to the reaction mixture at the same temperature, and the resultant mixture was gradually warmed to room temperature. After stirring for three days, CH_2Cl_2 was removed, and the residue was diluted with EtOAc. The solution was washed successively with 1 M of HCl, water, sat. aq $NaHCO_3$, and brine. The organic layer was dried over anhydrous Na_2SO_4 and concentrated in vacuo to give a crude product, which was purified by flash column chromatography on silica gel (40% EtOAc in *n*-hexane) to give **2** (72.1 mg, 58%) as a pale yellow oil. R_f = 0.58 (EtOAc). $[\alpha]_D^{20}$ −11.0 (*c* 1.00, $CHCl_3$). ^1H NMR (500 MHz, $CDCl_3$) δ: 7.38–7.17 (m, 1H), 5.98–5.76 (m, 2H), 5.34–5.09 (m, 4H), 4.72–4.56 (m, 1H), 4.42–4.25 (m, 1H), 4.11–4.05 (m, 1H), 4.05–3.84 (m, 4H), 3.73 (s, 0.6H), 3.72 (s, 2.4H), 3.63–3.39 (m, 4H), 2.65–2.41 (m, 1H), 2.27–1.96 (m, 3H), 1.48 (s, 9H). ^{13}C NMR (125 MHz, $CDCl_3$) δ: 172.5, 172.0, 171.0, 154.7, 134.5, 134.4, 134.34, 134.26, 117.3, 117.2, 117.1, 117.0, 80.9, 76.3, 72.04, 71.98, 69.6, 66.3, 66.0, 60.1, 52.7, 52.3, 52.1, 50.6, 50.4, 36.9, 35.6, 31.6, 28.3, 28.1. IR (film): 3385 (br), 2978, 2868, 1744, 1690 cm^{-1}. HRMS (ESI) *m*/*z*: $[M + Na]^+$ calcd. for $C_{21}H_{34}N_2O_7Na$, 449.2264; found, 449.2262.

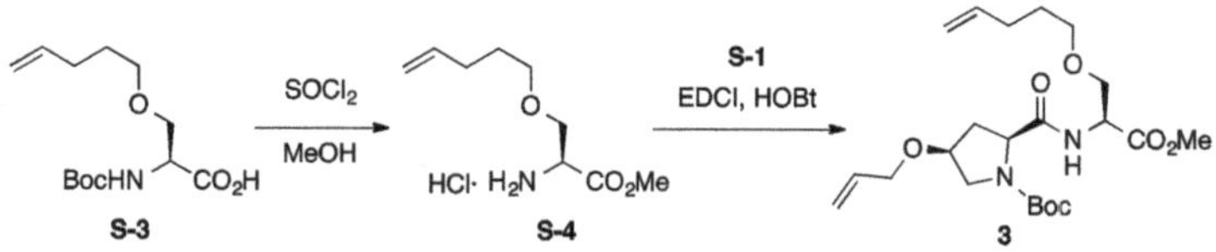

Boc-L-Hyp^{OAll}-L-Ser^{OPte}-OMe (**3**): to a solution of carboxylic acid **S-3** [51] (135 mg, 0.495 mmol) in MeOH (5 mL), thionyl chloride (0.143 mL, 1.98 mmol) was added dropwise at 0 °C. The reaction mixture was stirred at room temperature for 2 h and was concentrated to give H-L-Ser^{OPte}-OMe·HCl (**S-4**, R_f = 0.57 with 0.5% AcOH in EtOAc), which was used for the next step without further purification. To a mixture of H-L-Ser^{OPte}-OMe·HCl (**S-4**, 0.495 mmol) and Boc-L-Hyp^{OAll}-OH (**S-1**, 148 mg, 0.545 mmol) in CH_2Cl_2 (5 mL) were added EDCI·HCl (114 mg, 0.594 mmol), HOBt·H_2O (91.0 mg, 0.594 mmol), and DIPEA (0.253 mL, 1.49 mmol) at 0 °C, and the mixture was gradually warmed to room temperature. After stirring for 17 h, CH_2Cl_2 was removed under vacuum, and the residue was diluted with EtOAc. The resultant solution was washed successively with 1 M of HCl, water, sat. aq $NaHCO_3$, and brine. The organic layer was dried over anhydrous Na_2SO_4 and concentrated in vacuo to give a crude product, which was purified by flash column chromatography on silica gel (40% EtOAc in *n*-hexane) to give **3** (90.1 mg, 41% in 2 steps) as a pale yellow oil. R_f = 0.71 (EtOAc). $[\alpha]_D^{22}$ −2.6 (*c* 1.00, $CHCl_3$). ^1H NMR (500 MHz, $CDCl_3$) δ: 7.29 (br s, 1H), 7.11–6.90 (m, 1H), 5.98–5.71 (m, 2H), 5.34–5.22 (m, 1H), 5.21–5.11 (m, 1H), 5.05–4.92 (m, 2H), 4.77–4.63 (m, 1H), 4.44–4.28 (m, 1H), 4.11–3.92 (m, 2H), 3.92–3.78 (m, 2H), 3.75 (s, 3H), 3.66–3.48 (m, 3H), 3.47–3.35 (m, 2H), 2.67–2.45 (m, 1H), 2.24–2.02 (m, 3H), 1.67–1.57 (m, 2H), 1.49 (s, 9H). ^{13}C NMR (125 MHz, $CDCl_3$) δ: 172.1, 171.2, 170.7, 170.4, 154.7, 138.0, 134.4, 134.2, 117.2, 116.9, 114.79, 114.75, 81.0, 76.1, 72.0, 70.7, 70.6, 70.31, 70.27, 69.4, 65.9, 60.0, 52.8, 52.51, 52.45, 52.39, 52.2, 36.8, 35.3, 30.01, 29.99, 28.4, 28.2, 28.1. IR (film): 3428 (br), 2978, 2918, 1753, 1692 cm^{-1}. HRMS (ESI) *m*/*z*: $[M + Na]^+$ calcd. for $C_{22}H_{36}N_2O_7Na$, 463.2420; found, 463.2418.

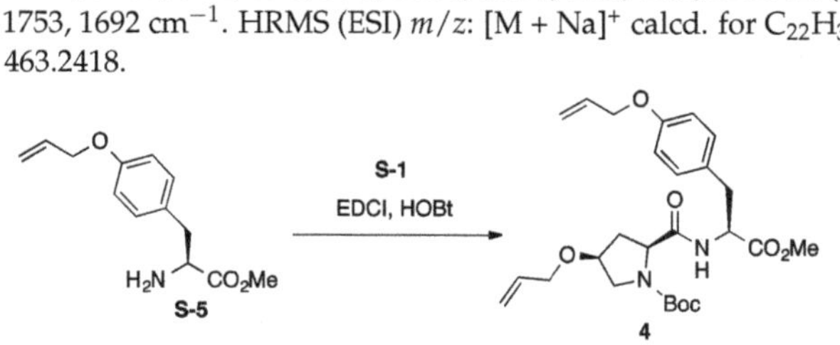

Boc-L-Hyp^{OAll}-L-Tyr^{OAll}-OMe (**4**): to a solution of Boc-L-Hyp^{OAll}-OH (**S-1**, 445 mg, 1.64 mmol) in CH_2Cl_2 (8 mL) were added EDCI·HCl (314 mg, 1.64 mmol) and HOBt·H_2O (301 mg, 1.97 mmol) at 0 °C, and the reaction mixture was stirred for 30 min at 0 °C. Then, a solution of *O*-allyl-L-tyrosine methyl ester (H-L-Tyr^{OAll}-OMe, **S-5** [52], 386 mg, 1.64 mmol) in CH_2Cl_2 (3 mL) was added to the reaction mixture at the same temperature, and the resultant mixture was gradually warmed to room temperature. After stirring for 35 h, CH_2Cl_2 was removed in vacuo, and the residue was diluted with EtOAc. The resultant solution was washed successively with 1 M of HCl, water, sat. aq $NaHCO_3$, and brine. The organic layer was dried over anhydrous Na_2SO_4 and concentrated in vacuo to give a crude product, which was purified by flash column chromatography on silica gel (50% EtOAc in *n*-hexane) to give **4** (562 mg, 70%) as a pale yellow oil. R_f = 0.75 (EtOAc). $[\alpha]_D^{23}$ +0.90 (*c* 1.00, $CHCl_3$). ^{1}H NMR (500 MHz, $CDCl_3$) δ: 7.10–6.98 (m, 2H), 6.87–6.70 (m, 3H), 6.10–5.99 (m, 1H), 5.90–5.77 (m, 1H), 5.40 (dp, *J* = 17.2, 1.7 Hz, 1H), 5.31–5.20 (m, 2H), 5.19–5.10 (m, 1H), 4.89–4.76 (m, 1H), 4.54–4.45 (m, 2H), 4.42–4.21 (m, 1H), 4.12–4.02 (m, 1H), 4.00–3.91 (m, 1H), 3.91–3.83 (m, 1H), 3.65 (s, 3H), 3.55 (br s, 2H), 3.13–2.99 (m, 1H), 2.94 (br s, 1H), 2.53–2.41 (m, 1H), 2.21–1.95 (m, 1H), 1.38 (s, 9H). ^{13}C NMR (125 MHz, $CDCl_3$) δ: 171.9, 171.6, 171.5, 171.1, 171.0, 157.6, 155.4, 154.5, 134.3, 134.1, 133.22, 133.18, 130.5, 130.2, 128.0, 127.8, 117.6, 117.5, 117.3, 117.2, 114.7, 114.6, 114.4, 81.0, 76.1, 72.0, 69.5, 68.69, 68.67, 65.9, 60.1, 59.3, 53.7, 53.3, 53.1, 52.9, 52.2, 52.0, 37.3, 37.2, 36.9, 36.8, 35.0, 32.5, 28.2, 28.0. IR (film): 3424 (br), 2978, 2934, 1744, 1665 cm^{-1}. HRMS (ESI) *m/z*: [M + Na]$^+$ calcd. for $C_{26}H_{36}N_2O_7Na$, 511.2420; found, 511.2422.

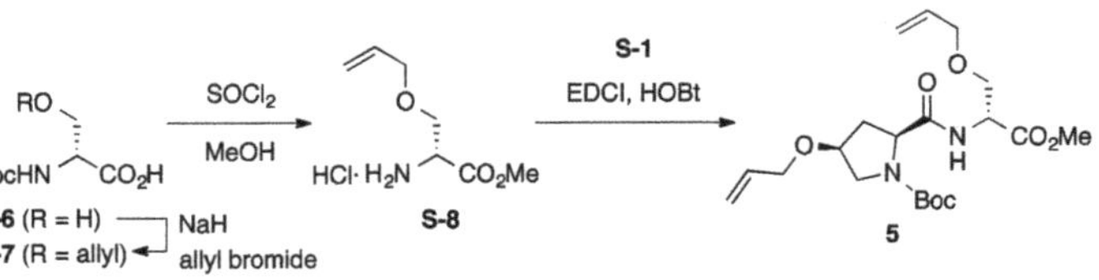

RO / BocHN–CO2H: S-6 (R = H) → S-7 (R = allyl) (NaH, allyl bromide); SOCl2, MeOH → HCl·H2N–CO2Me S-8; S-1, EDCI, HOBt → 5

Boc-L-*Hyp*^{OAll}-D-*Ser*^{OAll}-*OMe* (**5**): to a solution of Boc-D-Ser-OH (**S-6**, 2.05 g, 10.0 mmol) in DMF (35 mL) was added sodium hydride (60% in mineral oil, 880 mg, 22.0 mmol) portionwise at −15 °C, and the reaction mixture was stirred at the same temperature for 2 h. To the above suspension, allyl bromide (0.952 mL, 11.0 mmol) was added dropwise at −15 °C, and the reaction mixture was stirred at room temperature for 14 h. The reaction mixture was quenched by adding water and washed twice with Et_2O. The aqueous phase was acidified with 1 M of HCl, which was extracted with EtOAc three times. The combined organic layers were washed with water and brine, dried over anhydrous Na_2SO_4, and concentrated under vacuum. The residue was purified by flash column chromatography on silica gel (40% EtOAc in *n*-hexane) to give Boc-D-Ser^{OAll}-OH (**S-7**, 1.68 g, 69%, R_f = 0.28 with 10% MeOH in EtOAc) as a pale yellow oil. To a solution of **S-7** (123 mg, 0.500 mmol) in MeOH (5 mL) was added thionyl chloride (0.145 mL, 2.00 mmol) dropwise at 0 °C. The reaction mixture was stirred at room temperature for 2 h and was concentrated under vacuum to give crude H-D-Ser^{OAll}-OMe·HCl (**S-8**, R_f = 0.57 with 0.5% AcOH in EtOAc), which was used for the next step without further purification. To a mixture of H-D-Ser^{OAll}-OMe·HCl (**S-8**, 0.500 mmol) and Boc-L-Hyp^{OAll}-OH (**S-1**, 149 mg, 0.550 mmol) in CH_2Cl_2 (5 mL) were added EDCI·HCl (115 mg, 0.600 mmol), HOBt·H_2O (91.9 mg, 0.600 mmol), and DIPEA (0.255 mL, 1.50 mmol) at 0 °C, and the reaction mixture was gradually warmed to room temperature. After stirring for 17 h, CH_2Cl_2 was removed under vacuum, and the residue was diluted with EtOAc. The organic solution was washed successively with 1 M of HCl, water, sat. aq $NaHCO_3$, and brine. The organic layer was dried over anhydrous Na_2SO_4 and concentrated in vacuo to give a crude product, which was purified by flash column chromatography on silica gel (40% EtOAc in *n*-hexane) to give **5** (84.6 mg, 41% in 2 steps) as a pale yellow oil. R_f = 0.66 (EtOAc). $[\alpha]_D^{23}$ −22.6 (*c* 1.00, $CHCl_3$). ^{1}H NMR (500 MHz, $CDCl_3$) δ: 7.43–7.06 (m, 1H), 5.99–5.75 (m, 2H), 5.35–5.08 (m, 4H), 4.80–4.60 (m, 1H), 4.42–4.22 (m, 1H), 4.14–3.81 (m, 6H), 3.75 (s, 3H), 3.72–3.41 (m, 3H), 2.60–2.36 (m, 1H), 2.32–2.07 (m, 1H), 1.65–1.24 (m, 9H). ^{13}C NMR (125 MHz, $CDCl_3$) δ: 172.6, 171.2, 170.7, 170.3, 154.6, 134.4, 134.2, 134.0, 133.9, 117.4, 117.3, 117.2, 116.9, 80.7, 75.8, 72.14, 72.12, 72.0, 69.61, 69.59, 69.4, 65.9, 60.3, 59.7, 53.2, 52.6, 52.52, 52.46, 52.3, 36.8, 35.2, 33.4, 28.2. IR (film): 3325 (br), 2978, 2932, 1753, 1692 cm^{-1}. HRMS (ESI) *m/z*: [M + Na]$^+$ calcd. for $C_{20}H_{32}N_2O_7Na$, 435.2107; found, 435.2106.

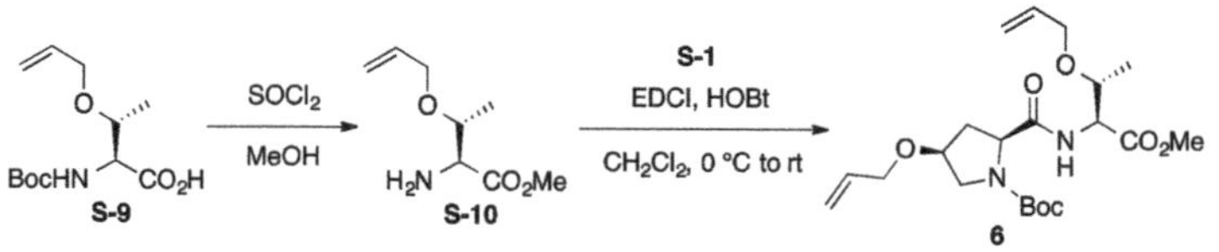

*Boc-*L*-Hyp*^{OAll}*-*L*-Thr*^{OAll}*-OMe* (**6**): to a solution of *N-tert*-butoxycarbonyl *O*-allyl-L-threonine (Boc-L-Thr^{OAll}-OH, **S-9**; 130 mg, 0.500 mmol) in MeOH (2.5 mL) was added thionyl chloride (0.144 mL, 2.00 mmol) dropwise at 0 °C. The reaction mixture was stirred at room temperature for 3 h prior to the addition of sat. $NaHCO_3$ aq. After removal of MeOH by evaporation, the aqueous residue was extracted with $CHCl_3$ (five times) and the combined organics were dried over anhydrous Na_2SO_4. Concentration of the solution gave H-L-Thr^{OAll}-OMe (**S-10**, 41.9 mg, 48%), which was used for the next step without further purification. To a solution of Boc-L-Hyp^{OAll}-OH (**S-1**, 72.1 mg, 0.266 mmol) in CH_2Cl_2 (0.8 mL) were added EDCI·HCl (51.0 mg, 0.266 mmol) and HOBt·H_2O (48.2 mg, 0.315 mmol) at 0 °C, and the solution was stirred for 30 min at 0 °C. Then, a solution of H-L-Thr^{OAll}-OMe (**S-10**, 41.9 mg, 0.242 mmol) in CH_2Cl_2 (0.8 mL) was added to the reaction mixture at the same temperature, and the resultant mixture was gradually warmed to room temperature. After stirring for 42 h, CH_2Cl_2 was removed, and the residue was diluted with EtOAc. The solution was washed successively with 1 M of HCl, water, sat. aq $NaHCO_3$, and brine. The organic layer was dried over anhydrous Na_2SO_4 and concentrated in vacuo to give a crude product, which was purified by flash column chromatography on silica gel (40% EtOAc in *n*-hexane) to give **6** (57.6 mg, 56%) as a pale yellow oil. R_f = 0.58 (EtOAc). $[\alpha]_D^{23}$ −11.2 (*c* 1.00, $CHCl_3$). ^{1}H NMR (400 MHz, $CDCl_3$) δ: 6.92 (s, 1H), 5.98–5.70 (m, 2H), 5.35–5.05 (m, 4H), 4.71–4.55 (m, 1H), 4.45–4.29 (m, 1H), 4.15–3.79 (m, 6H), 3.74 (s, 0.6H), 3.73 (s, 2.4H), 3.63–3.46 (m, 2H), 2.70–2.40 (m, 1H), 2.30–2.10 (m, 1H), 1.49 (s, 9H), 1.18 (d, *J* = 6.3 Hz, 0.6H), 1.13 (d, *J* = 6.4 Hz, 2.4H). ^{13}C NMR (100 MHz, $CDCl_3$) δ: 172.5, 171.8, 171.1, 170.6, 154.9, 134.5, 134.2, 117.2, 117.0, 116.8, 81.0, 76.0, 74.4, 74.2, 72.0, 69.8, 69.7, 66.0, 60.2, 56.4, 52.7, 52.2, 52.1, 36.9, 35.6, 28.1, 16.3, 16.1. IR (film): 3441 (br), 2978, 2934, 1753, 1703 cm^{-1}. HRMS (DART) *m*/*z*: $[M + H]^+$ calcd. for $C_{21}H_{35}N_2O_7$, 427.2444; found, 427.2437.

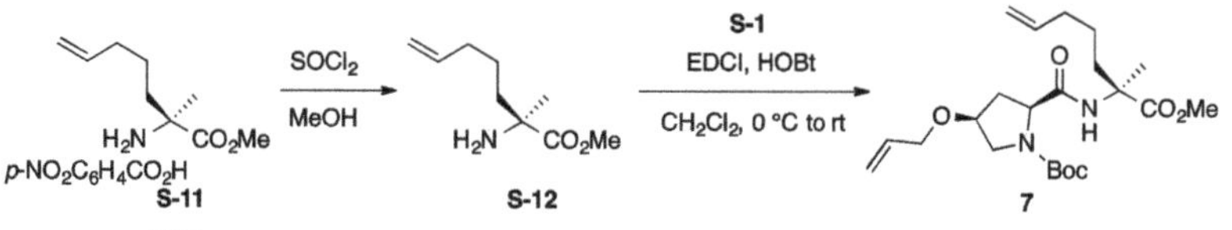

*Boc-*L*-Hyp*^{OAll}*-(S)-Ala(4-Pte)-OMe* (**7**): to a solution of *p*-nitrobenzoic acid salt of (*S*)-(4-pentenyl)alanine *tert*-butyl ester (H-(*S*)-Ala(4-Pte)-O^{t}Bu·*p*-$NO_2C_6H_4CO_2H$, **S-11**; 100 mg, 0.263 mmol) in MeOH (3 mL) was added thionyl chloride (0.152 mL, 2.10 mmol) dropwise at 0 °C. The reaction mixture was stirred at 65 °C for 69 h prior to the addition of sat. $NaHCO_3$ aq. After removal of MeOH by evaporation, the aqueous residue was extracted with $CHCl_3$ (five times) and the combined organics were dried over anhydrous Na_2SO_4. Concentration of the solution gave H-(*S*)-Ala(4-Pte)-OMe (**S-12**) contaminated with *p*-$NO_2C_6H_4CO_2Me$, which was used for the next step without further purification. To a solution of Boc-L-Hyp^{OAll}-OH (**S-1**, 60.5 mg, 0.223 mmol) in CH_2Cl_2 (1.5 mL) were added EDCI·HCl (42.8 mg, 0.223 mmol) and HOBt·H_2O (40.4 mg, 0.264 mmol) at 0 °C, and the solution was stirred for 30 min at 0 °C. Then, a solution of H-(*S*)-Ala(4-Pte)-OMe (**S-12**) in CH_2Cl_2 (0.5 mL) was added to the reaction mixture at the same temperature, and the resultant mixture was gradually warmed to room temperature. After stirring at room temperature for 3 d, CH_2Cl_2 was removed, and the residue was diluted with EtOAc. The solution was washed successively with 1 M of HCl, water, sat. aq $NaHCO_3$, and brine. The organic layer was dried over anhydrous Na_2SO_4 and concentrated in vacuo to give a crude product, which was purified by flash column chromatography on silica gel (40% EtOAc in *n*-hexane) to give **7** (46.4 mg, 42% in 2 steps) as a pale yellow oil. R_f = 0.58 (EtOAc). $[\alpha]_D^{23}$ −16.2 (*c* 1.00, $CHCl_3$). ^{1}H NMR (500 MHz, $CDCl_3$) δ: 7.28–6.82 (m, 1H), 5.99–5.79 (m, 1H),

5.79–5.67 (m, 1H), 5.35–5.22 (m, 1H), 5.22–5.13 (m, 1H), 5.03–4.90 (m, 2H), 4.33–4.13 (m, 1H), 4.10–3.84 (m, 3H), 3.77–3.71 (m, 3H), 3.71–3.61 (m, 1H), 3.56–3.46 (m, 1H), 2.50–2.32 (m, 1H), 2.30–2.06 (m, 2H), 2.05–1.98 (m, 2H), 1.83–1.74 (m, 1H), 1.58 (s, 1H), 1.53 (s, 2H), 1.48 (s, 9H), 1.43–1.34 (m, 1H), 1.23–1.13 (m, 1H). ^{13}C NMR (125 MHz, $CDCl_3$) δ: 174.8, 174.4, 171.4, 170.6, 154.9, 138.11, 138.06, 134.4, 134.2, 117.31, 117.25, 114.9, 114.8, 80.8, 80.6, 76.1, 72.1, 69.7, 66.2, 60.7, 60.1, 59.7, 52.9, 52.6, 52.4, 37.6, 36.5, 36.1, 35.5, 33.5, 33.4, 28.2, 23.4, 23.1, 23.0, 22.6. IR (film): 3393 (br), 2978, 2936, 1740, 1692 cm^{-1}. HRMS (DART) m/z: [M + H]$^+$ calcd. for $C_{22}H_{37}N_2O_6$, 425.2652; found, 425.2651.

3.3. Synthesis of Stapled Dipeptides **1′–7′**

Boc-L-Hyp^{OX}-L-Ser^{OX}-OMe (**1′**; X = *n*-but-2-enyl tether): to a solution of unstapled peptide **1** [20] (20.5 mg, 0.0500 mmol) in degassed CH_2Cl_2 (10 mL) was added second-generation Grubbs catalyst (8.5 mg, 0.010 mmol) at room temperature under an argon atmosphere. The reaction mixture was stirred at the same temperature for 2 h and then passed through a short plug of amino silica gel/silica gel, which was eluted with EtOAc. After removal of the solvent, the residue was purified by flash column chromatography on silica gel (70% EtOAc in *n*-hexane) to give **1′** (14.6 mg, 76%) as a colorless oil. R_f = 0.32 (EtOAc). ^{1}H NMR (500 MHz, $CDCl_3$) δ: 7.23–7.00 (m, 1H), 5.86–5.73 (m, 1H), 5.68 (dt, J = 11.9, 6.5 Hz, 1H), 4.87–4.57 (m, 1H), 4.45–4.15 (m, 2H), 4.10–3.78 (m, 6H), 3.76 (s, 3H), 3.72–3.62 (m, 1H), 3.45 (dd, J = 12.1, 3.9 Hz, 1H), 2.66–2.47 (m, 1H), 2.26–2.11 (m, 1H), 1.60–1.39 (m, 9H). ^{13}C NMR (125 MHz, $CDCl_3$) δ: 172.2, 171.4, 170.2, 154.9, 130.4, 130.0, 129.4, 81.0, 80.8, 78.6, 77.9, 77.3, 67.7, 67.2, 66.4, 66.0, 65.1, 60.2, 59.8, 54.0, 53.2, 52.5, 52.5, 35.7, 34.2, 30.9, 29.7, 28.2. HRMS (DART) m/z: [M + H]$^+$ calcd. for $C_{18}H_{29}N_2O_7$, 385.1975; found, 385.1970.

Boc-L-Hyp^{OX}-L-Hse^{OX}-OMe (**2′**; X = *n*-but-2-enyl tether): compound **2′** (15.0 mg, 75%) was obtained from compound **2** (21.3 mg, 0.0500 mmol) in a similar manner to that described for the synthesis of **1′**. Colorless oil. Eluent for column: 70% EtOAc/*n*-hexane. R_f = 0.34 (EtOAc). ^{1}H NMR (500 MHz, $CDCl_3$) δ: 6.78 (s, 1/3H), 6.41 (s, 2/3H), 5.86 (ddd, J = 10.8, 8.6, 6.6 Hz, 1/3H), 5.75 (dt, J = 15.7, 5.9 Hz, 2/3H), 5.74–5.59 (m, 1H), 4.78–4.66 (m, 1H), 4.35 (d, J = 10.0 Hz, 1H), 4.24–3.95 (m, 3H), 3.89–3.74 (m, 2H), 3.73 (s, 1H), 3.72 (s, 2H), 3.69–3.58 (m, 1H), 3.57–3.36 (m, 3H), 2.28–1.99 (m, 2H), 1.87–1.64 (m, 2H), 1.49 (s, 9H). ^{13}C NMR (125 MHz, $CDCl_3$) δ: 172.8, 172.2, 131.9, 131.7, 131.3, 128.7, 81.2, 75.9, 69.5, 68.5, 66.4, 63.5, 60.6, 60.2, 53.6, 52.3, 52.2, 49.4, 48.8, 34.7, 32.3, 28.2. HRMS (ESI) m/z: [M + Na]$^+$ calcd. for $C_{19}H_{30}N_2O_7Na$, 421.1951; found, 421.1954.

Boc-L-Hyp^{OX}-L-Ser^{OX}-OMe (**3′**; X = *n*-hex-2-enyl tether): compound **3′** (18.8 mg, 91%) was obtained from compound **3** (22.0 mg, 0.0500 mmol) in a manner similar to that described for the synthesis of **1′**. Colorless oil. Eluent for column: 60% EtOAc/*n*-hexane. R_f = 0.42 (EtOAc). ^{1}H NMR (500 MHz, $CDCl_3$) δ: 7.85 (d, J = 8.1 Hz, 0.2H), 7.26–7.09 (m, 0.8H), 5.93–5.81 (m, 0.8H), 5.66 (dt, J = 14.9, 5.9 Hz, 0.1H), 5.60 (ddd, J = 9.1, 7.4, 6.2 Hz, 0.2H), 5.47 (dt, J = 15.3, 4.2 Hz, 0.7H), 5.42 (td, J = 10.1, 10.0, 5.1 Hz, 0.2H), 4.83–4.64 (m, 1H), 4.43–4.24 (m, 1H), 4.12 (t, J = 4.4 Hz, 0.2H), 4.04 (t, J = 3.8 Hz, 0.8H), 3.92–3.77 (m, 3H), 3.76 (s, 0.6H), 3.74 (s, 2.4H), 3.72–3.53 (m, 3H), 3.46–3.36 (m, 1H), 3.30 (td, J = 9.6, 3.4 Hz, 1H), 2.61–1.99 (m, 4H), 1.78–1.56 (m, 2H), 1.55–1.37 (m, 9H). ^{13}C NMR (125 MHz, $CDCl_3$) δ: 172.7, 172.1, 170.8, 170.1, 155.2, 136.4, 133.8, 132.9, 131.9, 125.5, 124.3, 124.1, 80.9, 77.3, 72.7, 72.2, 70.7, 70.4, 70.3, 68.6, 68.1, 65.7, 60.5, 53.0, 52.8, 52.6, 52.4, 52.3, 36.4, 35.8, 32.3, 31.7, 28.6, 28.3, 28.1, 28.0, 27.9, 22.7. HRMS (ESI) m/z: [M + Na]$^+$ calcd. for $C_{20}H_{32}N_2O_7Na$, 435.2107; found, 435.2117.

Boc-L-Hyp^{OX}-L-Tyr^{OX}-OMe (**4′**; X = *n*-but-2-enyl tether): compound **4′** (5.4 mg, 23%) was obtained from compound **4** (24.4 mg, 0.0500 mmol) in a similar manner to that described for the synthesis of **1′**. White solid. Eluent for column: 50% EtOAc/*n*-hexane. R_f = 0.61 (EtOAc). ^{1}H NMR (500 MHz, $CDCl_3$) δ: 7.17 (d, J = 8.5 Hz, 1H), 6.96–6.82 (m, 2H), 6.76 (s, 1H), 6.25 (s, 1H), 5.58 (dt, J = 15.1, 4.9 Hz, 1H), 5.48 (dt, J = 15.1, 6.6, 5.6 Hz, 1H), 4.94 (ddd,

J = 10.9, 8.9, 4.5 Hz, 1H), 4.63 (d, J = 5.1 Hz, 2H), 4.18–3.82 (m, 3H), 3.79 (s, 3H), 3.77–3.54 (m, 2H), 3.37 (dd, J = 14.1, 4.5 Hz, 1H), 3.09 (s, 1H), 2.67 (t, J = 12.4 Hz, 1H), 2.39–1.82 (m, 2H), 1.46 (s, 9H). HRMS (ESI) m/z: $[M + Na]^+$ calcd. for $C_{24}H_{32}N_2O_7Na$, 483.2107; found, 483.2105.

Boc-L-Hyp^{OX}-D-Ser^{OX}-OMe (**5′**; X = n-but-2-enyl tether): compound **5′** (4.4 mg, 21%) was obtained from compound **2** (20.6 mg, 0.0500 mmol) in a similar manner to that described for the synthesis of **1′**. Colorless oil. Eluent for column: 70% EtOAc/n-hexane. R_f = 0.39 (EtOAc). ^{1}H NMR (500 MHz, $CDCl_3$) δ: 7.41 (br s, 0.6H), 7.17 (br s, 0.4H), 5.95–5.85 (m, 1H), 5.85–5.66 (m, 1H), 4.55–4.18 (m, 4H), 4.14 (t, J = 4.1 Hz, 1H), 4.02–3.94 (m, 1H), 3.93–3.84 (m, 2H), 3.82 (m, 1.2H), 3.77 (s, 1.8H), 3.73–3.56 (m, 2H), 3.45 (dd, J = 12.2, 4.1 Hz, 1H), 2.62 (d, J = 15.2 Hz, 0.6H), 2.50 (d, J = 14.0 Hz, 0.4H), 2.29–2.12 (m, 1H), 1.45 (d, J = 8.9 Hz, 9H). HRMS (ESI) m/z: $[M + Na]^+$ calcd. for $C_{18}H_{28}N_2O_7Na$, 407.1794; found, 407.1790.

Boc-L-Hyp^{OX}-L-Thr^{OX}-OMe (**6′**; X = n-but-2-enyl tether): compound **6′** (7.8 mg, 39%) was obtained from compound **6** (21.3 mg, 0.0500 mmol) in a similar manner to that described for the synthesis of **1′**. Colorless oil. Eluent for column: 70% EtOAc/n-hexane. R_f = 0.32 (EtOAc). ^{1}H NMR (500 MHz, $CDCl_3$) δ: 7.02 (d, J = 8.2 Hz, 1H), 5.86 (dt, J = 11.5, 6.7 Hz, 1H), 5.78 (dt, J = 11.5, 6.2 Hz, 1H), 4.72–4.54 (m, 1H), 4.39–4.21 (m, 2H), 4.16 (dd, J = 11.8, 6.5 Hz, 1H), 4.12–4.06 (m, 1H), 4.01 (dd, J = 11.8, 6.2 Hz, 1H), 3.89–3.76 (m, 2H), 3.72 (s, 3H), 3.76–3.62 (m, 1H), 3.45 (dd, J = 12.0, 3.3 Hz, 1H), 2.51 (d, J = 14.8 Hz, 1H), 2.27–2.14 (m, 1H), 1.43 (s, 9H), 1.21 (d, J = 6.3 Hz, 3H). HRMS (ESI) m/z: $[M + Na]^+$ calcd. for $C_{19}H_{30}N_2O_7Na$, 421.1951; found, 421.1958.

Boc-L-Hyp^{OX}-(S)-Ala(Et^X)-OMe (**7′**; X = n-but-2-enyl tether): compound **7′** (8.5 mg, 43%) was obtained from compound **7** (21.2 mg, 0.0500 mmol) in a similar manner to that described for the synthesis of **1′**. Colorless oil. Eluent for column: 5% MeOH in $CHCl_3$. R_f = 0.52 (10% MeOH in $CHCl_3$). ^{1}H NMR (500 MHz, $CDCl_3$) δ: 7.52 (s, 1H), 5.90–5.77 (m, 1H), 5.65 (dt, J = 10.5, 7.3 Hz, 1H), 4.42–4.22 (m, 1H), 4.15–3.92 (m, 2H), 3.74 (s, 3H), 3.70–3.43 (m, 3H), 2.83–2.64 (m, 1H), 2.62–2.48 (m, 1H), 2.22–1.86 (m, 4H), 1.64 (s, 3H), 1.52–1.43 (m, 9H), 1.51–1.43 (m, 1H), 1.22–1.12 (m, 1H). HRMS (ESI) m/z: $[M + Na]^+$ calcd. for $C_{20}H_{32}N_2O_6Na$, 419.2158; found, 419.2166.

3.4. *Synthesis of Stapled Octapeptides* **9** *and* **10**

Boc-L-Hyp^{OX}-L-Ser^{OX}-[(L-Leu)$_2$-Ac$_5$c]$_2$-OMe (**9**; X = n-but-2-enyl tether): compound **9** (10.2 mg, 52%) was obtained from compound **8** [19] (20.0 mg, 0.0184 mmol) in a similar manner to that described for the synthesis of **1′**. Eluent for column: 80% EtOAc/n-hexane. White amorphous. R_f = 0.26 (EtOAc). ^{1}H NMR (500 MHz, $CDCl_3$) δ: 7.73 (d, J = 2.3 Hz, 1H), 7.43 (d, J = 8.0 Hz, 1H), 7.37 (d, J = 4.8 Hz, 1H), 7.27–7.24 (m, 1H), 7.24–7.18 (m, 3H), 6.08 (dd, J = 10.9, 6.6 Hz, 1H), 6.04 (dd, J = 10.9, 5.8 Hz, 1H), 4.46 (ddd, J = 11.7, 5.3, 2.3 Hz, 1H), 4.39–4.30 (m, 2H), 4.26–4.15 (m, 5H), 4.01 (dd, J = 11.4, 5.3 Hz, 1H), 3.96–3.88 (m, 2H), 3.74 (dd, J = 10.0, 5.3 Hz, 1H), 3.71–3.68 (m, 1H), 3.67 (s, 3H), 3.66–3.60 (m, 1H), 3.46 (dd, J = 12.1, 3.4 Hz, 1H), 2.66 (dt, J = 13.5, 8.1 Hz, 1H), 2.45–2.30 (m, 2H), 2.26 (ddd, J = 13.6, 8.5, 6.7 Hz, 1H), 2.22–2.10 (m, 3H), 2.10–2.02 (m, 1H), 1.97–1.55 (m, 28H), 1.49 (s, 9H), 1.00–0.83 (m, 24H). HRMS (ESI) m/z: $[M + Na]^+$ calcd. for $C_{54}H_{90}N_8O_{13}Na$, 1081.6525; found, 1081.6536.

Boc-L-Hyp^{OX}-L-Ser^{OX}-[(L-Leu)$_2$-Ac$_5$c]$_2$-OMe (**10**; X = n-butyl tether): to a solution of peptide **9** (10.2 mg, 0.00963 mmol) in MeOH (2 mL) was added 10% Pd/C (10 mg) at room temperature and the reaction mixture was stirred at room temperature overnight. The resultant dark suspension was filtered through a short plug of celite (MeOH), and the organics were concentrated under vacuum. The crude material was purified by preparative TLC (EtOAc) to give **10** (8.5 mg, 83%) as white amorphous. R_f = 0.31 (EtOAc). ^{1}H NMR (500 MHz, $CDCl_3$) δ: 7.70 (s, 1H), 7.46 (d, J = 5.0 Hz, 1H), 7.44 (d, J = 8.0 Hz, 1H), 7.26–7.24 (m, 2H), 7.23 (d, J = 5.6 Hz, 2H), 4.35 (ddd, J = 11.4, 8.1, 3.0 Hz, 1H), 4.28 (ddd, J = 11.0, 4.8, 1.5 Hz,

1H), 4.24–4.16 (m, 2H), 4.14 (d, J = 10.9 Hz, 1H), 4.04 (t, J = 3.5 Hz, 1H), 3.98 (dd, J = 11.1, 5.1 Hz, 1H), 3.93 (dt, J = 9.6, 4.5 Hz, 1H), 3.83 (dd, J = 11.9, 2.2 Hz, 1H), 3.70 (dd, J = 9.3, 1.6 Hz, 1H), 3.67 (s, 3H), 3.63 (dd, J = 9.3, 2.7 Hz, 1H), 3.58 (dt, J = 9.3, 3.2 Hz, 1H), 3.54 (t, J = 11.2 Hz, 1H), 3.44–3.36 (m, 2H), 2.65 (dt, J = 13.6, 8.3 Hz, 1H), 2.38 (ddd, J = 15.1, 11.1, 4.3 Hz, 1H), 2.27 (dd, J = 13.8, 7.5 Hz, 1H), 2.24–2.03 (m, 5H), 1.96–1.66 (m, 22H), 1.66–1.57 (m, 4H), 1.52 (s, 9H), 0.99–0.93 (m, 9H), 0.92–0.85 (m, 15H). X-ray crystallographic data and CIF file of compound **10** are provided in the Supplementary Materials.

Supplementary Materials: The following are available online at https://www.mdpi.com/article/10.3390/ijms22105364/s1: ^{1}H and ^{13}C NMR spectra of compounds **2–7**, **1′–7′**, **9**, and **10**; X-ray crystallographic data of compound **10**, and CIF file of compound **10**.

Author Contributions: Conceptualization, A.U. and M.T.; methodology, A.U. and M.T.; validation, Y.M., A.U. and T.K.; formal analysis, Y.M., A.U., T.K., A.I., M.H., M.D. and M.T.; investigation, Y.M., A.U., T.K., A.I. and M.H.; writing—original draft preparation, A.U. and M.T.; writing—review and editing, Y.M., A.U., T.K., A.I., M.H., M.D. and M.T.; visualization, Y.M., A.U. and T.K.; supervision, M.D. and M.T.; project administration, A.U.; funding acquisition, A.U. and M.T. All authors have read and agreed to the published version of the manuscript.

Funding: This research was funded by JSPS KAKENHI Grant Numbers JP17H03998 (M.T.), JP18K14870 (A.U.), and JP20K06967 (A.U.), the Ube Industries Foundation (A.U.), and Shionogi Award in Synthetic Organic Chemistry, Japan (A.U.).

Institutional Review Board Statement: The study did not involve humans or animals.

Informed Consent Statement: The study did not involve humans.

Data Availability Statement: The data presented in this study are available on request from the corresponding author.

Acknowledgments: Y.M. is grateful for a fellowship from the Tokyo Biochemical Research Foundation. This work was the result of using research equipment shared in MEXT Project for promoting the public utilization of advanced research infrastructure (program for supporting introduction of the new sharing system), grant number JPMXS0422500320.

Conflicts of Interest: The authors declare no conflict of interest.

References

1. Moretto, A.; Crisma, M.; Formaggio, F.; Toniolo, C. Building a bridge between peptide chemistry and organic chemistry: Intramolecular macrocyclization reactions and supramolecular chemistry with helical peptide substrates. *Biopolymers* **2010**, *94*, 721–732. [CrossRef] [PubMed]
2. Verdine, G.L.; Hilinski, G.J. Stapled peptides for intracellular drug targets. *Methods Enzymol.* **2012**, *503*, 3–33. [CrossRef] [PubMed]
3. Cromm, P.M.; Spiegel, J.; Grossmann, T.N. Hydrocarbon stapled peptides as modulators of biological function. *ACS Chem. Biol.* **2015**, *10*, 1362–1375. [CrossRef] [PubMed]
4. Moiola, M.; Memeo, M.G.; Quadrelli, P. Stapled Peptides—A Useful Improvement for Peptide-Based Drugs. *Molecules* **2019**, *24*, 3654. [CrossRef] [PubMed]
5. Bozovičar, K.; Bratkovič, T. Small and Simple, yet Sturdy: Conformationally Constrained Peptides with Remarkable Properties. *Int. J. Mol. Sci.* **2021**, *22*, 1611. [CrossRef] [PubMed]
6. Blackwell, H.E.; Grubbs, R.H. Highly efficient synthesis of covalently cross-linked peptide helices by ring-closing metathesis. *Angew. Chem. Int. Ed.* **1998**, *37*, 3281–3284. [CrossRef]
7. Blackwell, H.E.; Sadowsky, J.D.; Howard, R.J.; Sampson, J.N.; Chao, J.A.; Steinmetz, W.E.; O'Leary, D.J.; Grubbs, R.H. Ring-closing metathesis of olefinic peptides: Design, synthesis, and structural characterization of macrocyclic helical peptides. *J. Org. Chem.* **2001**, *66*, 5291–5302. [CrossRef]
8. Schafmeister, C.E.; Po, J.; Verdine, G.L. An all-hydrocarbon cross-linking system for enhancing the helicity and metabolic stability of peptides. *J. Am. Chem. Soc.* **2000**, *122*, 5891–5892. [CrossRef]
9. Sawyer, T.K.; Partridge, A.W.; Kaan, H.Y.K.; Juang, Y.C.; Lim, S.; Johannes, C.; Yuen, T.Y.; Verma, C.; Kannan, S.; Aronica, P.; et al. Macrocyclic alpha helical peptide therapeutic modality: A perspective of learnings and challenges. *Bioorg. Med. Chem.* **2018**, *26*, 2807–2815. [CrossRef]
10. Ali, A.M.; Atmaj, J.; Oosterwijk, N.V.; Groves, M.R.; Dömling, A. Stapled peptides inhibitors: A new window for target drug discovery. *Comput. Struct. Biotechnol. J.* **2019**, *17*, 263–281. [CrossRef] [PubMed]

11. Hirano, M.; Saito, C.; Yokoo, H.; Goto, C.; Kawano, R.; Misawa, T.; Demizu, Y. Development of Antimicrobial Stapled Peptides Based on Magainin 2 Sequence. *Molecules* **2021**, *26*, 444. [CrossRef] [PubMed]
12. Jochim, A.L.; Arora, P.S. Systematic analysis of helical protein interfaces reveals targets for synthetic inhibitors. *ACS Chem. Biol.* **2010**, *5*, 919–923. [CrossRef] [PubMed]
13. Boal, A.K.; Guryanov, I.; Moretto, A.; Crisma, M.; Lanni, E.L.; Toniolo, C.; Grubbs, R.H.; O'Leary, D.J. Facile and *E*-selective intramolecular ring-closing metathesis reactions in 3_{10}-helical peptides: A 3D structural study. *J. Am. Chem. Soc.* **2007**, *129*, 6986–6987. [CrossRef] [PubMed]
14. Kim, Y.-W.; Kutchukian, P.S.; Verdine, G.L. Introduction of all-hydrocarbon *i,i*+3 staples into α-helices via ring-closing olefin metathesis. *Org. Lett.* **2010**, *12*, 3046–3049. [CrossRef] [PubMed]
15. Mangold, S.L.; O'Leary, D.J.; Grubbs, R.H. Z-Selective olefin metathesis on peptides: Investigation of side-chain influence, preorganization, and guidelines in substrate selection. *J. Am. Chem. Soc.* **2014**, *136*, 12469–12478. [CrossRef]
16. Mangold, S.L.; Grubbs, R.H. Stereoselective synthesis of macrocyclic peptides via a dual olefin metathesis and ethenolysis approach. *Chem. Sci.* **2015**, *6*, 4561–4569. [CrossRef]
17. Creighton, C.J.; Reitz, A.B. Synthesis of an eight-membered cyclic pseudo-dipeptide using ring closing metathesis. *Org. Lett.* **2001**, *3*, 893–895. [CrossRef]
18. Islam, N.M.; Islam, M.S.; Hoque, M.A.; Kato, T.; Nishino, N.; Ito, A.; Yoshida, M. Bicyclic tetrapeptides as potent HDAC inhibitors: Effect of aliphatic loop position and hydrophobicity on inhibitory activity. *Bioorg. Med. Chem.* **2014**, *22*, 3862–3870. [CrossRef]
19. Ueda, A.; Higuchi, M.; Sato, K.; Umeno, T.; Tanaka, M. Design and synthesis of helical *N*-terminal L-prolyl oligopeptides possessing hydrocarbon stapling. *Molecules* **2020**, *25*, 4667. [CrossRef]
20. Hill, T.A.; Shepherd, N.E.; Diness, F.; Fairlie, D.P. Constraining cyclic peptides to mimic protein structure motifs. *Angew. Chem. Int. Ed.* **2014**, *53*, 13020–13041. [CrossRef]
21. Paul, P.K.C.; Sukumar, M.; Bardi, R.; Piazzesi, A.M.; Valle, G.; Toniolo, C.; Balaram, P. Stereochemically constrained peptides. Theoretical and experimental studies on the conformations of peptides containing 1-aminocyclohexanecarboxylic acid. *J. Am. Chem. Soc.* **1986**, *108*, 6363–6370. [CrossRef]
22. Royo, S.; de Borggraeve, W.M.; Peggion, C.; Formaggio, F.; Crisma, M.; Jimeéez, A.I.; Cativiela, C.; Toniolo, C. Turn and helical peptide handedness governed exclusively by side-chain chiral centers. *J. Am. Chem. Soc.* **2005**, *127*, 2036–2037. [CrossRef] [PubMed]
23. Tanaka, M.; Demizu, Y.; Doi, M.; Kurihara, M.; Suemune, H. Chiral centers in the side chains of α-amino acids control the helical screw sense of peptides. *Angew. Chem. Int. Ed.* **2004**, *43*, 5360–5363. [CrossRef] [PubMed]
24. Nagano, M.; Tanaka, M.; Doi, M.; Demizu, Y.; Kurihara, M.; Suemune, H. Helical-screw directions of diastereoisomeric cyclic α-amino acid oligomers. *Org. Lett.* **2009**, *11*, 1135–1137. [CrossRef]
25. Oba, M.; Ishikawa, N.; Demizu, Y.; Kurihara, M.; Suemune, H.; Tanaka, M. Helical oligomers with a changeable chiral acetal moiety. *Eur. J. Org. Chem.* **2013**, 7679–7682. [CrossRef]
26. Hirata, T.; Ueda, A.; Oba, M.; Doi, M.; Demizu, Y.; Kurihara, M.; Nagano, M.; Suemune, H.; Tanaka, M. Amino equatorial effect of a six-membered ring amino acid on its peptide 3_{10}- and α-helices. *Tetrahedron* **2015**, *71*, 2409–2420. [CrossRef]
27. Crisma, M.; Toniolo, C. Helical screw-sense preferences of peptides based on chiral, C^{α}-tetrasubstituted α-amino acids. *Biopolymers* **2015**, *104*, 46. [CrossRef]
28. Koba, Y.; Hirata, Y.; Ueda, A.; Oba, M.; Doi, M.; Demizu, Y.; Kurihara, M.; Tanaka, M. Synthesis of chiral five-membered carbocyclic ring amino acids with an acetal moiety and helical conformations of its homo-chiral homopeptides. *Biopolymers* **2016**, *106*, 555–562. [CrossRef]
29. Eto, R.; Oba, M.; Ueda, A.; Uku, T.; Doi, M.; Matsuo, Y.; Tanaka, T.; Demizu, Y.; Kurihara, M.; Tanaka, M. Diastereomeric right- and left-handed helical structures with fourteen (*R*)-chiral centers. *Chem. Eur. J.* **2017**, *23*, 18120–18124. [CrossRef]
30. Koba, Y.; Ueda, A.; Oba, M.; Doi, M.; Kato, T.; Demizu, Y.; Tanaka, M. Left-handed helix of three-membered ring amino acid homopeptide interrupted by an N–H·ethereal O-type hydrogen bond. *Org. Lett.* **2018**, *20*, 7830–7834. [CrossRef]
31. Demizu, Y.; Tanaka, M.; Nagano, M.; Kurihara, M.; Doi, M.; Maruyama, T.; Suemune, H. Controlling 3_{10}-helix and α-helix of short peptides in the solid state. *Chem. Pharm. Bull.* **2007**, *55*, 840–842. [CrossRef]
32. Umeno, T.; Ueda, A.; Oba, M.; Doi, M.; Hirata, T.; Suemune, H.; Tanaka, M. Helical structures of L-Leu-based peptides having chiral six-membered ring amino acids. *Tetrahedron* **2016**, *72*, 3124–3131. [CrossRef]
33. Koba, Y.; Ueda, A.; Oba, M.; Doi, M.; Demizu, Y.; Kurihara, M.; Tanaka, M. Helical L-Leu-based peptides having chiral five-membered carbocyclic ring amino acids with an ethylene acetal moiety. *Chem. Select.* **2017**, *2*, 8108–8114. [CrossRef]
34. The Cambridge Crystallographic Data Centre (CCDC) Home Page. Available online: http://www.ccdc.cam.ac.uk/conts/retrieving.html (accessed on 18 May 2021).
35. Tanaka, M. Design and synthesis of chiral α,α-disubstituted amino acids and conformational study of their oligopeptides. *Chem. Pharm. Bull.* **2007**, *55*, 349–358. [CrossRef] [PubMed]
36. Wolf, W.M.; Stasiak, M.; Leplawy, M.T.; Bianco, A.; Formaggio, F.; Crisma, M.; Toniolo, C. Destabilization of the 3_{10}-helix in peptides based on C^{α}-tetrasubstituted α-amino acids by main-chain to side-chain hydrogen bonds. *J. Am. Chem. Soc.* **1998**, *120*, 11558–11566. [CrossRef]
37. Tanda, K.; Eto, R.; Kato, K.; Oba, M.; Ueda, A.; Suemune, H.; Doi, M.; Demizu, Y.; Kurihara, M.; Tanaka, M. Peptide foldamers composed of six-membered ring α, α-disubstituted α-amino acids with two changeable chiral acetal moieties. *Tetrahedron* **2015**, *71*, 3909–3914. [CrossRef]

38. Ueda, A.; Umeno, T.; Doi, M.; Akagawa, K.; Kudo, K.; Tanaka, M. Helical-peptide-catalyzed enantioselective Michael addition reactions and their mechanistic insights. *J. Org. Chem.* **2016**, *81*, 5864–5871. [CrossRef]
39. Ueda, A.; Higuchi, M.; Umeno, T.; Tanaka, M. Enantioselective synthesis of 2,4,5-trisubstituted tetrahydropyrans via peptide-catalyzed Michael addition followed by Kishi's reductive cyclization. *Heterocycles* **2019**, *99*, 989–1002. [CrossRef]
40. Umeno, T.; Ueda, A.; Doi, M.; Kato, T.; Oba, M.; Tanaka, M. Helical foldamer-catalyzed enantioselective 1,4-addition reaction of dialkyl malonates to cyclic enones. *Tetrahedron Lett.* **2019**, *60*, 151301. [CrossRef]
41. Ueda, A.; Ikeda, M.; Kasae, T.; Doi, M.; Demizu, Y.; Oba, M.; Tanaka, M. Synthesis of chiral α-trifluoromethyl α, α-disubstituted α-amino acids and conformational analysis of L-Leu-based peptides with (*R*)- or (*S*)-α-trifluoromethylalanine. *Chem. Select.* **2020**, *5*, 10882–10886. [CrossRef]
42. Oba, M. Cell-penetrating peptide foldamers: Drug-delivery tools. *ChemBioChem* **2019**, *20*, 2041–2045. [CrossRef] [PubMed]
43. Oba, M.; Kunitake, M.; Kato, T.; Ueda, A.; Tanaka, M. Enhanced and prolonged cell-penetrating abilities of arginine-rich peptides by introducing cyclic α, α-disubstituted α-amino acids with stapling. *Bioconj. Chem.* **2017**, *28*, 1801–1806. [CrossRef] [PubMed]
44. Kato, T.; Oba, M.; Nishida, K.; Tanaka, M. Cell-penetrating peptides using cyclic α, α-disubstituted α-amino acids with basic functional groups. *ACS Biomater. Sci. Eng.* **2018**, *4*, 1368–1376. [CrossRef] [PubMed]
45. Furukawa, K.; Tanaka, M.; Oba, M. siRNA delivery using amphipathic cell-penetrating peptides into human hepatoma cells. *Bioorg. Med. Chem.* **2020**, *28*, 115402. [CrossRef] [PubMed]
46. Kato, T.; Kita, Y.; Iwanari, K.; Asano, A.; Oba, M.; Tanaka, M.; Doi, M. Synthesis of six-membered carbocyclic ring α, α-disubstituted amino acids and arginine-rich peptides to investigate the effect of ring size on the properties of the peptide. *Bioorg. Med. Chem.* **2021**, *38*, 116111. [CrossRef] [PubMed]
47. Peters, C.; Bacher, M.; Buenemann, C.L.; Kricek, F.; Rondeau, J.-M.; Weigand, K. Conformationally constrained mimics of the membrane-proximal domain of FcεRIα. *ChemBioChem* **2007**, *8*, 1785–1789. [CrossRef]
48. Bortolini, O.; Cavazzini, A.; Giovannini, P.P.; Greco, R.; Marchetti, N.; Massi, A.; Pasti, L. A combined kinetic and thermodynamic approach for the interpretation of continuous-flow heterogeneous catalytic processes. *Chem. Eur. J.* **2013**, *19*, 7802–7808. [CrossRef]
49. Yamagata, N.; Demizu, Y.; Sato, Y.; Doi, M.; Tanaka, M.; Nagasawa, K.; Okuda, H.; Kurihara, M. Design of a stabilized short helical peptide and its application to catalytic enantioselective epoxidation of (*E*)-chalcone. *Tetrahedron Lett.* **2011**, *52*, 798–801. [CrossRef]
50. Demizu, Y.; Yamagata, N.; Nagoya, S.; Sato, Y.; Doi, M.; Tanaka, M.; Nagasawa, K.; Okuda, H.; Kurihara, M. Enantioselective epoxidation of α, β-unsaturated ketones catalyzed by stapled helical L-Leu-based peptides. *Tetrahedron* **2011**, *67*, 6155–6165. [CrossRef]
51. Tang, H.; Yin, L.; Lu, H.; Cheng, J. Water-soluble poly(L-serine)s with elongated and charged side- chains: Synthesis, conformations, and cell-penetrating properties. *Biomacromolecules* **2012**, *13*, 2609–2615. [CrossRef]
52. Bayardon, J.; Sinou, D. Synthesis of two new chiral fluorous bis(oxazolines) and their applications as ligands in catalytic asymmetric reactions. *Tetrahedron Asymmetry* **2005**, *16*, 2965–2972. [CrossRef]

International Journal of
Molecular Sciences

MDPI

Article

[68Ga]Ga-DFO-c(RGDyK): Synthesis and Evaluation of Its Potential for Tumor Imaging in Mice

Sona Krajcovicova [1], Andrea Daniskova [2], Katerina Bendova [2], Zbynek Novy [2], Miroslav Soural [1,2,*] and Milos Petrik [2,*]

1 Department of Organic Chemistry, Faculty of Science, Palacky University, 77900 Olomouc, Czech Republic; sona.krajcovicova@upol.cz

2 Institute of Molecular and Translational Medicine, Faculty of Medicine and Dentistry, Palacky University, 77900 Olomouc, Czech Republic; andrea.daniskova01@upol.cz (A.D.); katerina.bendova01@upol.cz (K.B.); zbynek.novy@upol.cz (Z.N.)

* Correspondence: miroslav.soural@upol.cz (M.S.); milos.petrik@upol.cz (M.P.); Tel.: +42-058-563-21-96 (M.S.); +42-058-563-21-26 (M.P.)

Abstract: Angiogenesis has a pivotal role in tumor growth and the metastatic process. Molecular imaging was shown to be useful for imaging of tumor-induced angiogenesis. A great variety of radiolabeled peptides have been developed to target $\alpha v\beta 3$ integrin, a target structure involved in the tumor-induced angiogenic process. The presented study aimed to synthesize deferoxamine (DFO)-based c(RGD) peptide conjugate for radiolabeling with gallium-68 and perform its basic preclinical characterization including testing of its tumor-imaging potential. DFO-c(RGDyK) was labeled with gallium-68 with high radiochemical purity. In vitro characterization including stability, partition coefficient, protein binding determination, tumor cell uptake assays, and ex vivo biodistribution as well as PET/CT imaging was performed. [^{68}Ga]Ga-DFO-c(RGDyK) showed hydrophilic properties, high stability in PBS and human serum, and specific uptake in U-87 MG and M21 tumor cell lines in vitro and in vivo. We have shown here that [^{68}Ga]Ga-DFO-c(RGDyK) can be used for $\alpha v\beta 3$ integrin targeting, allowing imaging of tumor-induced angiogenesis by positron emission tomography.

Keywords: deferoxamine; RGD peptides; integrins; radiodiagnostics; PET imaging

Citation: Krajcovicova, S.; Daniskova, A.; Bendova, K.; Novy, Z.; Soural, M.; Petrik, M. [68Ga]Ga-DFO-c(RGDyK): Synthesis and Evaluation of Its Potential for Tumor Imaging in Mice. *Int. J. Mol. Sci.* **2021**, *22*, 7391. https://doi.org/10.3390/ijms22147391

Academic Editors: Menotti Ruvo and Nunzianna Doti

Received: 14 May 2021
Accepted: 6 July 2021
Published: 9 July 2021

1. Introduction

Over the last 30 years, many radiolabeled peptides have been evaluated as promising radiotracers for imaging tumors by means of positron emission tomography (PET) or single photon emission computerized tomography (SPECT) [1,2]. The biological effects of such peptides are mediated via the high affinity targeting of specific receptors. These receptors are often massively overexpressed in numerous cancers, compared to their relatively low density in physiological organs, which is the main principle allowing molecular imaging and therapy of tumors with radiopeptides [3]. Based on the success of studies with ^{111}In-labeled somatostatin analogue octreotide (OctreoScan™, Curium, London, UK), the pioneering radiopeptide for tumor imaging, many other receptor-targeting peptides are currently under development or undergoing clinical trials, including arginine-glycine-aspartic acid (RGD)-based peptides [4].

RGD-based family of peptides preferentially bind to the receptors of the integrin superfamily. Integrins are heterodimeric transmembrane receptors interacting with a diverse groups of extracellular ligands [5]. They regulate cellular growth, proliferation, migration, signaling, and cytokine activation and release and thereby play important roles in cell proliferation and migration, apoptosis, tissue repair, as well as in all processes critical to inflammation, infection, and angiogenesis [6]. Among the 24 human integrin subtypes known to date, eight integrin dimers, i.e., $\alpha v\beta 1$, $\alpha v\beta 3$, $\alpha v\beta 5$, $\alpha v\beta 6$, $\alpha v\beta 8$, $\alpha 5\beta 1$, $\alpha 8\beta 1$, and $\alpha IIb\beta 3$, recognize the tripeptide RGD motif within extracellular matrix proteins and represent the most important integrin receptor subfamily involved in cancer and the metastatic process [7]. Of these, the integrin αvβ3 has been studied most extensively for its role in tumor angiogenesis using nuclear medicine imaging techniques [8]. A large variety of radiotracers based on RGD peptides have been developed and tested for targeting αvβ3 integrin in both preclinical and clinical settings.

Integrin-specific radiotracers are obtained by radiolabeling of RGD-based precursors, which can be used for imaging with scintigraphy and SPECT (by using gamma emitters like, e.g., Tc-99m and In-111) or PET (by using positron-emitting radionuclides, e.g., F-18, Ga-68, or Cu-64). For labeling of peptide-precursors with the use of a radiometal (e.g., Ga-68, Cu-64, Tc-99m, In-111), a specific metal chelating agent has to be introduced to the precursor's structure [9]. The choice of the chelating agent is largely determined by the nature and oxidation state of the radiometal to be used for labeling [10]. Among various chelating agents for radiometal labeling of RGD peptides, 1,4,7,10-tetraazacyclododecane-1,4,7,10-tetraacetic acid (DOTA), 1,4,7-triazacyclononane-1,4,7-triacetic acid (NOTA) and their derivatives are the most widely used [11].

Many previous works have shown that deferoxamine (DFO), a hexadentate hydroxamate siderophore, is a common and suitable chelator for labeling with radiometals such as Ga-67, Zr-89, In-111, and Ga-68 [12–16]. Although a series of near-infrared fluorescent conjugates containing DFO and multi-RGD peptide moieties were designed, synthesized, and affinity to αvβ3 integrin was evaluated in vitro [17], radiolabeled DFO-RGD peptide conjugates were not studied neither in vitro nor in vivo. Herein, we report the preparation of cyclic c(RGDyK) pentapeptide conjugated with *p*-SCN-Bn derivatized DFO to obtain the DFO-based c(RGDyK) conjugate, which was then labeled with Ga-68. Binding properties of [^{68}Ga]Ga-DFO-c(RGDyK) towards αvβ3 integrin were studied in vitro and in vivo, including PET/CT imaging in a mouse tumor model.

2. Results and Discussion

2.1. Conjugate Synthesis

To improve the pharmacokinetic profile, especially in vivo stability against enzymatic cleavage of RGD peptides alongside maintenance of high activity and specificity for αvβ3 integrin, several approaches have been developed in the past [18]. The enrichment of the pentapeptide chain with amino acids in unnatural *D*-configuration and subsequent cyclization were demonstrated to be significant in vivo stability improvements and are routinely used in the development of RGD peptides nowadays [19]. In our study, the c(RGDyK) sequence for conjugation with DFO chelating agent was chosen, as it is one of the most prominent structures for the development of molecular imaging compounds in order to determine αvβ3 expression. It was extensively studied both preclinically and clinically with different radiometal chelators for tumor imaging [20–26]. Although DFO is well-accessible and an established chelating agent in nuclear medicine for peptide and particularly antibody radiolabeling, to our knowledge, DFO-based RGD peptides for radiometal labeling have not been studied yet. DFO is a molecule known for its ability to bind many different metal ions for years [27]. In 1968, it was also approved for medical use by the FDA under the brand name Desferal® (Novartis, Basel, Switzerland) and became a well-established clinically used medication [16]. In nuclear medicine, DFO-based compounds offer possibilities for labeling with different radiometals as mentioned above, allowing a wide range of diagnostic as well as therapeutic applications [12,15,17].

The proposed synthetic pathway started from the commercially available 2-chlorotrityl chloride polystyrene resin **1**, which was acylated with Fmoc-Gly-OH in the presence of *N,N′*-diisopropylethylamine (DIPEA) as a base (Scheme 1) to give **2**. For the cleavage of Fmoc protecting group, a non-nucleophilic strong base 1,8-diazabicyclo[5.4.0]undec-7-ene (DBU) in CH_2Cl_2 was used, as we observed lower crude purities within the reaction sequence when the traditional cleavage protocol with piperidine in dimethylformamide (DMF) was applied. Construction of oligopeptide was accomplished by the conventional solid-phase peptide synthesis with *N,N′*-diisopropylcarbodiimide (DIC) and hydroxybenzotriazole

(HOBt) as the activating agents. Notably, we decided to incorporate *D*-tyrosine (y) amino acid into the linear sequence that should result in enhanced stability of the RGD peptide, as was mentioned above. The corresponding intermediates **3–6** were obtained in high crude purities (according to spectral data, see Supplementary Materials). The loading of pentapeptide **6** was quantified to 0.7 mmol/g. It is worth mentioning that the 4,4-dimethyl-2,6-dioxocyclohex-1-ylidene (Dde) protecting group of the lysine side chain was chosen due to its orthogonality with acid-labile protecting groups (Pbf, *tert*-butyl) as well as the Fmoc protecting group [28].

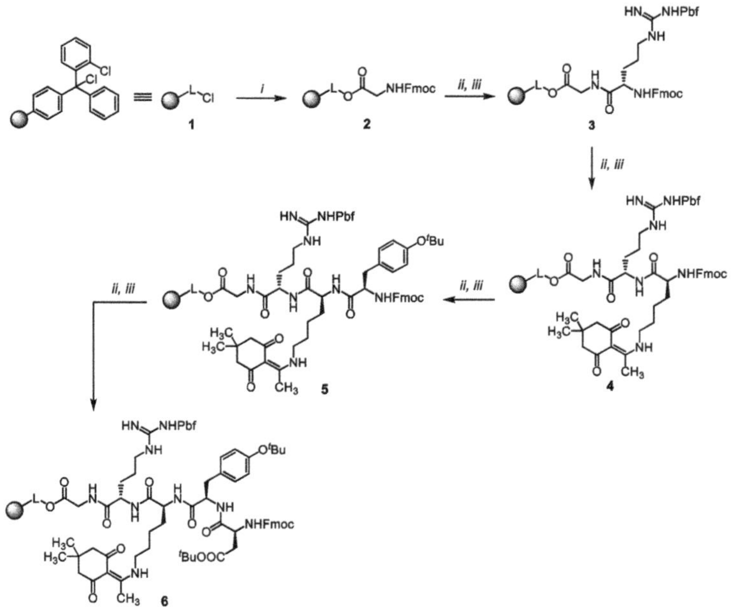

Scheme 1. Synthesis of linear pentapeptide. Reagents and conditions: (**i**) Fmoc-Gly-OH, DIPEA, DMF/CH_2Cl_2 (1:1), room temperature (r.t.), 16 h; (**ii**) DBU/CH_2Cl_2 (1:1), r.t., 10 min; (**iii**) Fmoc-amino acid (Fmoc-Arg(Pbf)-OH for **3**, Fmoc-Lys(Dde)-OH for **4**, Fmoc-D-Tyr(^{t}Bu)-OH for **5**, Fmoc-Asp(^{t}Bu)-OH for **6**), DIC, HOBt, DMF, r.t., 4–16 h.

Prior to the cyclization step, the linear peptide **6** was liberated from the resin (Scheme 2). Chemoselective cleavage (to maintain residual protecting groups) was performed with hexafluoroisopropanol (HFiP) in CH_2Cl_2. Following cyclization of **7** using benzotriazol-1-yl-oxytripyrrolidinophosphonium hexafluorophosphate (PyBOP) yielded the protected cyclized peptide **8**. The Dde protecting group was then cleaved with hydrazine (2%) which furnished the key intermediate **8** in excellent crude purity applicable for further modification with DFO (Scheme 2).

For the attachment of DFO, commercially available *p*-SCN-Bn-deferoxamine was applied. The reaction proceeded smoothly and with high crude purity of the corresponding intermediate. Following acid-mediated cleavage of residual protecting groups in trifuoroacetic acid (TFA) yielded the final DFO-based c(RGDyK) conjugate **9** (Scheme 3), which was purified using semipreparative reversed-phase high-performance liquid chromatography (RP-HPLC) and fully characterized (see Supplementary Materials).

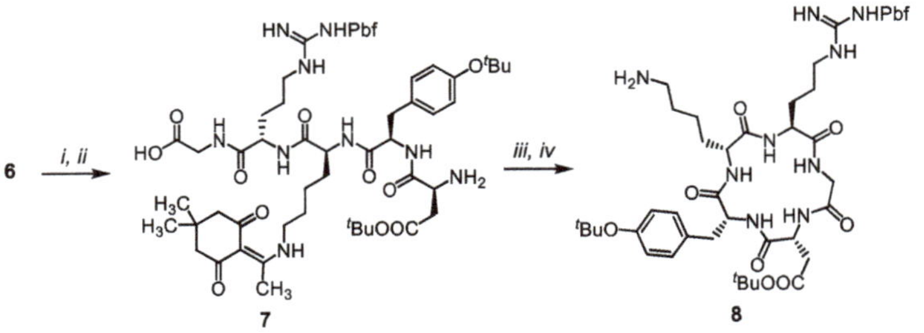

Scheme 2. Synthesis of cyclic pentapeptide. Reagents and conditions: (**i**) DBU/CH_2Cl_2 (1:1), room temperature (r.t.), 10 min; (**ii**) HFiP/CH_2Cl_2 (1:4), r.t., 3 h; (**iii**) PyBOP, DIPEA, DMF, r.t., 24 h; (**iv**) 2% $NH_2NH_2{\cdot}OH$ in DMF, r.t., 3 h.

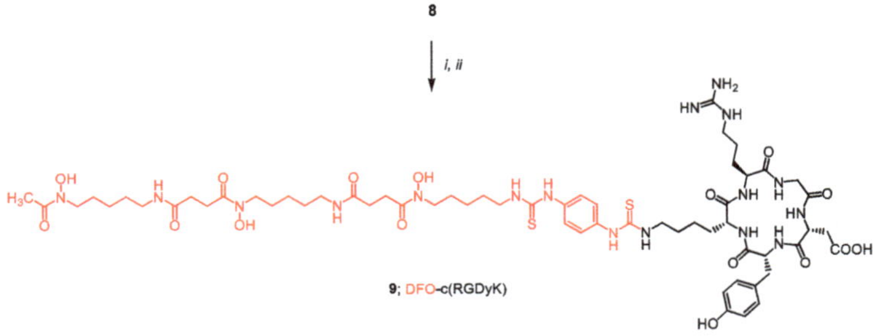

Scheme 3. Synthesis of final conjugate with deferoxamine. Reagents and conditions: (**i**) *p*-SCN-Bn-deferoxamine, DIPEA, DMSO/DMF 1:1, room temperature (r.t.), 1 h; (**ii**) trifluoroacetic acid (TFA)/CH_2Cl_2 1:1, r.t., 2 h.

2.2. Radiolabeling and In Vitro Characterization

DFO-c(RGDyK) **9** was radiolabeled with gallium-68 with molar activity of up to 6 GBq/µmol and radiochemical purity >98% in 5 min at 85 °C, confirmed by RP-HPLC (corresponding radiochromatogram is shown in Figure 1). The ^{68}Ga-labeled tracer was used without further purification for all the experiments. Gallium-68 is a positron emitter that decays with a half-life of 67.7 min and positron branching of 89.1%, emitting high-energy positrons of ca. 1.9 MeV [29,30]. The relatively short half-life can be a limitation, therefore, the short reaction time and high radiochemical yield and/or purity preventing any further purification or post-processing steps are important factors for the preparation of ^{68}Ga-labeled radiopharmaceuticals. In recent years, gallium-68 has attracted increasing interest in the field of nuclear medicine, currently being most often utilized in radiopharmaceuticals for oncology diagnostics [31].

[^{68}Ga]Ga-DFO-c(RGDyK) showed hydrophilic properties (log P = -2.01 ± 0.08) with plasma protein binding approximately 30% after 120 min of incubation at 37 °C in human serum. The in vitro stability of [^{68}Ga]Ga-DFO-c(RGDyK) was high in human serum and PBS (>97% in all tested time points), while in the presence of high excess of a competing metal and chelator, the stability of [^{68}Ga]Ga-DFO-c(RGDyK) decreased rapidly (see Table 1). The obtained in vitro data are in a good agreement with previously published data on [^{68}Ga]Ga-DFO [16]. However, [^{68}Ga]Ga-DFO-c(RGDyK) revealed higher lipophilicity, plasma protein binding and instability in the presence of a competing metal and chelator as compared with those of [^{68}Ga]Ga-NODAGA-c(RGDyK) [24], which subsequently influenced the in vivo behavior of the studied radiotracer.

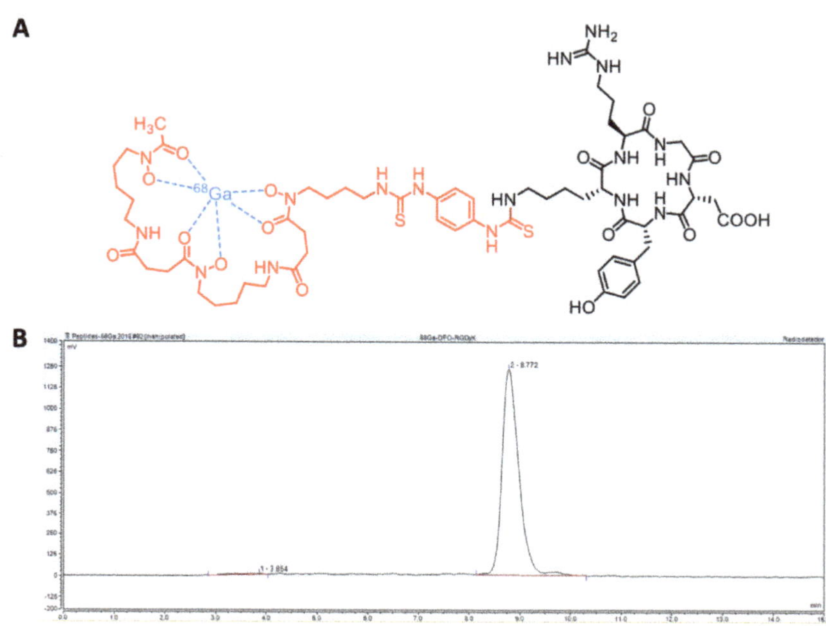

Figure 1. [^{68}Ga]Ga-DFO-c(RGDyK) structure (**A**) and representative radiochromatogram (**B**).

Table 1. In vitro characterization of [^{68}Ga]Ga-DFO-c(RGDyK). Log P, protein binding (expressed as % of protein bound activity of the total activity used) and stability in human serum, PBS, 0.1 M $FeCl_3$, and 6 mM DTPA.

Log P (n = 6)	Incubation Time (min)	Protein Binding (%) (n = 3)	Stability in Human Serum (%) (n = 3)	Stability in PBS (%) (n = 3)	Stability in Iron Solution (%) (n = 3)	Stability in DTPA Solution (%) (n = 3)
	30	30.7 ± 1.40	97.5 ± 0.46	97.6 ± 0.60	0.13 ± 0.12	75.4 ± 1.55
−2.01 ± 0.08	60	29.3 ± 1.78	98.1 ± 0.76	98.9 ± 0.27	0.47 ± 0.28	61.4 ± 5.53
	120	30.1 ± 3.32	98.3 ± 1.44	97.5 ± 0.81	0.22 ± 0.14	40.4 ± 4.39

In vitro uptake assays in tumor cell lines (U-87 MG, M21 and M21-L) showed specific uptake of [^{68}Ga]Ga-DFO-c(RGDyK) by cell lines expressing αvβ3 integrin (U-87 MG and M21), which could be blocked with an excess of appropriate competing inhibitor (NODAGA-c(RGDyK)). In the uptake assay using U-87 MG cells, the uptake of [^{68}Ga]Ga-DFO-c(RGDyK) increased with the incubation time and could be blocked with NODAGA-c(RGDyK) by about half (Figure 2A). This is consistent with the findings of Novy et al. [24], who showed similar in vitro uptake behavior of [^{68}Ga]Ga-NODAGA-c(RGDyK) in a U-87 MG cell line. In the cell uptake assay using M21 and M21-L cells, the uptake of [^{68}Ga]Ga-DFO-c(RGDyK) could only be blocked for the αvβ3-positive cells (M21), whereas for the αvβ3-negative control cells (M21-L), the uptake was approximately one half of the amount for the receptor-positive cells. The binding of [^{68}Ga]Ga-DFO-c(RGDyK) to the M21-L cells was very similar under blocked and unblocked conditions (Figure 2B). These results are in accordance with the data published by Knetsch et al. [32], who examined the in vitro uptake of [^{68}Ga]Ga-NODAGA-c(RGDfK) in M21 and M21-L cell lines. It also confirms that the small change in the peptide sequence from c(RGDyK) to c(RGDfK) does not have any significant impact on the αvβ3 integrin binding affinity [19].

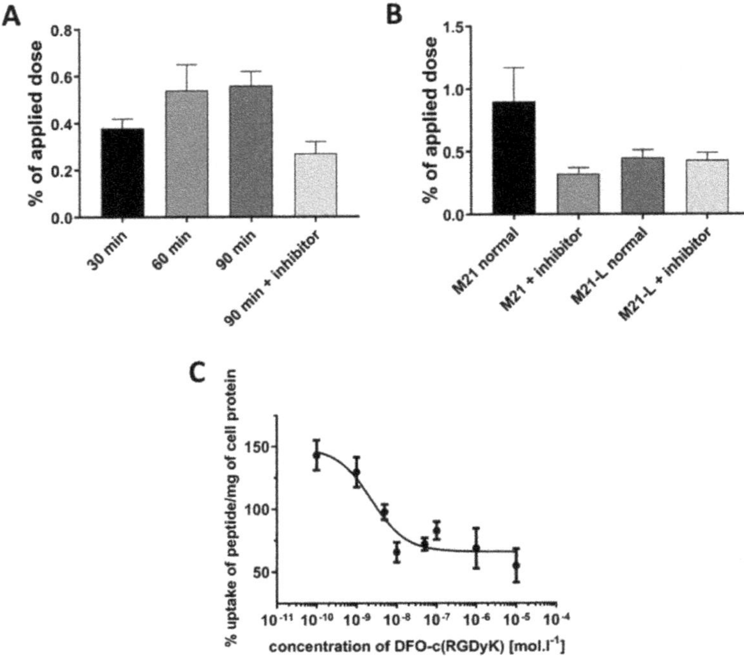

Figure 2. In vitro cell assays. The uptake of [^{68}Ga]Ga-DFO-c(RGDyK) in a U-87 MG cell line at various time points (30, 60, and 90 min) at 37 °C, and the uptake inhibition by co-incubation with 500-fold excess of inhibitor (cold NODAGA-c(RGDyK)) (**A**). The uptake of [^{68}Ga]Ga-DFO-c(RGDyK) in M21 ($\alpha v\beta 3$-positive) and M21-L ($\alpha v\beta 3$-negative) cell lines without or with inhibitor (500-fold excess of cold NODAGA-c(RGDyK)) for 60 min at 37 °C (**B**). Inhibition curve of [^{68}Ga]Ga-NODAGA-c(RGDyK) uptake obtained by incubation with increasing concentrations of cold DFO-c(RGDyK) **9** in M21 cells for 60 min at 37 °C. Data are presented as mean values ± s.d. (n = 6) (**C**).

The in vitro competition assays using increasing amounts of cold DFO-c(RGDyK) **9** showed that the inhibitory peptide was able to suppress the binding of [^{68}Ga]Ga-NODAGA-c(RGDyK) to the $\alpha v\beta 3$ integrin expressing M21 cells and that the binding kinetics followed a classic sigmoid inhibition curve. The IC_{50} value found for DFO-c(RGDyK) **9** was 2.35 ± 1.48 nM (Figure 2C). Although we did not use the gold standard radioligand [^{125}I]I-echistatin and observed relatively high nonspecific binding in the competition assay, which could affect the accuracy of IC_{50} calculations, the determined IC_{50} value is in accordance with the IC_{50} values of similar RGD-based conjugates published by Kapp et al. [19]. Other than monomeric RGD-based peptides, multimeric compounds presenting more than one RGD motif have also been introduced [18]. This "multimerisation" approach may result in improved target affinity and prolonged target retention, however, it may have a significant, and in some cases undesired, impact on in vivo behavior of the RGD conjugates [18,33].

2.3. In Vivo Characterization

Biodistribution studies were performed 30 and 90 min after [^{68}Ga]Ga-DFO-c(RGDyK) injection in healthy Balb/c and tumor-bearing (U-87 MG) Balb/c nude mice. [^{68}Ga]Ga-DFO-c(RGDyK) displayed relatively rapid excretion mainly via the renal system and showed minimal retention in blood and other organs in healthy animals 90 min after injection (Figure 3). The highest activity concentration in the organs at a later time point (90 min p.i.) was found in the kidneys (5.31 ± 0.12% ID/g), intestines (1.99 ± 0.01% ID/g), and the stomach (1.78 ± 0.05% ID/g) in healthy animals. The ex vivo biodistribution data were consistent with the results obtained from PET/CT imaging 30 and 90 min p.i. (Figure 3).

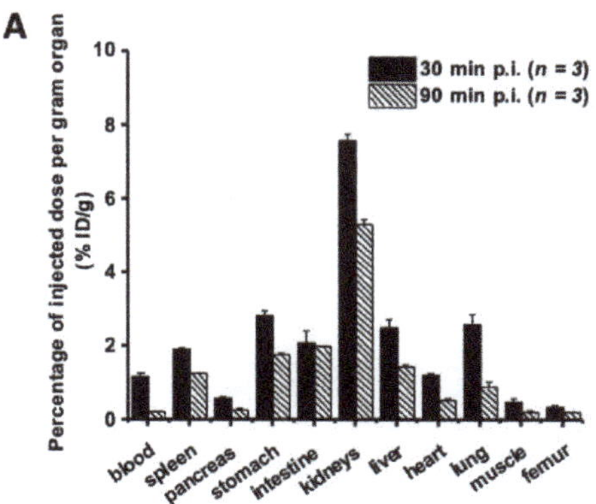

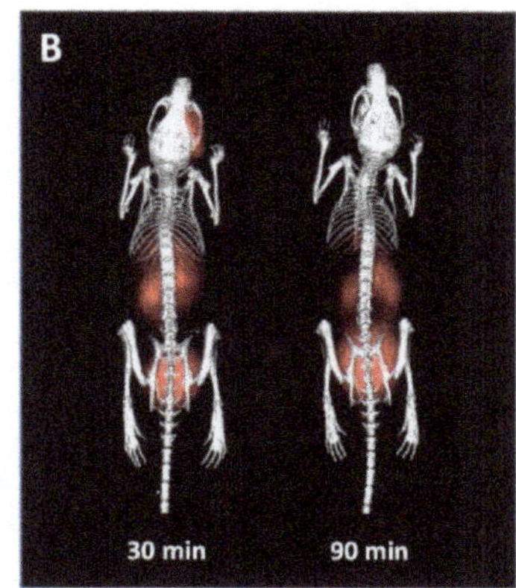

Figure 3. Ex vivo biodistribution of [^{68}Ga]Ga-DFO-c(RGDyK) in healthy Balb/c mice 30 and 90 min after r.o. injection (**A**). Static PET/CT imaging (3D volume rendered images) of [^{68}Ga]Ga-DFO-c(RGDyK) in healthy Balb/c mice 30 and 90 min after r.o. injection (**B**).

The results of biodistribution studies of [^{68}Ga]Ga-DFO-c(RGDyK) in U-87 MG xenografted Balb/c nude mice are presented in Figure 4. [^{68}Ga]Ga-DFO-c(RGDyK) in tumor-bearing animals showed in vivo behavior similar to that in healthy mice. In addition, [^{68}Ga]Ga-DFO-c(RGDyK) was accumulated in tumor tissue 30 min p.i. (3.03 ± 0.62% ID/g), and tumor washout was rather slow, as a significant amount of radioactivity (1.54 ± 0.56% ID/g), compared to that in other organs, was still present in tumor tissue after 90 min. In general, this is in good agreement with reports on other radiolabeled RGD peptides being evaluated on the similar integrin $\alpha v\beta 3$-expressing tumor model [20,24,33,34]. [^{68}Ga]Ga-DFO-c(RGDyK) displayed lower U-87 MG tumor uptake in vivo compared to that of multimeric RGD peptide counterparts and confirmed that multimerization usually improves tumor-targeting capability of radiolabeled peptides [18,35], including RDG-based peptides [36,37]. However, compared to monomeric conjugates, e.g., [^{68}Ga]Ga-DOTA-c(RGDyK) [23], [^{68}Ga]Ga-NOTA-c(RGDyK) [20], and [^{68}Ga]Ga-NODAGA-c(RGDyK) [24], [^{68}Ga]Ga-DFO-c(RGDyK) showed slightly slower pharmacokinetics and higher U-87 tumor accumulation and retention, which is in full agreement with the obtained in vitro data. Considering not only tumor uptake but also the pharmacokinetics and tissue distribution, monomeric or dimeric RGD-based peptide conjugates seem to be the most promising candidates for in vivo imaging of integrins-expressing tumors [20,33,37].

Animal PET/CT imaging studies were conducted in healthy mice and in the same tumor mouse model (U-87 MG) as biodistribution studies. Moreover, PET/CT imaging was also performed in M21($\alpha v\beta 3$-positive) and M21-L ($\alpha v\beta 3$-negative) xenografted Balb/c nude mice to evaluate the specificity of [^{68}Ga]Ga-DFO-c(RGDyK) uptake in the tumors with high and low $\alpha v\beta 3$ integrin expression in vivo. Static PET/CT images of [^{68}Ga]Ga-DFO-c(RGDyK) in U-87 MG tumor mice (Figure 5A) confirmed the results from ex vivo biodistribution showing the uptake in $\alpha v\beta 3$ integrin-overexpressing tissue and clearly visualizing the tumor 30 and 90 min p.i. with relatively decent target-to-organ contrast observed for both time points. PET/CT imaging of [^{68}Ga]Ga-DFO-c(RGDyK) in M21 and M21-L tumor-bearing mice (Figure 5B) displayed specific uptake of the radiotracer in $\alpha v\beta 3$ integrin-positive (M21) tumor tissue, while no uptake was observed in $\alpha v\beta 3$ integrin-negative (M21-L) tumor tissue both 30 and 90 min p.i. Animal PET/CT imaging of [^{68}Ga]Ga-DFO-c(RGDyK) in both tumor mouse models (U-87 MG and M21) confirmed that it has similar in vivo behavior as other analogical radiolabeled RGD-based peptides [20,23,24] and can be used for PET imaging of tumor angiogenesis. Moreover, DFO allows labeling with different radionuclides [12,15,38,39], which could open ways of the studied DFO-c(RGDyK) for SPECT imaging and theranostic applications [18,40].

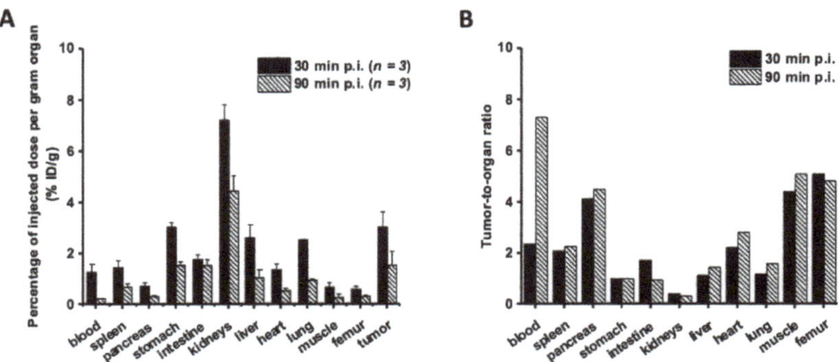

Figure 4. Ex vivo biodistribution of [^{68}Ga]Ga-DFO-c(RGDyK) in tumor-bearing (U-87 MG) Balb/c nude mice (**A**) and corresponding tumor-to-organ ratios (**B**) 30 and 90 min after r.o. injection.

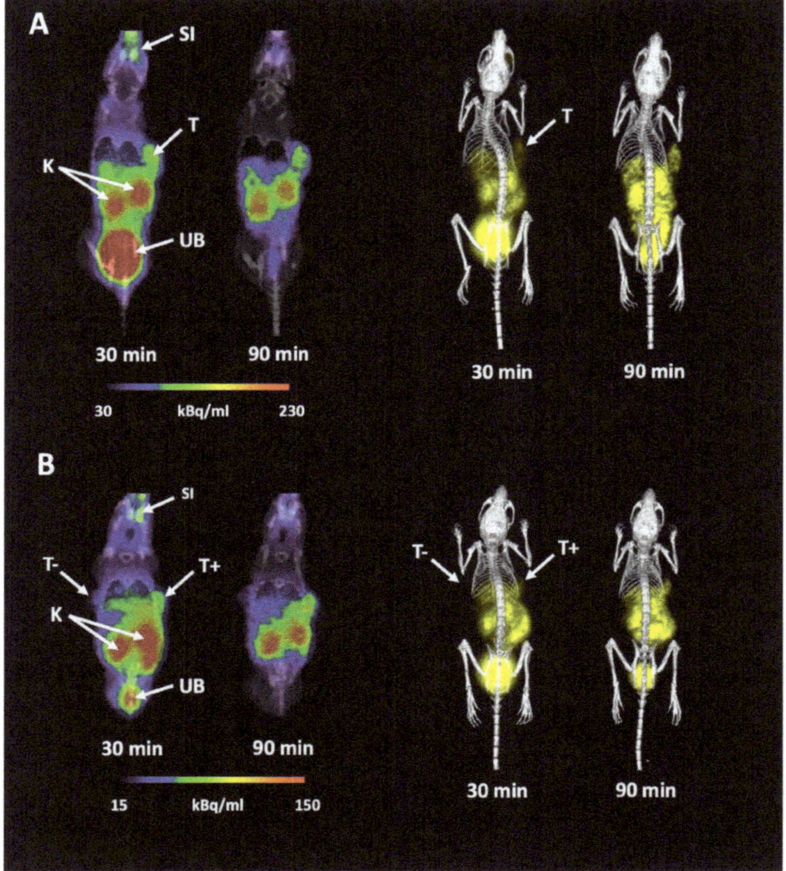

Figure 5. Static PET/CT imaging (coronal slices and 3D volume rendered images) of [^{68}Ga]Ga-DFO-c(RGDyK) in tumor-bearing (U-87 MG (**A**) or M21 and M21-L (**B**)) Balb/c nude mice 30 and 90 min after r.o. injection; K = kidneys, SI = site of injection, T = U-87 MG tumor, T+ = M21 tumor, T− = M21-L tumor, UB = urinary bladder.

3. Materials and Methods

3.1. Chemicals

All reagents were purchased from commercial sources as reagent or analytical grade and used without further purification. Solvents and chemicals were purchased from Sigma-Aldrich (St. Louis, MO, USA), Acros Organics (Geel, Belgium), AAPPTec (Louisville, KY, USA), or Fluorochem (Hadfield, UK). *p*-SCN-Bn-Deferoxamine was purchased from Macrocyclics (Plano, TX, USA). Anhydrous solvents were dried over 4 Å molecular sieves or stored as received from commercial suppliers. $^{68}GaCl_3$ for radiolabeling was eluted from a $^{68}Ge/^{68}Ga$-generator (Eckert & Ziegler Eurotope GmbH, Berlin, Germany) with 0.1 N HCl using a fractionated elution approach.

3.2. Conjugate Preparation

Reactions were performed in plastic reaction vessels (syringes, each equipped with a porous disk) using a manually operated synthesizer (Torviq, Tuscon, AZ, USA) or in ace-pressure tubes, unless stated otherwise. The volume of wash solvent was 10 mL per 1 g of resin. For washing, resin slurry was shaken with the fresh solvent for at least 1 min before changing the solvent. Resin-bound intermediates were dried under a stream of nitrogen for prolonged storage and/or quantitative analysis.

3.2.1. Procedure for Acylation with Fmoc-Gly-OH

2-Chlorotrityl chloride resin **1** (300 mg, loading 0.85 mmol/g) was added to the polypropylene fritted syringe and then washed with CH_2Cl_2 (3 × 5 mL). The solution of Fmoc-Gly-OH (267 mg, 0.9 mmol) and DIPEA (148 µL, 0.9 mmol) in a mixture of DMF/CH_2Cl_2 (1:1, 3 mL, *v*/*v*) was added to resin **1**. The reaction slurry was shaken at ambient temperature for 16 h, followed by washing with CH_2Cl_2 (5 × 5 mL), DMF (5 × 5 mL), and CH_2Cl_2 (5 × 5 mL). For capping, the solution of CH_2Cl_2/methanol/DIPEA (17:2:1, 10 mL, *v*/*v*) was added to the resin, and the slurry was shaken for an additional 1 h. Then, the resin was washed again with CH_2Cl_2 (3 × 5 mL). Subsequent cleavage from the resin (according to General procedure C) confirmed full conversion to product **2**.

Loading after this step was determined as follows: the sample of resin **2** (~30 mg) was washed with CH_2Cl_2 (5 × 3 mL) and MeOH (3 × 3 mL), dried under a stream of nitrogen and divided into two portions (2 × 12 mg). Both samples were treated with CH_2Cl_2/TFA (1:1, 1 mL, *v*/*v*) for 1 h, after which the cleavage cocktail was evaporated under a stream of nitrogen. Cleaved compounds were dissolved in CH_3CN/H_2O (1:1, 1 mL, *v*/*v*), diluted four times, and analyzed by ultra-high performance liquid chromatography coupled with mass spectrometry and ultraviolet detection (UHPLC/MS/UV). Loading of the resin was calculated with the use of an external standard (Fmoc-Ala-OH, 0.5 mg/mL).

3.2.2. General Procedure A for Deprotection of Fmoc

Corresponding resin (300 mg) was swollen in CH_2Cl_2 (5 mL) for 30 min. Solution of DBU/CH_2Cl_2 (1:1, 5 mL, *v*/*v*) was added to the resin, and the reaction slurry was shaken for an additional 10 min, then washed again with CH_2Cl_2 (3 × 5 mL) and DMF (3 × 5 mL). The resin was used in the next step without further analysis.

3.2.3. General Procedure B for Preparation of Pentapeptide

The corresponding resin (300 mg) was swollen in CH_2Cl_2 (5 mL) for 30 min and then washed with CH_2Cl_2 (3 × 5 mL) and DMF (3 × 5 mL). The Fmoc-protected amino acid (0.2 M) and HOBt (91 mg, 0.6 mmol) were dissolved in DMF (3 mL), and DIC (94 µL, 0.6 mmol) was added. The reaction mixture was added to the corresponding Fmoc-deprotected resin (according to General procedure A) and shaken for 4–16 h, followed by washing with CH_2Cl_2 (5 × 5 mL), DMF (5 × 5 mL), and CH_2Cl_2 (5 × 5 mL). Subsequent cleavage from the resin (according to General procedure C) confirmed the full conversion to products **3**–**6**. The loading of **6** was determined according to the above-mentioned procedure.

3.2.4. Procedure for HFiP Mediated Cleavage from the Resin (with Maintaining Protecting Groups)

Resin **6** (300 mg) was swollen in CH_2Cl_2 (5 mL) for 30 min and then washed with CH_2Cl_2 (3 × 5 mL) and DMF (3 × 5 mL). Fmoc protecting group was cleaved according to General procedure A. The solution of 1,1,1,3,3,3-hexafluoro-2-propanol (HFiP) in CH_2Cl_2 (1:4, 3 mL, *v*/*v*) was added to the resin, and the reaction slurry was shaken at ambient temperature for 3 h. Then, the resin was washed with a HFiP/CH_2Cl_2 (1:4, 3 × 4 mL, *v*/*v*) mixture, organic extracts were combined, and residual solvents were evaporated under reduced pressure. UHPLC/MS analysis confirmed quantitative cleavage of **6** from the resin, and the crude product **7** was used directly in the next step without further purification.

3.2.5. Procedure for Peptide Cyclization

Crude product **7** was dissolved in DMF (3 mL/300 mg of **6**) with subsequent addition of PyBOP (156 mg, 0.3 mmol) and DIPEA (200 μL, 1.2 mmol). The reaction mixture was stirred at ambient temperature for 24 h. UHPLC/MS analysis confirmed formation of the cyclized product (ESI−: 1147) which was used directly in the next step.

3.2.6. Procedure for Deprotection of Dde

$NH_2NH_2 \cdot OH$ (60 μL) was added into the reaction mixture with cyclized peptide, and the reaction mixture was stirred for an additional 3 h. Subsequent UHPLC/MS analysis confirmed the formation of **8**. The residual solvents were evaporated on high vacuum, and crude product **8** was purified by semipreparative HPLC.

3.2.7. Procedure for Acylation with *p*-SCN-Bn-Deferoxamine and Final Deprotection of Pbf and Tert-Butyl Protecting Groups

Compound **8** (15 mg, 0.01 mmol) was dissolved in anhydrous DMF (500 μL). The solution of *p*-SCN-Bn-deferoxamine (12 mg, 0.01 mmol) and DIPEA (10 μL, 0.05 mmol) in anhydrous dimethyl sulfoxide (DMSO; 500 μL) was added, and the resulting mixture was stirred at ambient temperature for 1 h, after which UHPLC/MS analysis confirmed the formation of conjugate. The residual solvents were evaporated under high vacuum, the solution of CH_2Cl_2/TFA (1:1, 1 mL, *v*/*v*) was added, and the reaction mixture was stirred at ambient temperature for an additional 2 h. Residual solvents were evaporated in a stream of nitrogen, and the final product **9** was purified by semipreparative HPLC.

3.2.8. Instrumentation and Analytics

For the UHPLC-MS analysis, a sample of resin (~5 mg) was treated with CH_2Cl_2/TFA (1:1, 1 mL, *v*/*v*), the cleavage cocktail was evaporated under a stream of nitrogen, and cleaved compounds extracted into CH_3CN/H_2O (1:1, 1 mL, *v*/*v*). Prior to HPLC separation (column Phenomenex Gemini, 50 × 2.00 mm, 3 μm particles, C18), the samples were injected by direct infusion into the mass spectrometer using an autosampler. Mobile phase was isocratic 80% CH_3CN and 20% 0.01 M ammonium acetate in H_2O or 95% methanol + 5% H_2O + 0.1% formic acid and flow of 0.3 mL/min.

Liquid chromatography coupled with mass spectrometry (LC/MS) analyses were carried out on a UHPLC-MS system consisting of a UHPLC chromatograph Acquity with a photodiode array detector and a single quadrupole mass spectrometer (Waters, Milford, MA, USA), using X-Select C18 column at 30 °C and flow rate of 0.6 mL/min. Mobile phase was (A) 0.01 M ammonium acetate in H_2O and (B) CH_3CN, linearly programmed from 10% A to 80% B over 2.5 min, kept for 1.5 min. The column was re-equilibrated with 10% of solution B for 1 min. The ESI source operated at a discharge current of 5 μA, vaporizer temperature of 350 °C, and capillary temperature of 200 °C. HPLC purification was carried out on a C18 reverse phase column (YMC Pack ODS-A (YMC, Kyoto, Japan), 20 × 100 mm, 5 μm particles), gradient was formed from CH_3CN and 0.01 M ammonium acetate in H_2O, flow rate of 15 mL/min. For lyophilization of residual solvents at −110 °C, a ScanVac Coolsafe 110-4 (Labogene, Lillerod, Denmark) was used.

High-resolution mass spectrometry (HRMS) analyses were performed using an LC chromatograph (Dionex Ultimate 3000, Thermo Fischer Scientific, Waltham, MA, USA) and an Exactive Plus Orbitrap high-resolution mass spectrometer (Thermo Fischer Scientific, Waltham, MA, USA) operating at positive full scan mode (120,000 FWMH) in the range of 100–1000 m/z. The settings for electrospray ionization were as follows: oven temperature of 150 °C and source voltage of 3.6 kV. The acquired data were internally calibrated with phthalate as a contaminant in methanol (m/z 297.15909). The samples were diluted to a final concentration of 0.1 mg/mL in CH_3CN/H_2O (9:1, v/v).

Nuclear magnetic resonance (NMR) spectra were recorded on a JEOL ECX500 spectrometer (JEOL, Tokyo, Japan) at magnetic field strengths of 11.75 T with operating frequencies 500.16 MHz (for ^{1}H), and 125.77 MHz (for ^{13}C) at 27 °C. Chemical shifts (δ) are reported in parts per million (ppm) and coupling constants (J) are reported in Hertz (Hz). The ^{1}H and ^{13}C NMR chemical shifts (δ in ppm) were referenced to the residual signals of DMSO-d_6 [2.50 (^{1}H) and 39.52 (^{13}C)]. The residual signal of ammonium acetate (from HPLC purification) exhibited a signal at 1.90 ppm (^{1}H) and at 21.3 ppm and 172.0 ppm (^{13}C).

Abbreviations in NMR spectra: br s—broad singlet, d—doublet, dd—doublet of doublets, m—multiplet, s—singlet.

3.3. Radiolabeling and Quality Control

For radiolabeling of DFO-c(RGDyK) **9**, $^{68}GaCl_3$ was obtained by eluting a commercial ^{68}Ge/^{68}Ga-generator with 0.1 N HCl. The fractionated elution method was used to increase the radioactivity to volume ratio to its maximum. DFO-c(RGDyK) **9** (0.1–50 μg) was incubated with 300 μL of ^{68}Ge/^{68}Ga-generator eluate (20–40 MBq), and pH was adjusted to 4–5 by adding 30 μL of 1.14 M $CH_3COONa \times 3H_2O$. After the incubation (1–30 min) at 85 °C and the adjustment of pH to 6–7 with 100 μL of 1.14 M $CH_3COONa \times 3H_2O$, the samples were analyzed by RP-HPLC with radiodetection.

Radiochemical purity (RCP) of [^{68}Ga]Ga-DFO-c(RGDyK) was determined by RP-HPLC using the gradient method. RP-HPLC analysis was performed with a Dionex UltiMate 3000 UHPLC system (Thermo Fisher Scientific, Waltham, MA, USA) consisting of an UltiMate 3000 RS pump, an UltiMate 3000 autosampler, an Ultimate 3000 column compartment (25 °C oven temperature), an UltiMate 3000 variable wavelength detector (UV detection at λ = 220 and 280 nm), and a GABI Star radiometric detector (Raytest GmbH, Straubenhardt, Germany). A Nucleosil 120-5 C18 250 × 40 mm column (WATREX, Prague, Czech Republic) with 1 mL/min flow rate was used with the following gradient: CH_3CN/H_2O/0.1% trifluoroacetic acid (TFA): 0–3.0 min 0% CH_3CN, 3.1–6.0 min 0–50% CH_3CN, 6.1–10.0 min 50% CH_3CN, 10.1–13.0 min 80% CH_3CN, 13.1–15 min 0% CH_3CN.

3.4. Partition Coefficient, In Vitro Stability, and Protein Binding

Partition coefficient (log P) of [^{68}Ga]Ga-DFO-c(RGDyK) was determined as follows: Radiolabeled DFO-c(RGDyK) was dissolved with phosphate-buffered saline (PBS) pH = 7.4 to 1 mL (~7 μM). Aliquots of 50 μL were added to 450 μL PBS and 500 μL octanol, and the mixture was vigorously vortexed for 20 min at 1500 rpm. The aqueous and organic solvents were separated by centrifugation (2 min at 2000× g), and 50 μL aliquots of both layers were collected and measured in a gamma counter (PerkinElmer, Waltham, MA, USA). Log p values were calculated from the obtained data in Microsoft Office Excel 2010 (Microsoft, Redmond WA, USA) (mean of n = 6).

In vitro stability of [^{68}Ga]Ga-DFO-c(RGDyK) was studied in different media, including PBS, human serum (Sigma Aldrich, St. Louis, MO, USA), competing metal (0.1 M $FeCl_3$), and chelator solution (6 mM diethylenetriaminepentaacetic acid (DTPA)). [^{68}Ga]Ga-DFO-c(RGDyK) was directly mixed with PBS and human serum at a 1:10 ratio, and with $FeCl_3$ and DTPA solutions at a 1:1 ratio. Thus, the prepared reaction mixtures were incubated for 30, 60, and 120 min, respectively, at 37 °C. After incubation, human serum samples were precipitated with acetonitrile and centrifuged (3 min, 2000× g). The supernatant was analyzed by RP-HPLC. Samples containing PBS, DTPA, and $FeCl_3$ were analyzed directly. The stability is reported as % RCP of [^{68}Ga]Ga-DFO-c(RGDyK) (n = 3).

For protein binding measurement, 10 µL of [^{68}Ga]Ga-DFO-c(RGDyK) was mixed with 190 µL of human serum or PBS as a control and incubated at 37 °C up to 120 min. The samples were collected at selected time points (30, 60, and 120 min after incubation) and analyzed by size exclusion chromatography using MicroSpin G-50 Columns (GE Healthcare, Buckinghamshire, UK) in triplicates. First, the columns were centrifuged at 2000× *g* for 1 min to remove the storage buffer. After adding 25 µL of tested sample, the columns were centrifuged again for 2 min with 2000× *g*. Column and eluate were measured in the gamma counter, and percentages of non-protein-bound radiolabeled peptide (column) and protein-bound radiotracer (eluate) were calculated.

3.5. Cell Culture and In Vitro Cell Assays

Human glioblastoma multiforme U-87 MG cells (ATCC, Manassas, VA, USA) were cultured in Dulbecco's Modified Eagle Medium (Merck, Darmstadt, Germany) supplemented with 10% fetal bovine serum, 0.1 mM non-essential amino acids, and 1.0 mM sodium pyruvate at 37 °C in a 5% carbon dioxide humidified incubator. Human melanoma cell lines M21 (fluorescence-activated cell sorting selected clone, which stably express $\alpha v\beta 3$ integrin; $\alpha v\beta 3$ integrin-positive) and M21-L ($\alpha v\beta 3$ integrin-negative) originated from prof. Cheresh, Departments of Immunology and Vascular Biology, The Scripps Research Institute, La Jolla, CA, USA and were a kind gift of prof. Decristoforo, Department of Nuclear Medicine, Medical University Innsbruck, Austria. Both M21 and M21-L cells were cultured in Dulbecco's Modified Eagle Medium (Merck, Darmstadt, Germany) supplemented with 10% fetal bovine serum. All the cells were subcultured and used for experiments at a confluency of 70–90%.

Cell uptake assays were performed in 24-well plates cell at cellular confluency of 70–90%. In the case of U-87 MG, collagen-coated plates (Waltham, MA, USA) were employed due to the insufficient adherence of this cell line. The uptake assay itself was carried out using a dedicated buffer instead of the regular cell culture media. This buffer consisted of 25 mM Tris/HCl, 5.4 mM KCl, 1.8 mM $CaCl_2$, 0.8 mM $MgSO_4$, 5 mM glucose, and 140 mM NaCl in H_2O [41]. The cells were incubated with [^{68}Ga]Ga-DFO-c(RGDyK) (7 nM) alone and also with an excess (3.5 µM) of cold NODAGA-c(RGDyK) (ABX, Radeberg, Germany) as the uptake inhibitor. This incubation was performed in the buffer described above. The incubation times with ^{68}Ga-labeled ligand were 30–90 min in the case of U-87 MG cells and 60 min in the case of M21 and M21-L cells at 37 °C. Afterwards, the 24-well plates were rinsed with PBS and the cells were lysed by 0.1 M NaOH and measured for radioactivity in the gamma counter. The uptake of [^{68}Ga]Ga-DFO-c(RGDyK) was calculated as the mean of the percentage of the applied dose ± standard deviation.

The determination of the IC_{50} value of [^{68}Ga]Ga-DFO-c(RGDyK) in vitro was carried out using an M21 cell line. This competition assay was performed in standard 24-well plates with the cells seeded one day before the experiment. The assay itself employed the same incubation buffer as in the above-described uptake assay. Briefly, the M21 cells were washed with PBS buffer, and cold DFO-c(RGDyK) **9** was added (triplicate wells) in nine different concentrations to cover a concentration range from 1×10^{-10} to 5×10^{-5} M. The cells were then incubated for 30 min at 37 °C. Next, [^{68}Ga]Ga-NODAGA-c(RGDyK) (15 nM) was added, followed by 60 min of incubation at 37 °C. Thereafter, the uptake was interrupted by buffer removal. The cells were rinsed with ice-cold PBS buffer and lysed with 0.1 M NaOH. The lysed cell were collected for radioactivity and protein content measurement. The protein content was determined using a standard BCA protein assay (Pierce™ BCA Protein Assay Kit, Thermo Fisher Scientific, Waltham, MA, USA). Radioactivity of the cell samples was measured in the gamma counter.

The cellular uptake of [^{68}Ga]Ga-NODAGA-c(RGDyK) was calculated as the percentage of applied dose per milligram of cell protein. These uptake values were used to plot a classical sigmoidal dose-response curve in GraphPad Prism (GraphPad Software,

San Diego, CA, USA). The IC_{50} value was determined from this curve using fitting analysis in the above-mentioned software.

3.6. Animal Experiments

The animal studies were performed using female 8–10-week-old Balb/c and athymic Balb/c nude mice (Envigo, Horst, The Netherlands). The animals were acclimatized to laboratory conditions for 1 week prior to experimental use and housed in a specific-pathogen-free animal facility with free access to animal chow and water. During the experiments, general health and body weight of the animals were monitored. The number of animals was reduced as much as possible (n = 3 per group and time point) for all in vivo experiments. The tracer injection as well as small animal imaging was carried out under 2% isoflurane anesthesia (FORANE, Abbott Laboratories, Abbott Park, IL, USA) to minimize animal suffering and to prevent animal motion.

For animal tumor models, athymic Balb/c nude mice were subcutaneously injected in the right flank with 5×10^6 U-87 MG cells mixed with Matrigel Matrix (Corning, NY, USA) at a 1:1 ratio or in the right and left flank with 5×10^6 M21 and M21-L cells. Tumor growth was continuously monitored by caliperation. When the tumor volume reached around 0.1–0.3 cm^3 (i.e., 6–8 weeks after the inoculation of tumor cells), the mice were used for ex vivo biodistribution studies or PET/CT imaging.

To evaluate pharmacokinetics and biodistribution of [^{68}Ga]Ga-DFO-c(RGDyK) in healthy and tumor-bearing (U-87 MG cell line) animals ex vivo, a group of three Balb/c or athymic Balb/c nude mice per time point were retro-orbitally (r.o.) injected with [^{68}Ga]Ga-DFO-c(RGDyK) (1–2 MBq/mouse, 1 µg DFO-c(RGDyK)). The animals were sacrificed by cervical dislocation at 30 and 90 min post-injection (p.i.). Organs and tissues of interest (blood, spleen, pancreas, stomach, intestines, kidneys, liver, heart, lung, muscle, bone, and tumor) were collected, weighed, and measured in the gamma counter. The results were expressed as percentage of injected dose per gram organ (% ID/g).

PET/CT imaging of the experimental animals was performed with an Albira PET/SPECT/CT small animal imaging system (Bruker Biospin Corporation, Woodbridge, CT, USA). The animals were r.o. injected with [^{68}Ga]Ga-DFO-c(RGDyK) at a dose of 5–7 MBq corresponding to ~2 µg of DFO-c(RGDyK) **9** per animal. Anesthetized animals were placed in a prone position in the Albira system before the start of imaging. Static PET/CT images were acquired over 30 min, starting 30 and 90 min after injection for both healthy and tumor-bearing (U-87 MG and M21 vs. M21-L) mice. A 10-min PET scan (axial FOV 148 mm) was performed, followed by a double CT scan (axial FOV 110 mm, 45 kVp, 400 µA, at 400 projections). ThesScans were reconstructed with Albira software (Bruker Biospin Corporation, Woodbridge, CT, USA) using the maximum likelihood expectation maximization (MLEM) and filtered backprojection (FBP) algorithms. After reconstruction, the acquired data was viewed and analyzed with the appropriate software (PMOD software, PMOD Technologies Ltd., Zurich, Switzerland and VolView software, Kitware, Clifton Park, NY, USA).

4. Conclusions

We have developed synthetic protocols for simple preparation of RGD peptides conjugated with deferoxamine. Based on the combination of solid-phase and solution-phase synthesis (post-cleavage oligopeptide modification), the conjugation via lysin side chain was regioselectively achieved due to a high level of orthogonality in the structure of corresponding intermediates. Furthermore, with respect to high crude purities, only two purification steps within the reaction sequence were required. The protocols can be applied to the preparation of various analogical conjugates bearing lysine in a peptide moiety.

Moreover, here we report for the first time, to our best knowledge, a DFO-based RGD peptide for radiometal labeling. The deferoxamine-based c(RGDyK) conjugate **9** could be easily radiolabeled with Ga-68 with high radiochemical purity, thus not requiring further purification steps. [^{68}Ga]Ga-DFO-c(RGDyK) showed excellent in vitro stability in human serum and PBS, high affinity for αvβ3 integrin, rapid predominantly renal elimination, and good tumor-to-background ratios, indicating that it may be applicable for imaging of tumors expressing αvβ3 integrins. Additionally, the use of DFO as a chelating moiety also allows labeling of the studied DFO-c(RGDyK) with different radiometals.

Supplementary Materials: The following are available online at https://www.mdpi.com/article/10.3390/ijms22147391/s1.

Author Contributions: Conceptualization, M.P. and M.S.; data curation, S.K., A.D., Z.N., K.B. and M.P.; investigation, S.K., A.D., Z.N. and K.B.; methodology, M.P. and M.S.; project administration, M.P.; chemistry, S.K. and M.S.; animal experiments, A.D., Z.N., K.B. and M.P.; writing—original draft, S.K. and M.P.; writing—review and editing, S.K., Z.N., M.S. and M.P. All authors have read and agreed to the published version of the manuscript.

Funding: This research was funded by the European Regional Development Fund–Project ENOCH (No. CZ.02.1.01/0.0/0.0/16_019/0000868) to M.P.

Institutional Review Board Statement: Animal studies were carried out in accordance with regulations and guidelines of the Czech Animal Protection Act (No. 246/1992) and with the approval of the Czech Ministry of Education, Youth, and Sports (MSMT-16402/2012-30 and MSMT-41830/2018-7) and the institutional Animal Welfare Committee of the Faculty of Medicine and Dentistry, Palacky University in Olomouc.

Informed Consent Statement: Not applicable.

Data Availability Statement: The data presented in this study are available in the article or Supplementary Materials. The raw datasets are available from the corresponding authors on reasonable request.

Acknowledgments: We thank the staff from the Animal Facilities of the Institute of Molecular and Translational Medicine, Faculty of Medicine and Dentistry, Palacky University in Olomouc for taking care of the animals.

Conflicts of Interest: The authors declare no conflict of interest.

References

1. Fani, M.; Maecke, H.R.; Okarvi, S.M. Radiolabeled peptides: Valuable tools for the detection and treatment of cancer. *Theranostics* **2012**, *2*, 481–501. [CrossRef]
2. Rangger, C.; Haubner, R. Radiolabelled peptides for positron emission tomography and endoradiotherapy in oncology. *Pharmaceuticals* **2020**, *13*, 22. [CrossRef]
3. Fani, M.; Maecke, H.R. Radiopharmaceutical development of radiolabelled peptides. *Eur. J. Nucl. Med. Mol. Imaging* **2012**, *39*, S11–S30. [CrossRef]
4. Laverman, P.; Sosabowski, J.K.; Boerman, O.C.; Oyen, W.J.G. Radiolabelled peptides for oncological diagnosis. *Eur. J. Nucl. Med. Mol. Imaging* **2012**, *39*, S78–S92. [CrossRef]
5. Leitha, D.; Izard, T. Roles of membrane domains in integrin-mediated cell adhesion. *Int. J. Mol. Sci.* **2020**, *21*, 5531. [CrossRef] [PubMed]
6. Mezu-Ndubuisi, O.J.; Maheshwari, A. The role of integrins in inflammation and angiogenesis. *Pediatr. Res.* **2020**, *89*, 1619–1626. [CrossRef] [PubMed]
7. Nieberler, M.; Reuning, U.; Reichart, F.; Notni, J.; Wester, H.J.; Schwaiger, M.; Weinmüller, M.; Räder, A.; Steiger, K.; Kessler, H. Exploring the role of RGD-recognizing integrins in cancer. *Cancers* **2017**, *9*, 116. [CrossRef] [PubMed]
8. Gaertner, F.C.; Kessler, H.; Wester, H.J.; Schwaiger, M.; Beer, A.J. Radiolabelled RGD peptides for imaging and therapy. *Eur. J. Nucl. Med. Mol. Imaging* **2012**, *39*, S126–S138. [CrossRef] [PubMed]
9. Jamous, M.; Haberkorn, U.; Mier, W. Synthesis of peptide radiopharmaceuticals for the therapy and diagnosis of tumor diseases. *Molecules* **2013**, *18*, 3379–3409. [CrossRef]
10. Liu, S. Bifunctional coupling agents for radiolabeling of biomolecules and target-specific delivery of metallic radionuclides. *Adv. Drug Deliv. Rev.* **2008**, *60*, 1347–1370. [CrossRef]
11. Shi, J.; Wang, F.; Liu, S. Radiolabeled cyclic RGD peptides as radiotracers for tumor imaging. *Biophys. Rep.* **2016**, *2*, 1–20. [CrossRef]
12. Ioppolo, J.A.; Caldwell, D.; Beiraghi, O.; Llano, L.; Blacker, M.; Valliant, J.F.; Berti, P.J. 67Ga-labeled deferoxamine derivatives for imaging bacterial infection: Preparation and screening of functionalized siderophore complexes. *Nucl. Med. Biol.* **2017**, *52*, 32–41. [CrossRef]

13. Oroujeni, M.; Garousi, J.; Andersson, K.G.; Löfblom, J.; Mitran, B.; Orlova, A.; Tolmachev, V. Preclinical evaluation of [^{68}Ga]Ga-DFO-ZEGFR:2377: A promising affibody-based probe for noninvasive PET imaging of EGFR expression in tumors. *Cells* **2018**, *7*, 141. [CrossRef] [PubMed]
14. Kaeppeli, S.A.M.; Schibli, R.; Mindt, T.L.; Behe, M. Comparison of desferrioxamine and NODAGA for the gallium-68 labeling of exendin-4. *EJNMMI Radiopharm. Chem.* **2019**, *4*, 9. [CrossRef] [PubMed]
15. Raavé, R.; Sandker, R.; Adumeau, P.; Jacobsen, C.B.; Mangin, F.; Meyer, M.; Moreau, M.; Bernhard, C.; Da Costa, L.; Dubois, A.; et al. Direct comparison of the in vitro and in vivo stability of DFO, DFO* and DFOcyclo* for 89 Zr-immunoPET. *Eur. J. Nucl. Med. Mol. Imaging* **2019**, *46*, 1966–1977. [CrossRef]
16. Petrik, M.; Umlaufova, E.; Raclavsky, V.; Palyzova, A.; Havlicek, V.; Pfister, J.; Mair, C.; Novy, Z.; Popper, M.; Hajduch, M.; et al. ^{68}Ga-labelled desferrioxamine-B for bacterial infection imaging. *Eur. J. Nucl. Med. Mol. Imaging* **2021**, *48*, 372–382. [CrossRef] [PubMed]
17. Ye, Y.; Bloch, S.; Xu, B.; Achilefu, S. A novel near-infrared fluorescent integrin targeted DFO analog. *Bioconjug. Chem.* **2008**, *19*, 225–234. [CrossRef] [PubMed]
18. Haubner, R.; Decristoforo, C. Radiolabelled RGD peptides and peptidomimetics for tumour targeting. *Front. Biosci.* **2009**, *14*, 872–886. [CrossRef]
19. Kapp, T.G.; Rechenmacher, F.; Neubauer, S.; Maltsev, O.V.; Cavalcanti-Adam, E.A.; Zarka, R.; Reuning, U.; Notni, J.; Wester, H.J.; Mas-Moruno, C.; et al. A comprehensive evaluation of the activity and selectivity profile of ligands for RGD-binding integrins. *Sci. Rep.* **2017**, *7*, 39805. [CrossRef]
20. Li, Z.B.; Chen, K.; Chen, X. (68)Ga-labeled multimeric RGD peptides for microPET imaging of integrin alpha(v)beta (3) expression. *Eur. J. Nucl. Med. Mol. Imaging* **2008**, *35*, 1100–1108. [CrossRef]
21. Jeong, J.M.; Hong, M.K.; Chang, Y.S.; Lee, Y.S.; Kim, Y.J.; Cheon, G.J.; Lee, D.S.; Chung, J.K.; Lee, M.C. Preparation of a promising angiogenesis PET imaging agent: ^{68}Ga-labeled c(RGDyK)-isothiocyanatobenzyl-1,4,7-triazacyclononane-1,4,7-triacetic acid and feasibility studies in mice. *J. Nucl. Med.* **2008**, *49*, 830–836. [CrossRef]
22. Oxboel, J.; Schjoeth-Eskesen, C.; El-Ali, H.H.; Madsen, J.; Kjaer, A. 64Cu-NODAGA-c(RGDyK) is a promising new angiogenesis PET tracer: Correlation between tumor uptake and integrin $\alpha V\beta 3$ expression in human neuroendocrine tumor xenografts. *Int. J. Mol. Imaging* **2012**, *2012*, 379807. [CrossRef]
23. Shin, U.C.; Jung, K.H.; Lee, J.W.; Lee, K.C.; Lee, Y.J.; Park, J.A.; Kim, J.Y.; Kang, J.H.; An, G.I.; Ryu, Y.H.; et al. Preliminary evaluation of new ^{68}Ga-labeled cyclic RGD peptides by PET imaging. *J. Radiopharm. Mol. Probes.* **2016**, *2*, 118–122.
24. Novy, Z.; Stepankova, J.; Hola, M.; Flasarova, D.; Popper, M.; Petrik, M. Preclinical evaluation of radiolabeled peptides for PET imaging of glioblastoma multiforme. *Molecules* **2019**, *24*, 2496. [CrossRef] [PubMed]
25. Van Der Gucht, A.; Pomoni, A.; Jreige, M.; Allemann, P.; Prior, J.O. ^{68}Ga-NODAGA-RGDyK PET/CT imaging in esophageal cancer: First-in-human imaging. *Clin. Nucl. Med.* **2016**, *41*, 491–492. [CrossRef] [PubMed]
26. Durante, S.; Dunet, V.; Gorostidi, F.; Mitsakis, P.; Schaefer, N.; Delage, J.; Prior, J.O. Head and neck tumors angiogenesis imaging with ^{68}Ga-NODAGA-RGD in comparison to 18F-FDG PET/CT: A pilot study. *EJNMMI Res.* **2020**, *10*, 47. [CrossRef]
27. Belotti, D.; Remelli, M. Deferoxamine B: A natural, excellent and versatile metal chelator. *Molecules* **2021**, *26*, 3255. [CrossRef]
28. Díaz-Mochón, J.J.; Bialy, L.; Bradley, M. Full Orthogonality between Dde and Fmoc: The direct yynthesis of PNA-peptide conjugates. *Org. Lett.* **2004**, *6*, 1127–1129. [CrossRef]
29. Fani, M.; Andre, J.P.; Maecke, H.R. ^{68}Ga-PET: A powerful generator-based alternative to cyclotron-based PET radiopharmaceuticals. *Contrast Media Mol. Imaging* **2008**, *3*, 67–77. [CrossRef]
30. Decristoforo, C. Gallium-68—A new opportunity for PET available from a long shelf-life generator—Automation and applications. *Curr. Radiopharm.* **2012**, *5*, 212–220. [CrossRef]
31. Velikyan, I. ^{68}Ga-Based radiopharmaceuticals: Production and application relationship. *Molecules* **2015**, *20*, 12913–12943. [CrossRef] [PubMed]
32. Knetsch, P.; Petrik, M.; Griessinger, C.M.; Rangger, C.; Fani, M.; Kesenheimer, C.; von Guggenberg, E.; Pichler, B.J.; Virgolini, I.; Decristoforo, C.; et al. [^{68}Ga]NODAGA-RGD for imaging $\alpha v\beta 3$ integrin expression. *Eur. J. Nucl. Med. Mol. Imaging* **2011**, *38*, 1303–1312. [CrossRef] [PubMed]
33. Kaeopookum, P.; Petrik, M.; Summer, D.; Klingler, M.; Zhai, C.; Rangger, C.; Haubner, R.; Haas, H.; Hajduch, M.; Decristoforo, C. Comparison of ^{68}Ga-labeled RGD mono- and multimers based on a clickable siderophore-based scaffold. *Nucl. Med. Biol.* **2019**, *78–79*, 1–10. [CrossRef] [PubMed]
34. Summer, D.; Grossrubatscher, L.; Petrik, M.; Michalcikova, T.; Novy, Z.; Rangger, C.; Klingler, M.; Haas, H.; Kaeopookum, P.; von Guggenberg, E.; et al. Developing targeted hybrid imaging probes by chelator scaffolding. *Bioconjug. Chem.* **2017**, *28*, 1722–1733. [CrossRef]
35. Carlucci, G.; Ananias, H.J.K.; Yu, Z.; Van de Wiele, C.; Dierckx, R.A.; de Jong, I.J.; Elsinga, P.H. Multimerization improves targeting of peptide radio-pharmaceuticals. *Curr. Pharm. Des.* **2012**, *18*, 2501–2516. [CrossRef] [PubMed]
36. Notni, J.; Pohle, C.; Wester, H.J. Be spoilt for choice with radiolabelled RGD peptides: Preclinical evaluation of ^{68}Ga-TRAP(RGD)$_3$. *Nucl. Med. Biol.* **2013**, *40*, 33–41. [CrossRef] [PubMed]
37. Liu, S. Radiolabeled cyclic RGD peptide bioconjugates as radiotracers targeting multiple integrins. *Bioconjug. Chem.* **2015**, *26*, 1413–1438. [CrossRef]

38. Chandra, R.; Pierno, C.; Braunstein, P. 111In Desferal: A new radiopharmaceutical for abscess detection. *Radiology* **1978**, *128*, 697–699. [CrossRef]
39. Govindan, S.V.; Michel, R.B.; Griffiths, G.L.; Goldenberg, D.M.; Mattes, M.J. Deferoxamine as a chelator for 67Ga in the preparation of antibody conjugates. *Nucl. Med. Biol.* **2005**, *32*, 513–519. [CrossRef] [PubMed]
40. Asati, S.; Pandey, V.; Soni, V. RGD peptide as a targeting moiety for theranostic purpose: An update study. *Int. J. Pept. Res. Ther.* **2019**, *25*, 49–65. [CrossRef]
41. Müller, S.A.; Holzapfel, K.; Seidl, C.; Treiber, U.; Krause, B.J.; Senekowitsch-Schmidtke, R. Characterization of choline uptake in prostate cancer cells following bicalutamide and docetaxel treatment. *Eur. J. Nucl. Med. Mol. Imaging* **2009**, *36*, 1434–1442. [CrossRef] [PubMed]

International Journal of
Molecular Sciences

MDPI

Review

Recent Applications of Retro-Inverso Peptides

Nunzianna Doti [1], Mario Mardirossian [2], Annamaria Sandomenico [1], Menotti Ruvo [1,*] and Andrea Caporale [3,*]

1 Institute of Biostructures and Bioimaging (IBB), National Research Council (CNR), 80134 Napoli, Italy; nunzianna.doti@cnr.it (N.D.); annamaria.sandomenico@cnr.it (A.S.)

2 Department of Medicine, Surgery and Health Sciences, University of Trieste, 34149 Trieste, Italy; mmardirossian@units.it

3 Institute of Crystallography (IC), National Research Council (CNR), 34149 Trieste, Italy

* Correspondence: menotti.ruvo@unina.it (M.R.); andrea.caporale@cnr.it (A.C.)

Abstract: Natural and *de novo* designed peptides are gaining an ever-growing interest as drugs against several diseases. Their use is however limited by the intrinsic low bioavailability and poor stability. To overcome these issues retro-inverso analogues have been investigated for decades as more stable surrogates of peptides composed of natural amino acids. Retro-inverso peptides possess reversed sequences and chirality compared to the parent molecules maintaining at the same time an identical array of side chains and in some cases similar structure. The inverted chirality renders them less prone to degradation by endogenous proteases conferring enhanced half-lives and an increased potential as new drugs. However, given their general incapability to adopt the 3D structure of the parent peptides their application should be careful evaluated and investigated case by case. Here, we review the application of retro-inverso peptides in anticancer therapies, in immunology, in neurodegenerative diseases, and as antimicrobials, analyzing pros and cons of this interesting subclass of molecules.

Keywords: retro-inverso peptides; anticancer peptides; drug delivery; peptide antigens; Aβ; IAPP; antimicrobial peptides

Citation: Doti, N.; Mardirossian, M.; Sandomenico, A.; Ruvo, M.; Caporale, A. Recent Applications of Retro-Inverso Peptides. *Int. J. Mol. Sci.* **2021**, *22*, 8677. https://doi.org/10.3390/ijms22168677

Academic Editor: Sonia Melino

Received: 28 July 2021
Accepted: 10 August 2021
Published: 12 August 2021

1. Topology, Structural Characteristics

As therapeutic agents, peptides have fascinating properties, such as very high specificity and binding affinity, generally low toxicity, and low risk of drug interactions [1]. Moreover, due to the high diversity which is sequence- and structure-dependent, they can be designed as potential drugs to target almost any disease. On the other hand, natural peptides, due to their low size and high sensitivity to most proteases, are quickly excreted or anyhow degraded, resulting in poor biodistribution, bioavailability, and rapid clearance [2], and thus limited therapeutic potential [2]. To increase peptide half-life, stability and bioavailability, many approaches have been proposed including PEGylation, backbone modifications, cyclization, side chain stapling, and lipidation [3,4]. Among these, modification of the backbone is one of the most invasive approaches, as it may profoundly affect the conformation of peptides, especially when it involves alteration of the residue's isomerization.

All amino acids (except glycine) possess chiral centers and occur in nature almost exclusively as L-enantiomers in proteins and natural peptides. D-amino acid-containing peptides are also found in nature, mainly in some frog species and bacteria, as a result of post-translational modifications [5].

Compared to L-amino acid peptides, D-peptides exhibit an innate resistance to enzymatic degradation and as such have acquired a special importance as potential biopharmaceuticals [6–8]. However, given the strong correlation between the structural properties of single residues and of the peptide molecule they belong to, partial or complete modification of peptide isomerism leads in most cases to reduction or even suppression of biological activity [9]. An elegant and often successful solution to translate biologically active L-peptides into D-analogues, is the use of retro-inverso (RI) peptides, which incorporate D-amino acids as stable surrogates of L-amino acids [10], but presented in a reverse (retro) order compared to the parent molecule [8]. RI analogues of all L-peptides, also known as retro-all-D or retro-enantio peptides, are thus peptides composed of D-amino acids introduced in the sequence in reverse direction. The importance of this subclass of peptides is that, when viewed in a fully extended conformation, the side-chains are superimposable with those of the parent L-peptide but with inverted amide bonds and N/C terminal groups [11]. Therefore, in those cases where activity is mostly associated to the array of side chains without any significant contribution from the backbone chemical groups and from the tridimensional organization, a RI analogue has the potential to achieve the same functions as the all-L parent peptide, but with superior stability toward proteolytic degradation [10]. In this regard, retro-inversion is commonly efficacious when one starts from unstructured peptides [12] which work by inductive adaptation on the interacting surfaces and whose array of side chains are more likely to adopt a topology similar to that of the parent peptide. This occurs most readily when the enthalpic contribution of the interactions established by the CO and NH groups of the backbone to stabilize preferential conformations and to bind the target, is relatively poor and can easily be overcome by rearrangements of the retro-inverso analogues which then gain access to the same conformational space of the parent peptide [13]. On these bases, this kind of quasi-neutral isomerization sometimes succeeds and sometimes fails [14–16], but in either case provides an interesting and valuable approach to understand the importance of structural elements in molecular recognition events. In particular, when the introduction of D-amino acids with reversal of the peptide backbone and of the C/N terminal groups leave unchanged the peptide binding properties and/or activities, clearly the role of the peptide backbone to the interaction is negligible. This observation was evidently demonstrated by Ruvo et al. [17] studying a very small bioactive peptide. They found that a close topochemical relationship existed between the parent tripeptide MYF-NH2 (three letters amino acids code: L-Met-L-Tyr-L-Phe), and its corresponding retro-inverso isomer (D-Phe-D-Tyr-D-Met) (Figure 1). In the two peptides, the amide bonds of the backbone were interchanged, whereas the 3D orientation of the side chains and the position of the amino group were identical. They demonstrated that the RI isomer retained the binding properties for the protein ligand, the receptor neurophysin II (NP II), and an affinity similar to that of the parent peptide [17]. In this case-study, the successful conversion toward the RI analogue was made possible due to the small size of the molecule analyzed in addition to regeneration of the amino group through a reduction reaction. In larger and structurally more complex peptides, the limitations described above must be taken into account in order to achieve equally effective translations. Indeed, the conversion from an all-L-parent peptide to its retro-inverso analogue implies that the values of the original φ and ψ dihedral angles are exchanged (retro conversion) and transformed to the identical negative values (chiral inversion) in all residues (Figure 1) [16,18,19]. As shown, an identical relationship holds between the inverso and the retro analogues which can be considered as the retro-inverso the one of the other. The angles of the parent peptide become the angles of the retro-peptide and *vice versa* and the same is true with the inverso and the retro analogues (Figure 1). As result, in the retro-inverso analogues, the direction of the peptide bonds is reversed while the side-chain orientation of the amino acid residues is retained. Since the bond lengths of the CO and NH groups are comparable, the positions of the side chains do not change significantly [13] and the two molecules appear as almost identical. Assuming that the activity of a peptide depends mainly on the interactions that the side chains establish with the surface of the target, the peptide functions can be therefore theoretically preserved [20]. Though, since recognition is also often mediated by backbone interactions and is governed by the molecule 3D organization, RI analogues are likely to successfully mimic the precursor molecule only in a restricted number of cases.

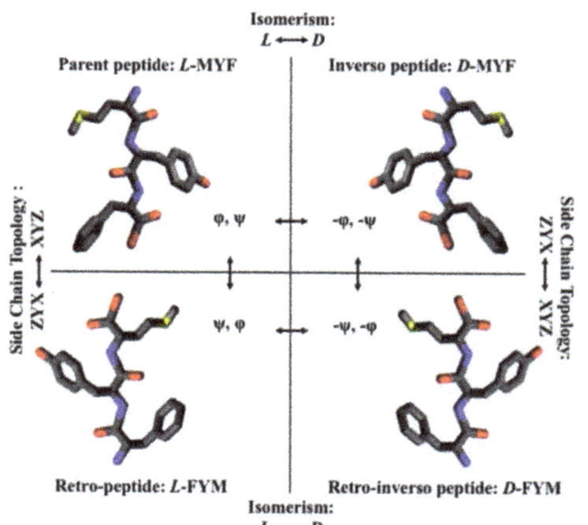

Figure 1. Topological relationship between a peptide and its inverso, retro, and retro-inverso analogues, illustrated for the example peptide MYF [17]. As shown, the topology of the side chains of the retro-inverso analogue, in the C-to-N orientation, is the same as the parent peptide in the N-to-C orientation (figure adapted by [13]).

The hydrogen bonds between the CO acceptors and NH donors generate a network of highly stabilizing interactions in peptides arranged as α-helices and β-sheets. If the network of stabilizing interactions is removed, the stability of the 3D structure will be severely compromised in the retro-inverso mimetics, largely affecting their activity [13]. From a topological point of view, the RI analogues of larger peptides [21,22] could adopt conformations similar to that of the parent peptide when the full-length protein or part of it predominantly contains structural elements whose energy in the Ramachandran map is not drastically changed during the conversion like in β-sheets and γ-turns. In this case, they are likely to be stabilized by similar side-chain-to-side-chain interactions. This observation is consistent with results reported in literature [13,17,23,24]. As an example, Peggion and coworkers in 2009 [25] proposed a structure-function relationship study on a mimetic peptide of the Parathyroid hormone (PTH) spanning residues 1-11 (PTH(1-11)). This peptide is a ligand of the PTH type-I receptor and was studied through the synthesis and characterization of all-D PTH retro-inverso analogues. The retro-inverso RI-PTH(1-11) analogues showed a reduced biological activity compared to the parent peptide, because of the absence of the α-helical structure which could be induced by introducing an Aib residues on the N-terminal position [26].

For the design of RI peptides two aspects should be therefore considered: (i) The importance of the interactions of the backbone amide groups of the parent peptide with the receptor are important [12,13,27]. (ii) Maintenance of the original 3D structure in the retro-inverted peptide, that means retention of most hydrogen bonds formed intra-backbone and those between the backbone and the side chains.

The current review focuses on the main applications of retro-inverso peptides as potential biotherapeutics with improved stability in vitro and in vivo. The interest around this fascinating subclass of molecules is driven by their potential use in a vast area of applications here reviewed, including diagnostics, cancer therapeutics, neurodegenerative diseases, and new antibiotics (as antimicrobial peptides).

2. Anticancer Applications—Diagnostic

In the cure of cancer, side effects following conventional drug treatments are currently on the rise. A growing number of studies indicate that peptides, more specifically anticancer peptides (ACPs), could be new valuable options in this field. Peptides have the advantage of exhibiting reduced immunogenicity, excellent tissue penetrability, and low-cost manufacturability compared to bigger molecules like proteins and antibodies. Also, they are easily modified to improve the in vivo stability and the biological activity, leading to an increased utility and versatility for cancer therapy. For these reasons, an ever-growing number of anticancer peptides (ACPs) are being evaluated at various stages of clinical trials [28,29]. Cancer development is characterized by a variety of processes that include migration from the primary tumor site, invasion through the basement membrane, invasion of metastatic cells into blood vessels, and finally localization at second sites [30]. During the proliferation of cancer cells, the core of the tumor becomes deficient in nutrients and oxygen. Therefore, this kind of cancer cells have to up-regulate the expression of pro-angiogenic factors to stimulate new blood vessels into and around the tumor to allow it to grow up [31,32]. Angiogenesis, which is the formation of new capillaries from pre-existing blood vessels, is the driving event in this physiological process that also regulates embryogenesis, postnatal growth, reproductive function, and wound healing [33]. The other side of the coin are the pathological mechanisms of angiogenesis which play important roles in numerous human diseases, particularly in the growth and spread of cancers [33–35]. The suppression of pathological angiogenesis, which is principally driven by vascular endothelial growth factors (VEGF) and by their interaction with the receptors VEGFR1 and VEGFR2, set on the surface of endothelial cells [34], is indeed an efficient and clinically validated approach in cancer therapy. Beyond antibodies and soluble receptors, also several peptides have been designed and tested as inhibitors of the VEGF-VEGFR kinase axes in the tumor angiogenic cascade, with many that have been approved by the regulatory agencies [35,36] or have entered various clinical trials [37,38] to block tumor growth and angiogenesis [37].

In this field, Vicari et al., in 2011 [38] designed peptides able to mimic the VEGF-binding site on VEGFR-2, reproducing the loop formed by the antiparallel β-sheets β5 and β6. The linear synthetic peptide VEGF-P3(NC) was cyclized by oxidation of two cysteines enabling the formation of the twisted peptide VEGF-P3(CYC) underpinning an anti-parallel β-sheet structure VEGF-P3(CYC) (Table 1). The cyclic peptide VEGF-P3(CYC) showed an increased affinity for VEGFR-2 and an improved capability to inhibit VEGF-dependent signaling pathways compared with the parent linear peptides, highlighting the close relationship between the structure and the activity of the molecule. In this case, the transition from the L peptide to its retro-inverso analogue was not successful. Indeed, the RI analogue, named VEGF-RI-P4(CYC) was not able to block VEGF-VEGFR-2 interaction, although VEGF-RI-P4(CYC) and VEGF-P3(CYC) showed a similar global conformation in solution as demonstrated by Circular Dichroism (CD). These results indicated that the RI analogue, despite the similar side chains disposition, was unable to expose its binding key residues in positions favorable for interacting with the receptor. Upon a more detailed structural characterization, the loss of activity was related to a different arrangement of the backbone as compared to the parent peptide [39].

Using a combinatorial screening on VEGF-activated endothelial cells, the retro-inverso peptide $_{D}$(LPR) was shown to target VEGFR1 and neuropilin-1 [40]. The motif $_{D}$(LPR) was successful validated as strong inhibitor of retinal angiogenesis in retinopathy models when administered in an eye-drop formulation [41], showing the effectiveness of the approach on short peptides.

Table 1. Names, peptide sequences reported in the review and their applications.

Name	Sequence [1]	Application	Ref
	Anticancer Applications—Diagnostic		
VEGF-P3(CYC)	$I^{76}TMQ^{79}CG^{92}IHQGQ$ $HPKIRMI^{80}CE^{93}MSF^{96}$ *	Inhibition angiogenesis	[38,39]
D(LPR)	D(Leu-Pro-Arg)	Inhibition retinal angiogenesis; Diagnostic	[41–43]
SP5	PRPSPKMGVSVS *	Drug delivery	[44,45]
$uPAR_{88–92}$	SRSRY *	Maintaining chemotactic activity and triggers directed cell migration and angiogenesis	[46,47]
RI-3	Ac-D(Tyr- Arg-Aib-Arg)-NH_2	Prevent extracellular invasion by tumor cells	[48]
D(RGD)	D(Asp-Gly-Arg)	Diagnostic	[49–51]
VS	SWFSRHRYSPFAVS *	Glioblastoma multiforme (GBM)	[52,53]
VAP	SNTRVAP *	Gliomas, glioma stem cells, vasculogenic mimicry and neovasculature	[54]
WSW	SYPGWSW *	Glioma cells and tumor neovasculature	[55]
BK	RPPGFSPFR *	Glioma cells	[56]
FP21	YTRDLVYGD PARPGIQGTGTF *	Ovarian cancer	[57–59]
T7	HAIYPRH *	Drug delivery	[60]
	Applications in Immunology		
TG19320	$(rty)_4K_2KG$	IgG binding	[61,62]
VSVp	RGYVYQGL *	antigen surface of hepatitis B virus	[63]
OVAp	SIINFEKL *	antigen surface of hepatitis B virus	[63]
PS1	HQLDPAFGANSTNPD *	antigen surface of hepatitis B virus	[63]
HAI	HAIYPRH *	Crossing BBB	[64]
THR	THRPPMWSPVWP *	Crossing BBB	[64]
InsB:9–23	HLVEALYLVCGERGG *	Analogue of diabetogenic islet peptide—prevents T-cell activation in humanized model mice	[65,66]
	Application in Neurodegenerative Diseases		
Amytrap	WKGEWTGR *	Blocking the oligomerization and aggregation of $A\beta_{1–42}$	[67–71]
$IAPP_{11–20}$	RLANFLVHSS *	Strong inhibitory effects on amylin aggregation in T2DM	[72]
β-$syn_{36–45}$	GVLYVGSKT *	Reduction of amyloid fibril and oligomer formation	[73]
	Application in Antimicrobial Antibiotics		
RI1018	rrwirvavilrv	Preventing formation of Biofilm	[74]
RI-JK6	rivwvrirrwqv	Preventing formation of Biofilm	[74]
RI-73	lwGvwrrvidwlr	Damaging the bacterial membrane	[75]
BMAP-28	GGLRSLGRKILRAWK KYGPIIVPIIRIG *	Broad antimicrobial activities	[76]

[1] The sequences reported are those of the parent peptides * (L-residues), unless otherwise indicated, like reporting D residues as lower-case letters or adding a "D" before the sequence.

Another example of the use of RI analogues against the VEGF-VEGFR complexes was shown by Calvanese et al. [42] who designed a series of cyclic peptides embedding the retro-inverso (RI) version of the consensus sequences RPL/LPPR, corresponding to peptides capable of preventing VEGF binding to VEGFR1 [41] and VEGFR2 [77], and to specifically inhibit human endothelial cells (EC) proliferation in vitro [77]. Direct binding experiments of the peptides to VEGFR1 and VEGFR2 identified a peptide that bound both receptors with a K_D in the low micromolar range but with a significant selectivity for VEGFR1 respect to VEGFR2 thus showing a potential as VEGFR1-selective diagnostic probe.

The important properties of the of LPR peptide's RI mimics were also recently demonstrated by Rezazadeh et al. [43]. Small size, high stability, high affinity for VEGF receptors and good distribution in tumor tissues were the characteristics of the $_D$(LPR) peptide which was suggested as a good candidate for use as SPECT probe for molecular imaging of cancer. $_D$(LPR) was indeed labeled with technetium-99m (^{99m}Tc), which is the first-choice radionuclide in diagnostic nuclear medicine. The authors prepared two different $_D$(LPR) analogues having sequences lprpK-HYNIC and HYNIC-Kplpr, both incorporating the HYNIC L-peptide (sequence L-His-L-Tyr-L-Asn-L-Ile-L-Cys) acting as a bifunctional chelating agent (BFCA) at the C- or N-termini of the targeting peptide. Their results demonstrated that $_D$(LPR) could be labeled not only with diagnostic radioisotopes but also with therapeutic radioisotopes for both imaging and curative purposes.

Another L-peptide, named SP5 (Table 1), that efficiently and specifically binds to the vasculature of tumors was discovered by in vivo phage display [44]. Its RI analogue, D-SP5 [45], was designed, prepared, and investigated and showed a stronger targeting ability to VEGF-stimulated HUVECs. Importantly, D-SP5 recognized the same binding site as the parent peptide, although the receptor was unknown. Authors also demonstrated that D-SP5 could be an effective agent for drug delivery for angiogenesis and a potential targeting vehicle for use in clinical cancer therapy. Also, D-SP5 was conjugated to micelles and loaded with doxorubicin (Dox), showing significantly stronger tumor inhibition efficiency compared to L-SP5 micelles/Dox, negative controls, and free Dox.

Another attractive approach for the clinical management of metastases arising from solid tumors is the control of cell motility. One major regulator of cell migration frequently overexpressed and commonly targeted by therapeutics is the urinary plasminogen activator receptor (uPAR), also known as the urokinase receptor which promotes the chemotactic activity, the migration and the angiogenesis of cancer many cells [78]. The minimal fragment of uPAR spanning residues 88–92 [46] (Table 1) was identified as able to maintain the uPAR chemotactic activity and to trigger cell migration and angiogenesis properties in vitro and in vivo [47]. The activity of this peptide, known as $uPAR_{88-92}$, was mediated by its direct interaction with the receptor FPR type 1 (FPR1), which is able to activate the vitronectin receptor [79]. Inhibition of the uPAR/FPR1 interaction represents an attractive target to inhibit the metastatic process in solid tumors. Carriero et al. [48] reported the tetrapeptide analogue RI-3, Ac-$_D$(Tyr-Arg-Aib-Arg)-NH_2 (Table 1), which at variance with the precursor peptide, was an antagonist of the uPAR/FPR1 interaction and as such could prevent in vivo extracellular matrix invasion, tumor cell infiltration into the blood, and capillary network formation.

Integrins [80,81] are other receptors playing important roles in cell–cell and cell–matrix interactions during developmental and pathological processes [80,82,83]. They are highly correlated with angiogenesis, tumorigenesis, metastasis, and drug resistance [84–86]. In particular, integrin $\alpha_v\beta_3$ isoform represents an interesting molecular target for many diagnostic and therapeutic applications [87–90]. $\alpha_v\beta_3$ specifically recognizes the consensus tripeptide Arg-Gly-Asp (RGD), derived from extracellular matrix proteins, which has been largely exploited as radiolabeled carrier for the early diagnosis of malignant tumors [91]. Recently, Karimi et al. [49] presented a HYNIC-$_D$(RGD) peptide labeled with the radioisotope ^{99m}Tc. The radiochemical purity of HYNIC-$_D$(RGD) was about 100%, showing the efficiency of this method to increase the quality of labeling compared to previous methods used for cyclic peptides [50]. At the same time, the retro-inverso portion of the peptide conferred higher stability in serum and higher affinity for integrins compared to the cyclic RGD parent peptide. Moreover, the peptide ^{99m}Tc-HYNIC-$_D$(RGD) showed radiochemical properties and in vitro targeting ability for human cancer cells similar to those reported in previous studies on other analogues [49]. Liu et al. [51] reported the in vitro investigation of several linear RI peptides based on the RGD motif conjugated to cell-penetrating peptides based on poly-arginines. Their results suggested that linear RI analogues were potentially useful as tumor targeting carriers with biological activity similar to RGD alone.

The peptide VS (Table 1) selected through the screening of a phage display library, also showed high binding affinity towards integrins, in particular against $\alpha_6\beta_1$ and $\alpha_v\beta_3$. The RI analogue [52] was designed, prepared and tested against glioblastoma multiforme (GBM), the most common and lethal tumor of the central nervous system [53]. Specifically, the RI variant of VS conjugated with PEG-PLA (poly-lactic acid) was used to prepare micelles which efficiently encapsulated doxorubicin (DOX), penetrated the tumor mass, and reduced its volume more efficiently compared to the control, the free drug, or other micelle formulations. These results showed that the Dox-loaded micelles functionalized with the RI-VS analogue had better anti-glioma effects in vivo, with fewer side effects compared with other formulations.

The peptide VAP (Table 1) was shown to have high binding affinity in vitro to GRP78 protein, which is overexpressed in gliomas, glioma stem cells, vasculogenic mimicry, and neovasculature [54]. The prediction of binding for the analogue RI-VAP to GRP78 was similar to that of the parent peptide and, in addition, remarkable tumor accumulation was observed experimentally by imaging in vivo. RI-VAP-modified paclitaxel-loaded polymeric micelles had better anti-tumor efficacy compared to free taxol, to paclitaxel-loaded simple micelles, and to micelles modified with parent peptide.

The short peptide WSW (Table 1) was reported to efficiently and selectively penetrate the blood–brain barrier (BBB) and blood–brain tumor barrier (BBTB) to reach glioma cells and tumor neovasculature, suggesting that it may be a suitable carrier for intracranial glioma targeting. Ran et al. [55] designed, synthesized and studied RI-WSW that exhibited higher endocytosis efficiency than the parent peptide. This property was explained by the higher targeting efficiency of the RI derivative and likely higher penetration efficiency. Moreover, micelles decorated on the surface with RI-WSW showed strong anti-angiogenesis and antitumor effects and increased penetration ability in vitro and in vivo toward tumor cells and angiogenic blood vessels. In a similar study Xie et al. [56] proposed a RI analogue of bradykinin, named RI-BK (Table 1), capable of crossing BBTB. The molecule was highly active and selective towards the bradykinin type 2 (B2) receptor, as also demonstrated by computational analyses. RI-BK was used to decorate paclitaxel (PTX)-loaded micelles whose accumulation was increased in glioma but not in normal brain. Co-administration of RI-BK increased the therapeutic efficiency of the drug-loaded nanocarriers in glioma. These results underscored the efficacy of glioma-targeted drug delivery, based on the use of micelles functionalized with retro-inverso peptides, in improving therapeutic efficacy for glioma treatment.

Follicle-stimulating hormone receptor (FSHR) expression is limited to the reproductive system [92,93] and might be targeted to deliver drugs against ovarian cancer with high selectivity and specificity. In particular, nanoparticles carrying the RI variant of the peptide FP21 showed to bind FSHR (Table 1). They were thus used as an ovarian cancer targeted delivery system [57–59], showing improved biostability compared to the parent peptide, with no degradation even after 12 h incubation with proteolytic enzymes. The data obtained on the RI peptide encouraged further developments and optimizations of the molecule for treating ovarian cancers expressing FSHR [57].

In another study, Zhang et al. demonstrated that the RI derivative of the same peptide FP21 conjugated to nanocarriers had significantly enhanced anti-tumor effects working by reducing the tumor volumes in nude mice from 33.3% to 58.5%. This effect was likely amplified by the high resistance of the ligand to hydrolysis [57–59]. Another tumor promoter is the Transferrin receptor (TfR), an important transmembrane glycoprotein involved in iron transport. TfR is overexpressed in tumors because of the increased demand for iron during tumor rapid growing [94,95]. Recently, the RI derivative of the TfR-targeting T7 peptide (Table 1) was shown to have enhanced serum stability and higher binding affinity to TfR [60] than the parent peptide. This property was efficiently exploited modifying the surface of liposomes (LIP) to realize a tumor selective drug delivery system. The RI-T7-LIP particles exhibited significantly higher accumulation in tumors than T7-LIP and Transferrin-LIP. A complete pharmacokinetic study was performed to further investigate the potential of RI-T7-LIP in vivo using Docetaxel-loaded RI-T7-LIP which induced markedly increased apoptotic and necrotic areas in the treated mouse models.

3. Applications in Immunology

As previously observed, a strong topological correlation is at the base of antigenic cross-recognition between linear antigens and the corresponding retro-inverso isomers [96–98]. In several studies, monospecific murine antibodies were used as conformational probes to demonstrate the existence of surface similarities between a cyclic peptide mimicking the CD4 surface, which was a synthetic analogue of the third complementarity-determining region (CDR3) of immunoglobulins, and its corresponding retro-inverso isomer [60,96]. Anti-CD4 antibodies have been used to inhibit in vivo the clinical symptoms associated to the CD4-dependent auto-immune disorder allergic encephalomyelitis [22]. On this ground, RI analogues of CD4 loops were hypothesized as potential synthetic vaccines, as immunodiagnostics, and for the development of new generations of immunomodulators for the treatment of various CD4-related diseases [99]. For these reasons the cross-recognition of peptide surfaces by anti-CD4 antibodies was investigated using RI-peptide mimetics showing a strong correlation between antibody recognition and the simple arrays of side chains, with minimal contributions from the backbone atoms. These observations anticipated later studies showing that even simple arrays of alternating side chains, obtained by glycine-alternated peptide sequences (sequence-simplified peptide antigens), exposed on the same face of a peptide ideally adopting a fully extended conformation, were sufficient to retain the binding to antibodies. These insights were obtained studying the surfaces of sequence-simplified variants of retro-, inverso-, and RI derivative of parent 15-mer peptide antigens [100]. A series of polyclonal antibodies was generated in rabbits against 15-residue chimeric peptides and RI analogues able to bind interleukin 2 and to inhibit its interaction with the p55 interleukin 2 receptor subunit [101].

In diseases of inflammatory origin, such as systemic lupus erythematosus (SLE), one major event leading to a pathological condition is the interaction of immunoglobulins (IgG) with the corresponding cellular receptors. The pathogenic mechanism in SLE is the production of autoantibodies. To inhibit the interaction between IgGs and their receptor, a proteolytically stable form of the tripeptide Arg-Thr-Tyr tetramerized on a multilysine scaffold and obtained by inverting the chiral centers of the tripeptide's amino acids (D-Arg-D-Thr-D-Tyr, TG19320) was used. The tripeptide bound the Fc portion of the antibodies [61], prevented the binding to Fc receptors and rescued from death transgenic mice harboring SLE-prone mutations [62].

Also, several negative examples of the use of RI peptides are reported in the current literature. Most parameters influencing the activity of these molecules have been described above. However, other possible reasons for their unsuccessful utilization are yet to be understood. Nair et al. carried out an interesting study to investigate the success or failure of retro-inverso isomers to mimic the corresponding all-L molecules in the case of antigenic epitopes (Table 1) [63]. They based their analysis on the T cell epitopes from vesicular stomatitis virus glycoprotein peptide (VSVp) and ovalbumin epitope (OVAp) and on the B cell epitope (PS1) derived from the antigen surface of hepatitis B virus. The parent VSVp and OVAp showed conformations similar to those of their corresponding RI analogues (both adopted extended conformations), and in the Ramachandran plots the distribution of ϕ and φ angles for the parent and the RI analogues occupied the same plot regions. On the contrary, in the case of the peptide PS1 and its RI derivative the two molecules showed distinct conformational propensities. Indeed, the parent peptide bound the antibody adopting a specific β-turn conformation that was not mimicked by the RI analogue. Although the plasticity of the epitope conformation in solution allowed a partial overlap of the angle values accessed by the RI analogue, the latter did not bind the anti-peptide antibody [63].

Recently, the RI analogues of a family of peptides capable to cross the blood–brain barrier (BBB), HAI and THR [64] were shown to possess improved protease-resistant properties and to maintain the original BBB shuttle activity of the parent peptide. However, the RI derivatives were much less immunogenic and as such provided an important improvement compared to the original molecules [102]. Another study reported the RI analogue of a peptide able to suppress T-cell activation in Type 1 diabetes mellitus (T1D) [103]. Currently, there are neither curative nor preventive treatments to block the auto-immune destruction of the islets of Langerhans (beta cells). A RI analogue of the diabetogenic islet peptide named InsB:9–23 (Table 1) [65,66], responsible of the auto-immune response, was shown to prevent T-cell activation in humanized model mice both ex vivo and in vivo. The peptide blocked the immune-mediated beta cells destruction, thereby suggesting a novel therapy for patients at earlier stages of T1D. The use of this molecule opened a highly positive clinical perspective since the treated animals showed a larger beta cells reserve compared to animals at later stages of the disease.

In this field, the use of bioinformatic platforms—now largely and freely accessible—is becoming an invaluable tool to quickly identify new peptides potentially suitable for developing synthetic vaccines and peptidomimetic therapeutics. Robson et al., for example, using bioinformatic tools identified the sequence KRSFIEDLLFNKV as a well conserved region around one of the known cleavage sites of the SARS coronavirus, used for cell entry by the virus itself. The authors proposed the use of a RI analogue and studied its conformational flexibility that might offer an advantage for the molecule's action in vivo because of the capacity to better adapt on the target binding site. According with preliminary studies using molecular modeling and docking, the proposed RI peptide was expected to bind to the angiotensin converting enzyme type 2 (ACE2) which is the target of SARS-CoV in lung cells, and to work as an inhibitor able to prevent the proteolysis required for activation of the S spike protein [104].

4. Applications in Neurodegenerative Diseases

Deposition of protein fibrils is one of the leading causes of pathological conditions associated to neurodegenerative diseases. Protein fibrils form when a protein in β-pleated sheet conformation self-associates, mainly through hydrogen bonds, precipitating and generating protein deposits which accumulate in many different organs and tissues [105]. It is currently debated whether the precipitated insoluble fibrils or actually soluble oligomers are the cytotoxic aggregative elements working as diseases etiologic agents [106]. However, blocking or slowing down the aggregating phenomena is believed to be a major therapeutic option in this field. In the case of Alzheimer's disease (AD), advanced approaches were initially based on direct immunization with $A\beta_{1-42}$, although the first clinical trials were stopped due to adverse effects involving detrimental T-cell mediated brain inflammation [107]. Also, no significant improvements in terms of reduction of symptoms and immunological adverse reactions in patients [108] were observed with bapineuzumab, a humanized anti-$A\beta_{1-42}$ monoclonal antibody (mAb). Very recently, aducanumab a mAb that binds only aggregated and soluble oligomers of Aβ has been approved for treating AD, strongly indicating that the anti-aggregation therapy with Aβ targeting molecules is an effective first line treatment for this disease. As alternative anti-aggregating agents, peptides capable to prevent $A\beta_{1-42}$ aggregation or to dissociate preformed $A\beta_{1-42}$ aggregates have been largely investigated. Recently, the synthesis of Amytrap, a tetrameric RI analogue of the all-L peptide WKGEWTGR has been reported. Amytrap was pegylated and conjugated to human serum albumin (HSA) to enhance its bioavailability [67] and as such displayed high affinity for the GSNKG region of $A\beta_{1-42}$, blocking oligomerization and aggregation of the full length polypeptide. Using this molecule, the authors observed a significant reduction of $A\beta_{1-42}$ levels in the brain as determined by immunohistochemical analyses of brain tissues. They also observed that Amytrap sequestered the soluble protein, shifting its ability to deposit into the brain. Previously, the same authors prepared the RI analogue of another peptide capable to block the oligomerization of $A\beta_{1-42}$. The parent molecule was a chimeric peptide obtained by conjugating the HIV-1 "TAT" sequence to the Aβ fragment 16–20 ($A\beta_{16-20}$). Its RI variant, named RI-OR2-TAT, was meant to act as a cell-permeable and brain-penetrant Aβ aggregation inhibitor [68]. With this molecule, the authors observed a rapid crossing of the BBB, an effective binding to the amyloid plaques and a reduction of the Aβ oligomers level of in the brain. Since RI-OR2-TAT inhibited Aβ aggregation at relatively high concentrations [68], recently it was covalently attached to nanoliposomes (NLs) using the 'click' chemistry. An efficient crossing of the BBB in in vitro models was observed and lower concentrations of this form of the peptide were enough to inhibit aggregation of Aβ. Also, protective effects towards the toxicity exerted by pre-aggregated Aβ on neuronal cells were observed in vivo, preventing memory loss in transgenic mice. The presence of NL improved the potency of RI-OR2-TAT due to the multivalent effect deriving from the presence of multiple copies of peptides decorating each liposome [69]. Morris et al. [70] reported a pre-clinical study of the same molecule labeled with ^{18}F, [^{18}F]RI-OR2-TAT, to demonstrate its in vivo stability and the hepatobiliary route as the primary excretion pathway of the intact peptide. These results were the base of a study where RI-OR2 was modified by replacing hydrophobic amino acids with non-natural building blocks. The final peptidomimetic, even more resistant to proteolytic degradation, retarded the aggregation of $A\beta_{1-42}$ and also partially dissolved newly aggregated oligomers [71].

Another approach proposed to combat Alzheimer's disease is blocking the initial cleavage of the amyloid-β protein precursor (AβPP) by the β-site AβPP cleaving enzyme 1 (BACE1). Some RI-analogues based on a fragment of AβPP were synthesized as chimeras with the TAT carrier to facilitate cell membrane permeation and crossing of the BBB. The authors observed a decrease of both $A\beta_{1-40}$ and $A\beta_{1-42}$ ($A\beta_{1-40/42}$) production without inducing cytotoxicity. Moreover, $A\beta_{1-40/42}$ levels decreased in plasma and brain, diminishing also the levels of soluble AβPP production and of insoluble Aβ following chronic treatments. These results suggested a possible use of the chimeric RI peptides as a selective disease-modifying therapy for AD [109].

A further application of retro-inverso peptides is to prepare nanocarriers and delivery systems as new tools in Alzheimer's disease. A recent approach was based on silencing BACE1 using RNA interference (RNAi). In particular using small interfering RNAs (siRNAs) some authors presented a "dual targeting" strategy based on nanoparticles (NP) built with a peptide component and a modified polyethylene glycol [110]. The peptide moiety was a BBB targeting peptide [111], more specifically it was the RI analogue of the all-L peptide TGNYKALHPHNG. The resulting NPs showed low toxicity and high transfection efficiency and were able to transfer siRNAs in the brain [112]. Using this system, the authors observed an increase of the BBB-penetration, of the neuron-targeting efficacy and higher neuroprotective effects reflected by improved cognitive performance. Also, the downregulation of the protein-tau phosphorylation level, the promotion of the axonal transport and the attenuation of microgliosis were observed in mice model [111] following treatment with these NPs.

Recently, molecular dynamics was used to study the formation of fibrils between the RI-$A\beta_{1-40/42}$ and the parent $A\beta_{1-40/42}$ to elucidate the mechanism of cross-fibril formation and the effect of RI-$A\beta_{1-40/42}$ on fibril stability. The resulting models indicated that $A\beta_{1-40/42}$ and RI-$A\beta_{1-40/42}$ generated a two-layer structure with similar stability. In particular, the dihedral angles were of opposite sign for the $A\beta_{1-40}$ fibrils and the extent of the twists was different. Furthermore, the twists of RI-$A\beta_{1-42}$ and of the parent peptide were close to zero. Analyzing the RI-Aβ fibrils, the authors observed that the number of hydrogen bonds connecting the chains within the fibril was lower compared to the parent peptide fibrils. The average number of missing hydrogen bonds was 7, mostly in the region around residues 23–29, and this strongly impacted on the different stability observed between the two fibrils. Data also suggested that the full-length RI-peptides could support fibril formation and their presence led to a decreased amount of soluble toxic Aβ oligomers [10]. These observations were in agreement with the experimental observations reviewed in this work.

In addition to AD, also type 2 diabetes mellitus (T2DM) and Parkinson's disease (PD) are related to amyloidogenesis. In T2DM, the Islet amyloid polypeptide (IAPP also known as amylin) aggregates into β-pleated sheet structures damaging pancreatic islet β-cells. The "hot spot" peptide segment encompassing residues 8–18 ($IAPP_{8-18}$, sequence ATQRLANFLVH) represents the "sticky" region of human IAPP, which is also able to assemble with $IAPP_{22-28}$, sequence NFGAIL [113,114]. In order to inhibit the early stages of IAPP hetero- and self-aggregation, a library of RI peptides covering the region 11–20 of IAPP (Table 1), was generated and studied evaluating their impact on the fibrillogenesis properties of full-length human IAPP [72]. The authors found a RI non-toxic analogue showing strong inhibitory effects on amylin aggregation, as confirmed by negative stain electron microscopy (TEM). Inquisitively, the RI-analogue alone aggregated already at low concentrations. The authors also introduced N-methylation as a way to prevent H-bond formation and avoid aggregation [115]. The new N-methylated RI variant showed a clear dose-dependent inhibition of fibril formation and was stable against an ample range of different proteolytic enzymes and in human plasma.

Shaltiel-Karyo and colleagues studied the inhibition of oligomerization of α-synuclein (α-syn) [116], a protein whose structural deformation is associated with PD. The isoform β-synuclein (β-syn) is a natural inhibitor of the aggregation of α-syn [73]. The entire sequence of β-syn was then systematically mapped using synthetic analogues to identify the domains able to mediate the molecular recognition between β-syn and α-syn. A synthetic RI-analogue of the 36–45 β-syn fragment (sequence GVLYVGSKTR) was able to reduce both amyloid fibril and soluble oligomer formation in vitro. The authors also tested the RI-analogue in a *Drosophila* model expressing a mutated α-syn in the nervous system and observed a reduction of α-syn accumulation in the brains of the flies, thus suggesting that this approach can pave the way for developing a novel class of therapeutic agents to treat PD in the future.

5. Application of RI Peptides as Antimicrobial Antibiotics

The systematic and widespread misuse and abuse of antibiotics has made antibiotic resistance a major medical complication following hospitalization [117,118]. The World Health Organization has identified a list of "priority pathogens", both Gram-positive and Gram-negative, which represent the biggest threat to human health caused by multidrug-resistant bacteria [119]. Among these microorganisms, those collected under the acronym "ESKAPE" (i.e., *Enterococcus faecium, Staphylococcus aureus, Klebsiella pneumoniae, Acinetobacter baumannii, Pseudomonas aeruginosa, Enterobacter* spp.) are those needing the urgent and prompt discovery of new antimicrobials. Many surgical procedures, or medical treatments that suppress immune system will become impracticable due to infections by antimicrobial resistant pathogens. Also, prophylactic treatments that are normally effective become inefficient. Today, the containment of infections by these microorganisms is very problematic [120] and new antibiotic drugs are urgently needed. An alternative to conventional antibiotics would be the use of antimicrobial peptides (AMPs) [121]. They are widely spread in nature being present in bacteria as well as in higher eukaryotes and play an important role in innate immunity and in both adaptive and non-adaptive immune responses [122]. Their antimicrobial action is based on multiple mechanisms that together contribute to eliminate pathogens [123,124]. However, despite the interesting and very promising antimicrobial effects displayed by many AMPs, so far only 10 peptide-based antimicrobials have reached the clinical use [120].

Indeed, several peptides have shown nephrotoxic or hemolytic side effects, strongly discouraging their use as drugs. Given their toxicity, the use of some approved AMPs, such as colistin, is relegated among the last treatment options against multi drug-resistant Gram-negative infections [125]. Nevertheless, novel AMPs can be obtained choosing among a vast repertoire of sequences and structures and novel peptides candidate as potential therapeutics are continuously developed at least at preclinical level. Starting from naturally occurring AMPs, synthetic derivatives have been rationally designed in order to maintain the antimicrobial pharmacophores, to improve the resistance to proteolysis, to reduce cytotoxicity and possibly improving the activity [126–128].

A recent application of the retro-inverso approach to antimicrobial peptides has been reported by Neubauer and colleagues, although this procedure not always results in enhanced antimicrobial activity [129]. The antimicrobial and hemolytic activities of a set of 6 AMPs were investigated together with their hydrophobicity, secondary structure content, and ability to self-associate. The antimicrobial peptides were aurein 1.2, CAMEL, citropin 1.1, omiganan, pexiganan, and temporin A together with their retro-inverso analogues. These peptides were selected for their broad-spectrum activity against fungi, bacteria and also for their possible anticancer applications (Table 2). Of interest, CAMEL, omiganan, pexiganan, and temporin A are in clinical trials, with potential uses against some ESKAPE bacteria strains (Tables 2 and 3). Of the compounds studied, the majority displayed antimicrobial activity (Table 3), although in most cases there was a decrease of the antibacterial potential with respect of the native molecule. In fact, only the RI omiganan displayed enhanced antimicrobial activity mainly against Gram-negative bacteria compared to parent peptide. Similarly, retro-inverso pexiganan exhibited a good activity towards *K. pneumoniae* and *P. aeruginosa*.

Table 2. Peptide-based antimicrobial compounds in clinical trials and their retro-inverso analogues, based on [118,129].

Peptide	Sequence	Net Charge	Helicity [a]		% ACN [b]	Application	Mechanism of Action	Status	Therapeutic Indication	Ref.
			SDS	DPC						
aurein 1.2	GLFDIIKK IAESF-NH_2 *	+1	<	>	43.93	Antimicrobial and anticancer properties	Prerequisite aggregation and carpet-like mechanism	in vitro [129]		[130]
RI-aurein 1.2	fseaikkiid flg-NH_2 **		>	=	37.10					
CAMEL	KWKLFKKIG AVLKVL-NH_2 *	+6	=	<	33.73	Broad spectrum antibacterial	Bacterial membrane disruption	Preclinical [119]	Bacterial infections [119]	[131,132]
RI-CAMEL	lvklvagikkf lkwK-NH_2 **		>	=	30.71					
citropin 1.1	GLFDVIKKVA SVIGGL-NH_2 *	+2	=	=	42.40	Broad spectrum antibacterial and anticancer properties	Prerequisite aggregation and carpet-like mechanism	in vitro [129]		[133]
RI-citropin 1.1	lggivsavkk ivdflg-NH_2 **		=	>	41.28					
Omiganan	ILRWPWWPW RRK-NH_2 *	+5	NO	NO	32.92	Broad spectrum anti-fungal, antibacterial	Bacterial membrane disruption	1. Phase III complete (discontinued) 2. Phase III complete 3. Phase III on going 4. Phase II complete 5. Phase II complete 6. Phase II complete 7. Phase II complete 8. Phase II on going [118]	1. Local catheter site infections 2. Topical skin antisepsis 3. Papulopustular rosacea 4. Acne vulgaris 5. Atopic dermatitis 6. Vulvar intraepithelial neoplasia 7. Condylomata acuminata (external genital warts) 8. Facial seborrhoeic dermatitis [118]	[134–136]
RI-omiganan	krrwpwwpwrli-NH_2 **		NO	NO	35.48					

Table 2. *Cont.*

Peptide	Sequence	Net Charge	Helicity [a]		% ACN [b]	Application	Mechanism of Action	Status	Therapeutic Indication	Ref.
			SDS	DPC						
Pexiganan	GIGKFLKKAKK FGKAFVKILK K-NH_2 *	+10	=	=	30.58	Broad spectrum antibacterial	Bacterial membrane disruption	Phase III complete; rejected, efficacy not superior to current therapies [118]	Infected diabetic foot ulcers [118]	[137,138]
RI-pexiganan	kklikvfakgfk kakklfk gig-NH_2 **		<	<	26.36					
temporin A	FLPLIGRVLS GIL-NH_2 *	+2	<	>	42.80	Gram-positive bacteria	Bacterial membrane disruption	Preclinical [119]	Bacterial infections [119]	[139,140]
RI-temporin A	ligslvrgil plf-NH2 **		<	<	38.91					

* All-L sequences are reported as capital letters. ** Lower case letters indicate amino acids in the D configuration. **Note**: [a]: The symbol =; >; < is referred to the helicity fraction calculated as in [141]. In particular, = means around 50%, > and < more or less 50%, respectively. Experimental conditions: CD spectra of the peptides were acquired in 10 mM phosphate buffer pH 7.4, containing SDS (sodium dodecyl sulfate) and DPC (dodecylphosphocholine) using a Jasco J-815 spectropolarimeter. All measurements were conducted using 0.15 mg/mL peptide solutions at 298 K [142]. [b]: Hydrophobicity was determined by HPLC and was expressed as the % *v*/*v* acetonitrile at the retention time of the peptides (tR) [129].

Table 3. MIC values (µg/mL) of anti-microbial peptides and of their retro-inverso analogues against reference strains of microorganisms [143]. Taken from reference [129].

	Gram-Positive			Gram-Negative		
Peptide	***E. faecalis*** [1] **PCM 2673**	***S. aureus*** [1] **ATCC 25923**	***S. pneumoniae*** **ATCC 49619**	***E. coli*** **ATCC 25922**	***K. pneumoniae*** [1] ***ATCC 700603***	***P. aeruginosa*** [1] ***ATCC 9027***
Aurein 1.2	64	128	64	128	16	256
RI-aurein 1.2	256	>256	256	256	128	>256
CAMEL	8	4	0.5	2	0.125	2
RI-CAMEL	64	128	128	128	2	8
citropin 1.1	32	16	32	32	16	128
RI-citropin 1.1	128	64	128	64	32	>256
Omiganan	16	16	8	16	8	16
RI-omiganan	16	8	8	8	4	4
Pexiganan	16	8	1	4	1	2
RI-pexiganan	64	128	4	8	0.125	2
Temporin A	64	4	>256	256	128	>256
RI-temporin A	256	64	>256	256	128	256

[1] These bacterial pathogens are comprised in the acronym ESKAPE, which are a group of Gram-positive and Gram-negative bacteria able to evade commonly used antibiotics due to their ever increasing multi-drug resistance (MDR) [117]. They represent the major cause of life-threatening nosocomial infections in immunocompromised and critically ill patients [144]. The acronym ESKAPE is based on the scientific names of six bacteria, *Enterococcus faecium*, *Staphylococcus aureus*, *Klebsiella pneumoniae*, *Acinetobacter baumannii*, *Pseudomonas aeruginosa*, and *Enterobacter* spp. In particular, *P. aeruginosa* and *S. aureus* are some of the most ubiquitous pathogens found in highly resistant biofilms [126,145].

In the list reported in Table 3, CAMEL is the only chimeric peptide, designed by Merrifield in 1995 [146], containing fragments of two peptides with different antimicrobial activities [147]. CAMEL, which is in preclinical trial, was one of the strongest antimicrobial peptides, but its RI analogue was only active toward *K. pneumoniae* and *P. aeruginosa*. Interestingly, the secondary structure of the peptide was not the prerequisite for establishing significant interactions between the peptide and its biological target and the antimicrobial activity was only due random interactions with the core lipidic membrane of the pathogen [146]. The differences in antimicrobial activity between the peptide and the RI analogue could therefore not be explained.

A major cause of antibiotic resistance is the formation of biofilms, which arise from bacteria growing on surfaces or at the air-liquid interfaces as a response to exogenous stresses. In biofilms, bacteria are encased in a protective extracellular matrix containing water, polysaccharides, proteins, extracellular DNA, and lipids [148]. de la Fuente-Núñez and colleagues reported the synthesis and analysis of a library of peptides and their RI-analogues to eradicate biofilms produced by *Pseudomonas aeruginosa* [149]. They observed that the RI-analogues (Table 1) named RI1018 and RI-JK6 were more potent at stimulating degradation and/or preventing accumulation of the stress-related second messenger nucleotide guanosine penta- and tetra-phosphate [(p)ppGpp] which plays an important role in biofilm development in many bacterial species [74]. They also demonstrated that these analogues killed bacteria growing as biofilms, which have a high adaptive resistance and are difficult to eradicate. Moreover, these peptides had synergic effect with common antibiotics, rendering biofilms more susceptible to their attack. Another AMP able to damage the bacterial membrane is a truncated and modified RI-analogue of Aurein 2.2 (RI-73, Table 1), which was recently used to eradicate preformed *Staphylococcus aureus* biofilms [75]. The antimicrobial activity of these analogues was increased 2- to 8-folds and when conjugated with biocompatible polyethylene glycol (PEG)-modified phospholipid micelles their toxicity toward human cells and aggregation were strongly reduced. Although RI-73 exhibited a good activity, the PEG-conjugated analogue showed a partially reduced activity.

A further public health problem in many countries throughout the world is represented by the insurgence of multi drug-resistance against protozoan parasites, such as *Leishmania*. Host defense peptides (HDPs) are becoming promising options for new therapies. HDPs have the advantage of their small size and their amphipathic and cationic character that is able to induce permeabilization of cell membranes. Cathelicidins, a family of HDPs, have shown significant antimicrobial activities against various parasites including *Leishmania* spp. [150]. In particular, a study was carried out using the bovine myeloid antimicrobial peptide 28 (BMAP-28, Table 1), a cathelicidin with broad antimicrobial activities, and its inversed and RI-analogues [76]. The study demonstrated that D- and RI-BMAP-28 were also effective antimicrobials against *Leishmania major*, working in a dose dependent manner with a mechanism leading to disruption of membrane integrity [151]. Thus, the protection conferred by RI-BMAP28, accompanied by a reduced toxicity and increased stability, could be exploited to develop effective antimicrobial therapeutics [152].

6. Conclusions and Future Perspectives

In the field of peptidomimetics, retro-inversion has been largely explored to improve peptide stability while retaining the parent molecule's activity. Changing the order of the amino acids and their configuration has been also a mean of introducing novelty and to overcome existing intellectual property claims [153]. The first examples of their use were reported by M. Goodman in the mid-1970s [154], who was interested in the study of stereochemical and conformational properties of retro–inverso (RI) amide bonds in linear peptides. Interesting examples were next reported by Merrifield with studies on the CAMEL peptide [146,147], which was a chimeric peptide derived from the merging of two AMP. Despite the amazing results reported in literature, the application of retro-inversion to generate peptidomimetics is still rare or however uncommon.

As also evidenced in this review, several studies have indeed reported that the general and straightforward process of retro-inversion becomes more likely effective with very short sequences where conformation plays a limited or no role and activity is mostly due to a simple array of side chains. For instance, Sakurai's results [155] suggested that the interaction between the RI analogue of VWRLLAPPFSNRLL and the ganglioside GM1, a glycolipid with high affinity for the cholera toxin subunit B (CTB), was mediated only by the peptide side chains while those of the backbone, whose direction was thus irrelevant, were completely negligible. One could thus expect that a RI analogue can better mimic the parent peptide when the free energy of interaction of the backbone with all other atoms is insignificant for the stability of the peptide 3D structure.

Beyond these basic rules applicable to short peptides or other specific examples, the reasons for the frequent failure of RI isomerization of longer molecules are still largely unclear, and definite instructions for possibly improving the success rate are unresolved. The reversal of the peptide backbone and the shift of the H-bond network it is involved into is a major alteration of the fine equilibrium of the forces that supports the conformation of a peptide having an organized 3D structure. Therefore, as for the folding of a natural molecule, the lack of one such important puzzle piece prevents the correct assembling of the structure although the side chains may potentially have access to the same conformational space of the parent molecule. We can thus conclude that the design of a successful retro-inverso analogue of a folded peptide has the same complications as for the *de novo* design of a new protein or peptide and one should thus proceed following the rules, still not well understood and codified, of protein folding, exploiting and using the geometrical and structural features of amino acids in D configuration. For example, the RI isomerization and structure reconstruction of the MDM2/MDMX peptide inhibitor stingin, which adopts an N-terminal loop and a C-terminal α-helix, lead to an isomer that partially retained binding (3.0–3.4 kcal/mol reduction) and showed a decreased ability to prevent the interaction with p53 [11]. These conformation and energy issues have been often discouraging because of the frequent loss of biological activity observed in larger molecules showing well-defined tridimensional organizations. Merrifield indeed soon observed that the efficiency of peptide retro-inversion was not only related to inversion of its chirality but to the global change of the 3D conformation [146]. These observations have been indirectly confirmed showing that retro-inverso analogues of unstructured peptides more often maintain or even increase the activity compared to the parent peptide [12].

On the other hands, peptides that assemble into β-sheets adopting extended conformations establish a large and well-organized network of interactions, mostly H-bonds, with the adjacent molecules. Also, the side chains are well packed each other. In this case, despite the strong backbone interactions, retro-inverso analogues have more chance to be successful if the registry of H-bonds and of side chain-to-side chain interactions is corrected to account for the inverted amide bonds. The molecular dynamic simulations of amyloid fibrils in AD [10] or amylin in T2D [156] indeed showed that the interactions of both side chains and backbone of RI peptides were re-aligned establishing different patterns of contacts and hydrogen bonding. Also, the twist of the RI analogue β-sheets was similar and the complex had only slightly lower stability compared to the parent peptides.

Computational approaches might be of great help and might open a new season in this field as suggested by Robson [104]. Despite their many limitations, we believe their use still has a place in the design of drugs based on bioactive peptides. This belief stems from the simplicity of the design, from the rapidity in making synthetic peptides and from the immediate benefits resulting when the molecules maintain their activity. Therefore, this review would be an incentive to continue working with these types of molecules, also to further investigate the conformational and topological space they need to occupy to fully mimic bioactive peptides with complex structure.

Author Contributions: Conceptualization, N.D., M.M., A.S., M.R. and A.C.; Writing—original draft preparation, M.R. and A.C.; Writing—review and editing, N.D., M.M., A.S., M.R. and A.C.; Supervision, M.R. The manuscript was written through contributions of all authors. All authors have read and agreed to the published version of the manuscript.

Funding: Authors acknowledge the support from Regione Campania for the projects: "Fighting Cancer resistance: Multidisciplinary integrated Platform for a technological Innovative Approach to Oncotherapies (Campania Oncotherapies)" and "Development of novel therapeutic approaches for treatment-resistant neoplastic diseases (SATIN)" project. The support from MUR for the project PRIN 2017M8R7N9 is also acknowledged.

Institutional Review Board Statement: Not applicable.

Informed Consent Statement: Not applicable.

Data Availability Statement: No datasets have been used for this study.

Conflicts of Interest: The authors declare no conflict of interest.

References

1. Lee, A.C.; Harris, J.L.; Khanna, K.K.; Hong, J.H. A Comprehensive Review on Current Advances in Peptide Drug Development and Design. *Int. J. Mol. Sci.* **2019**, *20*, 2383. [CrossRef]
2. Bruno, B.J.; Miller, G.D.; Lim, C.S. Basics and recent advances in peptide and protein drug delivery. *Ther. Deliv.* **2013**, *4*, 1443–1467. [CrossRef] [PubMed]
3. Corbi-Verge, C.; Garton, M.; Nim, S.; Kim, P.M. Strategies to Develop Inhibitors of Motif-Mediated Protein-Protein Interactions as Drug Leads. *Annu. Rev. Pharmacol. Toxicol.* **2017**, *57*, 39–60. [CrossRef]
4. Fosgerau, K.; Hoffmann, T. Peptide therapeutics: Current status and future directions. *Drug Discov. Today* **2015**, *20*, 122–128. [CrossRef] [PubMed]
5. Kreil, G. D-amino acids in animal peptides. *Annu. Rev. Biochem.* **1997**, *66*, 337–345. [CrossRef]
6. Di, L. Strategic Approaches to Optimizing Peptide ADME Properties. *AAPS J.* **2015**, *17*, 134–143. [CrossRef] [PubMed]
7. Fischer, P.M. The design, synthesis and application of stereochemical and directional peptide isomers: A critical review. *Curr. Protein Pept. Sci.* **2003**, *4*, 339–356. [CrossRef]
8. Fletcher, M.D.; Campbell, M.M. Partially Modified Retro-Inverso Peptides: Development, Synthesis, and Conformational Behavior. *Chem. Rev.* **1998**, *98*, 763–796. [CrossRef] [PubMed]
9. Grishin, D.V.; Zhdanov, D.D.; Pokrovskaya, M.V.; Sokolov, N.N. D-amino acids in nature, agriculture and biomedicine. *All Life* **2020**, *13*, 11–22. [CrossRef]
10. Xi, W.; Hansmann, U.H.E. The effect of retro-inverse D-amino acid Abeta-peptides on Abeta-fibril formation. *J. Chem. Phys.* **2019**, *150*, 095101. [CrossRef]
11. Li, C.; Zhan, C.; Zhao, L.; Chen, X.; Lu, W.Y.; Lu, W. Functional consequences of retro-inverso isomerization of a miniature protein inhibitor of the p53-MDM2 interaction. *Bioorg. Med. Chem.* **2013**, *21*, 4045–4050. [CrossRef]
12. Garton, M.; Nim, S.; Stone, T.A.; Wang, K.E.; Deber, C.M.; Kim, P.M. Method to generate highly stable D-amino acid analogs of bioactive helical peptides using a mirror image of the entire PDB. *Proc. Natl. Acad Sci. USA* **2018**, *115*, 1505–1510. [CrossRef] [PubMed]
13. Rai, J. Peptide and protein mimetics by retro and retroinverso analogs. *Chem. Biol. Drug Des.* **2019**, *93*, 724–736. [CrossRef]
14. Chorev, M. The partial retro-inverso modification: A road traveled together. *Biopolymers* **2005**, *80*, 67–84. [CrossRef]
15. Rai, J. Mini Heme-Proteins: Designability of Structure and Diversity of Functions. *Curr. Protein Pept. Sci.* **2017**, *18*, 1132–1140. [CrossRef]
16. Chorev, M.; Shavitz, R.; Goodman, M.; Minick, S.; Guillemin, R. Partially modified retro-inverso-enkephalinamides: Topochemical long-acting analogs in vitro and in vivo. *Science* **1979**, *204*, 1210–1212. [CrossRef] [PubMed]
17. Ruvo, M.; Fassina, G. End-group modified retro-inverso isomers of tripeptide oxytocin analogues: Binding to neurophysin II and enhancement of its self-association properties. *Int. J. Pept. Protein Res.* **1995**, *45*, 356–365. [CrossRef]
18. Sridhar, S.; Guruprasad, K. Can natural proteins designed with 'inverted' peptide sequences adopt native-like protein folds? *PLoS ONE* **2014**, *9*, e107647. [CrossRef] [PubMed]
19. Verdoliva, A.; Ruvo, M.; Cassani, G.; Fassina, G. Topological mimicry of cross-reacting enantiomeric peptide antigens. *J. Biol. Chem.* **1995**, *270*, 30422–30427. [CrossRef]
20. Chorev, M.; Goodman, M. Recent developments in retro peptides and proteins–an ongoing topochemical exploration. *Trends Biotechnol.* **1995**, *13*, 438–445. [CrossRef]
21. Brady, L.; Dodson, G. Drug design. Reflections on a peptide. *Nature* **1994**, *368*, 692–693. [CrossRef]
22. Jameson, B.A.; McDonnell, J.M.; Marini, J.C.; Korngold, R. A rationally designed CD4 analogue inhibits experimental allergic encephalomyelitis. *Nature* **1994**, *368*, 744–746. [CrossRef]
23. Taylor, E.M.; Otero, D.A.; Banks, W.A.; O'Brien, J.S. Retro-inverso prosaptide peptides retain bioactivity, are stable In vivo, and are blood-brain barrier permeable. *J. Pharmacol. Exp. Ther.* **2000**, *295*, 190–194. [PubMed]
24. Banerjee, A.; Raghothama, S.R.; Karle, I.L.; Balaram, P. Ambidextrous molecules: Cylindrical peptide structures formed by fusing left- and right-handed helices. *Biopolymers* **1996**, *39*, 279–285. [CrossRef]
25. Caporale, A.; Biondi, B.; Schievano, E.; Wittelsberger, A.; Mammi, S.; Peggion, E. Structure-function relationship studies of PTH(1-11) analogues containing D-amino acids. *Eur. J. Pharmacol.* **2009**, *611*, 1–7. [CrossRef]
26. Crisma, M.; Bisson, W.; Formaggio, F.; Broxterman, Q.B.; Toniolo, C. Factors governing 3(10)-helix vs alpha-helix formation in peptides: Percentage of C(alpha)-tetrasubstituted alpha-amino acid residues and sequence dependence. *Biopolymers* **2002**, *64*, 236–245. [CrossRef]
27. Wermuth, J.; Goodman, S.L.; Jonczyk, A.; Kessler, H. Stereoisomerism and biological activity of the selective and superactive alpha(v)beta(3) integrin inhibitor cyclo(-RGDfV-) and its retro-inverso peptide. *J. Am. Chem. Soc.* **1997**, *119*, 1328–1335. [CrossRef]
28. Yavari, B.; Mahjub, R.; Saidijam, M.; Raigani, M.; Soleimani, M. The Potential Use of Peptides in Cancer Treatment. *Curr. Protein Pept. Sci.* **2018**, *19*, 759–770. [CrossRef]
29. Chiangjong, W.; Chutipongtanate, S.; Hongeng, S. Anticancer peptide: Physicochemical property, functional aspect and trend in clinical application (Review). *Int. J. Oncol.* **2020**, *57*, 678–696. [CrossRef]
30. Wirtz, D.; Konstantopoulos, K.; Searson, P.C. The physics of cancer: The role of physical interactions and mechanical forces in metastasis. *Nat. Rev. Cancer* **2011**, *11*, 512–522. [CrossRef]
31. Carmeliet, P.; Jain, R.K. Angiogenesis in cancer and other diseases. *Nature* **2000**, *407*, 249–257. [CrossRef] [PubMed]

32. Nishida, N.; Yano, H.; Nishida, T.; Kamura, T.; Kojiro, M. Angiogenesis in cancer. *Vasc. Health Risk Manag.* **2006**, *2*, 213–219. [CrossRef]
33. Ferrara, J.L.M.; Cooke, K.R.; Teshima, T. The pathophysiology of acute graft-versus-host disease. *Int. J. Hematol.* **2003**, *78*, 181–187. [CrossRef]
34. Kaumaya, P.T.; Foy, K.C. Peptide vaccines and targeting HER and VEGF proteins may offer a potentially new paradigm in cancer immunotherapy. *Future Oncol.* **2012**, *8*, 961–987. [CrossRef]
35. Dass, C.R.; Tran, T.M.; Choong, P.F. Angiogenesis inhibitors and the need for anti-angiogenic therapeutics. *J. Dent. Res.* **2007**, *86*, 927–936. [CrossRef] [PubMed]
36. Nemeth, J.A.; Nakada, M.T.; Trikha, M.; Lang, Z.; Gordon, M.S.; Jayson, G.C.; Corringham, R.; Prabhakar, U.; Davis, H.M.; Beckman, R.A. Alpha-v integrins as therapeutic targets in oncology. *Cancer Investig.* **2007**, *25*, 632–646. [CrossRef]
37. Ferrara, N. The role of VEGF in the regulation of physiological and pathological angiogenesis. *EXS* **2005**, 209–231. [CrossRef]
38. Vicari, D.; Foy, K.C.; Liotta, E.M.; Kaumaya, P.T. Engineered conformation-dependent VEGF peptide mimics are effective in inhibiting VEGF signaling pathways. *J. Biol. Chem.* **2011**, *286*, 13612–13625. [CrossRef] [PubMed]
39. Foy, K.C.; Liu, Z.; Phillips, G.; Miller, M.; Kaumaya, P.T. Combination treatment with HER-2 and VEGF peptide mimics induces potent anti-tumor and anti-angiogenic responses in vitro and in vivo. *J. Biol. Chem.* **2011**, *286*, 13626–13637. [CrossRef]
40. Giordano, R.J.; Cardo-Vila, M.; Lahdenranta, J.; Pasqualini, R.; Arap, W. Biopanning and rapid analysis of selective interactive ligands. *Nat. Med.* **2001**, *7*, 1249–1253. [CrossRef]
41. Giordano, R.J.; Cardo-Vila, M.; Salameh, A.; Anobom, C.D.; Zeitlin, B.D.; Hawked, D.H.; Valente, A.P.; Almeida, F.C.L.; Nor, J.E.; Sidman, R.L.; et al. From combinatorial peptide selection to drug prototype (I): Targeting the vascular endothelial growth factor receptor pathway. *Proc. Natl. Acad. Sci. USA* **2010**, *107*, 5112–5117. [CrossRef] [PubMed]
42. Calvanese, L.; Caporale, A.; Foca, G.; Iaccarino, E.; Sandomenico, A.; Doti, N.; Apicella, I.; Incisivo, G.M.; De Falco, S.; Falcigno, L.; et al. Targeting VEGF receptors with non-neutralizing cyclopeptides for imaging applications. *Amino Acids* **2018**, *50*, 321–329. [CrossRef]
43. Rezazadeh, F.; Sadeghzadeh, N.; Abedi, S.M.; Abediankenari, S. Tc-99m labeled (D)(LPR): A novel retro-inverso peptide for VEGF receptor-1 targeted tumor imaging. *Nucl. Med. Biol.* **2018**, *62–63*, 54–62. [CrossRef]
44. Lee, T.Y.; Lin, C.T.; Kuo, S.Y.; Chang, D.K.; Wu, H.C. Peptide-mediated targeting to tumor blood vessels of lung cancer for drug delivery. *Cancer Res.* **2007**, *67*, 10958–10965. [CrossRef] [PubMed]
45. Li, Y.; Lei, Y.; Wagner, E.; Xie, C.; Lu, W.Y.; Zhu, J.H.; Shen, J.; Wang, J.; Liu, M. Potent Retro-Inverso D-Peptide for Simultaneous Targeting of Angiogenic Blood Vasculature and Tumor Cells. *Bioconjugate Chem.* **2013**, *24*, 133–143. [CrossRef]
46. Bifulco, K.; Longanesi-Cattani, I.; Gala, M.; Di Carluccio, G.; Masucci, M.T.; Pavone, V.; Lista, L.; Arra, C.; Stoppelli, M.P.; Carriero, M.V. The soluble form of urokinase receptor promotes angiogenesis through its Ser(88)-Arg-Ser-Arg-Tyr(92) chemotactic sequence. *J. Thromb. Haemost.* **2010**, *8*, 2789–2799. [CrossRef]
47. Resnati, M.; Pallavicini, I.; Wang, J.M.; Oppenheim, J.; Serhan, C.N.; Romano, M.; Blasi, F. The fibrinolytic receptor for urokinase activates the G protein-coupled chemotactic receptor FPRL1/LXA4R. *Proc. Natl. Acad. Sci. USA* **2002**, *99*, 1359–1364. [CrossRef]
48. Carriero, M.V.; Bifulco, K.; Ingangi, V.; Costantini, S.; Botti, G.; Ragone, C.; Minopoli, M.; Motti, M.L.; Rea, D.; Scognamiglio, G.; et al. Retro-inverso Urokinase Receptor Antagonists for the Treatment of Metastatic Sarcomas. *Sci. Rep.* **2017**, *7*, 1–17. [CrossRef]
49. Karimi, H.; Sadeghzadeh, N.; Abediankenari, S.; Rezazadeh, F.; Hallajian, F. Radiochemical Evaluation and In Vitro Assessment of the Targeting Ability of a Novel Tc-99m-HYNIC-RGD for U87MG Human Brain Cancer Cells. *Curr. Radiopharm.* **2017**, *10*, 139–144. [CrossRef]
50. Torabizadeh, S.A.; Abedi, S.M.; Noaparast, Z.; Hosseinimehr, S.J. Comparative assessment of a Tc-99m labeled H1299.2-HYNIC peptide bearing two different co-ligands for tumor-targeted imaging. *Bioorgan. Med. Chem.* **2017**, *25*, 2583–2592. [CrossRef]
51. Liu, Y.Y.; Mei, L.; Yu, Q.W.; Zhang, Q.Y.; Gao, H.L.; Zhang, Z.R.; He, Q. Integrin alpha(v)beta(3) targeting activity study of different retro-inverso sequences of RGD and their potentiality in the designing of tumor targeting peptides. *Amino Acids* **2015**, *47*, 2533–2539. [CrossRef]
52. Ren, Y.C.; Zhan, C.Y.; Gao, J.; Zhang, M.F.; Wei, X.L.; Ying, M.; Liu, Z.N.; Lu, W.Y. A D-Peptide Ligand of Integrins for Simultaneously Targeting Angiogenic Blood Vasculature and Glioma Cells. *Mol. Pharmaceut* **2018**, *15*, 592–601. [CrossRef]
53. van Ommeren, R.; Staudt, M.D.; Xu, H.; Hebb, M.O. Advances in HSP27 and HSP90-targeting strategies for glioblastoma. *J. Neuro Oncol.* **2016**, *127*, 209–219. [CrossRef] [PubMed]
54. Ran, D.N.; Mao, J.N.; Shen, Q.; Xie, C.; Zhan, C.Y.; Wang, R.F.; Lu, W.Y. GRP78 enabled micelle-based glioma targeted drug delivery. *J. Control. Release* **2017**, *255*, 120–131. [CrossRef]
55. Ran, D.N.; Mao, J.N.; Zhan, C.Y.; Xie, C.; Ruan, H.T.; Ying, M.; Zhou, J.F.; Lu, W.L.; Lu, W.Y. D-Retroenantiomer of Quorum-Sensing Peptide-Modified Polymeric Micelles for Brain Tumor-Targeted Drug Delivery. *ACS Appl. Mater. Inter.* **2017**, *9*, 25672–25682. [CrossRef]
56. Xie, Z.X.; Shen, Q.; Xie, C.; Lu, W.Y.; Peng, C.M.; Wei, X.L.; Li, X.; Su, B.X.; Gao, C.L.; Liu, M. Retro-inverso bradykinin opens the door of blood-brain tumor barrier for nanocarriers in glioma treatment. *Cancer Lett.* **2015**, *369*, 144–151. [CrossRef]
57. Zhang, M.Y.; Zhang, M.X.; Wang, J.; Cai, Q.Q.; Zhao, R.; Yu, Y.; Tai, H.Y.; Zhang, X.Y.; Xu, C.J. Retro-inverso follicle-stimulating hormone peptide-mediated polyethylenimine complexes for targeted ovarian cancer gene therapy. *Drug Deliv.* **2018**, *25*, 995–1003. [CrossRef]

58. Zhang, M.X.; Hong, S.S.; Cai, Q.Q.; Zhang, M.; Chen, J.; Zhang, X.Y.; Xu, C.J. Transcriptional control of the MUC16 promoter facilitates follicle-stimulating hormone peptide-conjugated shRNA nanoparticle-mediated inhibition of ovarian carcinoma in vivo. *Drug Deliv.* **2018**, *25*, 797–806. [CrossRef]
59. Hong, S.S.; Zhang, M.X.; Zhang, M.; Yu, Y.; Chen, J.; Zhang, X.Y.; Xu, C.J. Follicle-stimulating hormone peptide-conjugated nanoparticles for targeted shRNA delivery lead to effective gro-alpha silencing and antitumor activity against ovarian cancer. *Drug Deliv.* **2018**, *25*, 576–584. [CrossRef]
60. Tang, J.J.; Wang, Q.T.; Yu, Q.W.; Qiu, Y.; Mei, L.; Wan, D.D.; Wang, X.H.; Li, M.; He, Q. A stabilized retro-inverso peptide ligand of transferrin receptor for enhanced liposome-based hepatocellular carcinoma-targeted drug delivery. *Acta Biomater.* **2019**, *83*, 379–389. [CrossRef] [PubMed]
61. Moiani, D.; Salvalaglio, M.; Cavallotti, C.; Bujacz, A.; Redzynia, I.; Bujacz, G.; Dinon, F.; Pengo, P.; Fassina, G. Structural characterization of a Protein A mimetic peptide dendrimer bound to human IgG. *J. Phys. Chem. B* **2009**, *113*, 16268–16275. [CrossRef]
62. Marino, M.; Ruvo, M.; De Falco, S.; Fassina, G. Prevention of systemic lupus erythematosus in MRL/lpr mice by administration of an immunoglobulin-binding peptide. *Nat. Biotechnol.* **2000**, *18*, 735–739. [CrossRef] [PubMed]
63. Nair, D.T.; Kaur, K.J.; Singh, K.; Mukherjee, P.; Rajagopal, D.; George, A.; Bal, V.; Rath, S.; Rao, K.V.S.; Salunke, D.M. Mimicry of native peptide antigens by the corresponding retro-inverso analogs is dependent on their intrinsic structure and interaction propensities. *J. Immunol.* **2003**, *170*, 1362–1373. [CrossRef]
64. Lee, J.H.; Engler, J.A.; Collawn, J.F.; Moore, B.A. Receptor mediated uptake of peptides that bind the human transferrin receptor. *Eur. J. Biochem.* **2001**, *268*, 2004–2012. [CrossRef] [PubMed]
65. Lee, K.H.; Wucherpfennig, K.W.; Wiley, D.C. Structure of a human insulin peptide-HLA-DQ8 complex and susceptibility to type 1 diabetes. *Nat. Immunol.* **2001**, *2*, 501–507. [CrossRef]
66. Nakayama, M.; Abiru, N.; Moriyama, H.; Babaya, N.; Liu, E.; Miao, D.; Yu, L.; Wegmann, D.R.; Hutton, J.C.; Elliott, J.F.; et al. Prime role for an insulin epitope in the development of type 1 diabetes in NOD mice. *Nature* **2005**, *435*, 220–223. [CrossRef]
67. Gandbhir, O.; Sundaram, P. Pre-Clinical Safety and Efficacy Evaluation of Amytrap, a Novel Therapeutic to Treat Alzheimer's Disease. *J. Alzheimers Dis. Rep.* **2019**, *3*, 77–94. [CrossRef] [PubMed]
68. Parthsarathy, V.; McClean, P.L.; Holscher, C.; Taylor, M.; Tinker, C.; Jones, G.; Kolosov, O.; Salvati, E.; Gregori, M.; Masserini, M.; et al. A novel retro-inverso peptide inhibitor reduces amyloid deposition, oxidation and inflammation and stimulates neurogenesis in the APPswe/PS1DeltaE9 mouse model of Alzheimer's disease. *PLoS ONE* **2013**, *8*, e54769. [CrossRef]
69. Gregori, M.; Taylor, M.; Salvati, E.; Re, F.; Mancini, S.; Balducci, C.; Forloni, G.; Zambelli, V.; Sesana, S.; Michael, M.; et al. Retro-inverso peptide inhibitor nanoparticles as potent inhibitors of aggregation of the Alzheimer's Abeta peptide. *Nanomedicine* **2017**, *13*, 723–732. [CrossRef]
70. Morris, O.; Gregory, J.; Kadirvel, M.; Henderson, F.; Blykers, A.; McMahon, A.; Taylor, M.; Allsop, D.; Allan, S.; Grigg, J.; et al. Development & automation of a novel [(18)F]F prosthetic group, 2-[(18)F]-fluoro-3-pyridinecarboxaldehyde, and its application to an amino(oxy)-functionalised Abeta peptide. *Appl. Radiat. Isot.* **2016**, *116*, 120–127. [CrossRef]
71. Stark, T.; Lieblein, T.; Pohland, M.; Kalden, E.; Freund, P.; Zangl, R.; Grewal, R.; Heilemann, M.; Eckert, G.P.; Morgner, N.; et al. Peptidomimetics That Inhibit and Partially Reverse the Aggregation of $A\beta_{1-42}$. *Biochemistry* **2017**, *56*, 4840–4849. [CrossRef]
72. Obasse, I.; Taylor, M.; Fullwood, N.J.; Allsop, D. Development of proteolytically stable N-methylated peptide inhibitors of aggregation of the amylin peptide implicated in type 2 diabetes. *Interface Focus* **2017**, *7*, 20160127. [CrossRef]
73. Windisch, M.; Hutter-Paier, B.; Schreiner, E.; Wronski, R. Beta-Synuclein-derived peptides with neuroprotective activity: An alternative treatment of neurodegenerative disorders? *J. Mol. NeuroSci* **2004**, *24*, 155–165. [CrossRef]
74. de la Fuente-Nunez, C.; Reffuveille, F.; Haney, E.F.; Straus, S.K.; Hancock, R.E. Broad-spectrum anti-biofilm peptide that targets a cellular stress response. *PLoS Pathog* **2014**, *10*, e1004152. [CrossRef]
75. Kumar, P.; Pletzer, D.; Haney, E.F.; Rahanjam, N.; Cheng, J.T.J.; Yue, M.; Aljehani, W.; Hancock, R.E.W.; Kizhakkedathu, J.N.; Straus, S.K. Aurein-Derived Antimicrobial Peptides Formulated with Pegylated Phospholipid Micelles to Target Methicillin-Resistant Staphylococcus aureus Skin Infections. *ACS Infect. Dis.* **2019**, *5*, 443–453. [CrossRef]
76. Lynn, M.A.; Kindrachuk, J.; Marr, A.K.; Jenssen, H.; Pante, N.; Elliott, M.R.; Napper, S.; Hancock, R.E.; McMaster, W.R. Effect of BMAP-28 antimicrobial peptides on Leishmania major promastigote and amastigote growth: Role of leishmanolysin in parasite survival. *PLoS Negl. Trop. Dis.* **2011**, *5*, e1141. [CrossRef]
77. Binetruy-Tournaire, R.; Demangel, C.; Malavaud, B.; Vassy, R.; Rouyre, S.; Kraemer, M.; Plouet, J.; Derbin, C.; Perret, G.; Mazie, J.C. Identification of a peptide blocking vascular endothelial growth factor (VEGF)-mediated angiogenesis. *Embo. J.* **2000**, *19*, 1525–1533. [CrossRef]
78. Blasi, F. uPA, uPAR, PAI-I: Key intersection of proteolytic, adhesive and chemotactic highways? *Immunol. Today* **1997**, *18*, 415–417. [CrossRef]
79. Gargiulo, L.; Longanesi-Cattani, I.; Bifulco, K.; Franco, P.; Raiola, R.; Campiglia, P.; Grieco, P.; Peluso, G.; Stoppelli, M.P.; Carriero, M.V. Cross-talk between fMLP and vitronectin receptors triggered by urokinase receptor-derived SRSRY peptide. *J. Biol. Chem.* **2005**, *280*, 25225–25232. [CrossRef] [PubMed]
80. Barczyk, M.; Carracedo, S.; Gullberg, D. Integrins. *Cell Tissue Res.* **2010**, *339*, 269–280. [CrossRef]

81. Trabocchi, A.; Menchi, G.; Danieli, E.; Potenza, D.; Cini, N.; Bottoncetti, A.; Raspanti, S.; Pupi, A.; Guarna, A. Cyclic DGR-peptidomimetic containing a bicyclic reverse turn inducer as a selective alpha(v)beta(5) integrin ligand. *Amino Acids* **2010**, *38*, 329–337. [CrossRef] [PubMed]
82. Winograd-Katz, S.E.; Fassler, R.; Geiger, B.; Legate, K.R. The integrin adhesome: From genes and proteins to human disease. *Nat. Rev. Mol. Cell Biol.* **2014**, *15*, 273–288. [CrossRef] [PubMed]
83. Shattil, S.J.; Kim, C.; Ginsberg, M.H. The final steps of integrin activation: The end game. *Nat. Rev. Mol. Cell Biol.* **2010**, *11*, 288–300. [CrossRef] [PubMed]
84. Seguin, L.; Desgrosellier, J.S.; Weis, S.M.; Cheresh, D.A. Integrins and cancer: Regulators of cancer stemness, metastasis, and drug resistance. *Trends Cell Biol.* **2015**, *25*, 234–240. [CrossRef]
85. Leblanc, R.; Lee, S.C.; David, M.; Bordet, J.C.; Norman, D.D.; Patil, R.; Miller, D.; Sahay, D.; Ribeiro, J.; Clezardin, P.; et al. Interaction of platelet-derived autotaxin with tumor integrin alpha(V)beta(3) controls metastasis of breast cancer cells to bone. *Blood* **2014**, *124*, 3141–3150. [CrossRef]
86. Contois, L.W.; Akalu, A.; Caron, J.M.; Tweedie, E.; Cretu, A.; Henderson, T.; Liaw, L.; Friesel, R.; Vary, C.; Brooks, P.C. Inhibition of tumor-associated alpha v beta 3 integrin regulates the angiogenic switch by enhancing expression of IGFBP-4 leading to reduced melanoma growth and angiogenesis in vivo. *Angiogenesis* **2015**, *18*, 31–46. [CrossRef]
87. Kibria, G.; Hatakeyama, H.; Ohga, N.; Hida, K.; Harashima, H. Dual-ligand modification of PEGylated liposomes shows better cell selectivity and efficient gene delivery. *J. Control. Release* **2011**, *153*, 141–148. [CrossRef]
88. Guo, Z.M.; He, B.; Jin, H.W.; Zhang, H.R.; Dai, W.B.; Zhang, L.R.; Zhang, H.; Wang, X.Q.; Wang, J.C.; Zhang, X.; et al. Targeting efficiency of RGD-modified nanocarriers with different ligand intervals in response to integrin alpha v beta 3 clustering. *Biomaterials* **2014**, *35*, 6106–6117. [CrossRef]
89. Caporale, A.; Bolzati, C.; Incisivo, G.M.; Salvarese, N.; Grieco, P.; Ruvo, M. Improved synthesis on solid phase of dithiocarbamic cRGD-derivative and Tc-99m-radiolabelling. *J. Pept. Sci.* **2019**, *25*, e3140. [CrossRef]
90. Flechsig, P.; Lindner, T.; Loktev, A.; Roesch, S.; Mier, W.; Sauter, M.; Meister, M.; Herold-Mende, C.; Haberkorn, U.; Altmann, A. PET/CT Imaging of NSCLC with a alpha(v)beta(6) Integrin-Targeting Peptide. *Mol. Imaging Biol.* **2019**, *21*, 973–983. [CrossRef]
91. Evans, B.J.; King, A.T.; Katsifis, A.; Matesic, L.; Jamie, J.F. Methods to Enhance the Metabolic Stability of Peptide-Based PET Radiopharmaceuticals. *Molecules* **2020**, *25*, 2314. [CrossRef]
92. Papadimitriou, K.; Kountourakis, P.; Kottorou, A.E.; Antonacopoulou, A.G.; Rolfo, C.; Peeters, M.; Kalofonos, H.P. Follicle-Stimulating Hormone Receptor (FSHR): A Promising Tool in Oncology? *Mol. Diagn. Ther.* **2016**, *20*, 523–530. [CrossRef] [PubMed]
93. Perales-Puchalt, A.; Svoronos, N.; Rutkowski, M.R.; Allegrezza, M.J.; Tesone, A.J.; Payne, K.K.; Wickramasinghe, J.; Nguyen, J.M.; O'Brien, S.W.; Gumireddy, K.; et al. Follicle-Stimulating Hormone Receptor Is Expressed by Most Ovarian Cancer Subtypes and Is a Safe and Effective Immunotherapeutic Target. *Clin. Cancer Res.* **2017**, *23*, 441–453. [CrossRef]
94. Huang, R.Q.; Qu, Y.H.; Ke, W.L.; Zhu, J.H.; Pei, Y.Y.; Jiang, C. Efficient gene delivery targeted to the brain using a transferrin-conjugated polyethyleneglycol-modified polyamidoamine dendrimer. *FASEB J.* **2007**, *21*, 1117–1125. [CrossRef]
95. Wang, Y.; Chen, J.T.; Yan, X.P. Fabrication of Transferrin Functionalized Gold Nanoclusters/Graphene Oxide Nanocomposite for Turn-On Near-Infrared Fluorescent Bioimaging of Cancer Cells and Small Animals. *Anal. Chem.* **2013**, *85*, 2529–2535. [CrossRef]
96. Benkirane, N.; Friede, M.; Guichard, G.; Briand, J.P.; Vanregenmortel, M.H.V.; Muller, S. Antigenicity and Immunogenicity of Modified Synthetic Peptides Containing D-Amino-Acid Residues-Antibodies to a D-Enantiomer Do Recognize the Parent L-Hexapeptide and Reciprocally. *J. Biol. Chem.* **1993**, *268*, 26279–26285. [CrossRef]
97. Guichard, G.; Benkirane, N.; Zederlutz, G.; Vanregenmortel, M.H.V.; Briand, J.P.; Muller, S. Antigenic Mimicry of Natural L-Peptides with Retro-Inverso-Peptidomimetics. *Proc. Natl. Acad. Sci. USA* **1994**, *91*, 9765–9769. [CrossRef] [PubMed]
98. Verdoliva, A.; Ruvo, M.; Villain, M.; Cassani, G.; Fassina, G. Antigenicity of topochemically related peptides. *Biochim. Biophys. Acta* **1995**, *1253*, 57–62. [CrossRef]
99. Weiner, H.L. Oral Tolerance. *Proc. Natl. Acad. Sci. USA* **1994**, *91*, 10762–10765. [CrossRef]
100. Rossi, M.; Manfredi, V.; Ruvo, M.; Fassina, G.; Verdoliva, A. Sequence-simplification and chimeric assembly: New models of peptide antigen modification. *Mol. Immunol.* **2002**, *39*, 443–451. [CrossRef]
101. Fassina, G.; Cassani, G.; Gnocchi, P.; Fornasiero, M.C.; Isetta, A.M. Inhibition of interleukin-2/p55 receptor subunit interaction by complementary peptides. *Arch. Biochem. Biophys.* **1995**, *318*, 37–45. [CrossRef] [PubMed]
102. Arranz-Gibert, P.; Ciudad, S.; Seco, J.; Garcia, J.; Giralt, E.; Teixido, M. Immunosilencing peptides by stereochemical inversion and sequence reversal: Retro-D-peptides. *Sci. Rep.* **2018**, *8*, 6446. [CrossRef]
103. Lombardi, A.; Concepcion, E.; Hou, H.; Arib, H.; Mezei, M.; Osman, R.; Tomer, Y. Retro-inverso D-peptides as a novel targeted immunotherapy for Type 1 diabetes. *J. Autoimmun.* **2020**, *115*, 102543. [CrossRef] [PubMed]
104. Robson, B. Computers and viral diseases. Preliminary bioinformatics studies on the design of a synthetic vaccine and a preventative peptidomimetic antagonist against the SARS-CoV-2 (2019-nCoV, COVID-19) coronavirus. *Comput. Biol. Med.* **2020**, *119*, 103670. [CrossRef] [PubMed]
105. Soto, C.; Estrada, L.D. Protein misfolding and neurodegeneration. *Arch. Neurol.* **2008**, *65*, 184–189. [CrossRef] [PubMed]
106. Bloom, G.S. Amyloid-beta and tau: The trigger and bullet in Alzheimer disease pathogenesis. *JAMA Neurol.* **2014**, *71*, 505–508. [CrossRef]

107. Zotova, E.; Bharambe, V.; Cheaveau, M.; Morgan, W.; Holmes, C.; Harris, S.; Neal, J.W.; Love, S.; Nicoll, J.A.; Boche, D. Inflammatory components in human Alzheimer's disease and after active amyloid-beta42 immunization. *Brain* **2013**, *136*, 2677–2696. [CrossRef]
108. Salloway, S.; Sperling, R.; Fox, N.C.; Blennow, K.; Klunk, W.; Raskind, M.; Sabbagh, M.; Honig, L.S.; Porsteinsson, A.P.; Ferris, S.; et al. Two phase 3 trials of bapineuzumab in mild-to-moderate Alzheimer's disease. *N. Engl. J. Med.* **2014**, *370*, 322–333. [CrossRef]
109. Resende, R.; Ferreira-Marques, M.; Moreira, P.; Coimbra, J.R.M.; Baptista, S.J.; Isidoro, C.; Salvador, J.A.R.; Dinis, T.C.P.; Pereira, C.F.; Santos, A.E. New BACE1 Chimeric Peptide Inhibitors Selectively Prevent AbetaPP-beta Cleavage Decreasing Amyloid-beta Production and Accumulation in Alzheimer's Disease Models. *J. Alzheimers Dis.* **2020**, *76*, 1317–1337. [CrossRef]
110. Zheng, X.; Pang, X.; Yang, P.; Wan, X.; Wei, Y.; Guo, Q.; Zhang, Q.; Jiang, X. A hybrid siRNA delivery complex for enhanced brain penetration and precise amyloid plaque targeting in Alzheimer's disease mice. *Acta Biomater.* **2017**, *49*, 388–401. [CrossRef]
111. Guo, Q.; Xu, S.; Yang, P.; Wang, P.; Lu, S.; Sheng, D.; Qian, K.; Cao, J.; Lu, W.; Zhang, Q. A dual-ligand fusion peptide improves the brain-neuron targeting of nanocarriers in Alzheimer's disease mice. *J. Control. Release* **2020**, *320*, 347–362. [CrossRef]
112. Wang, P.; Zheng, X.; Guo, Q.; Yang, P.; Pang, X.; Qian, K.; Lu, W.; Zhang, Q.; Jiang, X. Systemic delivery of BACE1 siRNA through neuron-targeted nanocomplexes for treatment of Alzheimer's disease. *J. Control. Release* **2018**, *279*, 220–233. [CrossRef]
113. Andreetto, E.; Malideli, E.; Yan, L.M.; Kracklauer, M.; Farbiarz, K.; Tatarek-Nossol, M.; Rammes, G.; Prade, E.; Neumuller, T.; Caporale, A.; et al. A Hot-Segment-Based Approach for the Design of Cross-Amyloid Interaction Surface Mimics as Inhibitors of Amyloid Self-Assembly. *Angew. Chem. Int. Ed. Engl.* **2015**, *54*, 13095–13100. [CrossRef]
114. Bakou, M.; Hille, K.; Kracklauer, M.; Spanopoulou, A.; Frost, C.V.; Malideli, E.; Yan, L.M.; Caporale, A.; Zacharias, M.; Kapurniotu, A. Key aromatic/hydrophobic amino acids controlling a cross-amyloid peptide interaction versus amyloid self-assembly. *J. Biol. Chem.* **2017**, *292*, 14587–14602. [CrossRef]
115. Yan, L.M.; Velkova, A.; Tatarek-Nossol, M.; Andreetto, E.; Kapurniotu, A. IAPP mimic blocks Abeta cytotoxic self-assembly: Cross-suppression of amyloid toxicity of Abeta and IAPP suggests a molecular link between Alzheimer's disease and type II diabetes. *Angew. Chem. Int. Ed. Engl.* **2007**, *46*, 1246–1252. [CrossRef] [PubMed]
116. Shaltiel-Karyo, R.; Frenkel-Pinter, M.; Egoz-Matia, N.; Frydman-Marom, A.; Shalev, D.E.; Segal, D.; Gazit, E. Inhibiting alpha-synuclein oligomerization by stable cell-penetrating beta-synuclein fragments recovers phenotype of Parkinson's disease model flies. *PLoS ONE* **2010**, *5*, e13863. [CrossRef]
117. Mulani, M.S.; Kamble, E.E.; Kumkar, S.N.; Tawre, M.S.; Pardesi, K.R. Emerging Strategies to Combat ESKAPE Pathogens in the Era of Antimicrobial Resistance: A Review. *Front. Microbiol.* **2019**, *10*, 539. [CrossRef]
118. Mookherjee, N.; Anderson, M.A.; Haagsman, H.P.; Davidson, D.J. Antimicrobial host defence peptides: Functions and clinical potential. *Nat. Rev. Drug Discov.* **2020**, *19*, 311–332. [CrossRef] [PubMed]
119. De Oliveira, D.M.P.; Forde, B.M.; Kidd, T.J.; Harris, P.N.A.; Schembri, M.A.; Beatson, S.A.; Paterson, D.L.; Walker, M.J. Antimicrobial Resistance in ESKAPE Pathogens. *Clin. Microbiol. Rev.* **2020**, *33*, e00181-19. [CrossRef]
120. Browne, K.; Chakraborty, S.; Chen, R.; Willcox, M.D.; Black, D.S.; Walsh, W.R.; Kumar, N. A New Era of Antibiotics: The Clinical Potential of Antimicrobial Peptides. *Int. J. Mol. Sci.* **2020**, *21*, 7047. [CrossRef] [PubMed]
121. Divyashree, M.; Mani, M.K.; Reddy, D.; Kumavath, R.; Ghosh, P.; Azevedo, V.; Barh, D. Clinical Applications of Antimicrobial Peptides (AMPs): Where do we Stand Now? *Protein Pept. Lett.* **2020**, *27*, 120–134. [CrossRef]
122. Zhang, L.J.; Gallo, R.L. Antimicrobial peptides. *Curr. Biol.* **2016**, *26*, R14–R19. [CrossRef] [PubMed]
123. Zasloff, M. Antimicrobial peptides of multicellular organisms. *Nature* **2002**, *415*, 389–395. [CrossRef]
124. Roncevic, T.; Puizina, J.; Tossi, A. Antimicrobial Peptides as Anti-Infective Agents in Pre-Post-Antibiotic Era? *Int. J. Mol. Sci.* **2019**, *20*, 5713. [CrossRef] [PubMed]
125. El-Sayed Ahmed, M.A.E.-G.; Zhong, L.-L.; Shen, C.; Yang, Y.; Doi, Y.; Tian, G.-B. Colistin and its role in the Era of antibiotic resistance: An extended review (2000–2019). *Emerg. Microbes Infect.* **2020**, *9*, 868–885. [CrossRef]
126. de la Fuente-Nunez, C.; Cardoso, M.H.; de Souza Candido, E.; Franco, O.L.; Hancock, R.E. Synthetic antibiofilm peptides. *Biochim. Biophys. Acta* **2016**, *1858*, 1061–1069. [CrossRef]
127. Torres, M.D.T.; Sothiselvam, S.; Lu, T.K.; de la Fuente-Nunez, C. Peptide Design Principles for Antimicrobial Applications. *J. Mol. Biol.* **2019**, *431*, 3547–3567. [CrossRef]
128. Li, F.; Brimble, M. Using chemical synthesis to optimise antimicrobial peptides in the fight against antimicrobial resistance. *Pure Appl. Chem.* **2019**, *91*, 181–198. [CrossRef]
129. Neubauer, D.; Jaśkiewicz, M.; Migoń, D.; Bauer, M.; Sikora, K.; Sikorska, E.; Kamysz, E.; Kamysz, W. Retro analog concept: Comparative study on physico-chemical and biological properties of selected antimicrobial peptides. *Amino Acids* **2017**, *49*, 1755–1771. [CrossRef]
130. Baranska-Rybak, W.; Cirioni, O.; Dawgul, M.; Sokolowska-Wojdylo, M.; Naumiuk, L.; Szczerkowska-Dobosz, A.; Nowicki, R.; Roszkiewicz, J.; Kamysz, W. Activity of Antimicrobial Peptides and Conventional Antibiotics against Superantigen Positive Staphylococcus aureus Isolated from the Patients with Neoplastic and Inflammatory Erythrodermia. *Chemother. Res. Pract.* **2011**, *2011*, 270932. [CrossRef] [PubMed]
131. Sang, Y.; Blecha, F. Antimicrobial peptides and bacteriocins: Alternatives to traditional antibiotics. *Anim. Health Res. Rev.* **2008**, *9*, 227–235. [CrossRef]

132. Rodriguez-Hernandez, M.J.; Saugar, J.; Docobo-Perez, F.; de la Torre, B.G.; Pachon-Ibanez, M.E.; Garcia-Curiel, A.; Fernandez-Cuenca, F.; Andreu, D.; Rivas, L.; Pachon, J. Studies on the antimicrobial activity of cecropin A-melittin hybrid peptides in colistin-resistant clinical isolates of Acinetobacter baumannii. *J. Antimicrob. Chemother.* **2006**, *58*, 95–100. [CrossRef] [PubMed]
133. Sikorska, E.; Greber, K.; Rodziewicz-Motowidlo, S.; Szultka, L.; Lukasiak, J.; Kamysz, W. Synthesis and antimicrobial activity of truncated fragments and analogs of citropin 1.1: The solution structure of the SDS micelle-bound citropin-like peptides. *J. Struct. Biol.* **2009**, *168*, 250–258. [CrossRef] [PubMed]
134. Ng, S.M.S.; Teo, S.W.; Yong, Y.E.; Ng, F.M.; Lau, Q.Y.; Jureen, R.; Hill, J.; Chia, C.S.B. Preliminary investigations into developing all-D Omiganan for treating Mupirocin-resistant MRSA skin infections. Chem. *Biol. Drug Des.* **2017**, *90*, 1155–1160. [CrossRef]
135. Rubinchik, E.; Dugourd, D.; Algara, T.; Pasetka, C.; Friedland, H.D. Antimicrobial and antifungal activities of a novel cationic antimicrobial peptide, omiganan, in experimental skin colonisation models. *Int. J. Antimicrob. Agents* **2009**, *34*, 457–461. [CrossRef]
136. Melo, M.N.; Dugourd, D.; Castanho, M.A. Omiganan pentahydrochloride in the front line of clinical applications of antimicrobial peptides. *Recent Pat. Anti-Infect. Drug Discov.* **2006**, *1*, 201–207. [CrossRef] [PubMed]
137. Flamm, R.K.; Rhomberg, P.R.; Simpson, K.M.; Farrell, D.J.; Sader, H.S.; Jones, R.N. In vitro spectrum of pexiganan activity when tested against pathogens from diabetic foot infections and with selected resistance mechanisms. *Antimicrob. Agents Chemother.* **2015**, *59*, 1751–1754. [CrossRef]
138. Lopez-Medina, E.; Fan, D.; Coughlin, L.A.; Ho, E.X.; Lamont, I.L.; Reimmann, C.; Hooper, L.V.; Koh, A.Y. Candida albicans Inhibits Pseudomonas aeruginosa Virulence through Suppression of Pyochelin and Pyoverdine Biosynthesis. *PLoS Pathog.* **2015**, *11*, e1005129. [CrossRef]
139. Kim, J.B.; Iwamuro, S.; Knoop, F.C.; Conlon, J.M. Antimicrobial peptides from the skin of the Japanese mountain brown frog, Rana ornativentris. *J. Pept. Res.* **2001**, *58*, 349–356. [CrossRef] [PubMed]
140. Kamysz, W.; Mickiewicz, B.; Rodziewicz-Motowidlo, S.; Greber, K.; Okroj, M. Temporin A and its retro-analogues: Synthesis, conformational analysis and antimicrobial activities. *J. Pept. Sci.* **2006**, *12*, 533–537. [CrossRef] [PubMed]
141. Rao, T.; Ruiz-Gomez, G.; Hill, T.A.; Hoang, H.N.; Fairlie, D.P.; Mason, J.M. Truncated and helix-constrained peptides with high affinity and specificity for the cFos coiled-coil of AP-1. *PLoS ONE* **2013**, *8*, e59415. [CrossRef] [PubMed]
142. Ronga, L.; Langella, E.; Palladino, P.; Marasco, D.; Tizzano, B.; Saviano, M.; Pedone, C.; Improta, R.; Ruvo, M. Does tetracycline bind helix 2 of prion? An integrated spectroscopical and computational study of the interaction between the antibiotic and alpha helix 2 human prion protein fragments. *Proteins* **2007**, *66*, 707–715. [CrossRef] [PubMed]
143. Hughes, T.P.; Kaeda, J.; Branford, S.; Rudzki, Z.; Hochhaus, A.; Hensley, M.L.; Gathmann, I.; Bolton, A.E.; van Hoomissen, I.C.; Goldman, J.M.; et al. Frequency of major molecular responses to imatinib or interferon alfa plus cytarabine in newly diagnosed chronic myeloid leukemia. *N. Engl. J. Med.* **2003**, *349*, 1423–1432. [CrossRef]
144. Rice, L.B. Federal funding for the study of antimicrobial resistance in nosocomial pathogens: No ESKAPE. *J. Infect. Dis.* **2008**, *197*, 1079–1081. [CrossRef]
145. Khatoon, Z.; McTiernan, C.D.; Suuronen, E.J.; Mah, T.F.; Alarcon, E.I. Bacterial biofilm formation on implantable devices and approaches to its treatment and prevention. *Heliyon* **2018**, *4*, e01067. [CrossRef]
146. Merrifield, R.B.; Juvvadi, P.; Andreu, D.; Ubach, J.; Boman, A.; Boman, H.G. Retro and retroenantio analogs of cecropin-melittin hybrids. *Proc. Natl. Acad. Sci. USA* **1995**, *92*, 3449–3453. [CrossRef] [PubMed]
147. Andreu, D.; Ubach, J.; Boman, A.; Wahlin, B.; Wade, D.; Merrifield, R.B.; Boman, H.G. Shortened cecropin A-melittin hybrids. Significant size reduction retains potent antibiotic activity. *FEBS Lett.* **1992**, *296*, 190–194. [CrossRef]
148. de la Fuente-Nunez, C.; Reffuveille, F.; Fernandez, L.; Hancock, R.E. Bacterial biofilm development as a multicellular adaptation: Antibiotic resistance and new therapeutic strategies. *Curr. Opin. Microbiol.* **2013**, *16*, 580–589. [CrossRef]
149. de la Fuente-Nunez, C.; Reffuveille, F.; Mansour, S.C.; Reckseidler-Zenteno, S.L.; Hernandez, D.; Brackman, G.; Coenye, T.; Hancock, R.E. D-enantiomeric peptides that eradicate wild-type and multidrug-resistant biofilms and protect against lethal Pseudomonas aeruginosa infections. *Chem. Biol.* **2015**, *22*, 196–205. [CrossRef] [PubMed]
150. Kulkarni, M.M.; McMaster, W.R.; Kamysz, E.; Kamysz, W.; Engman, D.M.; McGwire, B.S. The major surface-metalloprotease of the parasitic protozoan, Leishmania, protects against antimicrobial peptide-induced apoptotic killing. *Mol. Microbiol.* **2006**, *62*, 1484–1497. [CrossRef]
151. Risso, A.; Braidot, E.; Sordano, M.C.; Vianello, A.; Macri, F.; Skerlavaj, B.; Zanetti, M.; Gennaro, R.; Bernardi, P. BMAP-28, an antibiotic peptide of innate immunity, induces cell death through opening of the mitochondrial permeability transition pore. *Mol. Cell Biol.* **2002**, *22*, 1926–1935. [CrossRef] [PubMed]
152. Kindrachuk, J.; Scruten, E.; Attah-Poku, S.; Bell, K.; Potter, A.; Babiuk, L.A.; Griebel, P.J.; Napper, S. Stability, toxicity, and biological activity of host defense peptide BMAP28 and its inversed and retro-inversed isomers. *Biopolymers* **2011**, *96*, 14–24. [CrossRef]
153. Fassina, G.; Verdoliva, A.; Ruvo, M. Antigenic Peptides. U.S. Patent US-5932692-A, 3 August 1999.
154. Goodman, M.; Chorev, M. On the concept of linear modified retro-peptide structures. *Acc. Chem. Res.* **1979**, *12*, 1–7. [CrossRef]
155. Sakurai, K. A Peptide–Glycolipid Interaction Probed by Retroinverso Peptide Analogues. *Chem. Pharm. Bull.* **2018**, *66*, 45–50. [CrossRef] [PubMed]
156. Pandey, P.; Nguyen, N.; Hansmann, U.H.E. d-Retro Inverso Amylin and the Stability of Amylin Fibrils. *J. Chem. Theory Comput.* **2020**, *16*, 5358–5368. [CrossRef] [PubMed]

International Journal of
Molecular Sciences

MDPI

Review

Role of Peptides in Diagnostics

Shashank Pandey [1,*], Gaurav Malviya [2] and Magdalena Chottova Dvorakova [3,4]

[1] Department of Pharmacology and Toxicology, Faculty of Medicine in Pilsen, Charles University, 32300 Pilsen, Czech Republic
[2] Cancer Research UK Beatson Institute, Garscube Estate, Switchback Road, Glasgow G611BD, UK; g.malviya@beatson.gla.ac.uk
[3] Department of Physiology, Faculty of Medicine in Pilsen, Charles University, 32300 Pilsen, Czech Republic; magdalena.dvorakova@lfp.cuni.cz
[4] Biomedical Center, Faculty of Medicine in Pilsen, Charles University, 32300 Pilsen, Czech Republic
* Correspondence: Shashank.Pandey@lfp.cuni.cz

Abstract: The specificity of a diagnostic assay depends upon the purity of the biomolecules used as a probe. To get specific and accurate information of a disease, the use of synthetic peptides in diagnostics have increased in the last few decades, because of their high purity profile and ability to get modified chemically. The discovered peptide probes are used either in imaging diagnostics or in non-imaging diagnostics. In non-imaging diagnostics, techniques such as Enzyme-Linked Immunosorbent Assay (ELISA), lateral flow devices (i.e., point-of-care testing), or microarray or LC-MS/MS are used for direct analysis of biofluids. Among all, peptide-based ELISA is considered to be the most preferred technology platform. Similarly, peptides can also be used as probes for imaging techniques, such as single-photon emission computed tomography (SPECT) and positron emission tomography (PET). The role of radiolabeled peptides, such as somatostatin receptors, interleukin 2 receptor, prostate specific membrane antigen, $\alpha\beta3$ integrin receptor, gastrin-releasing peptide, chemokine receptor 4, and urokinase-type plasminogen receptor, are well established tools for targeted molecular imaging ortumor receptor imaging. Low molecular weight peptides allow a rapid clearance from the blood and result in favorable target-to-non-target ratios. It also displays a good tissue penetration and non-immunogenicity. The only drawback of using peptides is their potential low metabolic stability. In this review article, we have discussed and evaluated the role of peptides in imaging and non-imaging diagnostics. The most popular non-imaging and imaging diagnostic platforms are discussed, categorized, and ranked, as per their scientific contribution on PUBMED. Moreover, the applicability of peptide-based diagnostics in deadly diseases, mainly COVID-19 and cancer, is also discussed in detail.

Keywords: peptides; diagnostic; ELISA; microarray; PET; SPECT; imaging diagnostic; non-imaging diagnostic

Citation: Pandey, S.; Malviya, G.; Chottova Dvorakova, M. Role of Peptides in Diagnostics. *Int. J. Mol. Sci.* **2021**, *22*, 8828. https://doi.org/10.3390/ijms22168828

Academic Editors: Menotti Ruvo and Nunzianna Doti

Received: 23 July 2021
Accepted: 13 August 2021
Published: 17 August 2021

1. Introduction

The development of accurate diagnostic methods is an urgent need in today's world. Due to the upsurge of various deadly diseases, rare diseases, and cancer, it is crucial to improve the diagnostic aspects, which will help the clinician to predict and examine therapeutic responses across a wide spectrum of diseases.

In the past few decades, immunodiagnostics has been an essential tool for clinical management and prognosis of a disease. To discover novel biomarkers, it is obligatory to understand the effect of a disease on the physiology of organisms, as well as their impact on genomic and proteomic patterns. In some scenarios, the development of new diagnostics is limited because of already known and well-characterized biomarkers. On the contrary, mapping of protein antigen for selection of linear epitopes by peptide scanning is a widely used technique [1,2]. Moreover, rapid development in peptide microarray technology has further advanced the screening platform for serological screenings [3]. To select, identify, and design immunodominant linear or continuous epitopes by scanning all the predicted protein sequences using bioinformatics approaches is easy for an effective, rapid, and inexpensive way to validate the diagnostic markers.

The first systematic method for identifying T- and B-cell epitopes was the PEPSCAN method [4–9]. However, the majority of diagnostic assays developed are based on antigen–antibody reactions, and diagnostic assays are limited to antigenic sites of antibodies; but, T-cell epitopes can also be defined equally well using similar methods [10].

Moreover, there are a number of methods to determine linear B-cell epitopes. Generally, epitopes are of two types: (1) continuous epitopes (i.e., epitopes are derived from the epitope-mapping experiments of antigenic protein sequences); and (2) discontinuous epitopes (i.e., epitopes are identified by screening of complex peptide libraries [5]. Common methods used for selection of linear epitopes are (1) prediction (by using algorithms); (2) epitope recognition; (3) mutation in antigenic sequence or 'escape mutants' of viruses; and (4) PEPSCAN, by which overlapping peptides are tested for their ability to bind the antibody. The most systematic and reliable method for identifying linear antigenic peptides among all four methods described earlier is PEPSCAN [7]. However, a linear epitope can also be selected by using Methods 1 to 3.

On the contrary, structural epitopes are also screened by using combinatorial peptide libraries, which can be comprised of myriad peptide variants of either chemical or biological origin. Phage display is a powerful strategy that includes three steps to create a peptide library to screen functional peptides and proteins for specific biological functions. To create a peptide library, random DNA sequences are inserted into genes encoding protein 3 (cPIII) or protein 8 (cPVIII) of the filamentous phages. The library is first screened negatively against non-specific ligands and then positively against the desired target in vitro and in vivo. The identified peptides will be chemically synthesized and validated. The detailed principles and practices have been excellently reviewed by Smith and Perenko [11].

Recently, Songprakhon and co-authors identified 11 different sequences of 12-mer peptides binding to dengue virus nonstructural protein 1 by using a phage-displayed peptide library [12].

In addition to phage display, an alternative strategy is combinatorial peptide libraries that generate functional peptides. Huge peptide libraries can be established by peptide synthesis techniques for screening of unique ligands. Moreover, combinatorial peptide libraries are advantageous because non-peptidic moieties, such as beta-amino acids, un-natural amino acid analog, and modified peptide residues (phosphorylated or glycosylated), can be incorporated into the peptide sequences. The detailed principles and practices have been excellently reviewed by Bozovičar and Bratkovič in 2019 [13], where the current trends of peptides in imaging and non-imagining diagnostics are described.

2. Role of Peptides in Diagnostics

To understand the role of peptide in diagnostics, we have thoroughly investigated the published literature of last decade (*w.e.f.* 1 January 2011 to 31 December 2020) on the PUBMED MEDLINE database using specific keywords such as "Diagnostic" along with two filters "protein" and "peptide". Although, the data acquired from these searches were based on algorithms and the results were dependent on the mapping of the articles/reviews/clinical trials and its match with specific words. However, many interesting facts were found during the scrutiny of the published data. In our search of the published articles in last decade (2011–2020) versus the total data published (1997–2022), we did not observe any big differences in the trend of using peptides versus proteins in diagnostics. Uses of peptides are always 2.5 times lower than proteins as per the published literature on PUBMED (Figure 1A1). We have also observed that use of peptides in diagnostics are constant and has been showing linear growth as per data published in 1 years, 5 years, and 10 years on PUBMED (Figure 1B1). The published literature on PUBMED for the last

1 year, 5 years, and 10 years has shown 18,963, 149,130, and 332,657 articles, respectively (Figure 1B2).

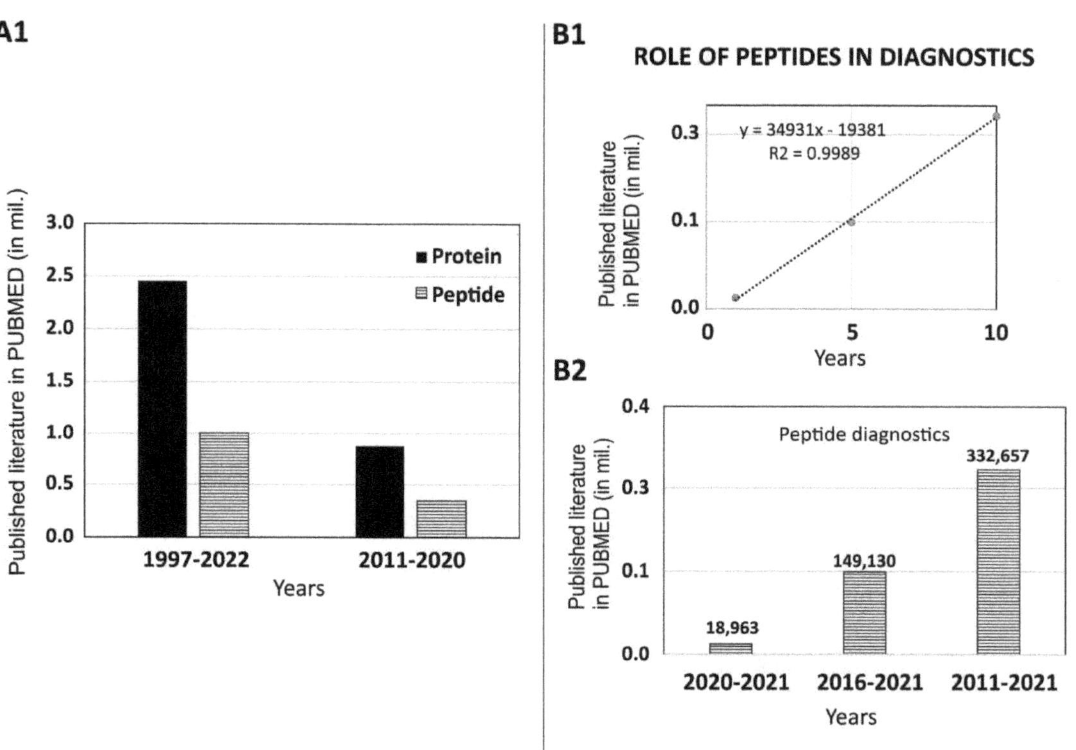

Figure 1. Role of peptides in diagnostics based on scientific research published on PUBMED: (**A1**) comparison of the published data of diagnostics using protein versus peptides-(**B1**,**B2**) exploring the role of peptides in diagnostics (1 year, 5 years, and 10 years).

To further understand the role of peptides in diagnostics and get a clear picture of the usage of peptides in diagnostics, we had critically analyzed our extracted data for the last decade (*w.e.f.* 1 January 2011 to 31 December 2020) on the PUBMED MEDLINE database using specific keywords, such as "Peptide" with three additional filters such as "Drug" or "Vaccine" or "Diagnostic". Data acquired from these searches were based on algorithms and the results were dependent on the mapping of the articles/reviews/clinical trials related to the keywords as mentioned above. It may contain some redundant data, due to the limitation of the analysis. However, some very interesting facts were found during the analysis, such as the total number of published scientific literature on PUBMED using the keyword "Peptide" along with additional filters such as "Drug" or "Vaccine" or "Diagnostic", which was 440,613, and 25,399, and 347,534, respectively. The data confirm that the use of peptides in drug was 1.26 times higher than peptides in diagnostics. However, peptides in diagnostics were 13.7 times higher than peptides in vaccines (Figure 2).

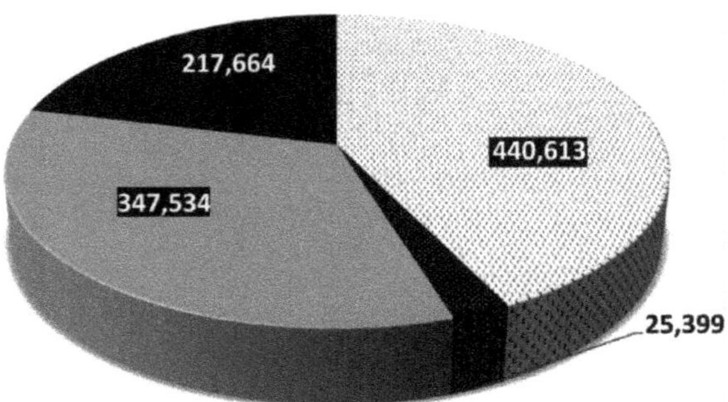

Figure 2. Based on scientific research published on PUBMED in the last decade (1 January 2011 to 31 December 2020).

3. Non-Imaging Diagnostics

Accurate and rapid detection of any diseases in humans has been a continuous challenge to diagnostic and epidemiological research. Efficient diagnosis is a crucial step, which helps in making an effective disease management strategy. A multitude of approaches have been attempted to identify pathogenic viruses and bacteria by using antigenic synthetic peptides in serological and molecular assays. Detection assays, which are based on peptides, have become increasingly substantial and indispensable for its advantages of using short synthetic peptides over conventional methods using recombinant proteins. Synthetic short peptide ligands with a length of more than eight amino acids have various advantages in the detection of specific antibodies [14].

To understand the role of peptides in non-imaging diagnostics, we have analyzed the published literature on PUBMED for last 5 decades (1 January 1970 to 31 December 2020). Non-imaging techniques such as ELISA, microarray, biosensors, microfluidics, and multiple Reaction monitoring were compared on PUBMED using keyword "peptide diagnostic". As per data published on PUBMED, we observed ELISA ranked 1st followed by microarray and biosensors (Figure 3).

3.1. ELISA

The ELISA technique was first developed by the Swiss scientists Engvall and Perlmann in 1971 by modifying the RIA method [15]. It is a quantitative analytical method that shows antigen–antibody reactions through a colorimetric assay, where an enzyme-linked conjugate and substrate are used to identify the presence of a specific concentration of the target molecule in biological fluids. In an ELISA assay, molecules such as peptides/proteins, hormones, vitamins, and drugs are coated in the polystyrene plate, which display a very high level of specificity against their cognate antibodies or antigens. Thus, ELISA assays are considered to be a very specific assay for quantification, where target antibodies or antigens can be measured in very low concentrations with hardly any risk of interference. Synthetic peptide-based ELISA can be developed by the three ways mentioned below: (1) target antibodies are immobilized in wells of the microtiter plate by using the adsorption procedure, wherein antibodies are immobilized in polystyrene plates; (2) species-specific anti-IgG or protein G-mediated immobilization, wherein anti-IgG or protein G are first coated in the plate and then target antibodies are captured by either anti-IgG or protein G; and (3) peptide-based capture, wherein peptides are directly immobilized in wells of the microtiter plate by the adsorption procedure. The assay is performed inthe solid phase of the microtiter plates, which is generally made up of rigid polystyrene, polyvinyl, and polypropylene materials. Synthetic peptides are first adsorbed in the microplates, followed by blocking with bovine serum albumin (BSA) for uncoated sites. The common enzymes that are employed with ELISA include peroxidase and alkaline phosphatase. These enzymes are conjugated with secondary antibodies. For alkaline phosphatase, P-nitro-phenyl phosphate (PNP) are used as substrates, which produce a yellow color in positive reactions. However, for the peroxidase conjugate, 5-amino salicylic acid and orthophenylenediamine (OPD) are used as the substrates, which produce a brown color in a positive reaction. The enzyme–substrate reaction is usually completed within 30–60 min. Sodium hydroxide (NaOH), hydrochloric acid (HCl), or sulfuric acid (H_2SO_4) are used to stop the reaction. The results are read at 400–600 nm on a spectrophotometer, as per the conjugate used. The technique is reviewed in more detail by Aydin in 2015 [16].

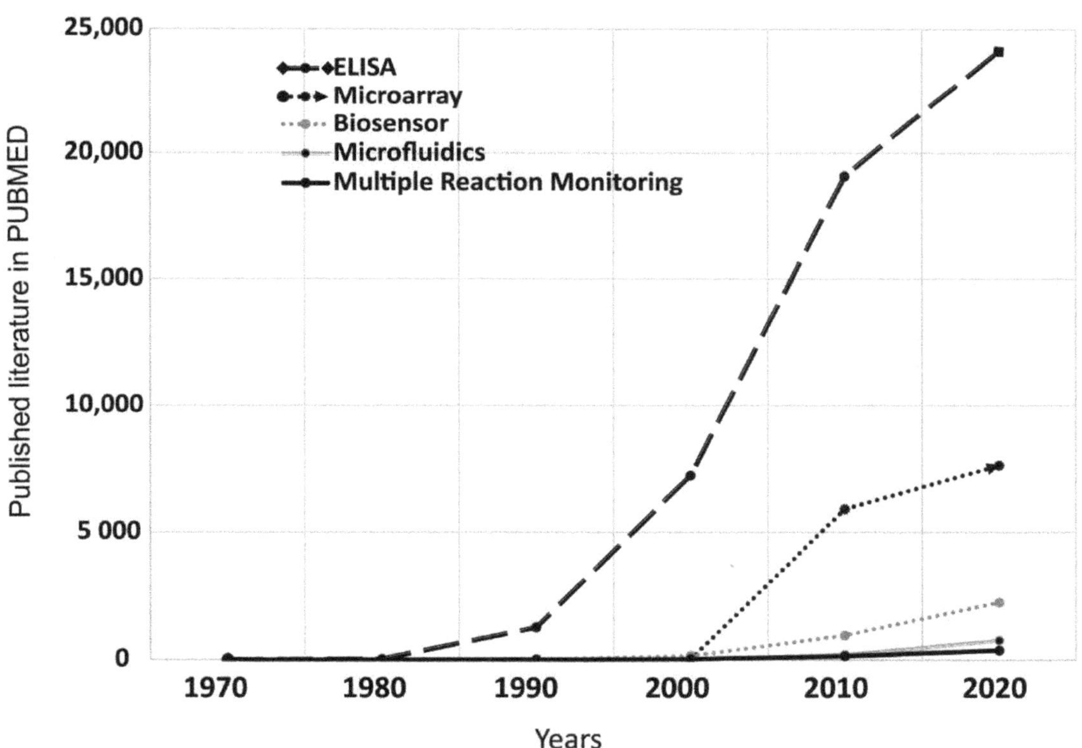

Figure 3. Understanding the role of peptides in diagnostics through the published literature on PUBMED in the last 5 decades (1 January 1970 to 31 December 2020). Comparison of the data published with the keywords "Peptide diagnostic" along with additional filters for ELISA, Microarray, Biosensors, Microfluidics, and Multiple Reaction monitoring. We observed ELISA ranked 1st followed by Microarray and Biosensors.

3.2. Microarray Technology

In late 1980, microarray technology was first developed [17]. Over time, it has become a valuable research tool for scientists and hold great promise in the field of diagnostics. Peptide microarrays are high-throughput, high-content miniature devices for immunoassays. Synthetic peptides are used as a probe in microarrays, wherein peptides are adsorbed on the surface of nitrocellulose-coated glass slides and are exposed to cellular extracts or serum or other specimens for molecular recognition events. The advantage of using microarray technology is the use of a number of different unique peptide biomarkers specific to the disease in real time. All probes can be immobilized in a random manner to ensure equal accessibility to all antibodies on the peptide microarray during epitope mapping, thus avoiding concentration-dependent effects on signal intensity. The technological concept of a peptide microarray is based on the substitution of linear epitopes of the protein with short overlapping synthetic peptides. These peptides typically consist of 10–15 amino acids and capture antigen-specific antibodies from serum samples [18].

3.3. Biosensors

In 1956, Leland C. Clark, Jr. has developed a biosensor to detect oxygen and later he was known as the 'father of biosensors'. His famous invention was later called by his name: the 'Clark electrode' [19]. Nowadays, biosensors are very common in clinical diagnosis and a number of point-of-care technologies (POCTs) have been developed for monitoring the disease diagnosis and its prognosis. In a general scenario, sensors are coupled with high-affinity biomolecules that allow selective detection of analytes. There are more than 84,000 indexed reports on the topic of 'biosensors' from 2005 to 2015 on 'Web of Science' [20]. A normal biosensor consists of five components:(1) an analyte, which can be any target molecule that needs to be detected by the biosensor; (2) a bioreceptor, which can be any molecule that specifically recognizes the analyte, such as a peptide, protein, cells, DNA, etc.; (3) a transducer, which is an element that converts one form of energy, such as bio-recognition, into another form of energy, such as optical or electrical signals; (4) an electronic circuit, which is a complex electronic circuit that performs amplification and conversion of signals to a digital form; and (5) a display, consisting of a user friendly system for interpretation of the results, such as the liquid crystal displays on computers or a direct printer that generates numbers or curves. It is a combination of hardware and software that generates the results of the biosensor in a user-friendly manner. A biosensor is a very sensitive device for measuring signal creating by biological or chemical reactions, which is proportional to the concentration of an analyte binding to its ligand. Biosensors are employed for disease monitoring, drug discovery, disease-causing micro-organisms, detection of pollutants, and presence of bio-markers indicating the disease stage in bodily fluids (blood, urine, saliva, and sweat). The technique is well reviewed by Bhalla et al. in 2016 [20].

3.4. Microfluidics

The field of science and technology that is associated with the control and manipulation of liquids at the microliter level is called microfluidics. Microfluidics is one of the powerful tools that is currently tying together with clinical diagnostics and generating a highly advanced version of POCT for precise and reproducibly results. It has revolutionized laboratory approaches for biological and chemical analysis from the bench-side to miniature chips. Moreover, these types of assays arecost effective and also do not require specific training to handle the device. Principally, the concept of microfluidics was associated with a framework of complexity and robustness in 1950s. The advantages of microfluidics are a reduced sample volume, scalability, laminar flow, and, hence, highly predictable fluid dynamics, a high resolution and sensitivity, and a short analysis time, leading to its low cost. The development in microfluidic technology has created a platform for genetic and proteomic analysis at the microscale level. This development is also associated with new advancement in technology along with their respective applications in pathogen detection to POCT devices, high-throughput combinatorial drug screening platforms, schemes for targeted drug delivery, advanced therapeutics, and novel biomaterials synthesis for tissue engineering.

Since the last two decades, microfluidics has started to show its impact in clinical diagnosis. The field of microfluidics is also evolving rapidly. The state of the art of microfluidic technologies is used to address the unmet challenges in diagnostics and can expand the horizons on clinical diagnostics, disease management, and patient care. Of the various microfluidic technologies that are available in the field, some are reliable and have been tested clinically. They can contribute to bridging the gap between this emerging technology and real-world applications [21].

Some advanced in vitro models, such as "organ-on-a-chip" technology, represents a new avenue in the field of scientific research and revolutionized the field of drug screening and toxicology studies [22]. Perestrelo et al. has reviewed interesting advancement in the field of microfluidic-based devices and its applications in the biomedical field, such as the body-on-a-chip concept [23].

3.5. Multiple Reaction Monitoring

In recent years, multiple reaction monitoring (MRM) has become more pivotal in clinical research for developing strategies for precision-based medicine or patient care. Thus, MRM is now used to evaluate proteomic/peptide biomarker verification with potential applications in medical screening. In this technique, high-quality tryptic peptides are selected and validated for quantitation of the proteins, its isoforms, and its post-translational modifications. The multiplexing of selected reaction monitoring (SRM) for targeting the number of proteins in a single run is known as MRM. It is a powerful technique based on a mass spectrometric approach for absolute and relative quantification of the proteins/peptides of interest in complex biological samples. MRM is a highly selective technique with a large capacity for multiplexing (~200 proteins per analysis per run). If the cost of transition is considered, it is rapid and cost-effective because the cost of the assay development to its deployment is low. For MRM assays, a triple quadrupole (QqQ) mass analyzer is required along with tandem quadrupole mass filters (Q1, Q3) and a collision cell. All the compartments are identical quadrupoles and may be used either to filter a specific mass-to-charge (m/z) ratio or to transmit a non-resolved ion of a specific range. Usually, the first quadrupole, Q1, is set to filter a specific precursor ion, which is passed through a collision cell and gets fragmented by the low-energy collision induced dissociation (CID) to create specific product ions. The specific product ion is detected by the Q3 analyzer for quantification. This process is referred to as the "transition process" and the technique is named "selected reaction monitoring"; the specific precursor/product ion pair is termed "transition" [24].

4. Peptides Application in Non-Imaging Diagnostics

In the 21st century, a number of peptide-based diagnostic systems has already been developed for commercial use or are on the verge of completion. There are a few examples of ELISAs with peptide-based diagnostic probes: C-peptide [25,26], gliadin [27], vasoactive intestinal peptide [28] diphtheria toxin (DTx) [29], *Chlamydia trachomatis* [30–34] human T-lymphotropic virus type I (HTLV-I) [35], human *C. pneumoniae* [31–33] and COVID 19 spike protein [36].

Moreover, rapid growth has been observed in peptide-based diagnostic systems mainly for diagnosis of cancer, heart disease, diabetes, Alzheimer disease, auto-immune disease, viral and bacterial infections, allergies, etc. (Figure 4). A few examples are quoted here for reference. Liu et al. has developed a novel affibody-based ELISA for detection of alpha-fetoprotein (AFP). AFP is an important biomarker associated with primary liver cancer. The peptide used in ELISA was a 58 amino acid peptide 'Affibody', which was derived from the Z domain of staphylococcal protein A. An affibody dimer $(Z_{AFP\ D2})_2$ showed higher binding affinity to AFP along with high thermal stability. The detection limit of the immunoassay using $(Z_{AFP\ D2})_2$ was 2 ng/mL [37]. Sahin et al. has selected and characterized the DE-Obs peptide HNDLFPSWYHNY by bio-panning of the phage display library on MKN-45 gastric cancer cells, which showed specific binding in MKN-45 cells [38]. Liu et al. has identified a 7-mer peptide that has the potential to be developed into a diagnostic test for residual hepatoma cells after trans-arterial chemoembolization [39]. Zhang et al. has demonstrated that the peptide sequence AADNAKTKSFPV has the potential to specifically recognize gastric cancer and discriminate neoplastic gastric mucosa from normal gastric mucosa. This can be used for early cancer detection during endoscopy [40]. Galvis-Jiménez et al. has developed an ELISA test to detect mammaglobin in blood samples from breast cancer patients vs. controls. Antibodies were generated in rabbits against four synthetic peptides of mammaglobin. All peptides showed immunogenicity and produced antibodies that were able to discriminate between the patients and controls. The results were obtained for an antiserum. B antiserum (against mammaglobin (31–39)) showed the best sensitivity (86.3%) and specificity (96%) [41].

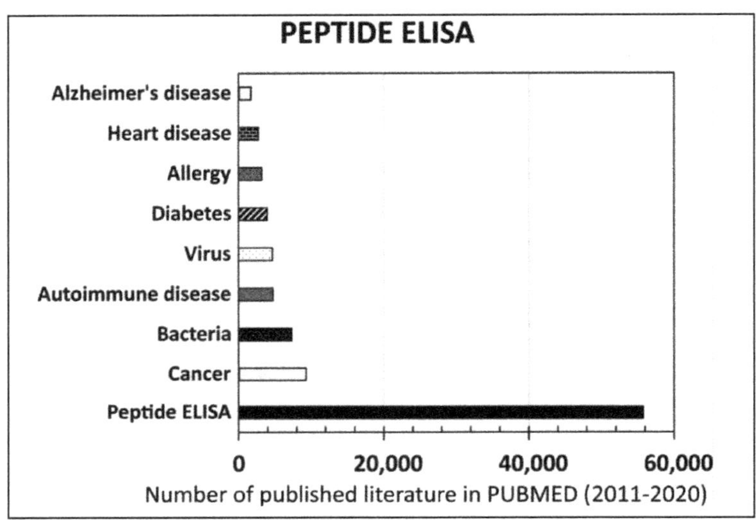

Figure 4. Understanding the role of peptides in ELISA in diagnostics through the published literature on PUBMED in the last decade (1 January 2011 to 31 December 2020). Comparison of the data published with the keyword peptide ELISA along with additional filters for Alzheimer's disease, Heart disease, Allergy, Diabetes, Virus, Autoimmune Disease, Bacteria, and Cancer, respectively. We observed Cancer was ranked 1st for peptide ELISA.

To understanding the role of GLP-1 in diabetes and its physiology, an accurate measurement of the GLP-1 metabolite is required. In 2017, Wewer Albrechtsen et al. developed an ELISA for measurement of the primary glucagon-like peptide-1 (GLP-1) metabolite, such as GLP-1 (7-36NH_2) and GLP-1 (9-36NH_2). The active form of GLP-1 is (7-36NH_2), which is rapidly degraded by the dipeptidyl peptidase 4 (DPP-4) enzyme and converts, by more than 90%, into an inactive form or to the primary metabolite (9-36NH_2) before reaching the target via the circulation. The developed ELISA could recognize both GLP-1 (9-36) NH_2 and nonamidated GLP-1 (9-37) [42–44]. The ADRB1-AB-immunogen-peptide (ESDEARRCYNDPK) impact of beta1-AAB on "myocardial recovery in patients with systolic heart failure" was published based on a peptide ELISA [45].

Increased C-peptide level is an important indicator for the diagnosis of diabetes. Lv et al. has developed an antibody sandwich ELISA for rapid detection of C-peptide in human urine of diabetic patients. Antibodies were developed in hen and rabbit by using PLL-C-peptide and BSA-C-peptide, respectively [46].

A peptide-ELISA was developed for detection of human H5N1 influenza viruses. ELISA was based on the antigenic H5 epitope (CNTKCQTP), which provides highly specific detection of antibodies to the H5N1 influenza viruses in human sera [47].

A rapid and accurate ELISA-based test was developed for HIV-1/2 antibody detection by using a peptide cocktail as an antigen. A novel peptide stretch, V3-I, covering the immunodominant epitope corresponding to the V3 hypervariable loop of gp120 antigens of selected Indian isolates, has been studied and incorporated in an antigenic cocktail of gp36, gp41, and rp24 of HIV-1/2. The peptide cocktail-based ELISA test showed 100% sensitivity and 99.3% specificity, with no cross reactivity [48]. A synthetic peptide of 11 amino acid was used to develop an ELISA for HIV-2. The peptide epitope in the ELISA was highly specific and sensitive towards anti-HIV-2 antibodies. The peptide ELISA showed 100% sensitivity with 94.9% specificity [47].

Lyme neuroborreliosis (LNB) is a disorder of the CNS caused by systemic infection of spirochetes. The diagnosis of LNB is a challenge to clinicians. Van Brugel et al. has demonstrated that the C6-peptide ELISA can be used for the diagnosis of LNB by using a patient's CSF. Serum–CSF pairs from LNB patients ($n = 59$), Lyme non-neuroborreliosis cases ($n = 36$), and neurological controls ($n = 74$) were tested in a C6-peptide ELISA, where the sensitivity of the C6-peptide ELISA for LNB patients in CSF was 95%, and the specificity was 83% in the Lyme non-neuroborreliosis patients, 96% in the infectious controls, and 97% in the neurological controls [49].

Davis has demonstrated that an ELISA can be developed to quantify cellular proteins, such as NGF, secreted into conditioned culture media. Neurotrophin is critical to neuronal viability, and has become a popular research focus for the treatment of neurodegenerative diseases [50].

5. Peptide Diagnostics and SARS-CoV-2

In 2019, a new coronavirus, SARS-CoV-2, which causes acute respiratory syndrome, began to spread around the world. The disease is known as COVID-19 (coronavirus disease 2019) and has so far caused the deaths of about 4 million people worldwide and more or less serious health problems for hundreds of millions more. It is clear that the need to establish the right diagnostic and therapeutic approach is critical. The basis for a successful fight against this pandemic is not only the determination of the most effective therapy but also prevention based on reliable testing and vaccination. Thousands of scientists immediately began to address this new problem using a variety of methodological approaches. Several of these methodologies are based on the use of peptides. Examples of the use of peptides in studying the properties of SARS-CoV-2 and research into the resultant COVID-19 can be found below.

5.1. Viral Epitope Profiling of SARS-CoV-2

Analysis of viral epitopes is crucial for understanding the immunogenicity of the viral proteome, while it is critical for improving the diagnostics and production of a functional vaccine. Peptides frequently and specifically recognized by COVID-19 patients were identified by VirScan-based serological profiling and used to create a Luminex assay predicting SARS-CoV-2 exposure with 90% sensitivity and 95% specificity [51].

5.2. Peptides Used for Antibody Diagnostics

Actually, a large variety of SARS-CoV-2 antibody diagnostic assays are used, including immunoassays based on the large recombinant protein or vice-versa specific epitope peptides identified from the whole antigen [52]. Using peptide epitopes would be beneficial with respect to assay specificity, while large recombinant proteins also include many cross-reactive epitopes that would react with low specificity antibodies, leading to a lower specificity of the test.

5.3. Peptides Used for Identification of SARS-CoV-2-Derived T Cell Epitopes

The identification of SARS-CoV-2-derived T cell epitopes is of critical importance for diagnostic tools as well as for peptides vaccines. One way how to identify them is using CD4+ and CD8+ T cell depletion assays and FACS-based analysis of activation markers. Results obtained by using these methods suggested that generation of effective adaptive immunity against SARS-CoV-2 requires the participation of both CD4+ and CD8+ T cells. This finding is very important for the preparation of a functional vaccine, as it is clear that it is necessary to incorporate both HLA-I-restricted and HLA-II-restricted epitopes in peptide-based vaccines to obtain optimal vaccination [53].

5.4. Peptides/Proteins as a Markers of COVID-19

D-dimer is produced during lysis of crosslinked fibrin. Results of some studies suggest that the D-dimer levels can be used as a prognostic marker in patients with COVID-19 [54–56]. Interferon gamma-induced protein 10 (IP-10) is a small cytokine secreted by endothelial cells, monocytes, and fibroblasts, which attracts activated Tcells to the site of inflammation. In COVID-19 patients, IP-10 was overexpressed in the acute phase of the disease regardless of other clinical characteristics; therefore, it has been suggested as a potential new biomarker for SARS-CoV-2 infection. In the study, SARS-CoV-2 peptide pools covering viral proteins were used in order to identify the immune biomarkers of SARS-CoV-2 infection [57].

6. Imaging Diagnostics

The most common targeted molecular imaging techniques, such as PET and SPECT, are playing a very important and essential role in modern diagnostics because the information provided by them are very specific, accurate, and shows disease distribution. On the contrary, using non-specific contrast agents has a low targeting efficiency, which can be superseded by using specific probes. Recent technological development has revealed various methodologies for designing specific, smart, and accurate probes. Among all the strategies, utilization of peptide-based probes has been the most successful. For discovery of specific peptide-based probes, the commonly used methods are phage display and combinatorial peptide chemistry. They have strongly impacted the use of available targeting peptides in an efficient and specific manner. The discovered peptides are either a specific target for a variety of disease-related receptors or surrogate biomarkers. These targeting peptides are either radiolabeled or coupled with the appropriate imaging moieties and used in imaging diagnostic. For this reason, labeled peptides have soon become a part of imaging diagnostic systems.

7. PET and SPECT Imaging

In 1951, Wrenn Jr et al. demonstrated the use of a positron-emitting radioisotope for brain tumor localization [58]; after this, in the 1960s and 1970s, PET gradually grew as a research imaging modality [59]. However, the clinical utility of PET in neurology and oncology patients was demonstrated in the 1980s and 1990s [60,61].

PET is a non-invasive molecular imaging modality that uses radioactive tracers in pico- and nanomolar amounts for visualization and quantitation of biological processes in vivo. In brief, PET starts with an intravenous injection of a radioactive probe (i.e., compound labelled with positron-emitting isotopes) that circulate throughout the body and accumulate in the inflammatory lesions, which results in the emission of a positron. The positron 'annihilates' with an electron and generates two 511 keV γ-photons that travels in the opposite direction at 180 degrees. This property is called as 'collinearity'. The two photons generated during the annihilation process are detected by the PET detector ring and known as 'coincidence detection'. Coincidence detection makes PET imaging more sensitive compared to SPECT imaging, where the γ-rays emitted from the target lesions are measured directly by the detectors. A detailed description about the PET and SPECT techniques and its applications are discussed in detail by Signore et al. in2010 [62].

Interestingly, labelled peptides were introduced into clinic more than three decades ago, since then these are increasingly being used in clinics for diagnosis of different diseases, staging, and evaluation of therapy response. These radiolabeled probes are utilized in the most sensitive molecular imaging techniques, i.e., PET and SPECT.

Human cells overexpress several peptide receptors in the diseased condition, which works as molecular targets, and radiolabeled peptides bind to these targets with high affinity and specificity, holding great potential for molecular diagnostic imaging.

To understand the role of peptides in imaging diagnostics mainly for PET and SPECT, we have analyzed the published literature on PUBMED for last 5 decades (1 January 1970 to 31 December 2020). As per data published on PUBMED, we observed that PET has grown drastically in last two decades (Figure 5).

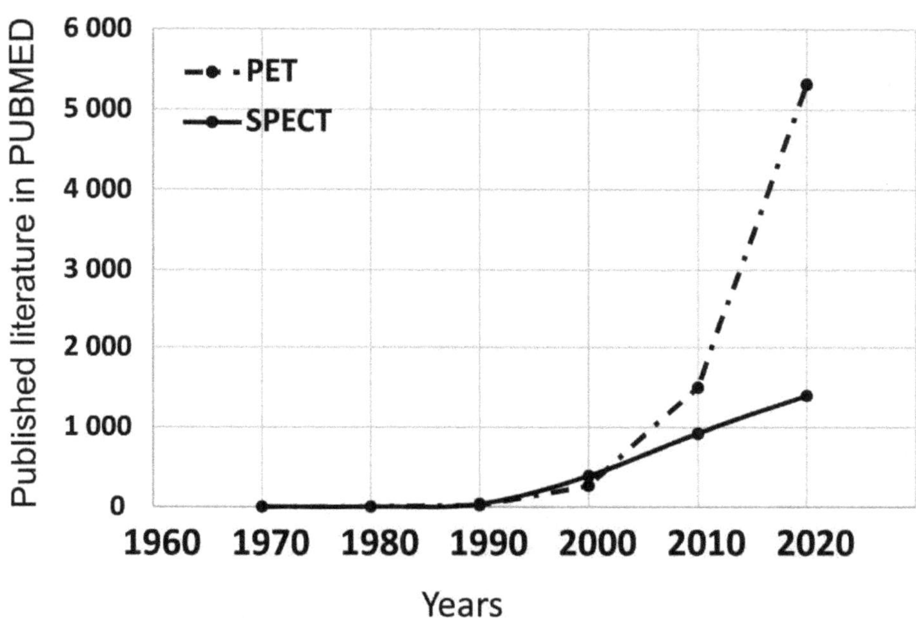

Figure 5. Understanding the role of peptides in imaging diagnostics (PET and SPECT) through the published literature on PUBMED in the last five decades (1 January1970 to 31 December 2020). Comparison of the data published with the keywords peptide diagnostic and sub-keywords (PET and SPECT). PET has grown drastically in last decade.

8. Peptides Application in Imaging Diagnostics

To understand the role of radiolabeled peptide in imagine diagnostics and get a clear picture of its usage in imaging diagnostics, we had critically analyzed the data for the last 10 years and 5 years on the PUBMED MEDLINE database. The specific keywords were either "PET" or "SPECT", used in two separate searches with seven additional filters of the most common radiolabeled peptides and their analogues used in PET and SPECT imaging, such as somatostatin receptors, interleukin 2 receptor, prostate specific membrane antigen, $\alpha\beta3$ integrin receptor, gastrin-releasing peptide, chemokine receptor 4, and urokinase-type plasminogen receptor. We observed that there is no big difference between the data published in 5 years versus 10 years for the peptides used in PET and approximately a similar amount of data published in 5years and 10 years, except for $\alpha\beta3$ integrin and urokinase-type plasminogen receptor. It shows that most of the research was done in last 5 years (Figure 6). Similarly, the data obtained from SPECT had shown a similar pattern as mentioned above for PET; but, interestingly, no published data was obtained for Interleukin 2 receptor and urokinase-type plasminogen receptor in the search of the last 5 years (Figure 7).

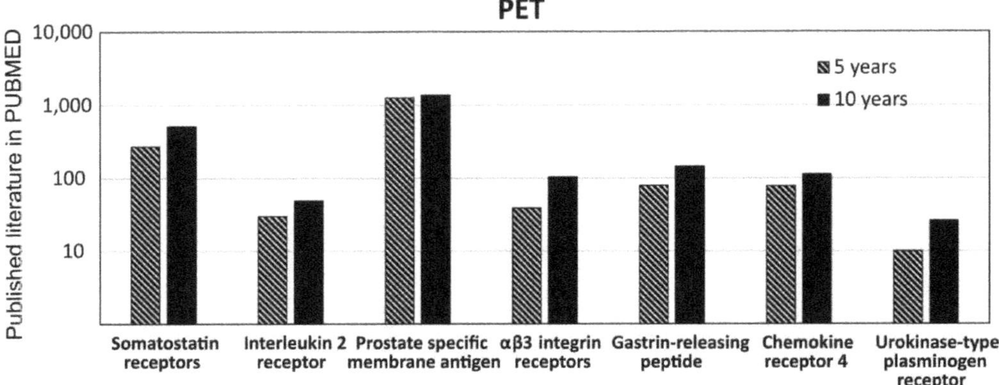

Figure 6. Understanding the role of radiolabeled peptide in PET through the published literature on PUBMED in the last 5 years and 10 years. Comparison of the data published with different keywords, such as somatostatin receptors, interleukin 2 receptor, prostate specific membrane antigen, αβ3 integrin receptor, gastrin-releasing peptide, chemokine receptor 4, and urokinase-type plasminogen receptor.

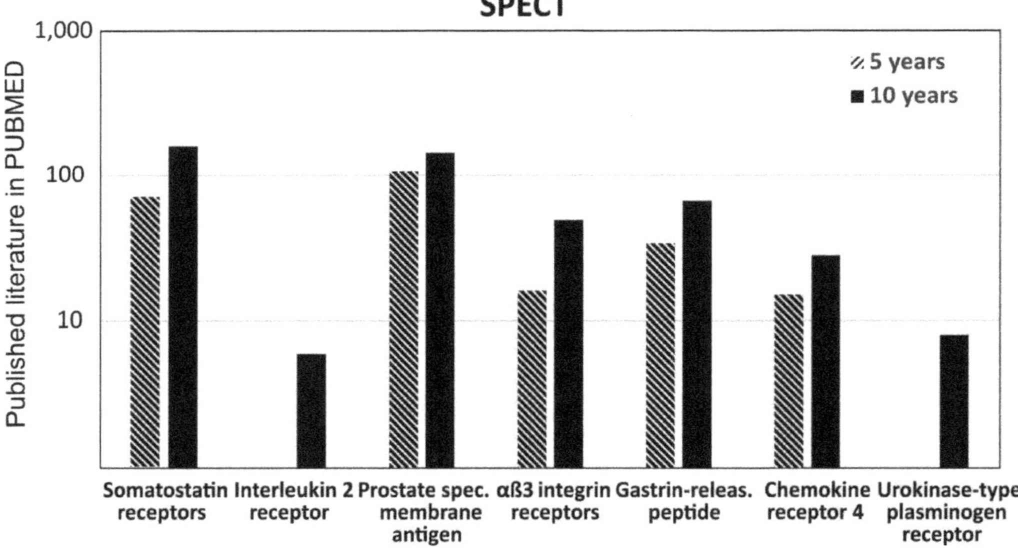

Figure 7. Understanding the role of radiolabeled peptide in SPECT through the published literature on PUBMED in the last 5 years and 10 years. Comparison of the data published with different keywords, such as somatostatin receptors, interleukin 2 receptor, prostate specific membrane antigen, αβ3 integrin receptor, gastrin-releasing peptide, chemokine receptor 4, and urokinase-type plasminogen receptor.

8.1. Somatostatin Receptors (SSTRs)

In 1973, Roger Guillemin's group first isolated somatostatin (SST) from an ovine hypothalamic extract and was characterized as tetradecapeptide [63]. SST is a regulatory and cyclic disulfide containing a peptide that is naturally present in 14 or 28 amino acids sequences. SST binds to somatostatin receptors (SSTRs) and regulates several physiological and cellular processes, which are expressed by many neuroendocrine cells, nerve cells, and inflammatory cells, such as lymphocytes, peripheral blood mononuclear cells, thymocytes, monocytes, and macrophages [64]. The function of SST is mediated by a family of G-protein-coupled receptors that includes five distinct subtypes, namely, SSTR1–SSTR5 [65].

Due to the wide range of interaction with SSTR in different inflammatory disease conditions and overexpression of SSTR in immune cells, inflammatory cells, and blood vessels, it was sensible to develop radiolabeled SST analogues with a different affinity for SSTRs for diagnostic PET and SPECT imaging in different oncological and inflammatory disease conditions, particularly in rheumatoid arthritis, Sjögren syndrome, and autoimmune thyroid diseases [66]. In brief, labelling of SST analogues were achieved by conjugation of the peptide with a bifunctional chelator, DOTA, NOTA, or DTPA, and afterwards labelling with a radionuclide, including ^{99m}Tc (EDDA/HYNIC-TOC) or ^{18}F or ^{68}Ga, ^{123}I, ^{111}In, or ^{64}Cu (DOTATATE) [67,68].

Several SST analogues have already been developed and assessed for clinical use; however, ^{111}In-labelled pentetreotide (also known as ^{111}In-octreotide or Octreoscan^{TM}), a DTPA-conjugate of octreotide, is extensively studied in diagnostic imaging. ^{111}In-pentetreotide has a high affinity for SSTR2 and SSTR5; it enters inside the cell by endocytosis, first taken by lysosomes and then moved to the nucleus. It is predominantly used for the assessment of neuroendocrine tumors and carcinoid tumors. Some other novel radiolabeled tracers for SSTRs also demonstrated decent affinity, including ^{68}Ga-DOTA-TOC (more selective to SSTR2 and SSTR5), ^{68}Ga-DOTA-TATE and ^{64}Cu-DOTA-TATE (affinity to SSTR2), ^{99m}Tc-EDDA/HYNIC-TOC (affinity to 2 and 5 type receptors), and ^{68}Ga-DOTA-NOC (affinity to 2, 3 and 5 type receptors) [69,70]. A study was performed by Yamaga et al. to compare the detection rate of^{68}Ga-DOTATATE PET/CT with ^{111}In-octreotide SPECT/CT in medullary thyroid carcinoma patients (n = 15) with increased calcitonin levels but negative conventional imaging after thyroidectomy. This study revealed a high sensitivity and accuracy, 100% and 93%, respectively, with ^{68}Ga-DOTATATE PET/CT, while ^{111}In-octreotide SPECT/CT showed a lower sensitivity and accuracy, 46% and 53%, respectively. In the same study, the authors also performed conventional imaging (CI) that was comparable with PET/CT scans with a sensitivity of 100% and accuracy of 93%, although ^{68}Ga-DOTATATE PET/CT demonstrated a higher detection rate compared to CI in detecting bone metastases [71].

Nevertheless, Johnbeck et al. performed a head-to-head comparison of the diagnostic accuracy of PET/CT scans of ^{64}Cu-DOTATATE with ^{68}Ga-DOTATOC in neuroendocrine tumor patients (n = 59). In this study, the authors found that 701 lesions were concordant in both PET/CT scans; however, only one of these scans detected an extra 68 lesions. The authors concluded that the patient-based sensitivity was the same for ^{64}Cu-DOTATATE and ^{68}Ga-DOTATOC PET/CT scans in these patients. However, ^{64}Cu-DOTATATE had a comparatively better lesion detection rate, as patient follow-up discovered that the majority of the extra lesions detected by ^{64}Cu-DOTATATE were true positive [72]. Nevertheless, the study also revealed thata >24 h shelf life and at least 3 h scanning window makethe ^{64}Cu-DOTATATE PET/CT scan more convenient to use in the clinical setting.

8.2. Interleukin-2 Receptor

Interleukin-2 (IL-2) has a very high affinity for interleukin-2 receptors (IL-2R), which is hetero trimers of the α, β, and γ subunit, named CD25, CD122, and CD132, respectively. The α-subunit, i.e., CD25, contains the key binding site for IL-2, which could be present as a soluble or transmembrane receptor [62]. High levels of IL-2R are expressed by activated lymphocytes during inflammatory processes, whileIL-2R expression is lower in resting immune cells; therefore, this receptor is an appropriate biomarker for the diagnosis of active inflammation in chronic inflammatory disease patients.

IL-2 is one of the most studied research radiotracer for imaging of infiltrating T cells in different inflammatory diseases, and radiolabeled with different radionuclides, including^{99m}Tc, ^{123}I, ^{125}I, ^{35}S, and, recently,^{18}F for PET imaging [62].

Signore et al. performed a study to evaluate in vivo the binding of ^{99m}Tc-IL2 with infiltrating lymphocytes in thirty patients with cutaneous lesions suspected of being melanoma [73]. In this study, histology revealed 21 melanoma lesions and 9 classified as benign. The authors reported ^{99m}Tc-IL2 uptake in 15 out of 21 (71%) melanomas lesions and 2 out of 9 (22%) benign cutaneous lesions. Additionally, in the ^{99m}Tc-IL2 scan, the target-to-background ratio significantly correlated with the number of IL2R-positive tumor-infiltrating lymphocytes. This study demonstrated that the ^{99m}Tc-IL2 scan might provide a tool for the in vivo assessment of tumor-infiltrating IL-2R-positive cells, which could be extremely beneficial for patient selection with unlabeled IL2 immunotherapy.

For PET imaging of IL-2, a novel method was published by Di Gialleonardo et al. on how to synthesize *N*-(4-Fluorobenzoyl)-interleukin-2 (FB-IL2) that specifically binds to IL-2R [74]. In addition, recently Khanapur et al. presented an improved synthesis of the same radiotracer [75]. To enable the use of FB-IL2 in clinical studies, a fully automated Good Manufacturing Practices (GMP)-compliant production process has been developed and published by Erik FJ de Vries's group at University Medical Centre Groningen, The Netherlands [76]. In addition, the same group performed a clinical trial (ClinicalTrials.gov; identifier NCT02922283) in patients with metastatic melanoma (stage IV). In this study, the researchers found FB-IL2 to be safe and feasible for human patient study without any side effects, although serial PET imaging was not able to detect a treatment-related immune response in this patient cohort [77].

8.3. Prostate-Specific Membrane Antigen (PSMA)

Prostate-specific membrane antigen (PSMA), also known as glutamate carboxypeptidase II (GCPII), or folate hydrolase, is an integral cell-surface membrane glycosylated metalloenzymeoverexpressed in prostate carcinomas. It has 19 amino acids (AA) intracellular N-terminal domain, 24 AA transmembrane helix, and a 707 AA extracellular C-terminal domain bearing 2 zinc ions and 2 binding pockets [78].

Mannweiler et al. investigated paraffin-embedded sections of patients with primary prostate carcinoma and distant metastases (n = 51). The immunohistochemistry data revealed that 96% of the primary tumors and 84% of metastases showed expression of PSMA, in the advanced prostate cancer cohort [79]. While only one case, i.e., 1.9%, was entirely negative for PSMA in both the primary and metastatic tissue. Therefore, PSMA is a suitable biomarker for diagnosis, staging, and therapy response monitoring in prostate cancer patients.

First images of prostate carcinoma patient detected by ^{68}Ga- labelled HBED-CC conjugate of the PSMA-specific pharmacophore Glu-NH-CO-NH-Lys (^{68}Ga-PSMA) was published by Afshar-Oromieh et al. [80]. In this study, an^{18}F-fluoroethylcholine (F-FECH) PET scan was unable to detect any lesions, while a ^{68}Ga-PSMA PET scan revealed a lesion adjacent to the urinary bladder matched with tumor relapse.

Recently, Grubmüller et al. performed a study to evaluate simultaneous [^{68}Ga]Ga-PSMA-11 PET/MRI for primary tumor-node-metastasis staging in prostate cancer patients (n = 122) prior to planned radical prostatectomy, compared with histology data. In this study, PSMA-PET/MRI correctly diagnosed prostate cancer in 119 of 122 patients (97.5%).The diagnostic accuracy for T staging was 82.5%, for T2 stage was 85%, for T3a stage was 79%, for T3b stage was 94%, and for N1 stage was 93% [81]. This study confirms the efficacy of [^{68}Ga]Ga-PSMA-11 PET/MRI for an accurate staging of newly diagnosed prostate cancer patients.

Another study was performed to compare the metabolic features of high-grade glioma (HGG) and low-grade glioma (LGG) tumors using ^{68}Ga-PSMA-617 and ^{18}F-FDG PET scans. In this study, the patients (n = 30) underwent both ^{68}Ga-PSMA-617 and ^{18}F-FDG PET scans over two consecutive days and then surgical treatment was performed. This study revealed that the ^{68}Ga-PSMA-617 PET scan is superior to the ^{18}F-FDG PET scan in differentiating HGG and LGG [82].

Nevertheless, several PET and SPECT studies with radiolabeled PSMAPET radiotracers were also performed in patients with thyroid cancer [83], hepatocellular carcinoma [84], prostate adenocarcinoma [85], glioblastoma [86], myeloma [87], sinonasal glomangiopericytoma [88], Sjögren syndrome [89], and bladder cancer [90], which shows PSMA-targeted radiotracers are now playing an increasing role in the diagnostic imaging of patients.

8.4. αvβ3 Integrin Receptors

Integrinsare a class of 24 heterodimeric transmembrane glycoproteins made up of different 18 α-subunits and 8 β-subunits, and play a key role in cellular interactions and transduction of signals between the extracellular matrix and interior of the cell [91]. The αvβ3 integrin, also referred as the vitronectin receptor, plays a key role in tumor metastasis and angiogenesis, so diagnostic examination with αvβ3 expression offers a great prospective strategy. On the surface of vitronectin, αvβ3-binding was mediated by RGD tripeptide, i.e., Arg-Gly-Asp, which acts as the core recognition motif [92]. In diagnostic PET and SPECT imaging, αvβ3 integrin is the most extensively studied integrinthat provides crucial information about the metastatic potential of tumor, and offers an optimal in vivo biomarker for angiogenesis.

Galacto-RGD is a radiolabeledαvβ3 antagonist that helps in the monitoring of αvβ3 expression with PET imaging and the first of its class studied in human patients. Haubner et al. radiolabeled glycosylated RGD-peptide (Galacto-RGD) using 4-nitrophenyl 2-fluoropropionate as a prosthetic group with a radiochemical yield of 85% and a high radiochemical purity of >98% [93]. Afterwards the same group performed a PET imaging study in nine patients, who suffered from either malignant melanoma with distant/lymph node metastasis, or chondrosarcoma, or soft tissue sarcoma, or osseous metastasis of renal cell carcinoma, or villonodular synovitis. Researchers selected these patients based on substantial evidence that these pathologies express αvβ3 [94]. This study demonstrated a 9-fold higher radiotracer accumulation in the tumor than in the muscle, which confirms the superior properties of Galacto-RGD for molecular imaging of αvβ3 integrin receptors.

A feasibility study is recently performed by Makowski et al. using Galacto-RGD PET/CT imaging in patients with acute myocardial infarction ($n = 12$) for αvβ3 expression assessment [95]. In this study, Galacto-RGD uptake significantly correlated with infarct size ($R = 0.73$). In addition, the authors found significant inverse correlation with restricted blood flow for all myocardial segments ($R = -0.39$) and in severely hypo-perfused areas ($R = -0.75$).

An SPECT tracer, NC100692, was evaluated for imaging αvβ3 expression in human breast cancer patients by Bach-Gansmo et al., where the authors were able to clearly detect 19 of 22 tumors using this tracer, which was safe and well tolerated by these patients [96].

Another novel integrin-targeted PET imaging radiotracer Fluciclatide (also known as AH111585) was evaluated for αvβ3 and αvβ5 imaging in melanoma and renal tumors. The authors demonstrated that an increased ^{18}F-fluciclatide uptake occurs at sites of acute myocardial infarction, in specific areas of subendocardial infarction, and hypokinesia associated with subsequent functional recovery [97]. Data from this study suggested that ^{18}F-fluciclatide is a potentially useful imaging biomarker for PET imaging of myocardial αvβ3 integrin expression.

8.5. Other Peptides for PET and SPECT Imaging

A number of radiolabeled peptides are already established in clinics or are being evaluated in different phases of clinical trials for PET and SPECT imaging of various inflammatory diseases. Apart from the above mentioned peptides, many other peptides are also under investigation, which includes but are not limited to peptides for bombesin (BBN) receptors, gastrin-releasing peptide (GRPR), chemokine receptor 4 (CXCR4), urokinase-type plasminogen receptor (uPAR), glucagon-like peptide receptor 1 (GLPR1), and caspase-3 imaging [98,99].

Recently, Kraus et al. evaluated the possibility of the C-X-C motif chemokine receptor 4 (CXCR4)-directed imaging with ^{68}Ga-Pentixafor PET/CT, to diagnose and quantify disease involvement in 12 myeloproliferative neoplasms patients, together with 5 non-oncologic control patients. Study data revealed that 12 out 12 patients were found positive in PET/CT, which was also confirmed by immunohistochemical staining [100]. This is the first data that shows the feasibility of CXCR4-directed imaging with ^{68}Ga-Pentixafor PET/CT in a myeloproliferative neoplasm patient cohort. In the beginning, CXCR4 was recognized as a co-receptor in human immunodeficiency virus-1 (HIV-1), which attracted the researchers' attention; afterwards, more investigation revealed overexpression of CXCR4 in 30 different cancers, including pancreatic, breast, lung, colorectal, prostate, ovarian, and skin cancers, lymphoma, and leukemia [101,102].

Zhang et al. published the first-in-human study of a ^{68}Ga-labeled heterodimeric peptide BBN-RGD [103]. This novel radiotracer, ^{68}Ga-BBN-RGD, targets $\alpha_v\beta_3$integrin as well as GRPR. In this study, the authors investigated the diagnostic accuracy and safety of ^{68}Ga-BBN-RGD PET scans in healthy volunteers (n = 5) and prostate cancer patients (n = 13) and comparedit with ^{68}Ga-labelled BBN. This study did not show any obvious side effect of ^{68}Ga-BBN-RGD administration in any healthy volunteer and/or patient. In patient scans, ^{68}Ga-BBN-RGD PET/CT diagnosed 3 of 4 primary tumors, while only 2 of 4 primary tumors were diagnosed with ^{68}Ga-BBN PET/CT. Interestingly, the authors found that ^{68}Ga-BBN was not able to detect lesions that were GPRR -ve and $\alpha_v\beta_3$ integrin +ve; however, ^{68}Ga-BBN-RGD was able to detect these lesions. Therefore, this novel approach for dual $\alpha_v\beta_3$ integrin and GRPR, targeting PET radiotracers, demonstrated a great potential in diagnosis and staging of primary prostate cancers as well as metastases lesions. Recently, another novel dual-targeting ^{68}Ga-NODAGA-LacN-E[c(RGDfK)]$_2$Glycopeptide has also been developed for PET imaging of cancer patients, which can diagnose integrin $\alpha_v\beta_3$ and galectin-3 expression in tumor and tumor endothelial cells [104]. As evident from the details above, enough novel radiolabeled peptides are now available in the clinical setting and many more peptides are currently in preclinical investigation and indifferent clinical trial stages; therefore, there is no doubt that it will have enormous clinical impact in diagnostics in the coming years.

9. Challenges in Peptide-Based Diagnostics

The major challenges in peptide-based diagnostics are the synthesis and purification of peptides. Fmoc/tBu strategies are widely used for solid-phase peptide synthesis (SPPS). The first amino acid is coupled to the resin. The first step is to deprotect the amine and then coupled with the free acid of the second amino acid. This cycle repeats until the desired sequences have been synthesized. SPPS cycles may also include capping steps, which block the ends of the unreacted amino acids from reacting. At the end of the synthesis, the crude peptide is cleaved from the solid support, while simultaneously removing all protecting groups using a strong acid reagent, such as trifluoroacetic acid or a nucleophile. The crude peptide can be precipitated from a non-polar solvent such as diethyl ether in order to remove soluble organic by-products. The crude peptide can be purified using reversed-phase HPLC [105,106].

To purify the longer peptides is very challenging, because the impurities of the by-products have somewhat similar sequences and shows the same retention time in HPLC purification. Kent and co-workers have proposed that some peptide sequences for intra-molecular or inter-molecular non-covalent interactions, which ultimately cause insoluble peptide aggregates, are not easy to solubilize for purification [107]. Moreover, synthesis of longer peptide sequences >50 amino acids has always been a challenging task for chemists, even when using well-advanced and automated peptide synthesis systems. Difficult peptide sequences are hydrophobic in nature and contain a large number of β-branched amino acids along with leucine, valine, phenylalanine, or isoleucine, etc. Peptide sequence with glycine may induce β-sheet packing. These type of peptide sequences can form β-sheet or α-helical structures within the molecule and therefore they have high aggregation potential and low solubility in aqueous or organic solvents. This results in generally difficult handling, synthesis, and purification [106].

Beside this, there are some challenges that depend upon the technique used for diagnosis. For example, non-imaging techniques, such as ELISA, microarray, and biosensor, are (1) a labor-intensive process—to prepare antibodies through a sophisticated cell culture technique; (2) high purity primary antibodies or synthetic peptides are required to set-up the experiments; (3) due to insufficient blocking of the surface onthe microtiter plate there is high possibility of getting false-positive or false-negative results; (4) antibodies instability during transportation and storage may cause false-negative results; (5) unavailability of specific antibodies or difficult peptide synthesis may cause a problem in setting up the process; and (6) it is a time-consuming process, requiring at least 24 h to complete [108].

The main limitation to set up the analysis is depended upon substances such as antigen-specific antibodies, a peptide epitope for specific antibodies, and its purity. Impure substances may cause either false-positive results or a low signal-to-noise ratio. However, imaging diagnostics are also having quite similar challenges but is seen to be more specific than non-imaging diagnostic because only labeled peptide probes are used.

As per the guideline for method validation ICH Q2 (R1), specificity is defined as follows: "Specificity is the ability to assess unequivocally the analyte in the presence of components, which may be expected to be present" (European Medicines Agency, ICH Topic Q 2 (R1), accessed on 12 August 2021) [109].

The specificity of the method is based on detection of a single or specific analyte, showing no cross reactivity to other molecules. This type of approach can be achieved by using unique peptide probes discovered or designed for specific target recognitions. These are used as a ligand in non-imaging diagnostics and as a tracer in imaging diagnostic. However, in some cases high similarities in analytes are observed, where a selective approach is considered to be the best solution that can be achieved.

For an adequate target-to-background ratio, the probe should be highly specific towards its receptor and should show a high binding affinity. Moreover, it should be functionally stable in physiological conditions and be cleared quickly from non-targeted sites in order to provide high-quality results. Additionally, it is also necessary to check the toxicity and immunogenicity of the probe for clinical translation [110].

In conclusion, peptide-based diagnostics is an interdisciplinary approach, for which scientists are performing basic research to discover unique peptides for targeting specific receptors reflecting a disease state, organic chemists are developing and characterizing the peptide-based probes, and biophysicists are improving image quality. Finally, clinicians are reviewing the outcome and importance of the diagnostics methods developed.

Author Contributions: Conceptualization, S.P.; formal analysis, S.P.; data curation, S.P. and G.M.; writing—original draft preparation, S.P., G.M. and M.C.D.; writing—review and editing, S.P., G.M. and M.C.D. All authors have read and agreed to the published version of the manuscript.

Funding: The work was supported by the Charles University Research Fund [Progres Q39].

Institutional Review Board Statement: Not applicable.

Informed Consent Statement: Not applicable.

Data Availability Statement: Not applicable.

Conflicts of Interest: The authors declare no conflict of interest.

Abbreviations

POCT	point-of-care technology;
ELISA	Enzyme-Linked Immunosorbent Assay;
MRM	multiple reaction monitoring;
PET	Positron Emission Tomography;
SPECT	Single Photon Emission Computed Tomography;
SSTRs	Somatostatin receptors;
PSMA	Prostate-specific membrane antigen;
SPPS	Solid-phase peptide synthesis.

References

1. Andresen, H.; Bier, F.F. Peptide microarrays for serum antibody diagnostics. *Methods Mol. Biol.* **2009**, *509*, 123–134.
2. Vanniasinkam, T.; Barton, M.D.; Heuzenroeder, M.W. B-Cell epitope mapping of the VapA protein of Rhodococcus equi: Implications for early detection of R. equi disease in foals. *J. Clin. Microbiol.* **2001**, *39*, 1633–1637. [CrossRef] [PubMed]
3. Pellois, J.P.; Zhou, X.; Srivannavit, O.; Zhou, T.; Gulari, E.; Gao, X. Individually addressable parallel peptide synthesis on microchips. *Nat. Biotechnol.* **2002**, *20*, 922–926. [CrossRef]
4. Carter, J.M. Epitope mapping of a protein using the Geysen (PEPSCAN) procedure. *Methods Mol. Biol.* **1994**, *36*, 207–223.
5. Van der Zee, R.; van Eden, W.; Meloen, R.H.; Noordzij, A.; van Embden, J.D. Efficient mapping and characterization of a T cell epitope by the simultaneous synthesis of multiple peptides. *Eur. J. Immunol.* **1989**, *19*, 43–47.
6. Geysen, H.M.; Rodda, S.J.; Mason, T.J.; Tribbick, G.; Schoofs, P.G. Strategies for epitope analysis using peptide synthesis. *J. Immunol. Methods* **1987**, *102*, 259–274. [CrossRef]
7. Geysen, H.M.; Rodda, S.J.; Mason, T.J. The delineation of peptides able to mimic assembled epitopes. In *Ciba Foundation Symposium 119—Synthetic Peptides as Antigens: Synthetic Peptides as Antigens, Volume 119*; Ciba Foundation: Glendale, CA, USA, 2007.
8. Geysen, H.M.; Barteling, S.J.; Meloen, R.H. Small peptides induce antibodies with a sequence and structural requirement for binding antigen comparable to antibodies raised against the native protein. *Proc. Natl. Acad. Sci. USA* **1985**, *82*, 178–182. [CrossRef]
9. Geysen, H.M.; Meloen, R.H.; Barteling, S.J. Use of peptide synthesis to probe viral antigens for epitopes to a resolution of a single amino acid. *Proc. Natl. Acad. Sci. USA* **1984**, *81*, 3998–4002. [CrossRef] [PubMed]
10. Meloen, R.H.; Langedijk, J.P.; Langeveld, J.P. Synthetic peptides for diagnostic use. *Vet. Q.* **1997**, *19*, 122–126. [CrossRef]
11. Barlow, D.J.; Edwards, M.S.; Thornton, J.M. Continuous and discontinuous protein antigenic determinants. *Nature* **1986**, *322*, 747–748. [CrossRef] [PubMed]
12. Songprakhon, P.; Thaingtamtanha, T.; Limjindaporn, T.; Puttikhunt, C.; Srisawat, C.; Luangaram, P.; Dechtawewat, T.; Uthaipibull, C.; Thongsima, S.; Yenchitsomanus, P.T.; et al. Peptides targeting dengue viral nonstructural protein 1 inhibit dengue virus production. *Sci. Rep.* **2020**, *10*, 12933. [CrossRef]
13. Bozovičar, K.; Bratkovič, T. Evolving a Peptide: Library Platforms and Diversification Strategies. *Int. J. Mol. Sci.* **2019**, *21*, 215. [CrossRef]
14. Brown, L.; Westby, M.; Souberbielle, B.E.; Szawlowski, P.W.; Kemp, G.; Hay, P.; Dalgleish, A.G. Optimisation of a peptide-based indirect ELISA for the detection of antibody in the serum of HIV-1 seropositive patients. *J. Immunol. Methods* **1997**, *200*, 79–88. [CrossRef]
15. Engvall, E.; Perlmann, P. Enzyme-linked Immunosorbent Assay (ELISA). Quantitative assay of immunoglobulin G. *Immunochemistry* **1971**, *8*, 871–874. [CrossRef]
16. Aydin, S. A short history, principles, and types of ELISA, and our laboratory experience with peptide/protein analyses using ELISA. *Peptides* **2015**, *72*, 4–15. [CrossRef]
17. Ekins, R.; Chu, F.; Micallef, J. High specific activity chemiluminescent and fluorescent markers: Their potential application to high sensitivity and 'multi-analyte' immunoassays. *J. Biolumin. Chemilumin.* **1989**, *4*, 59–78. [CrossRef]
18. Angenendt, P. Progress in protein and antibody microarray technology. *Drug Discov. Today* **2005**, *10*, 503–511. [CrossRef]
19. Newman, J.D.; Setford, S.J. Enzymatic biosensors. *Mol. Biotechnol.* **2006**, *32*, 249–268. [CrossRef]
20. Bhalla, N.; Jolly, P.; Formisano, N.; Estrela, P. Introduction to biosensors. *Essays Biochem.* **2016**, *60*, 1–8.
21. Sachdeva, S.; Davis, R.W.; Saha, A.K. Microfluidic Point-of-Care Testing: Commercial Landscape and Future Directions. *Front. Bioeng. Biotechnol.* **2020**, *8*, 602659. [CrossRef]
22. Pandey, S.; Dvorakova, M.C. Future Perspective of Diabetic Animal Models. *Endocr. Metab. Immune Disord. Drug Targets* **2020**, *20*, 25. [CrossRef] [PubMed]
23. Perestrelo, A.R.; Águas, A.C.; Rainer, A.; Forte, G. Microfluidic Organ/Body-on-a-Chip Devices at the Convergence of Biology and Microengineering. *Sensors* **2015**, *15*, 31142–31170. [CrossRef]
24. Vidova, V.; Spacil, Z. A review on mass spectrometry-based quantitative proteomics: Targeted and data independent acquisition. *Anal. Chim. Acta* **2017**, *964*, 7–23. [CrossRef]
25. Gresch, S.C.; Mutch, L.A.; Janecek, J.L.; Hegstad-Davies, R.L.; Graham, M.L. Cross-validation of commercial enzyme-linked immunosorbent assay and radioimmunoassay for porcine C-peptide concentration measurements in non-human primate serum. *Xenotransplantation* **2017**, *24*, e12320. [CrossRef]
26. Graham, M.L.; Gresch, S.C.; Hardy, S.K.; Mutch, L.A.; Janecek, J.L.; Hegstad-Davies, R.L. Evaluation of commercial ELISA and RIA for measuring porcine C-peptide: Implications for research. *Xenotransplantation* **2015**, *22*, 62–69. [CrossRef]
27. Lau, M.S.; Mooney, P.D.; White, W.L.; Rees, M.A.; Wong, S.H.; Hadjivassiliou, M.; Green, P.H.R.; Lebwohl, B.; Sanders, D.S. Office-Based Point of Care Testing (IgA/IgG-Deamidated Gliadin Peptide) for Celiac Disease. *Am. J. Gastroenterol.* **2018**, *113*, 1238–1246. [CrossRef] [PubMed]
28. Liu, M.; Zhao, G.; Wei, B.F. Attenuated serum vasoactive intestinal peptide concentrations are correlated with disease severity of non-traumatic osteonecrosis of femoral head. *J. Orthop. Surg. Res.* **2021**, *16*, 325. [CrossRef]
29. De-Simone, S.G.; Gomes, L.R.; Napoleão-Pêgo, P.; Lechuga, G.C.; de Pina, J.S.; Epitope, F.R.D. Mapping of the Diphtheria Toxin and Development of an ELISA-Specific Diagnostic Assay. *Vaccines* **2021**, *9*, 313. [CrossRef] [PubMed]

30. Gupta, K.; Brown, L.; Bakshi, R.K.; Press, C.G.; Chi, X.; Gorwitz, R.J.; Papp, J.R.; Geisler, W.M. Performance of Chlamydia trachomatis OmcB Enzyme-Linked Immunosorbent Assay in Serodiagnosis of Chlamydia trachomatis Infection in Women. *J. Clin. Microbiol.* **2018**, *56*, e00275-18. [CrossRef] [PubMed]
31. Rahman, K.S.; Darville, T.; Russell, A.N.; O'Connell, C.M.; Wiesenfeld, H.C.; Hillier, S.L.; Chowdhury, E.U.; Juan, Y.C.; Kaltenboeck, B. Discovery of Human-Specific Immunodominant Chlamydia trachomatis B Cell Epitopes. *Msphere* **2018**, *3*, e00246-18. [CrossRef]
32. Rahman, K.S.; Darville, T.; Russell, A.N.; O'Connell, C.M.; Wiesenfeld, H.C.; Hillier, S.L.; Lee, D.E.; Kaltenboeck, B. Comprehensive Molecular Serology of Human Chlamydia trachomatis Infections by Peptide Enzyme-Linked Immunosorbent Assays. *Msphere* **2018**, *3*, e00253-18. [CrossRef] [PubMed]
33. Rahman, K.S.; Darville, T.; Wiesenfeld, H.C.; Hillier, S.L.; Kaltenboeck, B. Mixed Chlamydia trachomatis Peptide Antigens Provide a Specific and Sensitive Single-Well Colorimetric Enzyme-Linked Immunosorbent Assay for Detection of Human Anti-C. trachomatis Antibodies. *Msphere* **2018**, *3*, e00484-18. [CrossRef] [PubMed]
34. Van Kruiningen, H.J.; Helal, Z.; Leroyer, A.; Garmendia, A.; Gower-Rousseau, C. ELISA Serology for Antibodies Against Chlamydia trachomatis in Crohn's Disease. *Gastroenterol. Res.* **2017**, *10*, 334–338. [CrossRef]
35. Mosadeghi, P.; Heydari-Zarnagh, H. Development and Evaluation of a Novel ELISA for Detection of Antibodies against HTLV-I Using Chimeric Peptides. *Iran. J. Allergy Asthma Immunol.* **2018**, *17*, 144–150.
36. Li, Y.; Lai, D.Y.; Lei, Q.; Xu, Z.W.; Wang, F.; Hou, H.; Chen, L.; Wu, J.; Ren, Y.; Ma, M.L.; et al. Systematic evaluation of IgG responses to SARS-CoV-2 spike protein-derived peptides for monitoring COVID-19 patients. *Cell Mol. Immunol.* **2021**, *18*, 621–631. [CrossRef] [PubMed]
37. Liu, J.; Cui, D.; Jiang, Y.; Li, Y.; Liu, Z.; Tao, L.; Zhao, Q.; Diao, A. Selection and characterization of a novel affibody peptide and its application in a two-site ELISA for the detection of cancer biomarker alpha-fetoprotein. *Int. J. Biol. Macromol.* **2021**, *166*, 884–892. [CrossRef]
38. Sahin, D.; Taflan, S.O.; Yartas, G.; Ashktorab, H.; Smoot, D.T. Screening and Identification of Peptides Specifically Targeted to Gastric Cancer Cells from a Phage Display Peptide Library. *Asian Pac. J. Cancer Prev.* **2018**, *19*, 927–932.
39. Liu, Y.; Xia, X.; Wang, Y.; Li, X.; Zhou, G.; Liang, H.; Feng, G.; Zheng, C. Screening and identification of a specific peptide for targeting hypoxic hepatoma cells. *Mol. Cell Probes* **2016**, *30*, 246–253. [CrossRef]
40. Zhang, W.J.; Sui, Y.X.; Budha, A.; Zheng, J.B.; Sun, X.J.; Hou, Y.C.; Wang, T.D.; Lu, S.Y. Affinity peptide developed by phage display selection for targeting gastric cancer. *World J. Gastroenterol.* **2012**, *18*, 2053–2060. [CrossRef]
41. Galvis-Jiménez, J.M.; Curtidor, H.; Patarroyo, M.A.; Monterrey, P.; Ramírez-Clavijo, S.R. Mammaglobin peptide as a novel biomarker for breast cancer detection. *Cancer Biol. Ther.* **2013**, *14*, 327–332. [CrossRef] [PubMed]
42. Wettergren, A.; Wøjdemann, M.; Holst, J.J. The inhibitory effect of glucagon-like peptide-1 (7-36)amide on antral motility is antagonized by its N-terminally truncated primary metabolite GLP-1 (9-36)amide. *Peptides* **1998**, *19*, 877–882. [CrossRef]
43. Wewer Albrechtsen, N.J.; Asmar, A.; Jensen, F.; Törang, S.; Simonsen, L.; Kuhre, R.E.; Asmar, M.; Veedfald, S.; Plamboeck, A.; Knop, F.K.; et al. A sandwich ELISA for measurement of the primary glucagon-like peptide-1 metabolite. *Am. J. Physiol. Endocrinol. Metab.* **2017**, *313*, E284–E291. [CrossRef]
44. Wewer Albrechtsen, N.J.; Bak, M.J.; Hartmann, B.; Christensen, L.W.; Kuhre, R.E.; Deacon, C.F.; Holst, J.J. Stability of glucagon-like peptide 1 and glucagon in human plasma. *Endocr. Connect* **2015**, *4*, 50–57. [CrossRef]
45. Wenzel, K.; Schulze-Rothe, S.; Müller, J.; Wallukat, G.; Haberland, A. Difference between beta1-adrenoceptor autoantibodies of human and animal origin-Limitations detecting beta1-adrenoceptor autoantibodies using peptide based ELISA technology. *PLoS ONE* **2018**, *13*, e0192615. [CrossRef] [PubMed]
46. Lv, R.; Chen, Y.; Xia, N.; Liang, Y.; He, Q.; Li, M.; Qi, Z.; Lu, Y.; Zhao, S. Development of a double-antibody sandwich ELISA for rapid detection to C-peptide in human urine. *J. Pharm. Biomed. Anal.* **2019**, *162*, 179–184. [CrossRef]
47. Velumani, S.; Ho, H.T.; He, F.; Musthaq, S.; Prabakaran, M.; Kwang, J. A novel peptide ELISA for universal detection of antibodies to human H5N1 influenza viruses. *PLoS ONE* **2011**, *6*, e20737. [CrossRef] [PubMed]
48. Tiwari, R.P.; Jain, A.; Khan, Z.; Kumar, P.; Bhrigu, V.; Bisen, P.S. Designing of novel antigenic peptide cocktail for the detection of antibodies to HIV-1/2 by ELISA. *J. Immunol. Methods* **2013**, *387*, 157–166. [CrossRef] [PubMed]
49. Van Burgel, N.D.; Brandenburg, A.; Gerritsen, H.J.; Kroes, A.C.; van Dam, A.P. High sensitivity and specificity of the C6-peptide ELISA on cerebrospinal fluid in Lyme neuroborreliosis patients. *Clin. Microbiol. Infect.* **2011**, *17*, 1495–1500. [CrossRef] [PubMed]
50. Davis, J.B. ELISA for Monitoring Nerve Growth Factor. *Methods Mol. Biol.* **2017**, *1606*, 141–147. [PubMed]
51. Shrock, E.; Fujimura, E.; Kula, T.; Timms, R.T.; Lee, I.H.; Leng, Y.; Robinson, M.L.; Sie, B.M.; Li, M.Z.; Chen, Y.; et al. Viral epitope profiling of COVID-19 patients reveals cross-reactivity and correlates of severity. *Science* **2020**, *370*, eabd4250. [CrossRef]
52. Zhang, Y.; Yang, Z.; Tian, S.; Li, B.; Feng, T.; He, J.; Jiang, M.; Tang, X.; Mei, S.; Li, H.; et al. A newly identified linear epitope on non-RBD region of SARS-CoV-2 spike protein improves the serological detection rate of COVID-19 patients. *BMC Microbiol.* **2021**, *21*, 194. [CrossRef]
53. Ma, Y.; Liu, F.; Lin, T.; Chen, L.; Jiang, A.; Tian, G.; Nielsen, M.; Wang, M. Large-scale identification of T cell epitopes derived from SARS-CoV-2 for the development of peptide vaccines against COVID-19. *J. Infect Dis.* **2021**. [CrossRef]
54. Huang, C.; Wang, Y.; Li, X.; Ren, L.; Zhao, J.; Hu, Y.; Zhang, L.; Fan, G.; Xu, J.; Gu, X.; et al. Clinical features of patients infected with 2019 novel coronavirus in Wuhan, China. *Lancet* **2020**, *395*, 497–506. [CrossRef]

55. Dong, Y.; Zhou, H.; Li, M.; Zhang, Z.; Guo, W.; Yu, T.; Gui, Y.; Wang, Q.; Zhao, L.; Luo, S.; et al. A novel simple scoring model for predicting severity of patients with SARS-CoV-2 infection. *Transbound Emerg. Dis.* **2020**, *67*, 2823–2829. [CrossRef]
56. Zhang, L.; Yan, X.; Fan, Q.; Liu, H.; Liu, X.; Liu, Z.; Zhang, Z. D-dimer levels on admission to predict in-hospital mortality in patients with COVID-19. *J. Thromb. Haemost.* **2020**, *18*, 1324–1329. [CrossRef] [PubMed]
57. Petruccioli, E.; Fard, S.N.; Navarra, A.; Petrone, L.; Vanini, V.; Cuzzi, G.; Gualano, G.; Pierelli, L.; Bertoletti, A.; Nicastri, E.; et al. Exploratory analysis to identify the best antigen and the best immune biomarkers to study SARS-CoV-2 infection. *J. Transl. Med.* **2021**, *19*, 272. [CrossRef] [PubMed]
58. Wrenn, F.R., Jr.; Good, M.L.; Handler, P. The use of positron-emitting radioisotopes for the localization of brain tumors. *Science* **1951**, *113*, 525–527. [CrossRef]
59. Ter-Pogossian, M.M.; Phelps, M.E.; Hoffman, E.J.; Mullani, N.A. A positron-emission transaxial tomograph for nuclear imaging (PETT). *Radiology* **1975**, *114*, 89–98. [CrossRef]
60. Di Chiro, G. Positron emission tomography using [18F] fluorodeoxyglucose in brain tumors. A powerful diagnostic and prognostic tool. *Investig. Radiol.* **1987**, *22*, 360–371. [CrossRef]
61. Patz, E.F., Jr.; Lowe, V.J.; Hoffman, J.M.; Paine, S.S.; Burrowes, P.; Coleman, R.E.; Goodman, P.C. Focal pulmonary abnormalities: Evaluation with F-18 fluorodeoxyglucose PET scanning. *Radiology* **1993**, *188*, 487–490. [CrossRef] [PubMed]
62. Signore, A.; Mather, S.J.; Piaggio, G.; Malviya, G.; Dierckx, R.A. Molecular imaging of inflammation/infection: Nuclear medicine and optical imaging agents and methods. *Chem. Rev.* **2010**, *110*, 3112–3145. [CrossRef] [PubMed]
63. Brazeau, P.; Vale, W.; Burgus, R.; Ling, N.; Butcher, M.; Rivier, J.; Guillemin, R. Hypothalamic polypeptide that inhibits the secretion of immunoreactive pituitary growth hormone. *Science* **1973**, *179*, 77–79. [CrossRef] [PubMed]
64. Narayanan, S.; Kunz, P.L. Role of somatostatin analogues in the treatment of neuroendocrine tumors. *J. Natl. Compr. Cancer Netw.* **2015**, *13*, 109–117. [CrossRef]
65. Patel, Y.C. Somatostatin and its receptor family. *Front. Neuroendocrinol.* **1999**, *20*, 157–198. [CrossRef] [PubMed]
66. Anzola, L.K.; Glaudemans, A.; Dierckx, R.; Martinez, F.A.; Moreno, S.; Signore, A. Somatostatin receptor imaging by SPECT and PET in patients with chronic inflammatory disorders: A systematic review. *Eur. J. Nucl. Med. Mol. Imaging* **2019**, *46*, 2496–2513. [CrossRef]
67. Sosabowsky, J.; Melendez-Alafort, L.; Mather, S. Radiolabelling of peptides for diagnosis and therapy of non-oncological diseases. *Q. J. Nucl. Med.* **2003**, *47*, 223–237.
68. Rambaldi, P.F.; Cuccurullo, V.; Briganti, V.; Mansi, L. The present and future role of (111)In pentetreotide in the PET era. *Q. J. Nucl. Med. Mol. Imaging* **2005**, *49*, 225–235. [PubMed]
69. Cascini, G.L.; Cuccurullo, V.; Tamburrini, O.; Rotondo, A.; Mansi, L. Peptide imaging with somatostatin analogues: More than cancer probes. *Curr. Radiopharm.* **2013**, *6*, 36–40. [CrossRef]
70. Patel, M.; Tena, I.; Jha, A.; Taieb, D.; Pacak, K. Somatostatin Receptors and Analogs in Pheochromocytoma and Paraganglioma: Old Players in a New Precision Medicine World. *Front. Endocrinol. (Lausanne)* **2021**, *12*, 625312. [CrossRef]
71. Yamaga, L.Y.I.; Cunha, M.L.; Neto, G.C.C.; Garcia, M.R.T.; Yang, J.H.; Camacho, C.P.; Wagner, J.; Funari, M.B.G. (68)Ga-DOTATATE PET/CT in recurrent medullary thyroid carcinoma: A lesion-by-lesion comparison with (111)In-octreotide SPECT/CT and conventional imaging. *Eur. J. Nucl. Med. Mol. Imaging* **2017**, *44*, 1695–1701. [CrossRef]
72. Johnbeck, C.B.; Knigge, U.; Loft, A.; Berthelsen, A.K.; Mortensen, J.; Oturai, P.; Langer, S.W.; Elema, D.R.; Kjaer, A. Head-to-Head Comparison of (64)Cu-DOTATATE and (68)Ga-DOTATOC PET/CT: A Prospective Study of 59 Patients with Neuroendocrine Tumors. *J. Nucl. Med.* **2017**, *58*, 451–457. [CrossRef]
73. Signore, A.; Annovazzi, A.; Barone, R.; Bonanno, E.; D'Alessandria, C.; Chianelli, M.; Mather, S.J.; Bottoni, U.; Panetta, C.; Innocenzi, D.; et al. 99mTc-interleukin-2 scintigraphy as a potential tool for evaluating tumor-infiltrating lymphocytes in melanoma lesions: A validation study. *J. Nucl. Med.* **2004**, *45*, 1647–1652.
74. Di Gialleonardo, V.; Signore, A.; Willemsen, A.T.; Sijbesma, J.W.; Dierckx, R.A.; de Vries, E.F. Pharmacokinetic modelling of N-(4-[(18)F]fluorobenzoyl)interleukin-2 binding to activated lymphocytes in an xenograft model of inflammation. *Eur. J. Nucl. Med. Mol. Imaging* **2012**, *39*, 1551–1560. [CrossRef] [PubMed]
75. Khanapur, S.; Yong, F.F.; Hartimath, S.V.; Jiang, L.; Ramasamy, B.; Cheng, P.; Narayanaswamy, P.; Goggi, J.L.; Robins, E.G. An Improved Synthesis of N-(4-[(18)F]Fluorobenzoyl)-Interleukin-2 for the Preclinical PET Imaging of Tumour-Infiltrating T-cells in CT26 and MC38 Colon Cancer Models. *Molecules* **2021**, *26*, 1728. [CrossRef] [PubMed]
76. Van der Veen, E.L.; Antunes, I.F.; Maarsingh, P.; Hessels-Scheper, J.; Zijlma, R.; Boersma, H.H.; Jorritsma-Smit, A.; Hospers, G.A.P.; de Vries, E.G.E.; Hooge, M.N.L.; et al. Clinical-grade N-(4-[(18)F]fluorobenzoyl)-interleukin-2 for PET imaging of activated T-cells in humans. *EJNMMI Radiopharm. Chem.* **2019**, *4*, 15. [CrossRef]
77. Van de Donk, P.P.; Wind, T.T.; Hooiveld-Noeken, J.S.; van der Veen, E.L.; Glaudemans, A.; Diepstra, A.; Jalving, M.; de Vries, E.G.E.; de Vries, E.F.J.; Hospers, G.A.P. Interleukin-2 PET imaging in patients with metastatic melanoma before and during immune checkpoint inhibitor therapy. *Eur. J. Nucl. Med. Mol. Imaging* **2021**. [CrossRef]
78. Grauer, L.S.; Lawler, K.D.; Marignac, J.L.; Kumar, A.; Goel, A.S.; Wolfert, R.L. Identification, purification, and subcellular localization of prostate-specific membrane antigen PSM' protein in the LNCaP prostatic carcinoma cell line. *Cancer Res.* **1998**, *58*, 4787–4789. [PubMed]
79. Mannweiler, S.; Amersdorfer, P.; Trajanoski, S.; Terrett, J.A.; King, D.; Mehes, G. Heterogeneity of prostate-specific membrane antigen (PSMA) expression in prostate carcinoma with distant metastasis. *Pathol. Oncol. Res.* **2009**, *15*, 167–172. [CrossRef]

80. Afshar-Oromieh, A.; Haberkorn, U.; Eder, M.; Eisenhut, M.; Zechmann, C.M. [68Ga]Gallium-labelled PSMA ligand as superior PET tracer for the diagnosis of prostate cancer: Comparison with 18F-FECH. *Eur. J. Nucl. Med. Mol. Imaging* **2012**, *39*, 1085–1086. [CrossRef]
81. Grubmüller, B.; Baltzer, P.; Hartenbach, S.; D'Andrea, D.; Helbich, T.H.; Haug, A.R.; Goldner, G.M.; Wadsak, W.; Pfaff, S.; Mitterhauser, M.; et al. PSMA Ligand PET/MRI for Primary Prostate Cancer: Staging Performance and Clinical Impact. *Clin. Cancer Res.* **2018**, *24*, 6300–6307. [CrossRef]
82. Liu, D.; Cheng, G.; Ma, X.; Wang, S.; Zhao, X.; Zhang, W.; Yang, W.; Wang, J. PET/CT using (68) Ga-PSMA-617 versus (18) F-fluorodeoxyglucose to differentiate low- and high-grade gliomas. *J. Neuroimaging* **2021**, *31*, 733–742. [CrossRef] [PubMed]
83. Usmani, S.; Al-Turkait, D.; Al-Kandari, F.; Ahmed, N. Thyroid Cancer Detected on 68Ga-PMSA PET/CT. *J. Pak. Med. Assoc.* **2021**, *71*, 1511–1512. [PubMed]
84. Gündoğan, C.; Ergül, N.; Çakır, M.S.; Kılıçkesmez, Ö.; Gürsu, R.U.; Aksoy, T.; Çermik, T.F. (68)Ga-PSMA PET/CT Versus (18)F-FDG PET/CT for Imaging of Hepatocellular Carcinoma. *Mol. Imaging Radionucl. Ther.* **2021**, *30*, 79–85. [CrossRef]
85. Zhao, Q.; Yang, B.; Dong, A.; Zuo, C. 68Ga-PSMA-11 PET/CT in Isolated Bilateral Adrenal Metastases From Prostate Adenocarcinoma. *Clin. Nucl. Med.* **2021**. [CrossRef]
86. Holzgreve, A.; Biczok, A.; Ruf, V.C.; Liesche-Starnecker, F.; Steiger, K.; Kirchner, M.A.; Unterrainer, M.; Mittlmeier, L.; Herms, J.; Schlegel, J.; et al. PSMA Expression in Glioblastoma as a Basis for Theranostic Approaches: A Retrospective, Correlational Panel Study Including Immunohistochemistry, Clinical Parameters and PET Imaging. *Front. Oncol.* **2021**, *11*, 646387. [CrossRef]
87. Veerasuri, S.; Redman, S.; Graham, R.; Meehan, C.; Little, D. Non-prostate uptake on (18)F-PSMA-1007 PET/CT: A case of myeloma. *BJR Case Rep.* **2021**, *7*, 20200102. [PubMed]
88. Sakthivel, P.; Kumar, A.; Arunraj, S.T.; Singh, C.A.; Kumar, R. 68Ga-PSMA PET/CT Scan on Postoperative Assessment of Sinonasal Glomangiopericytoma. *Clin. Nucl. Med.* **2021**, *46*, e478–e479.
89. Li, R.; Li, D.; Li, X.; Zuo, C.; Cheng, C. The Appearance of Sjögren Syndrome on 68Ga-PSMA-11 PET/CT. *Clin. Nucl. Med.* **2021**, *46*, 517–519. [CrossRef]
90. Tumedei, M.M.; Ravaioli, S.; Matteucci, F.; Celli, M.; de Giorgi, U.; Gunelli, R.; Puccetti, M.; Paganelli, G.; Bravaccini, S. Spotlight on PSMA as a new theranostic biomarker for bladder cancer. *Sci. Rep.* **2021**, *11*, 9777. [CrossRef] [PubMed]
91. Zitzmann, S.; Ehemann, V.; Schwab, M. Arginine-glycine-aspartic acid (RGD)-peptide binds to both tumor and tumor-endothelial cells in vivo. *Cancer Res.* **2002**, *62*, 5139–5143.
92. Haubner, R.; Wester, H.J.; Reuning, U.; Senekowitsch-Schmidtke, R.; Diefenbach, B.; Kessler, H.; Stöcklin, G.; Schwaiger, M. Radiolabeled alpha(v)beta3 integrin antagonists: A new class of tracers for tumor targeting. *J. Nucl. Med.* **1999**, *40*, 1061–1071. [PubMed]
93. Haubner, R.; Kuhnast, B.; Mang, C.; Weber, W.A.; Kessler, H.; Wester, H.J.; Schwaiger, M. [18F]Galacto-RGD: Synthesis, radiolabeling, metabolic stability, and radiation dose estimates. *Bioconjug. Chem.* **2004**, *15*, 61–69. [CrossRef] [PubMed]
94. Haubner, R.; Weber, W.A.; Beer, A.J.; Vabuliene, E.; Reim, D.; Sarbia, M.; Becker, K.F.; Goebel, M.; Hein, R.; Wester, H.J.; et al. Noninvasive visualization of the activated alphavbeta3 integrin in cancer patients by positron emission tomography and [18F]Galacto-RGD. *PLoS Med.* **2005**, *2*, e70. [CrossRef]
95. Makowski, M.R.; Rischpler, C.; Ebersberger, U.; Keithahn, A.; Kasel, M.; Hoffmann, E.; Rassaf, T.; Kessler, H.; Wester, H.J.; Nekolla, S.G.; et al. Multiparametric PET and MRI of myocardial damage after myocardial infarction: Correlation of integrin $\alpha v\beta 3$ expression and myocardial blood flow. *Eur. J. Nucl. Med. Mol. Imaging* **2021**, *48*, 1070–1080. [CrossRef]
96. Bach-Gansmo, T.; Danielsson, R.; Saracco, A.; Wilczek, B.; Bogsrud, T.V.; Fangberget, A.; Tangerud, A.; Tobin, D. Integrin receptor imaging of breast cancer: A proof-of-concept study to evaluate 99mTc-NC100692. *J. Nucl. Med.* **2006**, *47*, 1434–1439. [PubMed]
97. Mena, E.; Owenius, R.; Turkbey, B.; Sherry, R.; Bratslavsky, G.; Macholl, S.; Miller, M.P.; Somer, E.J.; Lindenberg, L.; Adler, S.; et al. [^{18}F]fluciclatide in the in vivo evaluation of human melanoma and renal tumors expressing $\alpha v\beta$ 3 and α vβ 5 integrins. *Eur. J. Nucl. Med. Mol. Imaging* **2014**, *41*, 1879–1888. [CrossRef]
98. Chianelli, M.; Boerman, O.C.; Malviya, G.; Galli, F.; Oyen, W.J.; Signore, A. Receptor binding ligands to image infection. *Curr. Pharm. Des.* **2008**, *14*, 3316–3325. [CrossRef]
99. Elvas, F.; Berghe, T.V.; Adriaenssens, Y.; Vandenabeele, P.; Augustyns, K.; Staelens, S.; Stroobants, S.; van der Veken, P.; Wyffels, L. Caspase-3 probes for PET imaging of apoptotic tumor response to anticancer therapy. *Org. Biomol. Chem.* **2019**, *17*, 4801–4824. [CrossRef]
100. Kraus, S.; Dierks, A.; Rasche, L.; Kertels, O.; Kircher, M.; Schirbel, A.; Zovko, J.; Steinbrunn, T.; Tibes, R.; Wester, H.J.; et al. (68)Ga-Pentixafor-PET/CT imaging represents a novel approach to detect chemokine receptor CXCR4 expression in myeloproliferative neoplasms. *J. Nucl. Med.* **2021**, *121*, 262206.
101. Cojoc, M.; Peitzsch, C.; Trautmann, F.; Polishchuk, L.; Telegeev, G.D.; Dubrovska, A. Emerging targets in cancer management: Role of the CXCL12/CXCR4 axis. *Onco Targets Ther.* **2013**, *6*, 1347–1361.
102. Burger, J.A.; Peled, A. CXCR4 antagonists: Targeting the microenvironment in leukemia and other cancers. *Leukemia* **2009**, *23*, 43–52. [CrossRef]
103. Zhang, J.; Niu, G.; Lang, L.; Li, F.; Fan, X.; Yan, X.; Yao, S.; Yan, W.; Huo, L.; Chen, L.; et al. Clinical Translation of a Dual Integrin $\alpha v\beta 3$- and Gastrin-Releasing Peptide Receptor-Targeting PET Radiotracer, 68Ga-BBN-RGD. *J. Nucl. Med.* **2017**, *58*, 228–234. [CrossRef] [PubMed]

104. Gyuricza, B.; Szabó, J.P.; Arató, V.; Szücs, D.; Vágner, A.; Szikra, D.; Fekete, A. Synthesis of Novel, Dual-Targeting (68)Ga-NODAGA-LacN-E[c(RGDfK)](2) Glycopeptide as a PET Imaging Agent for Cancer Diagnosis. *Pharmaceutics* **2021**, *13*, 796. [CrossRef] [PubMed]
105. Fields, G.B. Introduction to peptide synthesis. *Curr. Protoc. Protein Sci.* **2002**, *69*, 18. [CrossRef] [PubMed]
106. Mueller, L.K.; Baumruck, A.C.; Zhdanova, H.; Tietze, A.A. Challenges and Perspectives in Chemical Synthesis of Highly Hydrophobic Peptides. *Front. Bioeng. Biotechnol.* **2020**, *8*, 162. [CrossRef]
107. Kochendoerfer, G.G.; Kent, S.B. Chemical protein synthesis. *Curr. Opin. Chem. Biol.* **1999**, *3*, 665–671. [CrossRef]
108. Sakamoto, S.; Putalun, W.; Vimolmangkang, S.; Phoolcharoen, W.; Shoyama, Y.; Tanaka, H.; Morimoto, S. Enzyme-linked immunosorbent assay for the quantitative/qualitative analysis of plant secondary metabolites. *J. Nat. Med.* **2018**, *72*, 32–42. [CrossRef]
109. European Medicines Agency, ICH Topic Q 2 (R1). Available online: https://www.ema.europa.eu/en/documents/scientific-guideline/ich-q-2-r1-validation-analytical-procedures-text-methodology-step-5_en.pdf (accessed on 12 August 2021).
110. Sun, X.; Li, Y.; Liu, T.; Li, Z.; Zhang, X.; Chen, X. Peptide-based imaging agents for cancer detection. *Adv. Drug Deliv. Rev.* **2017**, *110–111*, 38–51. [CrossRef] [PubMed]

International Journal of *Molecular Sciences*

MDPI

Review

Peptide-Assisted Nucleic Acid Delivery Systems on the Rise

Shabnam Tarvirdipour [1,2], Michal Skowicki [1,3], Cora-Ann Schoenenberger [1,3,*] and Cornelia G. Palivan [1,3,*]

1 Department of Chemistry, University of Basel, Mattenstrasse 24a, 4058 Basel, Switzerland; shabnam.tarvirdipour@unibas.ch (S.T.); michaljerzy.skowicki@unibas.ch (M.S.)
2 Department of Biosystem Science and Engineering, ETH Zurich, Mattenstrasse 26, 4058 Basel, Switzerland
3 NCCR-Molecular Systems Engineering, BPR1095, Mattenstrasse 24a, 4058 Basel, Switzerland
* Correspondence: cora-ann.schoenenberger@unibas.ch (C.-A.S.); cornelia.palivan@unibas.ch (C.G.P.)

Abstract: Concerns associated with nanocarriers' therapeutic efficacy and side effects have led to the development of strategies to advance them into targeted and responsive delivery systems. Owing to their bioactivity and biocompatibility, peptides play a key role in these strategies and, thus, have been extensively studied in nanomedicine. Peptide-based nanocarriers, in particular, have burgeoned with advances in purely peptidic structures and in combinations of peptides, both native and modified, with polymers, lipids, and inorganic nanoparticles. In this review, we summarize advances on peptides promoting gene delivery systems. The efficacy of nucleic acid therapies largely depends on cell internalization and the delivery to subcellular organelles. Hence, the review focuses on nanocarriers where peptides are pivotal in ferrying nucleic acids to their site of action, with a special emphasis on peptides that assist anionic, water-soluble nucleic acids in crossing the membrane barriers they encounter on their way to efficient function. In a second part, we address how peptides advance nanoassembly delivery tools, such that they navigate delivery barriers and release their nucleic acid cargo at specific sites in a controlled fashion.

Keywords: amphiphilic peptides; non-viral gene delivery; nanocarrier; peptide self-assemblies; stimuli responsive

Citation: Tarvirdipour, S.; Skowicki, M.; Schoenenberger, C.-A.; Palivan, C.G. Peptide-Assisted Nucleic Acid Delivery Systems on the Rise. *Int. J. Mol. Sci.* **2021**, *22*, 9092. https://doi.org/10.3390/ijms22169092

Academic Editors: Nunzianna Doti and Menotti Ruvo

Received: 20 July 2021
Accepted: 19 August 2021
Published: 23 August 2021

1. Introduction

Introducing exogenous nucleic acids into human target cells has been receiving a great deal of attention for the treatment of several human diseases, in particular cancer and other genetic disorders. Quite recently, a new treatment involving gene editing CRISPER has made a mark by using mRNA encoding Cas [1,2]. In face of the worldwide coronavirus pandemic, mRNA has moved into the limelight as vaccine and many companies are working on other mRNA vaccines and therapeutics [3,4]. Both vaccines and disease intervention involve delivering nucleic acids to intracellular locations on a path strewn with obstacles. To ultimately accomplish modification of protein expression by replacing or adding missing or defective genes, regulating gene expression at the RNA level (e.g., gene silencing by RNA interference, modification of RNA processing), controlling microRNA activity or by genome editing and reprogramming of cells, nucleic acids face a number of challenging barriers. Hence, despite a broad range of possible therapeutic approaches, the clinical success of gene therapy has yet to meet the expectations. The lack of efficacy and issues with clinical safety, in particular with viral vectors, which make up about 70% of vectors used in gene therapy, are the main reasons gene delivery systems fail in clinical trials [5,6]. This has led to the emergence of non-viral vector systems, such as liposomes and polymer supramolecular assemblies with better biological safety. However, their efficacy is predominantly hampered by insufficient localization of the therapeutic agents at the site of interest, both at the extracellular and intracellular level [7]. Owing to their remarkable potency, selectivity and low toxicity, peptides offer ideal alternatives to overcome these hurdles [8]. In addition, advancements in nanosystems continue to open new avenues for an efficient delivery of therapeutics and, thus, nanotechnology has become a favored tool in medicine [7].

Nanocarriers based on their size, shape, charge, and surface chemistry are internalized by target cells through different pathways including clathrin-mediated endocytosis, caveolae- or cholesterol-mediated endocytosis, phagocytosis, and macropinocytosis [9,10]. After entering cells by endocytosis, nanocarriers usually remain sequestered in corresponding transport vesicles and their fate depends on the endocytic pathway but also on the physicochemical properties of the nanocarriers. Endosomal sequestration consists of multiple membrane fusions, in which the endocytic vesicles sequentially merge with early and late endosomes, proceeding all the way to the lysosomal compartment [10]. A constant decrease in intravesicular pH and increase in digestive enzymatic content throughout this pathway have a major impact on the stability of payloads and, subsequently, on efficacy [10,11]. These limitations have led to the search for strategies that can properly protect the macromolecular drugs from degradation and specifically target the major subcellular compartments. Furthermore, a boost of discovery research for the better understanding of intracellular trafficking routes highlight the need for carriers that overcome the barriers associated with the delivery to the intracellular site of action [11].

The major shortcomings of most commonly used non-viral nucleic acid delivery systems, such as lipoplexes and polyplexes include nonspecific distribution, inefficient cytoplasmic delivery, and organelle targeting. In contrast, peptide-based nanocarriers, e.g., peptide nanoparticles, also called peptiplexes, or peptidic multicompartment micelles, and nano-assemblies equipped with peptides hold great promise as delivery platforms, since they can be tweaked to facilitate penetration of cell membranes and to localize to distinct subcellular compartments. In addition, peptides are easy to synthesize with a desired bioactivity, and, by multivalent presence, endow the nanocarrier with high avidity for the target [10,12]. Owing to the highly specific targeting capacity of corresponding peptides, therapeutic nanocarriers are able to pass through the cell membrane and reach the specific tissue and cells which results in enhanced intracellular distribution and extended therapeutic window [13]. Furthermore, smart delivery systems are promising options to provide solutions related to uncontrolled release of payloads: besides a biocompatible nanocarrier and suitable targeting moieties, these platforms include stimulus-responsive elements which endow them with triggered cargo release [14].

The concept of using peptides as targeting moieties for therapeutic and diagnostic purposes has created new avenues for modern pharmaceutical industries [13,15]. Although clinical progress in the application of peptides, alone or combined with nano-assemblies, is slowly moving forward, large investments and wide-ranging research efforts confirm their promising potential as a delivery platform for therapeutic systems. Increased interest in smart nanocarrier design with particular focus on, but not limited to, cancer therapy with the aim of precision medicine application has boosted this unique class of pharmaceutical compounds into high demand [16–18].

In this review, we discuss various types of membrane active and stimuli responsive peptides with regard to their role in refining different nanocarriers for gene delivery applications. As peptides take center stage, we do not cover predominantly lipidic nor inorganic nanoparticle gene delivery systems. We describe properties of peptides that promote site-specific localization of nucleic acids and of peptide-based nano-assemblies. Then, we lay the emphasis on peptide designs that confer stimuli-responsiveness upon nanosystems with the aim to control payload release. Targeting, controlling, and stimuli-responsive peptides advance nanosystems from non-specific carriers of nucleic acids to smart site-specific gene delivery systems.

2. Peptide-Guided Delivery of Nucleic Acids across Biological Barriers

Membrane active peptides interact with cellular membranes by traversing them, disrupting them or by residing at the membrane interface and fusing with them [19]. They are known to overcome site-specific delivery barriers and facilitate intracellular delivery of various bioactive cargos with low cytotoxicity [19,20]. Although there is a wide variety of membrane-active peptides, here we mainly discuss peptides for targeting nucleic acid delivery systems to specific cells and tissues, and peptides that assist in the delivery of nanocarriers across membrane barriers, such as cell penetrating peptides (CPPs), peptides facilitating endosomal escape and those that target nanocarriers to subcellular organelles (Figure 1).

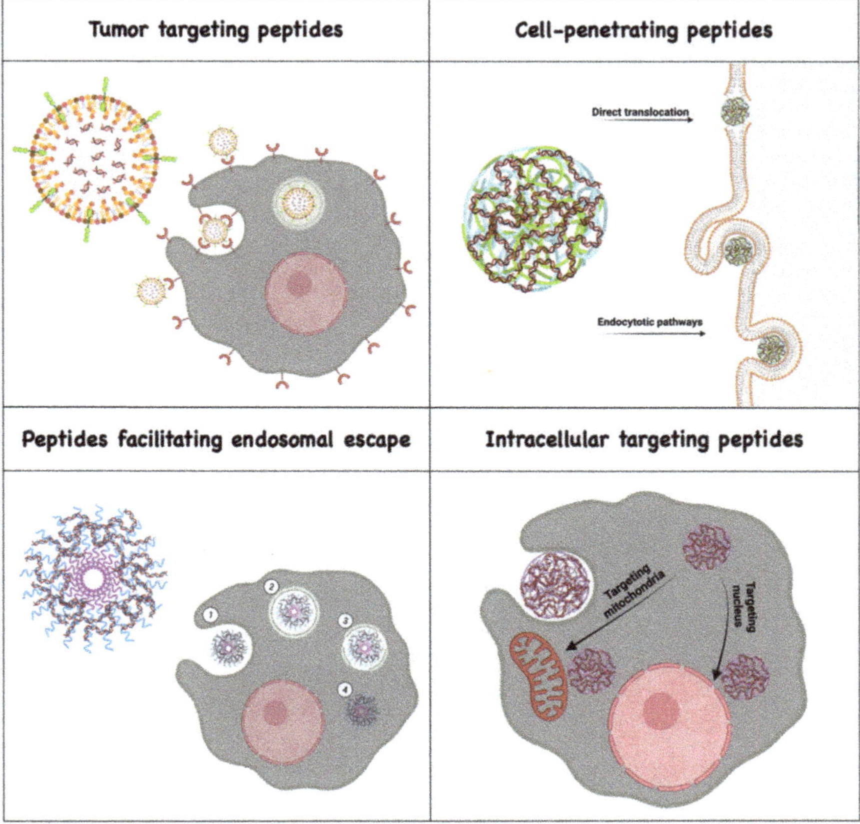

Figure 1. Classes of membrane active peptides facilitating the delivery of nucleic acid across biological barriers. Created with BioRender.com (Access to BioRender: June–July).

2.1. Tumor-Targeting Peptides

The ability of peptides to mediate translocation across membranes, traffic to desired sites, as well as executing many fundamental cellular functions made them promising candidates for targeting [21]. Owing to the high mortality related to cancer, substantial research investments have been made over the past decades in order to develop specific cancer diagnostics and treatments that improve survival rate [22]. The aberrant proliferation of tumor cells, accompanied by the up-regulation of their molecular markers result in high levels of specific receptors in the tumor and its microenvironment [23]. Thus, tumor-targeted delivery methods incorporate peptides or antibodies that are selective to the receptors overexpressed on the tumors [24]. Selective targeting of these tumor-associated markers promises the accurate targeting of signaling pathways that are dysregulated in the tumor [25].

Although the use of antibodies to target tumors has become highly successful both in tumor diagnosis and therapy, some deficiencies associated with antibodies, such as inadequate pharmacokinetics and limited tissue accessibility, as well as impaired interactions with the immune system limit their clinical application. Compared to antibodies and other tumor-targeting ligands, peptides offer better cell or tissue penetration, high affinity and targeting specificity, low immunogenicity, high stability, and improved pharmacokinetics by chemical modifications [26]. Tumor-targeting peptides, usually comprising less than 50 amino acids, are synthesized naturally or artificially [27,28]. For example, peptide sequences containing an arginine-glycine-aspartic acid (RGD) motif are among the most prominent targeting moieties for non-viral delivery systems [29]. The strong affinity of the RGD motif for integrin receptors expressed on vascular endothelial cells and overexpressed on many cancer cells [30] facilitates cell attachment and uptake of nanocarriers by receptor-mediated endocytosis (Figure 2) [31].

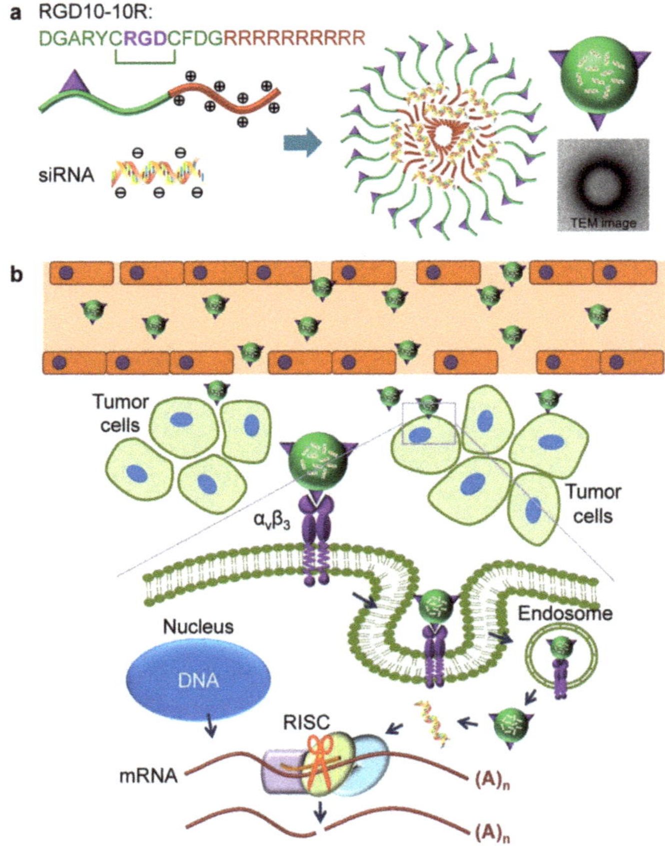

Figure 2. Schematic representation of (**a**) fabrication of the RGD10-10R/siRNA complex, (**b**) tumor-targeted siRNA delivery involving ligand/receptor interactions. siRNAs accumulated in the tumor tissue and then entered the tumor cells in a receptor (αvβ3)-mediated endocytosis (RME) manner in vitro. After being internalized by cells, peptide/siRNA complexes escaped from the endosomes/lysosomes. Then, siRNAs were released from the complexes and loaded by RNA-induced silencing complex (RISC). Targeted messenger RNA complementary to the guide strand (antisense strand) of siRNA was selected and cleaved by argonaute protein. Reprinted with permission from [31]. Copyright 2015 Springer Nature.

Likewise, the synthetic nonapeptide LyP-1 is an example of a tumor targeting peptide that can selectively bind to its primary receptor p32 protein overexpressed in various tumor-associated cells and atherosclerotic plaque macrophages [32]. Binding leads to proteolytic cleavage of LyP-1 into a truncated version whose exposed C-terminal CendR motif becomes active and triggers binding to NRP1 and/or NRP2 cell surface receptors [32,33]. This interaction promotes cellular internalization of LyP-1 and its bioconjugates. NRP1/2 also mediates transfer to the nucleus, which makes LyP 1-based delivery systems more effective in imaging and treatment of diseases [34]. An overview of different tumor-targeting peptides developed for cancer gene therapy is presented in Table 1.

Table 1. Examples of peptides used for targeting in cancer gene therapy.

Peptide Name	Cargo	Cancer Type	Ref.
RGD	siRNA	breast	[35]
cRGD	siRNA	brain	[36]
	siRNA	skin	[37]
iRGD	siRNA	pancreatic	[38]
	siRNA	lung	[39]
RGDfC	siRNA and doxorubicin	liver	[40]
CRGDK	siRNA and BAplatin	breast	[41]
CGKRK	siRNA	breast and brain	[42]
KTLLPTP	siRNA and paclitaxel	pancreatic	[43]
HAIYPRH	siRNA and doxorubicin	breast	[44]
LyP-1 and iRGD	siRNA	ovarian	[45]
YHWYGYTPQNVI	siRNA	liver	[46]
T7	pDNA	bone	[47]

2.2. Cell-Penetrating Peptides

Cell-penetrating peptides (CPPs) are short peptides (less than 30 amino acids) derived from naturally occurring proteins, designed de novo or a combination of both [48]. CPPs by virtue of their ability to permeate the cell membrane in an innocuous manner provided a means for successful cellular entry and intracellular trafficking of a wide variety of cargos including nucleic acids (Figure 3) [49–53]. In addition to sequence length, charge and amphipathicity are the main structural parameters determining internalization but also cargo interactions. Penetration of nucleic acids across the cell membranes is a key step in gene delivery and paves the way for an efficient gene therapy [52]. Nucleic acids can be conjugated to CPPs, either by non-covalent complex formation or by covalent bonds [54]. CPPs promote the intracellular distribution of these membrane-impermeable therapeutic molecules without destroying the integrity of cellular membranes and, thus, widen the therapeutic window of cargos [13].

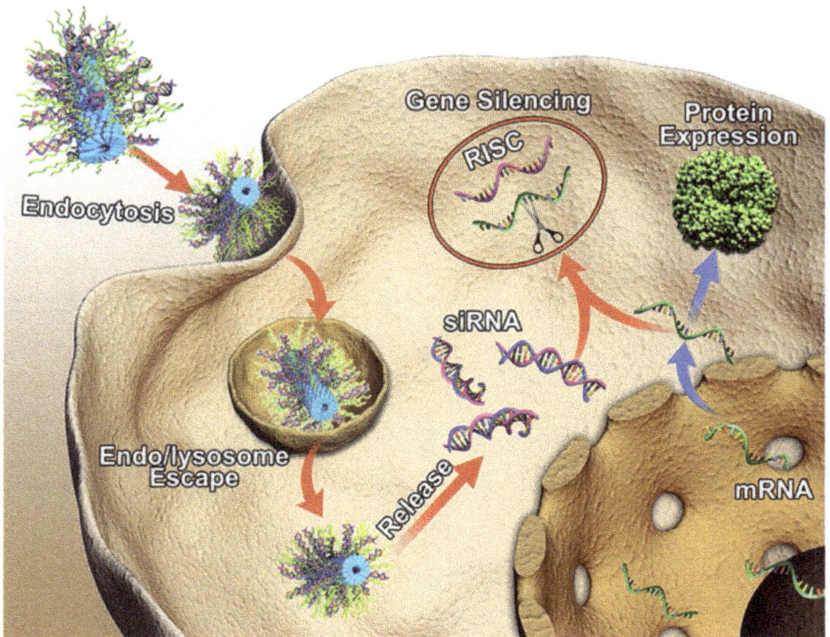

Figure 3. Schematic illustration of cell penetrating TAT peptides complexed with siRNA and integrated into modified tobacco mosaic virus (TMV) for virus-inspired gene silencing. Reprinted with permission from [52]. Copyright 2018 American Chemical Society.

CPPs can be classified according to their physicochemical properties as cationic, amphipathic, and hydrophobic, which largely impacts the type of cell-membrane interactions and uptake mechanism [55]. Extensive literature is available on the structure–activity relationship of CPPs [56–59]. Examples of CPPs classified according to their physicochemical properties and the genetic cargo they delivered are summarized in Table 2.

Table 2. CPP classification based on physicochemical properties.

Cell-Penetrating Peptides								
Cationic			**Amphipathic**			**Hydrophobic**		
Name(origin)	**Cargo**	**Ref.**	**Name(origin)**	**Cargo**	**Ref.**	**Name(origin)**	**Cargo**	**Ref.**
Diatos Peptide Vectors (DPV)	siRNA	[60]	MPG	pDNA siRNA	[61,62]	C105Y	pDNA	[63–66]
HIV-1 twinarginine translocation (TAT)	pDNA siRNA	[67–74]	Transportan	pDNA siRNA	[72,75,76]	K-FGF	pDNA	[77]
arginine-rich peptides	pDNA siRNA	[78–82]	NickFect (NF)	pDNA siRNA	[83–86]	Bip	pDNA	[87]
Polyarginine	pDNA	[75,76,88–90]	PepFect (PF)	pDNA mRNA	[91,92]	Melittin-derived peptides	siRNA	[93]
Penetratin	pDNA	[94–96]	MAP	siRNA	[97]			
L5a	pDNA	[98,99]	Crotamine	pDNA	[100–102]			
			VP22	pDNA	[103,104]			
Protamine	pDNA mRNA	[105–109]	Antennapedia (Antp)	AON siRNA	[110,111]			
			Pep-1	pDNA	[112,113]			
			CADY	siRNA	[114–116]			
			FGF	pDNA	[77]			
			pVEC	pDNA	[117,118]			

Cationic CPPs show a high affinity for negatively charged cell membranes because of electrostatic interactions and, thus, internalize into the cell through a receptor-independent mechanism. The key factors determining the activity of cationic CPPs are the number and position of positively charged amino acids in their structure [57]. TAT and penetratin, the first cationic CPPs discovered, have been widely used to promote cellular uptake and transfection efficiency of various lipid-, polymer-, and peptide-based nanocarriers [119–121]. Accordingly, several artificial homopolymers of arginine and lysine peptides have been developed to effectively translocate cargo across the membrane [122,123]. Notably, the rate of cell uptake and subsequently transfection efficiency was higher for arginine-rich peptides compared to polylysines [123–125].

Although most naturally occurring CPPs are cationic, the major class of CPPs is amphipathic [48]. Amphipathic CPPs consist of polar and non-polar (rich in hydrophobic) amino acid regions that are able to fold into α-helical and β-sheet-like structures. The secondary structure might change in response to different physiological conditions which, in turn, affects their penetration ability [57]. Prominent representatives of amphipathic CPPs are various variants of N-Methylpurine DNA Glycosylase or MPG, where amphiphilicity is a leading factor for their translocation across the membrane [126]. MPGs undergo a conformational transition from unordered into a folded state upon their interaction with membrane phospholipids mediated by polar residues. The resulting β-sheet conformation governed by the hydrophobic domain of MPG lead to transient pore-formation in the cell membrane, which in turn enable the MPG/cargo complexes direct penetration across the membrane independent of endocytosis [126,127]. In addition to MPGs' function in promoting cellular internalization, it is well known for its strong electrostatic interactions with oligonucleotides [128]. Consequently, MPG family members form stable noncovalent nanocomplexes with nucleic acids that enter cells independently of the endosomal pathway. Accordingly, MPG has shown to efficiently deliver small interfering RNA (siRNA) and plasmid DNA (pDNA) into cultured cell lines [129]. Transportan and its analogs NickFect and PepFect are other examples of amphipathic peptides that can condense pDNA and siRNA into stable nanocomplexes [130–132]. Although their hydrophobicity appears to be responsible for the nanocomplexes' stability, the pH-induced change of their charge plays a key role in promoting oligonucleotide condensation and high delivery efficiency.

Hydrophobic CPPs with low positive or negative net charge are less common and their uptake mechanism is not well understood. For example, natural C105Y, K-FGF, and Bip peptide belong to this group and their non-polar amino acids' affinity to the hydrophobic domain of cell membranes mediate their translocation [48].

2.3. *Peptides Facilitating Endosomal Escape*

Endosomal escape is a crucial step in improving intracellular delivery and efficiency of nucleic acids [133]. Following endocytosis as the major uptake route for many peptide-based nanocarriers, most internalized nanocarriers significantly suffer from their interaction with endosomal membrane which leads to their entrapment and eventually their enzymatic degradation in the lysosomal compartment [134,135]. Peptides are the most promising candidates for promoting endosomal escape [136,137]. In particular, pH-sensitive peptides that at physiological pH adopt a random coil structure, transform to an α-helical conformation able to induce membrane pore formation in the acidic environment of endosomes [138]. Similarly, studies on viruses escaping the lysosome have shown that some viral peptides change from a hydrophilic ring structure to a hydrophobic spiral structure which will target the core of the bilayer and destroy the stability of the membrane [139]. Peptide development aims at facilitating escape from the endosome via pore formation, the "proton sponge effect", or conformational changes.

2.3.1. Fusogenic Peptides

Fusogenic peptides (FPs) are short peptides with the potential to promote membrane destabilization and delivery of nucleic acids to the cytosol and/or the nucleus [140]. Fusogenic peptides consist of hydrophilic and hydrophobic domains that are able to form helical structures at endosomal pH. This allows for direct engagement with the endosomal membrane upon which further energetically favorable conformational changes induce pore formation in the membrane. Disruption of the bilayer eventually leads to endosomal escape of nanocarriers equipped with FPs and release of cargos to the cytoplasm (Figure 4). Overcoming the endosomal membrane barrier presents an important role in facilitating nucleic acids localization to distinct subcellular compartments as their site of action [54]. However, before integrating a fusogenic peptide into gene delivery systems, the cellular uptake mechanism should be considered. Since the fusogenic activity of these peptides is due to a pH-dependent shift in conformation, non-acidic endocytotic pathways, such as caveolae-mediated endocytosis and macropinocytosis will revoke their membrane lytic activity [141].

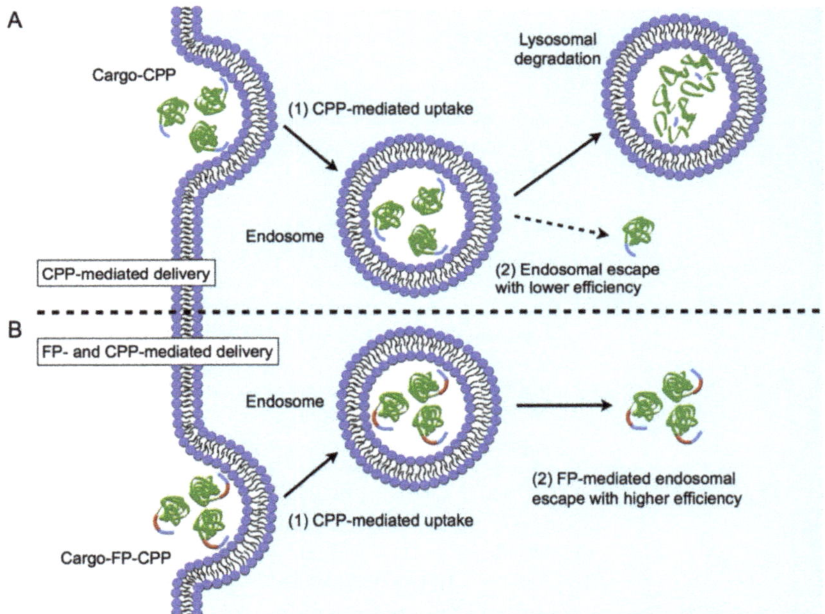

Figure 4. Schematic representation of FP and CPP-mediated delivery. (**A**) A conventional CPP-mediated delivery. (1) (**A**) cationic CPP (blue) interacts electrostatically with the anionic cell-surface, and a CPP-fused cargo (green) is internalized into the endosome by endocytosis. (2) Because the endosomal escape efficiency of CPP is low, the cargo-CPP is subjected to lysosomal degradation. (**B**) FP- and CPP-mediated delivery. (1) An FP (red)- and CPP-fused cargo is internalized into the endosome by endocytosis. (2) Because the efficiency of FP-mediated endosomal escape is relatively high, the cargo-FP-CPP is efficiently transferred from the endosome to the cytoplasm. Reprinted with permission from [142]. Copyright 2017 Elsevier.

Fusogenic peptides are either derived from the transduction domain of proteins that interact with cell membranes such as HA2, INF7, and melittin or are synthetic amphipathic peptides that can penetrate membranes [141,143].

Wild-type HA2(1–23) peptide and a glutamic acid-enriched analogue (INF7) from influenza virus hemagglutinin are the oldest and best studied fusogenic peptides used for gene delivery [144–149]. These peptides, based on the protonation of their acidic residues upon a decrease in pH, assume a helix structure and consequently promote the endosomal escape, which, in turn, results in enhanced transfection efficiency [150,151].

Melittin, a cationic amphipathic peptide composed of 26 amino acids, is derived from the venom of the honey bee Apis mellifera. Melittin with its predominantly hydrophobic 20 N-terminal amino acids and hydrophilic C-terminus acts mainly like a natural detergent on the membrane and is well known for its cytolytic activity [152]. Oligomerization of this peptide results in the formation of transmembrane channels which lead to osmotic cell lysis [153]. Owing to its cytotoxicity, the use of melittin as an agent to promote gene delivery in transfected mammalian cells is limited. More recently, less toxic melittin analogues that retained their ability to escape from the endosome were shown to enhance the efficiency of non-viral gene delivery systems [154–156].

A number of pH-responsive synthetic amphipathic peptides mimic the fusogenic activity of virus-derived peptides. The most prominent representatives consist of non-polar alanine-leucine-alanine repeating units with considerable repetitive content of either glutamic acid, lysine, or arginine, and are named GALA, KALA, or RALA, respectively [157–161]. Likewise, upon protonation at endosomal pH (5.0), they assume an amphipathic α-helical conformation which is associated with a significant affinity for binding to phospholipid membranes. As a consequence, pore formation, membrane fusion, and/or lysis are induced. Their membrane lytic activity explains extensive utilization of these fusogenic peptide in modulating non-viral gene delivery systems [76,162–166]. Similar to GALA, JTS-1, a negatively charged amphipathic peptide with strong nonpolar amino acids in the hydrophobic domain and glutamic acid residues in the hydrophilic domain is able to form an α-helical structure [167]. Owing to the endosomolytic capacity of JTS-1-modified carriers, improved transfection activity was reported in several studies [168,169].

Taking advantage of the fusogenic properties of peptides, alone or in combination with other advantageous attributes, promotes the efficacy of peptide-based nanocarriers for traversing membranes and, thereby, improves their therapeutic effects. Yet, there is a strong need for systematic study of different fusogenic peptides under similar conditions in order to elucidate the mechanisms and pin down the parameters that ultimately will maximize therapeutic outcome. This comprehensive comparison of fusogenic peptides will allow for designing a robust, widely applicable delivery system.

2.3.2. Histidine-Rich Peptides

The combination of being able to condense nucleic acids and at the same time promote endosomal escape spurred efforts to incorporate histidine-rich amphipathic peptides into various gene delivery systems [170]. As described for fusogenic peptides, histidine residues that become protonated during acidification of the endosome interact with negatively charged membrane lipids (Figure 5) and destabilize the membrane [171]. Histidylation of different non-viral vectors among which polylysine was the first example, was found to increase the buffering capacity of vectors [172]. Substituting several lysines with histidines turns polylysine into a successful gene delivery vector with enhanced transfection efficiency [173–178].

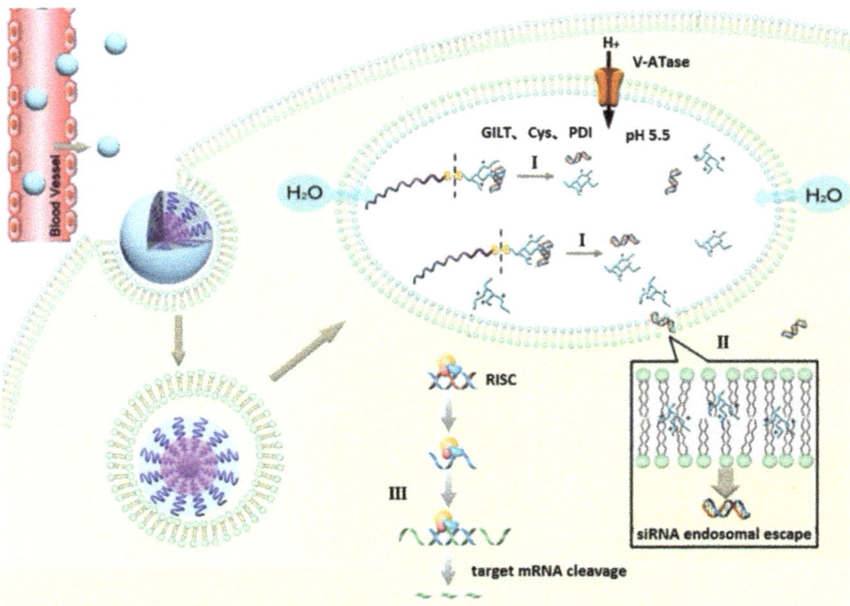

Figure 5. Schematic illustration of the intracellular trafficking of mPEG-b-PLA-Phis-ssPEI and siRNA complexes. After internalized by tumor cell, (**I**) the complexes rapidly disassemble to release siRNA and free polyethylenimine (PEI) molecules in response to the acidic and reductive microenvironment, (**II**) efficiently escape from the endosome, facilitated simultaneously by cleaved PEI chains inducing membrane destabilization, the "proton sponge effect" of polyhistidine and polyethylenimine, as well as the relative small size of after disassembly, (**III**) achieve efficient gene silencing by cytosolic target mRNA cleavage. Adapted with permission from [179]. Copyright 2018 Elsevier.

Influenza-derived, histidine-rich H5WYG peptide that is capable of traversing intracellular barriers to deliver nucleic acids, is another well-known example that raised great interest. This pH sensitive peptide has been extensively applied to improve the gene delivery efficacy of different polymeric, peptidic, and lipid-based carriers [180–186].

Cationic LAH4 is another widely studied histidine-rich peptide where the protonation of the imidazole groups invokes chloride ion, as well as proton influx into the endosome, creating a hypertonic environment. Although the so-called "proton sponge effect", i.e., the osmotic influx of water triggering endosome lysis, has been recognized as the primary route of endosomal escape, other mechanisms also exist [187,188]. Changes in pH modulate the amphipathicity and membrane topology of LAH4 and derivatives: they are transmembrane at neutral pH, whereas under acidic conditions, LAH4 peptides align parallel to the phospholipid bilayer surface [189,190]. The interactions of the peptides with the bilayer interface eventually result in pore-formation [191] and membrane lysis [192], thereby modulating nucleic acid delivery. Inspired by the ability to enhance the efficiency of several gene delivery systems, researchers in the past extensively used LAH4 peptides as targeting moiety [187,193–198]. By now, there is a growing number of peptidic, as well as polymeric and lipidic carriers that utilize other histidine-rich moieties, such as $O_{10}H_6$ [199], $MS(O_{10}H_6)$ [200,201], histidine-rich Tat peptide [202–204], or His6 RPCs [205,206] to develop new promising gene delivery strategies. The possibility of intracellular delivery of nucleic acids in a nontoxic manner, which is a necessary prerequisite for gene therapy, has opened interesting perspectives in non-viral gene delivery. Nevertheless, there are many unanswered questions regarding the precise capacity or trafficking routes involved in favoring endosomal escape [172]. Hence, a comprehensive screen and quantification of this step will provide further improvements for exploiting histidine-rich peptides in the field of gene therapy.

2.4. Peptides Assisting Delivery to Subcellular Organelles

A major focus in gene therapy is delivering nucleic acids to those intracellular compartments where they are therapeutically most effective. By escaping the endosome, siRNA and mRNA cargos arrive at their final destination, the cytosol, whereas DNA cargoes require translocation to the nucleus or to mitochondria. Intracellular targeting peptides serve a promising approach to specifically direct their cargo to the respective organelles and ensure membrane interactions that support delivery. Obviously, such peptides are particularly favorable candidates to be integrated into gene delivery systems [10,207]. The degree of translocation enhancement depends on the characteristics of both, delivery system and targeting peptide [208]. In the following sections, we address the mechanisms of intracellular nanoparticle trafficking and provide examples employing intracellular targeting peptides to ensure nuclear and mitochondrial targeting of nucleic acids.

2.4.1. Nuclear Localization Signals

Nucleocytoplasmic transport is major consideration for effective non-viral gene delivery [207,209]. Once inside the cell, most DNA must translocate into the nucleus where they can be either transcribed into the messenger RNA (mRNA) or interfere with transcription and RNA processing [210,211]. Nuclear localization signals (NLSs) are short peptide motifs rich in arginine, lysine, or proline that mediate nuclear translocation and when attached to foreign macromolecules or nanocarriers, deliver them to the nucleus [212]. For example, polymersomes, artificial vesicles resulting from self-assembly of amphiphilic copolymers, bypass the nuclear pore complexes (NPCs) that regulate transport into and out of the nucleus and deliver payloads directly into cell nuclei (Figure 6) [213]. Active transport of macromolecules to the nucleus is carried out by interactions of the NLS with importin receptors (karyopherins) and specific proteins of the NPC [214,215].

In view of the fact that the nuclear membrane is the main barrier restricting transgene expression of most non-viral carriers, gene therapy is the obvious field for the application of nuclear targeting peptides [216]. The significance of incorporating NLS peptides into non-viral delivery systems that can adequately favor the genetic materials release into the nucleus manifests itself by expanding case studies [207,217]. Hereby, positively charged NLS peptides either are attached to the negatively charged DNA via electrostatic interactions or are covalently coupled to the phosphate backbone of the DNA or to the condensing agent of the non-viral vector [216].

A frequently used NLS is derived from the large tumor antigen of Simian virus 40, SV40 (PKKKRKV). The positive charges of SV40 NLS peptide not only help in DNA condensation, but also mediate nuclear targeting [218]. Consistently, addition of the SV40 NLS peptide and its derivatives enhanced the transfection efficiency of many non-viral carriers [219–222].

Another interesting NLS is M9, a 38 amino acid peptide derived from heterogeneous ribonucleoprotein A1 (hnRNP A1) which is a major nuclear pre-mRNA binding protein. M9 is responsible for ferrying hnRNP A1 into the nucleus and also contains a nuclear export sequence (NES) [223]. Owing to its rather low positive charge, M9 is relatively poor at condensing DNA. On the other hand, it strongly interacts with its known receptor transportin 1 [224]. Therefore, utilizing the nuclear import effect of M9 in combination with positively charged biomaterials that condense the DNA offers great potential for gene therapy [142,225,226]. Furthermore, an NLS sequence derived from the HIV-1 viral protein (Vpr) promotes nuclear import through a karyopherin α–independent mechanism [227]. Examples for an enhanced transfection efficiency with Vpr-containing non-viral vectors are reviewed by Cartier and Reszka [217].

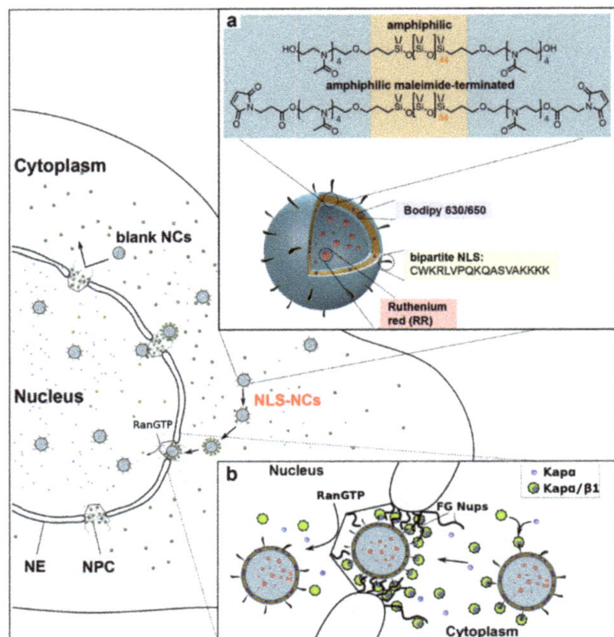

Figure 6. Organelle-specific targeting of polymersome NCs into the cell nucleus. (**a**) NLS-NCs self-assemble from amphiphilic PMOXA-PDMS-PMOXA triblock copolymers. Two model compounds are used to test for nuclear delivery: Ruthenium red (RR) that is encapsulated within the NLS-NC lumen, and Bodipy 630/650 that incorporates into its polymeric membrane. (**b**) The nuclear transport mechanism involves Kapα•Kapβ1 that (1) authenticates NLS-NCs for selective NPC transport, (2) binds to FG Nups, and (3) releases NLS-NCs into the nucleus upon binding RanGTP. Reprinted with permission from [213]. Copyright 2020 National Academy of Sciences.

Other examples of targeting sequences that facilitate nuclear transport of exogenous DNA include Xenopus protein nucleoplasmin [228,229], adenoviral peptide (Ad) [217,230], human T-cell leukaemia virus (HTLV) [224], Epstein–Barr virus nuclear antigen (EBNA)-1 [231]. By overcoming intracellular barriers, these peptides greatly expand the perspectives in non-viral gene delivery.

2.4.2. Mitochondrial Delivery

Mitochondria have their own genome whose mutations are associated with numerous disorders, such as cancer, diabetes, neurodegenerative diseases including Parkinson's disease, and more recently infectious and autoimmune diseases [225,226]. Given the link between mitochondrial dysfunction and disease, targeting of therapeutic interventions to these subcellular organelles is of vital importance [232,233]. Efficient mitochondrial gene therapy requires nanocarriers that, once inside the cell, target mitochondria and ferry nucleic acids across the outer (OMM), as well as inner mitochondrial membrane (IMM) [234]. The hydrophobicity and negative charge of MMs require that for efficient mitochondrial delivery, negatively charged DNA be shielded by carrier molecules, such as peptides that have amphiphilic and cationic properties.

Many natural and artificial short peptides and polypeptides have mitochondrial targeting ability [235,236]. Typically, these peptides comprise hydrophobic (phenylalanine, tyrosine, isoleucine) and positively charged (D-arginine, lysine) amino acids. The development of mitochondrion-targeted delivery strategies involving mitochondrial targeting sequences (MTSs), which are typically tens of amino acids in length, mostly takes advantage of MM properties including the high negative potential (−160 to −180 mV) and the intrinsic protein import machinery. Although MTSs vary in length, they have in common an α-helical structure with an amphiphilic surface that mediates internalization by endogenous transmembrane transporters. However, because of their rather large size, low solubility and insufficient permeability across the plasma membrane, MTSs by themselves are not suitable for delivering exogenous nucleic acids.

In addition to MTSs that are recognized by translocators, several smaller peptides consisting of 4–16 cationic and hydrophobic residues efficiently target and permeate mitochondrial double membranes [235,237]. These mitochondria-penetrating peptides (MPPs), also known as mitochondrial CPPs (mtCPPs), typically appear to penetrate cellular membranes directly rather than by endocytosis [238]. Consequently, nanocarriers targeted by MPPs circumvent endosome/lysosome segregation, which also increases the chance of (gene) delivery to the mitochondria. In addition, MPPs appear to have marginal effects on mitochondrial membrane potential [239]

Considering that cell and mitochondrial membrane barriers have distinct compositions and properties, a single peptide will not be able to mediate the crossing of both. Here, combining CPP activity and mitochondrial targeting can act synergistically to optimize delivery of peptide-based DNA nanoparticles to mitochondria (Figure 7) [233,240]. For example, a library of fusion peptides with mitochondria targeting (mtCPP1) and cell-penetrating properties (Pepfect14, a stearylated CPP forming ASO nanocomplexes with splice-correction activity in cells [241]) that self-assembled with antisense oligonucleotides (ASO) into complexes was successful in knocking down mitochondrial mRNA [242]. A combinatorial approach to develop mitochondrial gene expression was also pursued by incorporating an MTS into WRAP peptides (short tryptophan/arginine rich peptides; [243]) which formed nanocomplexes with plasmid DNA encoding the mitochondrial ND1 gene that were taken up by cells and targeted to mitochondria [244]. Systematic analysis of CPPs and MTSs revealed that while both types of peptides were rich in Ala and Arg, the latter included Leu, suggesting a role for Leu in targeting to mitochondria [245].

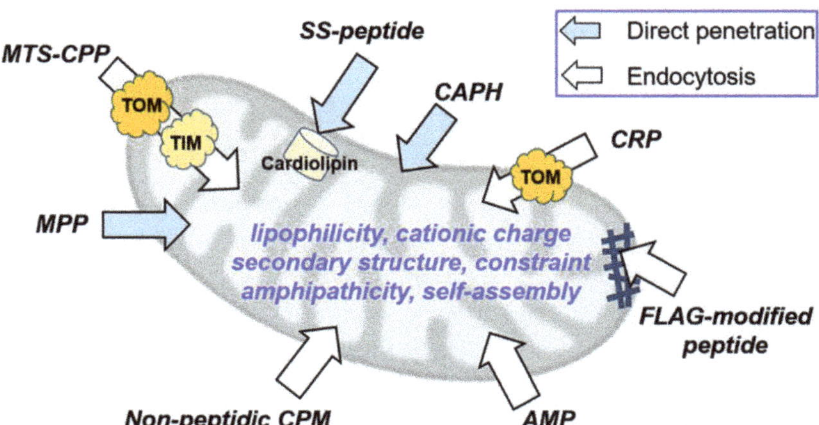

Figure 7. Mitochondrion-targeting peptides and peptidomimetics based on their structural classes and reported applications. Abbreviations: MTSCPP, MTS with cell-penetrating peptides; MPP, mitochondrion-penetrating peptides; SS-peptides, Szeto–Schiller peptides; CAPH, cationic amphiphilic polyproline helix; CRP, cysteine-rich peptides; FLAG-modified peptide, FLAG tag-based peptide that self-assembled into a nanofiber; AMP, peptides derived from antimicrobial peptides; CPM, nonpeptidic cell-penetrating motif; OMM, outer mitochondrial membrane; IMM, inner mitochondrial membrane; IMS, intermembrane space. Reprinted with modification from [233]. Copyright 2020 American Chemical Society.

3. Peptide-Related Nano-Assemblies for Nucleic Acid Delivery

Peptides have great potential as self-assembly building blocks on account of primary and secondary structure variability. Depending on the design, they form various supramolecular assemblies, such as vesicles [246], micelles [247], nanotubes [248], nanofibers [249], or nanoribbons [250]. Choosing corresponding peptide building blocks allows for tuning size and shape of the nano-assembly to obtain improved nanocarrier properties including cargo loading and delivery. Moreover, the weak interactions involved in peptide self-assembly are sensitive to environmental conditions, enabling nano-assemblies to exhibit specific functionalities in response to different external stimuli, such as temperature, pH, redox state, enzymes, or even light. We first focus on examples of supramolecular assemblies where peptides represent the predominant building block of the nanocarriers or are integrated into the nanocarriers to improve their targeting and gene delivery properties, and then discuss examples with stimuli-responsiveness.

Prominent structures that serve as nucleic acid carriers are micelles entrapping DNA during self-assembly, also called "micelleplexes" [251]. If purely peptidic, micelleplexes possess minimal cytotoxicity [20]. However, often peptides are used as a targeting or uptake-facilitating moieties associated with nanocarriers made out of different, less biocompatible materials. A micelle forming polymer-peptide conjugate used as an siRNA carrier has recently been reported as an effective tool in anti-metastasis cancer therapies (Figure 8) [252]. Methoxy-polyethylene glycol combined polycaprolactone conjugated with a cytoplasm-responsive peptide CH2R4H2C (MPEG-PCL-CH2R4H2C) was used to entrap anti-RelA siRNA (siRelA). RelA is a subunit of NF-κB involved in metastasis, especially cancer cell migration and invasion. The MPEG-PCL part of the conjugate was expected to improve blood retention and tumor accumulation and to facilitate micelle formation. Consistent with this notion, siRelA/MPEG-PCL-CH2R4H2C micelleplexes successfully delivered siRNA into cancer cells in a lung metastasis mouse model, causing inhibition of RelA accompanied by significant suppression of metastasis.

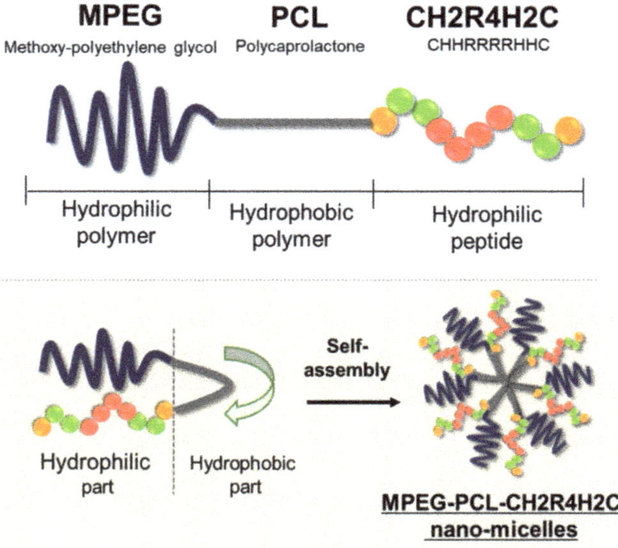

Figure 8. The structure of MPEG-PCL-CH2R4H2C and nano-micelle formation. Reprinted with permission from [252]. Copyright 2020 Multidisciplinary Digital Publishing Institute (MDPI).

More recently, peptide-assisted polymeric micelles were used for inhibiting the mitotic cycle of prostate cancer cells by siRNA delivery in cell lines and in tumor-burdened nude mice [253]. The hydrophilic segments of acetal-polyethylene oxide-*b*-polycaprolactone (A-PEO-PCL) copolymers were chemically modified with a TAT peptide and a ligand for prostate-specific membrane antigen (DCL) to enhance targeting and cell penetration of self-assembled micelles loaded with siRNA and docetaxel (anti-cancer drug).

Other small molecules, for example a palmitoyl chain conjugated to the N-terminus of GGGAAAKRK [254], proved useful in promoting self-assembly of peptides to distinct nanocarriers with a hydrophobic core. Accordingly, surfactant-like palmitoyl-GGGAAAKRK formed peptide nanofibers (PNFs) in the presence of siRNA specific for the down-regulation of BCL2 protein. Human SH-SY5Y cells showed significant uptake of PNF:siBCL2 constructs in vitro and silencing of *BCL2* in specific loci of rat brains demonstrated effective delivery of siRNA. In another example, a branched amphiphilic peptide comprising oligolysine segments with DNA binding properties formed different structures depending on the peptide/DNA ratio: at high peptide/DNA ratio, it coated the DNA surface forming nanofibers and at low peptide/DNA ratio, it condensed the DNA into nanometer-sized compacted structures [255]. Using pDNA encoding green fluorescent protein (GFP) as cargo, the peptide nanocarrier demonstrated higher transfection efficiency in HeLa cells compared to Lipofectin (commercial transfection agent) when the total number of transfectants alive was considered.

Another type of versatile and reproducible supramolecular nanocarrier with well-defined structure and composition are dendrimers [256]. Many types of dendrimers including peptide dendrimers (PPI; [257]), poly(L-lysine) dendrimers, and polyamidoamine (PAMAM) dendrimers that display electrostatic interactions with nucleic acids and protect the cargo from degradation, are particularly suited for gene delivery. In addition, dendrimers lend themselves to surface conjugation of peptide moieties that enhance gene delivery. Conjugation of TAT (HIV transactivator of transcription) peptide to PAMAM dendrimer formed nanometer-sized (105 nm–115 nm) 'dendriplexes' with GFP pDNA [258] that displayed an increased transfection efficiency in Vero cells compared to PAMAM without TAT.

Stimuli-Responsive Gene Delivery Systems

Peptide-based assemblies are particularly attractive for developing stimuli-responsive therapeutic nanocarriers for several reasons: they are biocompatible, readily degraded and then removed from the organism, but most importantly, highly sensitive to environmental conditions. Small changes in external factors, such as temperature or pH, can induce transformation of secondary structures (α-helices, β-sheets, and β-turns), thereby affecting the morphology and concomitantly the function and bio-activity of the polypeptides [259]. Moreover, control over cargo release is a highly sought-after feature in gene delivery systems and stimuli-responsive nanocarriers can deliver nucleic acids more efficiently by reducing unspecific release. To date, peptide vectors forming complexes with DNA by electrostatic interactions still make up the majority of peptide-based nanocarriers. As stimuli-responsiveness can be readily obtained by modifying the peptide sequence, these nanoparticles (peptiplexes) represent the main targets for a control of cargo release by external stimuli. However, nano-assemblies, based on their modularity offer not only increased DNA loading capacity but are also more susceptible towards environmental stimuli.

The pH-responsive peptides have been intensively investigated in delivery and diagnostic systems because pH variations are typical for many biological systems, intracellular compartments (lysosomes and endosomes), specific organs (gastrointestinal tract and vagina), and pathological conditions [260,261]. Particularly, the microenvironment of many tumor tissues has a lower pH (<6.5) compared to normal tissues (pH 7.4). Thus, polypeptides that change conformation in a pH-dependent fashion can find application in selective binding to cancer sites which can be exploited for tumor diagnosis and treatment. Nanocarriers taken up by endocytosis encounter acidification in the endosome, which is exploited by pH-responsive peptides to increase endosomal escape. For example, (Fmoc)2KH7-TAT, an amphiphilic, pH-responsive chimeric peptide [262], complexed with pGL-3 reporter plasmid mediated transfection of 293T and HeLa cells by promoting endosomal escape via protonation of KH residues. Moreover, co-delivery of p53 plasmid and doxorubicin using (Fmoc)2KH7-TAT self-assembled micelleplexes inhibited cell growth in vitro and tumor growth in vivo.

Nanocarriers with peptide-mediated redox-sensitivity have emerged as a fascinating type of biomedical material with potential for triggered gene and drug delivery inside cells. Sensitivity of peptides to the redox state is provided by disulphide bonds, diselenide bonds, succinimide-thioether linkage or by redox sensitive groups, such as "trimethyl-locked" benzoquinone [263]. In a reducing environment, redox-responsive nanocarriers undergo a change in conformation and release their cargo. A major advantage of redox-responsive nanocarriers is their stability in normal tissues which avoids cytotoxicity caused by the unwanted release of therapeutic cargo. Tumor tissues show 4-fold higher glutathione levels compared to healthy tissue and thus triggers redox-responsive cargo release [264].

A "smart", redox-sensitive peptide designed to trigger the assembly of gadolinium nanoparticles inside cells, was successfully applied in magnetic resonance imaging of tumors in a xenograft mouse model [265]. Acetyl-RVRR-C(StBu)-K(Gd-DOTA)-CBT contains an RVRR sequence which mediates cell membrane translocation but is also a cleavage site for intracellular furin, typically upregulated in many tumors, and a disulphided Cys motif. After entering the cell, the disulfide bond is reduced by intracellular glutathione (GSH) and subsequently, the RVRR motif is cleaved by furin in situ. The cleavage product quickly condenses to amphiphilic dimers that self-assemble via π-π stacking into Gd-containing nanoparticles.

Peptide structures and the weak interactions contributing to self-assembly of peptide-based nanocarriers are inherently sensitive to temperature [266,267]. An interesting example of a thermo-responsive, purely peptidic DNA nanocarrier are multi-compartment micellar nanoparticles (MCM-NPs) assembled from (HR)3gT peptide (Figure 9) [20].

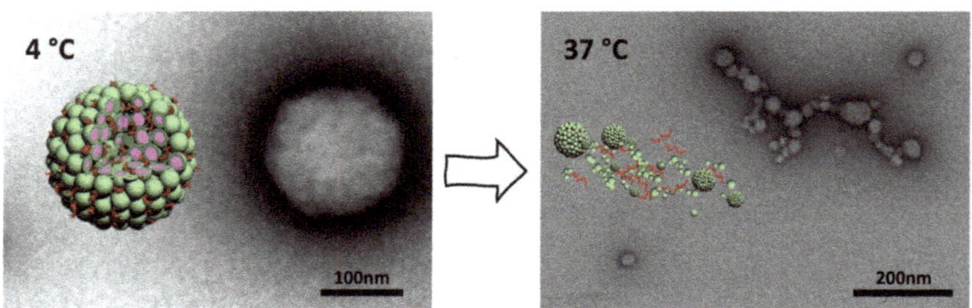

Figure 9. Schematic representation and TEM micrograph of the self-assembled (HR)3gT multi-compartment micellar nano-assembly (MCM) at 4 °C (*left*) and temperature-induced disassembly of MCMs into disperse or clustered smaller MCMs and individual micelles at 37 °C (*right*). Modified from [20] with permission from the Royal Society of Chemistry.

Although the multicompartment micellar structure of NPs was stable at 4 °C, increasing the temperature to 37 °C triggered structural changes that led to the disassembly into smaller MCMs and individual micelles after several hours. On account of the high cellular uptake efficiency and thermo-responsive disassembly at physiological temperature, MCM-NPs are a promising DNA delivery vehicle with great potential for application in vivo. Similar multicompartment micellar NPs assembled from H_3SSgT peptide bearing a disulfide functional group between hydrophilic and hydrophobic domain, were developed for redox-responsive codelivery of oligonucleotides and drugs [268]. The disulfide bond conferred responsiveness to physiological concentrations of reducing agent upon NPs, resulting in release of the incorporated cargo. The advantage of a supramolecular multicompartment structure over individual micelles lies in the increased capacity for oligonucleotide condensation [269]. Together with the ability to entrap various hydrophobic cargos, this makes MCM-NPs well-suited for biomedical applications.

Light has received much attention as an external stimulus, as it provides spatiotemporal control that can be triggered remotely. By crosslinking peptides with specific light-absorbing molecules it is possible to obtain photo-responsive conjugates that allow for light-stimulated assembly of nanostructures or light-induced release of cargo molecules. Such light-sensitive conjugates of peptides and photosensitizers can serve as light-controllable phototherapeutic agents [270].

Photo-crosslinking by UV (254 nm) of poly(ethylene glycol)-*b*-poly(l-glutamic acid) diblock copolymer was shown to convert the core of self-assembled core-shell micellar structures to nanogels that, depending on the composition of the copolymer, could release drug payload in a pH-dependent manner [271]. These results indicated the potential of nanogels fabricated by photo-crosslinking of polypeptide micelles as intelligent delivery systems.

Micelles with a photocleavable poly(S-(o-nitrobenzyl)-l-cysteine) (PNBC) core surrounded by a hydrophilic poly(ethylene glycol) (PEO) corona were also obtained by self-assembly of PNBC-b-PEO amphiphilic block copolymer [272]. UV irradiation (365 nm) of these micelles gradually removed nitrobenzyl groups from PNBC-b-PEO resulting in a shrinkage of the micelles. If micelles were prepared in the presence of doxorubicin, a photo-triggered release of the drug was observed in vitro. Since self-assembly is achieved in aqueous solution, photocleavable polypeptide-based block copolymers lend themselves to developing photoresponsive nanomedicines for anticancer therapy.

4. Combinatorial Approach for Advanced Nucleic Acid Delivery

In view of gene therapy, endowing nanocarries with targeting features and stimuli-responsiveness that provides site-specific, triggerable control over cargo release could optimize delivery efficacy, and, at the same time, minimize adverse effects.

For example, folate-receptor targeting, acid-sensitive polymeric micelles (F-ASPM) have been successfully applied to deliver siRNA to breast cancer cells (Figure 10) [273]. In this approach, poly (L-histidine)-polythelene-glycol (PEG-PHIS) and folate-conjugated PEG-PHIS amphiphilic block copolymers in the presence of CPP-coupled siRNA self-assembled into micelles where folate functioned as targeting ligand and histidine residues provided pH-responsiveness. As PEG-PHIS block copolymers have a pKb of 6.5–7.0, micelles formed by this copolymer dissociate at pH 6.5–7.0 which renders them suitable for constructing drug/DNA delivery systems that are sensitive to the extracellular tumor environment. In addition, c-myc silencing siRNA conjugation to a CPP was obtained by reduction-sensitive disulfide bonding which turned the resulting micelles into a dual responsive nanocarrier able to target tumor cells where release and delivery of siRNA are promoted by the respective peptide sequences.

The combination of targeting and stimuli-responsiveness is also provided by ROSE, a redox-sensitive, oligopeptide-guided, self-assembling, and efficiency-enhanced carrier system [274]. In ROSE, adamantyl-PEG chains with and without disulfide bonded SP94 targeting oligopeptide, mixed with hydroxypropyl-β-cyclodextrin formed supramolecular complexes condensing tumor-suppressor microRNA-34a (miR-34a). Oligopeptide-guided specificity for hepatocarcinoma cells and release of miRNA following disulfide cleavage in the reducing environment significantly improved the tumor-suppressing effect of ROSE/miR-34a over conventional gene delivery strategies.

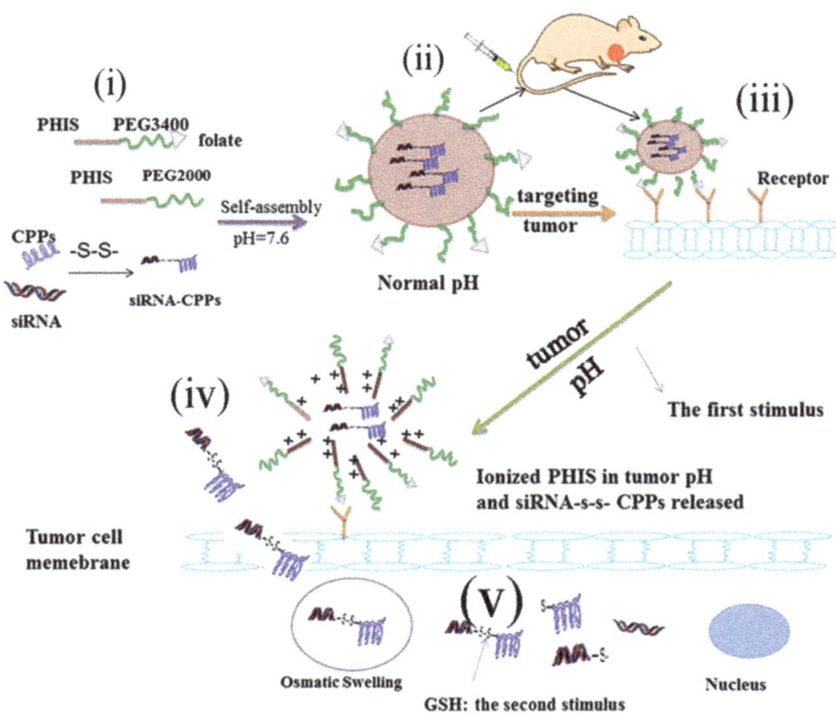

Figure 10. Diagram of F-ASPM formation and the mechanism of siRNA delivery into cancer cells. (**i**) synthesis of PEG-PHIS and F-PEG-PHIS. (**ii**) self-assembly of amphiphilic block copolymers in the presence of siRNA-CPP into acid-sensitive nanocarriers with active targeting ability (F-ASPM). (**iii**) binding of F-ASPM to cancer cells by folate receptor targeting. (**iv**) pH-stimulated release of siRNA-CPPs from F-ASPM. (**v**) GSH-mediated cleavage of disulfide bond of siRNA-CPPs leading to release free siRNA into cytosol. Reprinted with permission from [273]. Copyright 2016 John Wiley and Sons.

5. Conclusions

Since the introduction of peptides as potential delivery system for a variety of therapeutic cargos, extensive research has focused on their application in gene therapy. To be suitably tailored for gene therapy, peptide-based nanocarriers must comply with issues of targeting, cellular uptake, and intracellular trafficking, all of which involve biological membranes and how they can be overcome. A combinatorial approach, e.g., designer peptides composed of cationic cell-penetrating and hydrophobic endosomal escape domains in combination with a gene carrier peptide composed of targeting and cationic DNA-binding domains affording triggered, site-specific (cytosol, nucleus, mitochondria) release of nucleic acids, may offer some improvement of efficacy. Other properties, including, but not limited to, low cytotoxicity, target specificity, biodegradability, and cost and time efficiency of synthesis greatly contribute to the potential of peptides in nanomedicine. Nevertheless, to broadly realize bench to bedside translation of peptide-related gene delivery systems, innovative technologies need to be pursued to achieve peptide-based nanocarriers that more specifically and efficiently deliver nucleic acids or nucleic acid modifying systems to the desired sites. In many cases, such nanocarriers would further benefit from either sustained or triggered delivery options.

Advances in peptide development have made peptide-assisted gene delivery more efficient in vitro and, in some instances, in small animal models [275]. For example, cell and tissue selectivity could be greatly enhanced in the newest generation of CPPs [276]. Other advances which allow for improved performance with regard to targeting and delivery of nucleic acids include adapting peptide sequences to facilitate escape or release from intracellular vesicles or respond to environmental stimuli for a controlled release of cargo, and the development of composite, multivalent peptide-based, or peptide-coupled structures.

Intriguingly, while revolutionary and versatile peptide tools have inspired a great deal of hope regarding the treatment of genetic diseases, peptide nanocarriers are awaiting clinical translation. For example, none of the nanocarriers associated with CPPs have so far been approved for clinical studies. Evidently, besides overcoming membrane barriers, a string of challenges remain that need to be tackled for peptide nanocarriers to make a breakthrough in clinical application whereas lipid-based formulations, for all their drawbacks [277], are being used in the field of gene therapy and as a delivery vehicle in mRNA-based vaccines [278]. Short circulation half-lives, inadequate biodistribution, and poor chemical and physical serum stability, especially susceptibility to proteolytic degradation associated with off-target nucleic acid release, hamper clinical translation of peptide-based nanocarriers. Refining their preparation with regard to gene loading efficiency and product homogeneity would be a possible improvement. However, reaching the final target with high selectivity and adequate accumulation at the target site remains a major issue for in vivo applications. Here, peptide modifications (unnatural amino acids, cyclization) and conjugate molecules (PEGylation, hydrocarbon chains) that prolong the circulation time and enhance the structural stability of nanocarriers in the serum come to mind [279,280]. Peptidomimetics, often based on natural peptide sequences, that exhibit improved proteolytic stability or even new folds and morphologies designed to enhance bio availability, improve transport through the blood–brain barrier, or reduce the rate of clearance, are emerging. However, their modifications bear the risk of reducing potency or even introducing toxicity, for example D-amino acids [281]. Another alternative to clear abovementioned hurdles is developing advanced multifunctional carriers comprising various agents, each of which can overcome the barrier through distinct dictated functions. Undoubtedly, peptides offer the largest potential when it comes to nucleic acid condensation, targeting, endosomal escape, and subcellular localization as a part of multifunctional advanced delivery systems. But again, the combination of functionalities bears the danger of affecting the individual functions. Thus, extensive research on the development of stimuli-responsive purely peptidic systems with suitable physicochemical properties for nucleic acid delivery is being pursued at many levels.

A better understanding of the mechanisms by which peptide-based delivery systems use to overcome membrane but also other biological barriers, together with advancements in the synthesis of innovative materials tailored to environmental conditions and extensive in vivo studies herald a bright future for peptide-based delivery systems in gene therapy and in nanomedicine in general.

Funding: Swiss Nanoscience Institute (SNI), NCCR Molecular Systems Engineering, University of Basel, Switzerland.

Institutional Review Board Statement: Not applicable.

Informed Consent Statement: Not applicable.

Data Availability Statement: Not applicable.

Conflicts of Interest: The authors declare no conflict of interest.

References

1. Gillmore, J.D.; Gane, E.; Taubel, J.; Kao, J.; Fontana, M.; Maitland, M.L.; Seitzer, J.; O'Connell, D.; Walsh, K.R.; Wood, K.; et al. CRISPR-Cas9 In Vivo Gene Editing for Transthyretin Amyloidosis. *N. Engl. J. Med.* **2021**, *385*, 493–502. [CrossRef] [PubMed]
2. Rosenblum, D.; Gutkin, A.; Kedmi, R.; Ramishetti, S.; Veiga, N.; Jacobi, A.M.; Schubert, M.S.; Friedmann-Morvinski, D.; Cohen, Z.R.; Behlke, M.A.; et al. CRISPR-Cas9 Genome Editing Using Targeted Lipid Nanoparticles for Cancer Therapy. *Sci. Adv.* **2020**, *6*, eabc9450. [CrossRef] [PubMed]
3. Pardi, N.; Hogan, M.J.; Porter, F.W.; Weissman, D. MRNA Vaccines—A New Era in Vaccinology. *Nat. Rev. Drug Discov.* **2018**, *17*, 261–279. [CrossRef] [PubMed]

4. Jackson, L.A.; Anderson, E.J.; Rouphael, N.G.; Roberts, P.C.; Makhene, M.; Coler, R.N.; McCullough, M.P.; Chappell, J.D.; Denison, M.R.; Stevens, L.J.; et al. An MRNA Vaccine against SARS-CoV-2 - Preliminary Report. *N. Engl. J. Med.* **2020**, *383*, 1920–1931. [CrossRef]
5. Fogel, D.B. Factors Associated with Clinical Trials That Fail and Opportunities for Improving the Likelihood of Success: A Review. *Contemp. Clin. Trials Commun.* **2018**, *11*, 156–164. [CrossRef]
6. Seyhan, A.A. Lost in Translation: The Valley of Death across Preclinical and Clinical Divide—Identification of Problems and Overcoming Obstacles. *Transl. Med. Commun.* **2019**, *4*, 18. [CrossRef]
7. Blanco, E.; Shen, H.; Ferrari, M. Principles of Nanoparticle Design for Overcoming Biological Barriers to Drug Delivery. *Nat. Biotechnol.* **2015**, *33*, 941–951. [CrossRef] [PubMed]
8. Muttenthaler, M.; King, G.F.; Adams, D.J.; Alewood, P.F. Trends in Peptide Drug Discovery. *Nat. Rev. Drug Discov.* **2021**, *20*, 309–325. [CrossRef]
9. Kumari, S.; Mg, S.; Mayor, S. Endocytosis Unplugged: Multiple Ways to Enter the Cell. *Cell Res.* **2010**, *20*, 256–275. [CrossRef] [PubMed]
10. Parodi, A.; Corbo, C.; Cevenini, A.; Molinaro, R.; Palomba, R.; Pandolfi, L.; Agostini, M.; Salvatore, F.; Tasciotti, E. Enabling Cytoplasmic Delivery and Organelle Targeting by Surface Modification of Nanocarriers. *Nanomedicine* **2015**, *10*, 1923–1940. [CrossRef] [PubMed]
11. Petros, R.A.; DeSimone, J.M. Strategies in the Design of Nanoparticles for Therapeutic Applications. *Nat. Rev. Drug Discov.* **2010**, *9*, 615–627. [CrossRef]
12. Ruoslahti, E. Peptides as Targeting Elements and Tissue Penetration Devices for Nanoparticles. *Adv. Mater.* **2012**, *24*, 3747–3756. [CrossRef] [PubMed]
13. Jeong, W.; Bu, J.; Kubiatowicz, L.J.; Chen, S.S.; Kim, Y.; Hong, S. Peptide–Nanoparticle Conjugates: A next Generation of Diagnostic and Therapeutic Platforms? *Nano Converg.* **2018**, *5*, 38. [CrossRef] [PubMed]
14. Hossen, S.; Hossain, M.K.; Basher, M.K.; Mia, M.N.H.; Rahman, M.T.; Uddin, M.J. Smart Nanocarrier-Based Drug Delivery Systems for Cancer Therapy and Toxicity Studies: A Review. *J. Adv. Res.* **2019**, *15*, 1–18. [CrossRef] [PubMed]
15. Lee, A.C.-L.; Harris, J.L.; Khanna, K.K.; Hong, J.-H. A Comprehensive Review on Current Advances in Peptide Drug Development and Design. *Int. J. Mol. Sci.* **2019**, *20*, 2383. [CrossRef] [PubMed]
16. Cooper, B.M.; Iegre, J.; Donovan, D.H.O.; Halvarsson, M.Ö.; Spring, D.R. Peptides as a Platform for Targeted Therapeutics for Cancer: Peptide–Drug Conjugates (PDCs). *Chem. Soc. Rev.* **2021**, *50*, 1480–1494. [CrossRef]
17. Mitchell, M.J.; Billingsley, M.M.; Haley, R.M.; Wechsler, M.E.; Peppas, N.A.; Langer, R. Engineering Precision Nanoparticles for Drug Delivery. *Nat. Rev. Drug Discov.* **2021**, *20*, 101–124. [CrossRef]
18. Sandra, F.; Khaliq, N.U.; Sunna, A.; Care, A. Developing Protein-Based Nanoparticles as Versatile Delivery Systems for Cancer Therapy and Imaging. *Nanomaterials* **2019**, *9*, 1329. [CrossRef] [PubMed]
19. Avci, F.G.; Sariyar Akbulut, B.; Ozkirimli, E. Membrane Active Peptides and Their Biophysical Characterization. *Biomolecules* **2018**, *8*, 77. [CrossRef]
20. Tarvirdipour, S.; Schoenenberger, C.-A.; Benenson, Y.; Palivan, C.G. A Self-Assembling Amphiphilic Peptide Nanoparticle for the Efficient Entrapment of DNA Cargoes up to 100 Nucleotides in Length. *Soft Matter.* **2020**, *16*, 1678–1691. [CrossRef] [PubMed]
21. Sánchez-Navarro, M.; Teixidó, M.; Giralt, E. Jumping Hurdles: Peptides Able To Overcome Biological Barriers. *Acc. Chem. Res.* **2017**, *50*, 1847–1854. [CrossRef] [PubMed]
22. Cabral, B.P.; da Graça Derengowski Fonseca, M.; Mota, F.B. The Recent Landscape of Cancer Research Worldwide: A Bibliometric and Network Analysis. *Oncotarget* **2018**, *9*, 30474–30484. [CrossRef]
23. Ruoslahti, E. Specialization of Tumour Vasculature. *Nat. Rev. Cancer* **2002**, *2*, 83–90. [CrossRef]
24. Marelli, U.K.; Rechenmacher, F.; Sobahi, T.R.A.; Mas-Moruno, C.; Kessler, H. Tumor Targeting via Integrin Ligands. *Front. Oncol.* **2013**, *3*. [CrossRef]
25. Brown, K.C. Peptidic Tumor Targeting Agents: The Road from Phage Display Peptide Selections to Clinical Applications. *Curr. Pharm. Des.* **2010**, *16*, 1040–1054. [CrossRef]
26. Zhao, N.; Qin, Y.; Liu, H.; Cheng, Z. Tumor-Targeting Peptides: Ligands for Molecular Imaging and Therapy. *Anticancer Agents Med. Chem.* **2018**, *18*, 74–86. [CrossRef] [PubMed]
27. Li, Z.J.; Cho, C.H. Peptides as Targeting Probes against Tumor Vasculature for Diagnosis and Drug Delivery. *J. Transl. Med.* **2012**, *10*, S1. [CrossRef]
28. Ko, J.K.; Auyeung, K.K. Identification of Functional Peptides from Natural and Synthetic Products on Their Anticancer Activities by Tumor Targeting. *Curr. Med. Chem.* **2014**, *21*, 2346–2356. [CrossRef] [PubMed]
29. Zhao, J.; Santino, F.; Giacomini, D.; Gentilucci, L. Integrin-Targeting Peptides for the Design of Functional Cell-Responsive Biomaterials. *Biomedicines* **2020**, *8*, 307. [CrossRef] [PubMed]
30. Wu, P.-H.; Opadele, A.E.; Onodera, Y.; Nam, J.-M. Targeting Integrins in Cancer Nanomedicine: Applications in Cancer Diagnosis and Therapy. *Cancers* **2019**, *11*, 1783. [CrossRef] [PubMed]
31. Huang, Y.; Wang, X.; Huang, W.; Cheng, Q.; Zheng, S.; Guo, S.; Cao, H.; Liang, X.-J.; Du, Q.; Liang, Z. Systemic Administration of SiRNA via CRGD-Containing Peptide. *Sci. Rep.* **2015**, *5*, 12458. [CrossRef] [PubMed]
32. Roth, L.; Agemy, L.; Kotamraju, V.R.; Braun, G.; Teesalu, T.; Sugahara, K.N.; Hamzah, J.; Ruoslahti, E. Transtumoral Targeting Enabled by a Novel Neuropilin-Binding Peptide. *Oncogene* **2012**, *31*, 3754–3763. [CrossRef]

33. Jiang, Y.; Liu, S.; Zhang, Y.; Li, H.; He, H.; Dai, J.; Jiang, T.; Ji, W.; Geng, D.; Elzatahry, A.A.; et al. Magnetic Mesoporous Nanospheres Anchored with LyP-1 as an Efficient Pancreatic Cancer Probe. *Biomaterials* **2017**, *115*, 9–18. [CrossRef] [PubMed]
34. Song, N.; Zhao, L.; Zhu, M.; Zhao, J. Recent Progress in LyP-1-Based Strategies for Targeted Imaging and Therapy. *Drug Deliv.* **2019**, *26*, 363–375. [CrossRef] [PubMed]
35. Vaidya, A.M.; Sun, Z.; Ayat, N.; Schilb, A.; Liu, X.; Jiang, H.; Sun, D.; Scheidt, J.; Qian, V.; He, S.; et al. Systemic Delivery of Tumor-Targeting SiRNA Nanoparticles against an Oncogenic LncRNA Facilitates Effective Triple-Negative Breast Cancer Therapy. *Bioconjug. Chem.* **2019**, *30*, 907–919. [CrossRef] [PubMed]
36. He, S.; Cen, B.; Liao, L.; Wang, Z.; Qin, Y.; Wu, Z.; Liao, W.; Zhang, Z.; Ji, A. A Tumor-Targeting CRGD-EGFR SiRNA Conjugate and Its Anti-Tumor Effect on Glioblastoma in Vitro and in Vivo. *Drug Deliv.* **2017**, *24*, 471–481. [CrossRef]
37. Zhang, Y.; Li, S.; Zhou, X.; Sun, J.; Fan, X.; Guan, Z.; Zhang, L.; Yang, Z. Construction of a Targeting Nanoparticle of 3′,3″-Bis-Peptide-SiRNA Conjugate/Mixed Lipid with Postinserted DSPE-PEG2000-CRGD. *Mol. Pharm.* **2019**, *16*. [CrossRef]
38. Lo, J.H.; Hao, L.; Muzumdar, M.D.; Raghavan, S.; Kwon, E.J.; Pulver, E.M.; Hsu, F.; Aguirre, A.J.; Wolpin, B.M.; Fuchs, C.S.; et al. IRGD-Guided Tumor-Penetrating Nanocomplexes for Therapeutic SiRNA Delivery to Pancreatic Cancer. *Mol. Cancer Ther.* **2018**, *17*, 2377–2388. [CrossRef]
39. Zhou, Y.; Yuan, Y.; Liu, M.; Hu, X.; Quan, Y.; Chen, X. Tumor-Specific Delivery of KRAS SiRNA with IRGD-Exosomes Efficiently Inhibits Tumor Growth. *ExRNA* **2019**, *1*, 28. [CrossRef]
40. Xia, Y.; Xu, T.; Wang, C.; Li, Y.; Lin, Z.; Zhao, M.; Zhu, B. Novel Functionalized Nanoparticles for Tumor-Targeting Co-Delivery of Doxorubicin and SiRNA to Enhance Cancer Therapy. *Int. J. Nanomed.* **2017**, *13*, 143–159. [CrossRef]
41. Bai, Y.; Li, Z.; Liu, L.; Sun, T.; Fan, X.; Wang, T.; Gou, Z.; Tan, S. Tumor-Targeting Peptide for Redox-Responsive Pt Prodrug and Gene Codelivery and Synergistic Cancer Chemotherapy. *ACS Appl. Bio Mater.* **2019**, *2*, 1420–1426. [CrossRef]
42. Sharma, M.; El-Sayed, N.S.; Do, H.; Parang, K.; Tiwari, R.K.; Aliabadi, H.M. Tumor-Targeted Delivery of SiRNA Using Fatty Acyl-CGKRK Peptide Conjugates. *Sci. Rep.* **2017**, *7*, 6993. [CrossRef]
43. Li, Y.; Wang, H.; Wang, K.; Hu, Q.; Yao, Q.; Shen, Y.; Yu, G.; Tang, G. Targeted Co-Delivery of PTX and TR3 SiRNA by PTP Peptide Modified Dendrimer for the Treatment of Pancreatic Cancer. *Small* **2017**, *13*, 1602697. [CrossRef] [PubMed]
44. Wan, W.; Qu, C.; Zhou, Y.; Zhang, L.; Chen, M.; Liu, Y.; You, B.; Li, F.; Wang, D.; Zhang, X. Doxorubicin and SiRNA-PD-L1 Co-Delivery with T7 Modified ROS-Sensitive Nanoparticles for Tumor Chemoimmunotherapy. *Int. J. Pharm.* **2019**, *566*, 731–744. [CrossRef] [PubMed]
45. Kim, B.; Sun, S.; Varner, J.A.; Howell, S.B.; Ruoslahti, E.; Sailor, M.J. Securing the Payload, Finding the Cell, and Avoiding the Endosome: Peptide-Targeted, Fusogenic Porous Silicon Nanoparticles for Delivery of SiRNA. *Adv. Mater.* **2019**, *31*, 1902952. [CrossRef]
46. Liang, Y.; Peng, J.; Li, N.; Yu-Wai-Man, C.; Wang, Q.; Xu, Y.; Wang, H.; Tagalakis, A.D.; Du, Z. Smart Nanoparticles Assembled by Endogenous Molecules for SiRNA Delivery and Cancer Therapy via CD44 and EGFR Dual-Targeting. *Nanomed. Nanotechnol. Biol. Med.* **2019**, *15*, 208–217. [CrossRef]
47. Lu, Y.; Jiang, W.; Wu, X.; Huang, S.; Huang, Z.; Shi, Y.; Dai, Q.; Chen, J.; Ren, F.; Gao, S. Peptide T7-Modified Polypeptide with Disulfide Bonds for Targeted Delivery of Plasmid DNA for Gene Therapy of Prostate Cancer. *Int. J. Nanomed.* **2018**, *13*, 6913–6927. [CrossRef]
48. Desale, K.; Kuche, K.; Jain, S. Cell-Penetrating Peptides (CPPs): An Overview of Applications for Improving the Potential of Nanotherapeutics. *Biomater. Sci.* **2021**, *9*, 1153–1188. [CrossRef]
49. Torchilin, V.P. Cell Penetrating Peptide-Modified Pharmaceutical Nanocarriers for Intracellular Drug and Gene Delivery. *Pept. Sci.* **2008**, *90*, 604–610. [CrossRef]
50. Wang, F.; Wang, Y.; Zhang, X.; Zhang, W.; Guo, S.; Jin, F. Recent Progress of Cell-Penetrating Peptides as New Carriers for Intracellular Cargo Delivery. *J. Control. Release* **2014**, *174*, 126–136. [CrossRef]
51. Khan, M.M.; Filipczak, N.; Torchilin, V.P. Cell Penetrating Peptides: A Versatile Vector for Co-Delivery of Drug and Genes in Cancer. *J. Control. Release* **2021**, *330*, 1220–1228. [CrossRef] [PubMed]
52. Tian, Y.; Zhou, M.; Shi, H.; Gao, S.; Xie, G.; Zhu, M.; Wu, M.; Chen, J.; Niu, Z. Integration of Cell-Penetrating Peptides with Rod-like Bionanoparticles: Virus-Inspired Gene-Silencing Technology. *Nano Lett.* **2018**, *18*, 5453–5460. [CrossRef] [PubMed]
53. Conde, J.; Ambrosone, A.; Hernandez, Y.; Tian, F.; McCully, M.; Berry, C.C.; Baptista, P.V.; Tortiglione, C.; de la Fuente, J.M. 15 Years on SiRNA Delivery: Beyond the State-of-the-Art on Inorganic Nanoparticles for RNAi Therapeutics. *Nano Today* **2015**, *10*, 421–450. [CrossRef]
54. Kang, Z.; Meng, Q.; Liu, K. Peptide-Based Gene Delivery Vectors. *J. Mater. Chem. B* **2019**, *7*, 1824–1841. [CrossRef] [PubMed]
55. Madani, F.; Lindberg, S.; Langel, Ü.; Futaki, S.; Gräslund, A. Mechanisms of Cellular Uptake of Cell-Penetrating Peptides. *J. Biophys.* **2011**, *2011*, 1–10. [CrossRef] [PubMed]
56. Lindgren, M.; Langel, Ü. Classes and Prediction of Cell-Penetrating Peptides. In *Cell-Penetrating Peptides: Methods and Protocols*; Langel, Ü., Ed.; Methods in Molecular Biology; Humana Press: Totowa, NJ, USA, 2011; pp. 3–19, ISBN 978-1-60761-919-2.
57. Xie, J.; Bi, Y.; Zhang, H.; Dong, S.; Teng, L.; Lee, R.J.; Yang, Z. Cell-Penetrating Peptides in Diagnosis and Treatment of Human Diseases: From Preclinical Research to Clinical Application. *Front. Pharmacol.* **2020**, *11*. [CrossRef]
58. Trabulo, S.; Cardoso, A.L.; Mano, M.; De Lima, M.C.P. Cell-Penetrating Peptides—Mechanisms of Cellular Uptake and Generation of Delivery Systems. *Pharmaceuticals* **2010**, *3*, 961–993. [CrossRef]

59. Derakhshankhah, H.; Jafari, S. Cell Penetrating Peptides: A Concise Review with Emphasis on Biomedical Applications. *Biomed. Pharmacother.* **2018**, *108*, 1090–1096. [CrossRef]
60. Alluis, B.; Fruchart, J.-S. Cell Penetrating Peptide Conjugates for Delivering of Nucleic Acids into a Cell 2013. Patent Application No 13/659,395, 30 May 2013.
61. Simeoni, F.; Morris, M.C.; Heitz, F.; Divita, G. Insight into the Mechanism of the Peptide-based Gene Delivery System MPG: Implications for Delivery of SiRNA into Mammalian Cells. *Nucleic Acids Res.* **2003**, *31*, 2717–2724. [CrossRef]
62. Crombez, L.; Morris, M.C.; Dufort, S.; Aldrian-Herrada, G.; Nguyen, Q.; Mc Master, G.; Coll, J.-L.; Heitz, F.; Divita, G. Targeting Cyclin B1 through Peptide-Based Delivery of SiRNA Prevents Tumour Growth. *Nucleic Acids Res.* **2009**, *37*, 4559–4569. [CrossRef]
63. Rhee, M.; Davis, P. Mechanism of Uptake of C105Y, a Novel Cell-Penetrating Peptide. *J. Biol. Chem.* **2006**, *281*, 1233–1240. [CrossRef]
64. Sun, W.; Ziady, A.G. Real-Time Imaging of Gene Delivery and Expression with DNA Nanoparticle Technologies. In *Micro and Nano Technologies in Bioanalysis: Methods and Protocols*; Foote, R.S., Lee, J.W., Eds.; Methods in Molecular BiologyTM; Humana Press: Totowa, NJ, USA, 2009; pp. 525–546, ISBN 978-1-59745-483-4.
65. Saamoah-Moffatt, S.; Wiehle, S.; Christina, R.J. Enhanced Gene Delivery by a Novel Tumor-Specific Vector Containing Peptide and Nucleic Acid Based Nuclear Translocation Signals. *Mol. Ther.* **2004**, *9*, S137–S138. [CrossRef]
66. Lee, M.; Choi, J.S.; Ko, K.S. 202. DNA Delivery to the Mitochondria Sites Using Leader Peptide Conjugated Polyethylenimine. *Mol. Ther.* **2005**, *11*, S79–S80. [CrossRef]
67. Rajala, A.; Wang, Y.; Zhu, Y.; Ranjo-Bishop, M.; Ma, J.-X.; Mao, C.; Rajala, R.V.S. Nanoparticle-Assisted Targeted Delivery of Eye-Specific Genes to Eyes Significantly Improves the Vision of Blind Mice In Vivo. *Nano Lett.* **2014**, *14*, 5257–5263. [CrossRef] [PubMed]
68. Wu, Z.; Zhan, S.; Fan, W.; Ding, X.; Wu, X.; Zhang, W.; Fu, Y.; Huang, Y.; Huang, X.; Chen, R.; et al. Peptide-Mediated Tumor Targeting by a Degradable Nano Gene Delivery Vector Based on Pluronic-Modified Polyethylenimine. *Nanoscale Res. Lett.* **2016**, *11*, 122. [CrossRef]
69. Dong, S.; Zhou, X.; Yang, J. TAT Modified and Lipid–PEI Hybrid Nanoparticles for Co-Delivery of Docetaxel and PDNA. *Biomed. Pharmacother.* **2016**, *84*, 954–961. [CrossRef]
70. Ishiguro, S.; Alhakamy, N.A.; Uppalapati, D.; Delzeit, J.; Berkland, C.J.; Tamura, M. Combined Local Pulmonary and Systemic Delivery of AT2R Gene by Modified TAT Peptide Nanoparticles Attenuates Both Murine and Human Lung Carcinoma Xenografts in Mice. *J. Pharm. Sci.* **2017**, *106*, 385–394. [CrossRef]
71. Bahadoran, A.; Ebrahimi, M.; Yeap, S.K.; Safi, N.; Moeini, H.; Hair-Bejo, M.; Hussein, M.Z.; Omar, A.R. Induction of a Robust Immune Response against Avian Influenza Virus Following Transdermal Inoculation with H5-DNA Vaccine Formulated in Modified Dendrimer-Based Delivery System in Mouse Model. *Int. J. Nanomed.* **2017**, *12*, 8573–8585. [CrossRef] [PubMed]
72. Li, L.; Hu, S.; Chen, X. Non-Viral Delivery Systems for CRISPR/Cas9-Based Genome Editing: Challenges and Opportunities. *Biomaterials* **2018**, *171*, 207–218. [CrossRef]
73. Koizumi, K.; Nakamura, H.; Iijima, M.; Matsuzaki, T.; Somiya, M.; Kumasawa, K.; Kimura, T.; Kuroda, S. In Vivo Uterine Local Gene Delivery System Using TAT-Displaying Bionanocapsules. *J. Gene Med.* **2019**, *21*, e3140. [CrossRef]
74. Yi, A.; Sim, D.; Lee, Y.-J.; Sarangthem, V.; Park, R.-W. Development of Elastin-like Polypeptide for Targeted Specific Gene Delivery in Vivo. *J. Nanobiotechnol.* **2020**, *18*, 15. [CrossRef] [PubMed]
75. Yoon, J.Y.; Yang, K.-J.; Park, S.-N.; Kim, D.-K.; Kim, J.-D. The Effect of Dexamethasone/Cell-Penetrating Peptide Nanoparticles on Gene Delivery for Inner Ear Therapy. *Int. J. Nanomed.* **2016**, *11*, 6123–6134. [CrossRef]
76. Khalil, I.A.; Harashima, H. An Efficient PEGylated Gene Delivery System with Improved Targeting: Synergism between Octaarginine and a Fusogenic Peptide. *Int. J. Pharm.* **2018**, *538*, 179–187. [CrossRef] [PubMed]
77. Lee, J.; Jung, J.; Kim, Y.-J.; Lee, E.; Choi, J.S. Gene Delivery of PAMAM Dendrimer Conjugated with the Nuclear Localization Signal Peptide Originated from Fibroblast Growth Factor 3. *Int. J. Pharm.* **2014**, *459*, 10–18. [CrossRef] [PubMed]
78. Li, Y.; Li, Y.; Wang, X.; Lee, R.J.; Teng, L. Fatty Acid Modified Octa-Arginine for Delivery of SiRNA. *Int. J. Pharm.* **2015**, *495*, 527–535. [CrossRef] [PubMed]
79. Van Rossenberg, S.; van Keulen, A.; Drijfhout, J.-W.; Vasto, S.; Koerten, H.K.; Spies, F.; van 't Noordende, J.; van Berkel, T.; Biessen, E.a.L. Stable Polyplexes Based on Arginine-Containing Oligopeptides for in Vivo Gene Delivery. *Gene Ther.* **2004**, *11*, 457–464. [CrossRef] [PubMed]
80. Br, L.; Md, L.; Hj, C.; Hj, L. Arginine-Rich Cell-Penetrating Peptides Deliver Gene into Living Human Cells. *Gene* **2012**, *505*, 37–45. [CrossRef]
81. Liu, C.; Liu, X.; Rocchi, P.; Qu, F.; Iovanna, J.L.; Peng, L. Arginine-Terminated Generation 4 PAMAM Dendrimer as an Effective Nanovector for Functional SiRNA Delivery in Vitro and in Vivo. *Bioconjug. Chem.* **2014**, *25*, 521–532. [CrossRef]
82. Kato, T.; Yamashita, H.; Misawa, T.; Nishida, K.; Kurihara, M.; Tanaka, M.; Demizu, Y.; Oba, M. Plasmid DNA Delivery by Arginine-Rich Cell-Penetrating Peptides Containing Unnatural Amino Acids. *Bioorg. Med. Chem.* **2016**, *24*, 2681–2687. [CrossRef]
83. Arukuusk, P.; Pärnaste, L.; Hällbrink, M.; Langel, Ü. PepFects and NickFects for the Intracellular Delivery of Nucleic Acids. In *Cell-Penetrating Peptides: Methods and Protocols*; Langel, Ü., Ed.; Methods in Molecular Biology; Springer: New York, NY, USA, 2015; pp. 303–315, ISBN 978-1-4939-2806-4.
84. Margus, H.; Arukuusk, P.; Langel, Ü.; Pooga, M. Characteristics of Cell-Penetrating Peptide/Nucleic Acid Nanoparticles. *Mol. Pharm.* **2016**, *13*, 172–179. [CrossRef] [PubMed]

85. Freimann, K.; Arukuusk, P.; Kurrikoff, K.; Pärnaste, L.; Raid, R.; Piirsoo, A.; Pooga, M.; Langel, Ü. Formulation of Stable and Homogeneous Cell-Penetrating Peptide NF55 Nanoparticles for Efficient Gene Delivery In Vivo. *Mol. Ther. Nucleic Acids* **2018**, *10*, 28–35. [CrossRef]
86. Padari, K.; Porosk, L.; Arukuusk, P.; Pooga, M. Characterization of Peptide–Oligonucleotide Complexes Using Electron Microscopy, Dynamic Light Scattering, and Protease Resistance Assay. In *Oligonucleotide-Based Therapies: Methods and Protocols*; Gissberg, O., Zain, R., Lundin, K.E., Eds.; Methods in Molecular Biology; Springer: New York, NY, USA, 2019; pp. 127–139, ISBN 978-1-4939-9670-4.
87. Minchin, R.F.; Yang, S. Endosomal Disruptors in Non-Viral Gene Delivery. *Expert Opin. Drug Deliv.* **2010**, *7*, 331–339. [CrossRef] [PubMed]
88. Jiang, Q.-Y.; Lai, L.-H.; Shen, J.; Wang, Q.-Q.; Xu, F.-J.; Tang, G.-P. Gene Delivery to Tumor Cells by Cationic Polymeric Nanovectors Coupled to Folic Acid and the Cell-Penetrating Peptide Octaarginine. *Biomaterials* **2011**, *32*, 7253–7262. [CrossRef] [PubMed]
89. Oba, M.; Demizu, Y.; Yamashita, H.; Kurihara, M.; Tanaka, M. Plasmid DNA Delivery Using Fluorescein-Labeled Arginine-Rich Peptides. *Bioorg. Med. Chem.* **2015**, *23*, 4911–4918. [CrossRef]
90. Zhang, L.; Li, Z.; Sun, F.; Xu, Y.; Du, Z. Effect of Inserted Spacer in Hepatic Cell-Penetrating Multifunctional Peptide Component on the DNA Intracellular Delivery of Quaternary Complexes Based on Modular Design. *Int. J. Nanomed.* **2016**, *11*, 6283–6295. [CrossRef] [PubMed]
91. Van den Brand, D.; Gorris, M.A.J.; van Asbeck, A.H.; Palmen, E.; Ebisch, I.; Dolstra, H.; Hällbrink, M.; Massuger, L.F.A.G.; Brock, R. Peptide-Mediated Delivery of Therapeutic MRNA in Ovarian Cancer. *Eur. J. Pharm. Biopharm.* **2019**, *141*, 180–190. [CrossRef]
92. Kurrikoff, K.; Veiman, K.-L.; Künnapuu, K.; Peets, E.M.; Lehto, T.; Pärnaste, L.; Arukuusk, P.; Langel, Ü. Effective in Vivo Gene Delivery with Reduced Toxicity, Achieved by Charge and Fatty Acid -Modified Cell Penetrating Peptide. *Sci. Rep.* **2017**, *7*, 17056. [CrossRef] [PubMed]
93. Hou, K.K.; Pan, H.; Lanza, G.M.; Wickline, S.A. Melittin Derived Peptides for Nanoparticle Based SiRNA Transfection. *Biomaterials* **2013**, *34*, 3110–3119. [CrossRef]
94. Christiaens, B.; Dubruel, P.; Grooten, J.; Goethals, M.; Vandekerckhove, J.; Schacht, E.; Rosseneu, M. Enhancement of Polymethacrylate-Mediated Gene Delivery by Penetratin. *Eur. J. Pharm. Sci.* **2005**, *24*, 525–537. [CrossRef]
95. Zorko, M.; Langel, Ü. Cell-Penetrating Peptides: Mechanism and Kinetics of Cargo Delivery. *Adv. Drug Deliv. Rev.* **2005**, *57*, 529–545. [CrossRef]
96. Liu, C.; Jiang, K.; Tai, L.; Liu, Y.; Wei, G.; Lu, W.; Pan, W. Facile Noninvasive Retinal Gene Delivery Enabled by Penetratin. *ACS Appl. Mater. Interfaces* **2016**, *8*, 19256–19267. [CrossRef]
97. Mo, R.H.; Zaro, J.L.; Shen, W.-C. Comparison of Cationic and Amphipathic Cell Penetrating Peptides for SiRNA Delivery and Efficacy. *Mol. Pharm.* **2012**, *9*, 299–309. [CrossRef] [PubMed]
98. Liu, B.R.; Huang, Y.-W.; Aronstam, R.S.; Lee, H.-J. Identification of a Short Cell-Penetrating Peptide from Bovine Lactoferricin for Intracellular Delivery of DNA in Human A549 Cells. *PLoS ONE* **2016**, *11*, e0150439. [CrossRef] [PubMed]
99. Liu, B.R.; Huang, Y.-W.; Korivi, M.; Lo, S.-Y.; Aronstam, R.S.; Lee, H.-J. The Primary Mechanism of Cellular Internalization for a Short Cell- Penetrating Peptide as a Nano-Scale Delivery System. *Curr. Pharm. Biotechnol.* **2017**, *18*, 569–584. [CrossRef] [PubMed]
100. Freitas, V.J.F.; Campelo, I.S.; Silva, M.M.A.S.; Cavalcanti, C.M.; Teixeira, D.I.A.; Camargo, L.S.A.; Melo, L.M.; Rádis-Baptista, G. Disulphide-Less Crotamine Is Effective for Formation of DNA–Peptide Complex but Is Unable to Improve Bovine Embryo Transfection. *Zygote* **2020**, *28*, 72–79. [CrossRef]
101. Campeiro, J.D.; Dam, W.; Monte, G.G.; Porta, L.C.; de Oliveira, L.C.G.; Nering, M.B.; Viana, G.M.; Carapeto, F.C.; Oliveira, E.B.; van den Born, J.; et al. Long Term Safety of Targeted Internalization of Cell Penetrating Peptide Crotamine into Renal Proximal Tubular Epithelial Cells in Vivo. *Sci. Rep.* **2019**, *9*, 1–13. [CrossRef]
102. Nascimento, F.D.; Hayashi, M.A.F.; Kerkis, A.; Oliveira, V.; Oliveira, E.B.; Rádis-Baptista, G.; Nader, H.B.; Yamane, T.; dos Santos Tersariol, I.L.; Kerkis, I. Crotamine Mediates Gene Delivery into Cells through the Binding to Heparan Sulfate Proteoglycans. *J. Biol. Chem.* **2007**, *282*, 21349–21360. [CrossRef]
103. Zavaglia, D.; Favrot, M.-C.; Eymin, B.; Tenaud, C.; Coll, J.-L. Intercellular Trafficking and Enhanced in Vivo Antitumour Activity of a Non-Virally Delivered P27-VP22 Fusion Protein. *Gene Ther.* **2003**, *10*, 314–325. [CrossRef]
104. Chen, H.-C.G.; Chiou, S.-T.; Zheng, J.-Y.; Yang, S.-H.; Lai, S.-S.; Kuo, T.-Y. The Nuclear Localization Signal Sequence of Porcine Circovirus Type 2 ORF2 Enhances Intracellular Delivery of Plasmid DNA. *Arch. Virol.* **2011**, *156*, 803–815. [CrossRef]
105. Kim, N.H.; Provoda, C.; Lee, K.-D. Design and Characterization of Novel Recombinant Listeriolysin O–Protamine Fusion Proteins for Enhanced Gene Delivery. *Mol. Pharm.* **2015**, *12*, 342–350. [CrossRef]
106. Rezaee, M.; Oskuee, R.K.; Nassirli, H.; Malaekeh-Nikouei, B. Progress in the Development of Lipopolyplexes as Efficient Non-Viral Gene Delivery Systems. *J. Control. Release* **2016**, *236*, 1–14. [CrossRef]
107. Men, K.; Zhang, R.; Zhang, X.; Huang, R.; Zhu, G.; Tong, R.; Yang, L.; Wei, Y.; Duan, X. Delivery of Modified MRNA Encoding Vesicular Stomatitis Virus Matrix Protein for Colon Cancer Gene Therapy. *RSC Adv.* **2018**, *8*, 12104–12115. [CrossRef]
108. Limeres, M.J.; Suñé-Pou, M.; Prieto-Sánchez, S.; Moreno-Castro, C.; Nusblat, A.D.; Hernández-Munain, C.; Castro, G.R.; Suñé, C.; Suñé-Negre, J.M.; Cuestas, M.L. Development and Characterization of an Improved Formulation of Cholesteryl Oleate-Loaded Cationic Solid-Lipid Nanoparticles as an Efficient Non-Viral Gene Delivery System. *Colloids Surf. B Biointerfaces* **2019**, *184*, 110533. [CrossRef]

109. Zhang, F.; Li, H.-Y. Preparation of Lipid–Peptide–DNA (LPD) Nanoparticles and Their Use for Gene Transfection. In *Nanoparticles in Biology and Medicine: Methods and Protocols*; Ferrari, E., Soloviev, M., Eds.; Methods in Molecular Biology; Springer US: New York, NY, USA, 2020; pp. 91–98, ISBN 978-1-07-160319-2.
110. Astriab-Fisher, A.; Sergueev, D.; Fisher, M.; Ramsay Shaw, B.; Juliano, R.L. Conjugates of Antisense Oligonucleotides with the Tat and Antennapedia Cell-Penetrating Peptides: Effects on Cellular Uptake, Binding to Target Sequences, and Biologic Actions. *Pharm. Res.* **2002**, *19*, 744–754. [CrossRef]
111. Davidson, T.J.; Harel, S.; Arboleda, V.A.; Prunell, G.F.; Shelanski, M.L.; Greene, L.A.; Troy, C.M. Highly Efficient Small Interfering RNA Delivery to Primary Mammalian Neurons Induces MicroRNA-Like Effects before MRNA Degradation. *J. Neurosci.* **2004**, *24*, 10040–10046. [CrossRef] [PubMed]
112. Muñoz-Morris, M.A.; Heitz, F.; Divita, G.; Morris, M.C. The Peptide Carrier Pep-1 Forms Biologically Efficient Nanoparticle Complexes. *Biochem. Biophys. Res. Commun.* **2007**, *355*, 877–882. [CrossRef]
113. Chang, J.-C.; Liu, K.-H.; Chuang, C.-S.; Su, H.-L.; Wei, Y.-H.; Kuo, S.-J.; Liu, C.-S. Treatment of Human Cells Derived from MERRF Syndrome by Peptide-Mediated Mitochondrial Delivery. *Cytotherapy* **2013**, *15*, 1580–1596. [CrossRef] [PubMed]
114. Crowet, J.-M.; Lins, L.; Deshayes, S.; Divita, G.; Morris, M.; Brasseur, R.; Thomas, A. Modeling of Non-Covalent Complexes of the Cell-Penetrating Peptide CADY and Its SiRNA Cargo. *Biochim. Biophys. Acta BBA Biomembr.* **2013**, *1828*, 499–509. [CrossRef]
115. Rydström, A.; Deshayes, S.; Konate, K.; Crombez, L.; Padari, K.; Boukhaddaoui, H.; Aldrian, G.; Pooga, M.; Divita, G. Direct Translocation as Major Cellular Uptake for CADY Self-Assembling Peptide-Based Nanoparticles. *PLoS ONE* **2011**, *6*, e25924. [CrossRef]
116. Konate, K.; Lindberg, M.F.; Vaissiere, A.; Jourdan, C.; Aldrian, G.; Margeat, E.; Deshayes, S.; Boisguerin, P. Optimisation of Vectorisation Property: A Comparative Study for a Secondary Amphipathic Peptide. *Int. J. Pharm.* **2016**, *509*, 71–84. [CrossRef]
117. Elmquist, A.; Lindgren, M.; Bartfai, T.; Langel, Ü. VE-Cadherin-Derived Cell-Penetrating Peptide, PVEC, with Carrier Functions. *Exp. Cell Res.* **2001**, *269*, 237–244. [CrossRef] [PubMed]
118. Rajpal; Mann, A.; Khanduri, R.; Naik, R.J.; Ganguli, M. Structural Rearrangements and Chemical Modifications in Known Cell Penetrating Peptide Strongly Enhance DNA Delivery Efficiency. *J. Control. Release* **2012**, *157*, 260–271. [CrossRef] [PubMed]
119. Patel, S.G.; Sayers, E.J.; He, L.; Narayan, R.; Williams, T.L.; Mills, E.M.; Allemann, R.K.; Luk, L.Y.P.; Jones, A.T.; Tsai, Y.-H. Cell-Penetrating Peptide Sequence and Modification Dependent Uptake and Subcellular Distribution of Green Florescent Protein in Different Cell Lines. *Sci. Rep.* **2019**, *9*, 6298. [CrossRef]
120. Zhang, D.; Wang, J.; Xu, D. Cell-Penetrating Peptides as Noninvasive Transmembrane Vectors for the Development of Novel Multifunctional Drug-Delivery Systems. *J. Control. Release* **2016**, *229*, 130–139. [CrossRef] [PubMed]
121. Gessner, I.; Neundorf, I. Nanoparticles Modified with Cell-Penetrating Peptides: Conjugation Mechanisms, Physicochemical Properties, and Application in Cancer Diagnosis and Therapy. *Int. J. Mol. Sci.* **2020**, *21*, 2536. [CrossRef]
122. Fischer, R.; Fotin-Mleczek, M.; Hufnagel, H.; Brock, R. Break on through to the Other Side—Biophysics and Cell Biology Shed Light on Cell-Penetrating Peptides. *ChemBioChem* **2005**, *6*, 2126–2142. [CrossRef]
123. Zhou, Y.; Han, S.; Liang, Z.; Zhao, M.; Liu, G.; Wu, J. Progress in Arginine-Based Gene Delivery Systems. *J. Mater. Chem. B* **2020**. [CrossRef]
124. Nam, H.Y.; Nam, K.; Hahn, H.J.; Kim, B.H.; Lim, H.J.; Kim, H.J.; Choi, J.S.; Park, J.-S. Biodegradable PAMAM Ester for Enhanced Transfection Efficiency with Low Cytotoxicity. *Biomaterials* **2009**, *30*, 665–673. [CrossRef] [PubMed]
125. Tünnemann, G.; Ter-Avetisyan, G.; Martin, R.M.; Stöckl, M.; Herrmann, A.; Cardoso, M.C. Live-Cell Analysis of Cell Penetration Ability and Toxicity of Oligo-Arginines. *J. Pept. Sci.* **2008**, *14*, 469–476. [CrossRef]
126. Singh, T.; Murthy, A.S.N.; Yang, H.-J.; Im, J. Versatility of Cell-Penetrating Peptides for Intracellular Delivery of SiRNA. *Drug Deliv.* **2018**, *25*, 2005–2015. [CrossRef]
127. Ruseska, I.; Zimmer, A. Internalization Mechanisms of Cell-Penetrating Peptides. *Beilstein J. Nanotechnol.* **2020**, *11*, 101–123. [CrossRef]
128. Guidotti, G.; Brambilla, L.; Rossi, D. Cell-Penetrating Peptides: From Basic Research to Clinics. *Trends Pharmacol. Sci.* **2017**, *38*, 406–424. [CrossRef]
129. Munyendo, W.L.; Lv, H.; Benza-Ingoula, H.; Baraza, L.D.; Zhou, J. Cell Penetrating Peptides in the Delivery of Biopharmaceuticals. *Biomolecules* **2012**, *2*, 187–202. [CrossRef] [PubMed]
130. Veiman, K.-L.; Mäger, I.; Ezzat, K.; Margus, H.; Lehto, T.; Langel, K.; Kurrikoff, K.; Arukuusk, P.; Suhorutšenko, J.; Padari, K.; et al. PepFect14 Peptide Vector for Efficient Gene Delivery in Cell Cultures. *Mol. Pharm.* **2013**, *10*, 199–210. [CrossRef] [PubMed]
131. Pärnaste, L.; Arukuusk, P.; Langel, K.; Tenson, T.; Langel, Ü. The Formation of Nanoparticles between Small Interfering RNA and Amphipathic Cell-Penetrating Peptides. *Mol. Ther. Nucleic Acids* **2017**, *7*, 1–10. [CrossRef]
132. Carreras-Badosa, G.; Maslovskaja, J.; Periyasamy, K.; Urgard, E.; Padari, K.; Vaher, H.; Tserel, L.; Gestin, M.; Kisand, K.; Arukuusk, P.; et al. NickFect Type of Cell-Penetrating Peptides Present Enhanced Efficiency for MicroRNA-146a Delivery into Dendritic Cells and during Skin Inflammation. *Biomaterials* **2020**, *262*, 120316. [CrossRef] [PubMed]
133. Varkouhi, A.K.; Scholte, M.; Storm, G.; Haisma, H.J. Endosomal Escape Pathways for Delivery of Biologicals. *J. Control. Release* **2011**, *151*, 220–228. [CrossRef] [PubMed]
134. Mosquera, J.; García, I.; Liz-Marzán, L.M. Cellular Uptake of Nanoparticles versus Small Molecules: A Matter of Size. *Acc. Chem. Res.* **2018**, *51*, 2305–2313. [CrossRef] [PubMed]

135. Behzadi, S.; Serpooshan, V.; Tao, W.; Hamaly, M.A.; Alkawareek, M.Y.; Dreaden, E.C.; Brown, D.; Alkilany, A.M.; Farokhzad, O.C.; Mahmoudi, M. Cellular Uptake of Nanoparticles: Journey Inside the Cell. *Chem. Soc. Rev.* **2017**, *46*, 4218–4244. [CrossRef]
136. Lönn, P.; Kacsinta, A.D.; Cui, X.-S.; Hamil, A.S.; Kaulich, M.; Gogoi, K.; Dowdy, S.F. Enhancing Endosomal Escape for Intracellular Delivery of Macromolecular Biologic Therapeutics. *Sci. Rep.* **2016**, *6*, 1–9. [CrossRef]
137. Ahmad, A.; Khan, J.M.; Haque, S. Strategies in the Design of Endosomolytic Agents for Facilitating Endosomal Escape in Nanoparticles. *Biochimie* **2019**, *160*, 61–75. [CrossRef]
138. Erazo-Oliveras, A.; Muthukrishnan, N.; Baker, R.; Wang, T.-Y.; Pellois, J.-P. Improving the Endosomal Escape of Cell-Penetrating Peptides and Their Cargos: Strategies and Challenges. *Pharmaceuticals* **2012**, *5*, 1177–1209. [CrossRef]
139. Daussy, C.F.; Wodrich, H. "Repair Me If You Can": Membrane Damage, Response, and Control from the Viral Perspective. *Cells* **2020**, *9*, 2042. [CrossRef]
140. Alhakamy, N.A.; Nigatu, A.S.; Berkland, C.J.; Ramsey, J.D. Noncovalently Associated -Penetrating Peptides for Gene Delivery Applications. *Ther. Deliv.* **2013**, *4*, 741–757. [CrossRef]
141. Loughran, S.P.; McCrudden, C.M.; McCarthy, H.O. Designer Peptide Delivery Systems for Gene Therapy. *Eur. J. Nanomed.* **2015**, *7*, 85–96. [CrossRef]
142. Sudo, K.; Niikura, K.; Iwaki, K.; Kohyama, S.; Fujiwara, K.; Doi, N. Human-Derived Fusogenic Peptides for the Intracellular Delivery of Proteins. *J. Control. Release* **2017**, *255*, 1–11. [CrossRef] [PubMed]
143. Ferrer-Miralles, N.; Vázquez, E.; Villaverde, A. Membrane-Active Peptides for Non-Viral Gene Therapy: Making the Safest Easier. *Trends Biotechnol.* **2008**, *26*, 267–275. [CrossRef] [PubMed]
144. Wadia, J.S.; Stan, R.V.; Dowdy, S.F. Transducible TAT-HA Fusogenic Peptide Enhances Escape of TAT-Fusion Proteins after Lipid Raft Macropinocytosis. *Nat. Med.* **2004**, *10*, 310–315. [CrossRef] [PubMed]
145. Ye, S.; Tian, M.; Wang, T.; Ren, L.; Wang, D.; Shen, L.; Shang, T. Synergistic Effects of Cell-Penetrating Peptide Tat and Fusogenic Peptide HA2-Enhanced Cellular Internalization and Gene Transduction of Organosilica Nanoparticles. *Nanomed. Nanotechnol. Biol. Med.* **2012**, *8*, 833–841. [CrossRef] [PubMed]
146. Karjoo, Z.; McCarthy, H.O.; Patel, P.; Nouri, F.S.; Hatefi, A. Systematic Engineering of Uniform, Highly Efficient, Targeted and Shielded Viral-Mimetic Nanoparticles. *Small* **2013**, *9*, 2774–2783. [CrossRef]
147. Zhao, X.; Wang, J.; Tao, S.; Ye, T.; Kong, X.; Ren, L. In Vivo Bio-Distribution and Efficient Tumor Targeting of Gelatin/Silica Nanoparticles for Gene Delivery. *Nanoscale Res. Lett.* **2016**, *11*, 195. [CrossRef]
148. Golan, M.; Feinshtein, V.; David, A. Conjugates of HA2 with Octaarginine-Grafted HPMA Copolymer Offer Effective SiRNA Delivery and Gene Silencing in Cancer Cells. *Eur. J. Pharm. Biopharm.* **2016**, *109*, 103–112. [CrossRef]
149. Cantini, L.; Attaway, C.C.; Butler, B.; Andino, L.M.; Sokolosky, M.L.; Jakymiw, A. Fusogenic-Oligoarginine Peptide-Mediated Delivery of SiRNAs Targeting the CIP2A Oncogene into Oral Cancer Cells. *PLoS ONE* **2013**, *8*, e73348. [CrossRef]
150. Funhoff, A.M.; van Nostrum, C.F.; Koning, G.A.; Schuurmans-Nieuwenbroek, N.M.E.; Crommelin, D.J.A.; Hennink, W.E. Endosomal Escape of Polymeric Gene Delivery Complexes Is Not Always Enhanced by Polymers Buffering at Low PH. *Biomacromolecules* **2004**, *5*, 32–39. [CrossRef]
151. Wang, Y.; Mangipudi, S.S.; Canine, B.F.; Hatefi, A. A Designer Biomimetic Vector with a Chimeric Architecture for Targeted Gene Transfer. *J. Control. Release Off. J. Control. Release Soc.* **2009**, *137*, 46–53. [CrossRef]
152. Hou, K.K.; Pan, H.; Schlesinger, P.H.; Wickline, S.A. A Role for Peptides in Overcoming Endosomal Entrapment in SiRNA Delivery—A Focus on Melittin. *Biotechnol. Adv.* **2015**, *33*, 931–940. [CrossRef] [PubMed]
153. Raghuraman, H.; Chattopadhyay, A. Melittin: A Membrane-Active Peptide with Diverse Functions. *Biosci. Rep.* **2007**, *27*, 189–223. [CrossRef]
154. Paray, B.A.; Ahmad, A.; Khan, J.M.; Taufiq, F.; Pathan, A.; Malik, A.; Ahmed, M.Z. The Role of the Multifunctional Antimicrobial Peptide Melittin in Gene Delivery. *Drug Discov. Today* **2021**, *26*, 1053–1059. [CrossRef] [PubMed]
155. Schellinger, J.G.; Pahang, J.A.; Johnson, R.N.; Chu, D.S.H.; Sellers, D.L.; Maris, D.O.; Convertine, A.J.; Stayton, P.S.; Horner, P.J.; Pun, S.H. Melittin-Grafted HPMA-Oligolysine Based Copolymers for Gene Delivery. *Biomaterials* **2013**, *34*, 2318–2326. [CrossRef] [PubMed]
156. Sun, Y.; Yang, Z.; Wang, C.; Yang, T.; Cai, C.; Zhao, X.; Yang, L.; Ding, P. Exploring the Role of Peptides in Polymer-Based Gene Delivery. *Acta Biomater.* **2017**, *60*, 23–37. [CrossRef] [PubMed]
157. Nouri, F.S.; Wang, X.; Dorrani, M.; Karjoo, Z.; Hatefi, A. A Recombinant Biopolymeric Platform for Reliable Evaluation of the Activity of PH-Responsive Amphiphile Fusogenic Peptides. *Biomacromolecules* **2013**, *14*, 2033–2040. [CrossRef] [PubMed]
158. Lee, H.; Jeong, J.H.; Park, T.G. PEG Grafted Polylysine with Fusogenic Peptide for Gene Delivery: High Transfection Efficiency with Low Cytotoxicity. *J. Control. Release* **2002**, *79*, 283–291. [CrossRef]
159. Bennett, R.; Yakkundi, A.; McKeen, H.D.; McClements, L.; McKeogh, T.J.; McCrudden, C.M.; Arthur, K.; Robson, T.; McCarthy, H.O. RALA-Mediated Delivery of FKBPL Nucleic Acid Therapeutics. *Nanomedicine* **2015**, *10*, 2989–3001. [CrossRef]
160. Mulholland, E.J.; Ali, A.; Robson, T.; Dunne, N.J.; McCarthy, H.O. Delivery of RALA/SiFKBPL Nanoparticles via Electrospun Bilayer Nanofibres: An Innovative Angiogenic Therapy for Wound Repair. *J. Control. Release* **2019**, *316*, 53–65. [CrossRef] [PubMed]
161. Yan, L.-P.; Castaño, I.M.; Sridharan, R.; Kelly, D.; Lemoine, M.; Cavanagh, B.L.; Dunne, N.J.; McCarthy, H.O.; O'Brien, F.J. Collagen/GAG Scaffolds Activated by RALA-SiMMP-9 Complexes with Potential for Improved Diabetic Foot Ulcer Healing. *Mater. Sci. Eng. C* **2020**, *114*, 111022. [CrossRef]

162. Akita, H.; Masuda, T.; Nishio, T.; Niikura, K.; Ijiro, K.; Harashima, H. Improving in Vivo Hepatic Transfection Activity by Controlling Intracellular Trafficking: The Function of GALA and Maltotriose. *Mol. Pharm.* **2011**, *8*, 1436–1442. [CrossRef]
163. Li, X.; Chen, Y.; Wang, M.; Ma, Y.; Xia, W.; Gu, H. A Mesoporous Silica Nanoparticle–PEI–Fusogenic Peptide System for SiRNA Delivery in Cancer Therapy. *Biomaterials* **2013**, *34*, 1391–1401. [CrossRef]
164. Ali, A.A.; McCrudden, C.M.; McCaffrey, J.; McBride, J.W.; Cole, G.; Dunne, N.J.; Robson, T.; Kissenpfennig, A.; Donnelly, R.F.; McCarthy, H.O. DNA Vaccination for Cervical Cancer; a Novel Technology Platform of RALA Mediated Gene Delivery via Polymeric Microneedles. *Nanomed. Nanotechnol. Biol. Med.* **2017**, *13*, 921–932. [CrossRef]
165. McCrudden, C.M.; McBride, J.W.; McCaffrey, J.; McErlean, E.M.; Dunne, N.J.; Kett, V.L.; Coulter, J.A.; Robson, T.; McCarthy, H.O. Gene Therapy with RALA/INOS Composite Nanoparticles Significantly Enhances Survival in a Model of Metastatic Prostate Cancer. *Cancer Nanotechnol.* **2018**, *9*, 5. [CrossRef]
166. Sousa, Â.; Almeida, A.M.; Faria, R.; Konate, K.; Boisguerin, P.; Queiroz, J.A.; Costa, D. Optimization of Peptide-Plasmid DNA Vectors Formulation for Gene Delivery in Cancer Therapy Exploring Design of Experiments. *Colloids Surf. B Biointerfaces* **2019**, *183*, 110417. [CrossRef]
167. Gottschalk, S.; Sparrow, J.T.; Hauer, J.; Mims, M.P.; Leland, F.E.; Woo, S.L.-Y.; Smith, L.C. A Novel DNA-Peptide Complex for Efficient Gene Transfer and Expression in Mammalian Cells. *Gene Ther.* **1996**, *3*, 448–457.
168. Guryanov, I.A.; Vlasov, G.P.; Lesina, E.A.; Kiselev, A.V.; Baranov, V.S.; Avdeeva, E.V.; Vorob'ev, V.I. Cationic Oligopeptides Modified with Lipophilic Fragments: Use for DNA Delivery to Cells. *Russ. J. Bioorg. Chem.* **2005**, *31*, 18–26. [CrossRef]
169. Van Rossenberg, S.M.W.; Sliedregt-Bol, K.M.; Meeuwenoord, N.J.; van Berkel, T.J.C.; van Boom, J.H.; van der Marel, G.A.; Biessen, E.A.L. Targeted Lysosome Disruptive Elements for Improvement of Parenchymal Liver Cell-Specific Gene Delivery. *J. Biol. Chem.* **2002**, *277*, 45803–45810. [CrossRef] [PubMed]
170. Meng, Z.; Luan, L.; Kang, Z.; Feng, S.; Meng, Q.; Liu, K. Histidine-Enriched Multifunctional Peptide Vectors with Enhanced Cellular Uptake and Endosomal Escape for Gene Delivery. *J. Mater. Chem. B* **2017**, *5*, 74–84. [CrossRef]
171. Kichler, A.; Mason, A.J.; Bechinger, B. Cationic Amphipathic Histidine-Rich Peptides for Gene Delivery. *Biochim. Biophys. Acta BBA Biomembr.* **2006**, *1758*, 301–307. [CrossRef] [PubMed]
172. Midoux, P.; Pichon, C.; Yaouanc, J.-J.; Jaffrès, P.-A. Chemical Vectors for Gene Delivery: A Current Review on Polymers, Peptides and Lipids Containing Histidine or Imidazole as Nucleic Acids Carriers. *Br. J. Pharmacol.* **2009**, *157*, 166–178. [CrossRef]
173. Chen, Q.-R.; Zhang, L.; Stass, S.A.; Mixson, A.J. Co-Polymer of Histidine and Lysine Markedly Enhances Transfection Efficiency of Liposomes. *Gene Ther.* **2000**, *7*, 1698–1705. [CrossRef]
174. Pichon, C.; Gonçalves, C.; Midoux, P. Histidine-Rich Peptides and Polymers for Nucleic Acids Delivery. *Adv. Drug Deliv. Rev.* **2001**, *53*, 75–94. [CrossRef]
175. Perche, F.; Gosset, D.; Mével, M.; Miramon, M.-L.; Yaouanc, J.-J.; Pichon, C.; Benvegnu, T.; Jaffrès, P.-A.; Midoux, P. Selective Gene Delivery in Dendritic Cells with Mannosylated and Histidylated Lipopolyplexes. *J. Drug Target.* **2011**, *19*, 315–325. [CrossRef]
176. Perche, F.; Benvegnu, T.; Berchel, M.; Lebegue, L.; Pichon, C.; Jaffrès, P.-A.; Midoux, P. Enhancement of Dendritic Cells Transfection in Vivo and of Vaccination against B16F10 Melanoma with Mannosylated Histidylated Lipopolyplexes Loaded with Tumor Antigen Messenger RNA. *Nanomed. Nanotechnol. Biol. Med.* **2011**, *7*, 445–453. [CrossRef]
177. Perche, F.; Lambert, O.; Berchel, M.; Jaffrès, P.-A.; Pichon, C.; Midoux, P. Gene Transfer by Histidylated Lipopolyplexes: A Dehydration Method Allowing Preservation of Their Physicochemical Parameters and Transfection Efficiency. *Int. J. Pharm.* **2012**, *423*, 144–150. [CrossRef] [PubMed]
178. Zhu, H.; Dong, C.; Dong, H.; Ren, T.; Wen, X.; Su, J.; Li, Y. Cleavable PEGylation and Hydrophobic Histidylation of Polylysine for SiRNA Delivery and Tumor Gene Therapy. *ACS Appl. Mater. Interfaces* **2014**, *6*, 10393–10407. [CrossRef] [PubMed]
179. Zhu, J.; Qiao, M.; Wang, Q.; Ye, Y.; Ba, S.; Ma, J.; Hu, H.; Zhao, X.; Chen, D. Dual-Responsive Polyplexes with Enhanced Disassembly and Endosomal Escape for Efficient Delivery of SiRNA. *Biomaterials* **2018**, *162*, 47–59. [CrossRef]
180. Sadeghian, F.; Hosseinkhani, S.; Alizadeh, A.; Hatefi, A. Design, Engineering and Preparation of a Multi-Domain Fusion Vector for Gene Delivery. *Int. J. Pharm.* **2012**, *427*, 393–399. [CrossRef] [PubMed]
181. Asseline, U.; Gonçalves, C.; Pichon, C.; Midoux, P. Improved Nuclear Delivery of Antisense 2′-Ome RNA by Conjugation with the Histidine-Rich Peptide H5WYG. *J. Gene Med.* **2014**, *16*, 157–165. [CrossRef]
182. McErlean, E.M.; McCrudden, C.M.; McCarthy, H.O. Delivery of Nucleic Acids for Cancer Gene Therapy: Overcoming Extra- and Intra-Cellular Barriers. *Ther. Deliv.* **2016**, *7*, 619–637. [CrossRef]
183. McBride, J.W.; Massey, A.S.; McCaffrey, J.; McCrudden, C.M.; Coulter, J.A.; Dunne, N.J.; Robson, T.; McCarthy, H.O. Development of TMTP-1 Targeted Designer Biopolymers for Gene Delivery to Prostate Cancer. *Int. J. Pharm.* **2016**, *500*, 144–153. [CrossRef]
184. Alipour, M.; Hosseinkhani, S.; Sheikhnejad, R.; Cheraghi, R. Nano-Biomimetic Carriers Are Implicated in Mechanistic Evaluation of Intracellular Gene Delivery. *Sci. Rep.* **2017**, *7*, 1–13. [CrossRef]
185. Alipour, M.; Majidi, A.; Molaabasi, F.; Sheikhnejad, R.; Hosseinkhani, S. In Vivo Tumor Gene Delivery Using Novel Peptideticles: PH-Responsive and Ligand Targeted Core–Shell Nanoassembly. *Int. J. Cancer* **2018**, *143*, 2017–2028. [CrossRef]
186. Thapa, R.K.; Sullivan, M.O. Gene Delivery by Peptide-Assisted Transport. *Curr. Opin. Biomed. Eng.* **2018**, *7*, 71–82. [CrossRef]
187. Liu, J.; Guo, N.; Gao, C.; Liu, N.; Zheng, X.; Tan, Y.; Lei, J.; Hao, Y.; Chen, L.; Zhang, X. Effective Gene Silencing Mediated by Polypeptide Nanoparticles LAH4-L1-SiMDR1 in Multi-Drug Resistant Human Breast Cancer. *J. Biomed. Nanotechnol.* **2019**, *15*, 531–543. [CrossRef]

188. Liang, W.; Lam, J.K.W. Endosomal Escape Pathways for Non-Viral Nucleic Acid Delivery Systems. Intech Open: Rijeka, Croatia, 2012; ISBN 978-953-51-0662-3.
189. Bechinger, B. Towards Membrane Protein Design: PH-Sensitive Topology of Histidine-Containing Polypeptides. *J. Mol. Biol.* **1996**, *263*, 768–775. [CrossRef]
190. Perrone, B.; Miles, A.J.; Salnikov, E.S.; Wallace, B.A.; Bechinger, B. Lipid Interactions of LAH4, a Peptide with Antimicrobial and Nucleic Acid Transfection Activities. *Eur. Biophys. J.* **2014**, *43*, 499–507. [CrossRef]
191. Marquette, A.; Mason, A.J.; Bechinger, B. Aggregation and Membrane Permeabilizing Properties of Designed Histidine-Containing Cationic Linear Peptide Antibiotics. *J. Pept. Sci. Off. Publ. Eur. Pept. Soc.* **2008**, *14*, 488–495. [CrossRef] [PubMed]
192. Wolf, J.; Aisenbrey, C.; Harmouche, N.; Raya, J.; Bertani, P.; Voievoda, N.; Süss, R.; Bechinger, B. PH-Dependent Membrane Interactions of the Histidine-Rich Cell-Penetrating Peptide LAH4-L1. *Biophys. J.* **2017**, *113*, 1290–1300. [CrossRef]
193. Langlet-Bertin, B.; Leborgne, C.; Scherman, D.; Bechinger, B.; Mason, A.J.; Kichler, A. Design and Evaluation of Histidine-Rich Amphipathic Peptides for SiRNA Delivery. *Pharm. Res.* **2010**, *27*, 1426–1436. [CrossRef] [PubMed]
194. Xu, Y.; Liang, W.; Qiu, Y.; Cespi, M.; Palmieri, G.F.; Mason, A.J.; Lam, J.K.W. Incorporation of a Nuclear Localization Signal in PH Responsive LAH4-L1 Peptide Enhances Transfection and Nuclear Uptake of Plasmid DNA. *Mol. Pharm.* **2016**, *13*, 3141–3152. [CrossRef] [PubMed]
195. Moulay, G.; Leborgne, C.; Mason, A.J.; Aisenbrey, C.; Kichler, A.; Bechinger, B. Histidine-Rich Designer Peptides of the LAH4 Family Promote Cell Delivery of a Multitude of Cargo. *J. Pept. Sci.* **2017**, *23*, 320–328. [CrossRef]
196. Liu, N.; Bechinger, B.; Süss, R. The Histidine-Rich Peptide LAH4-L1 Strongly Promotes PAMAM-Mediated Transfection at Low Nitrogen to Phosphorus Ratios in the Presence of Serum. *Sci. Rep.* **2017**, *7*, 1–12. [CrossRef]
197. Kichler, A.; Mason, A.J.; Marquette, A.; Bechinger, B. Histidine-Rich Cationic Cell-Penetrating Peptides for Plasmid DNA and siRNA Delivery. In *Nanotechnology for Nucleic Acid Delivery: Methods and Protocols*; Ogris, M., Sami, H., Eds.; Methods in Molecular Biology; Springer: New York, NY, USA, 2019; pp. 39–59, ISBN 978-1-4939-9092-4.
198. Marquette, A.; Leborgne, C.; Schartner, V.; Salnikov, E.; Bechinger, B.; Kichler, A. Peptides Derived from the C-Terminal Domain of HIV-1 Viral Protein R in Lipid Bilayers: Structure, Membrane Positioning and Gene Delivery. *Biochim. Biophys. Acta BBA Biomembr.* **2020**, *1862*, 183149. [CrossRef]
199. Chamarthy, S.P.; Kovacs, J.R.; McClelland, E.; Gattens, D.; Meng, W.S. A Cationic Peptide Consists of Ornithine and Histidine Repeats Augments Gene Transfer in Dendritic Cells. *Mol. Immunol.* **2003**, *40*, 483–490. [CrossRef] [PubMed]
200. Kovacs, J.R.; Zheng, Y.; Shen, H.; Meng, W.S. Polymeric Microspheres as Stabilizing Anchors for Oligonucleotide Delivery to Dendritic Cells. *Biomaterials* **2005**, *26*, 6754–6761. [CrossRef] [PubMed]
201. Jia, L.; Kovacs, J.R.; Zheng, Y.; Gawalt, E.S.; Shen, H.; Meng, W.S. Attenuated Alloreactivity of Dendritic Cells Engineered with Surface-Modified Microspheres Carrying a Plasmid Encoding Interleukin-10. *Biomaterials* **2006**, *27*, 2076–2082. [CrossRef]
202. Lo, S.L.; Wang, S. An Endosomolytic Tat Peptide Produced by Incorporation of Histidine and Cysteine Residues as a Nonviral Vector for DNA Transfection. *Biomaterials* **2008**, *29*, 2408–2414. [CrossRef] [PubMed]
203. Hao, X.; Li, Q.; Ali, H.; Zaidi, S.S.A.; Guo, J.; Ren, X.; Shi, C.; Xia, S.; Zhang, W.; Feng, Y. POSS-Cored and Peptide Functionalized Ternary Gene Delivery Systems with Enhanced Endosomal Escape Ability for Efficient Intracellular Delivery of Plasmid DNA. *J. Mater. Chem. B* **2018**, *6*, 4251–4263. [CrossRef]
204. Li, Q.; Hao, X.; Zaidi, S.S.A.; Guo, J.; Ren, X.; Shi, C.; Zhang, W.; Feng, Y. Oligohistidine and Targeting Peptide Functionalized TAT-NLS for Enhancing Cellular Uptake and Promoting Angiogenesis in Vivo. *J. Nanobiotechnol.* **2018**, *16*, 29. [CrossRef] [PubMed]
205. Read, M.L.; Singh, S.; Ahmed, Z.; Stevenson, M.; Briggs, S.S.; Oupicky, D.; Barrett, L.B.; Spice, R.; Kendall, M.; Berry, M.; et al. A Versatile Reducible Polycation-Based System for Efficient Delivery of a Broad Range of Nucleic Acids. *Nucleic Acids Res.* **2005**, *33*, e86. [CrossRef]
206. Stevenson, M.; Ramos-Perez, V.; Singh, S.; Soliman, M.; Preece, J.A.; Briggs, S.S.; Read, M.L.; Seymour, L.W. Delivery of SiRNA Mediated by Histidine-Containing Reducible Polycations. *J. Control. Release Off. J. Control. Release Soc.* **2008**, *130*, 46–56. [CrossRef]
207. Durymanov, M.; Reineke, J. Non-Viral Delivery of Nucleic Acids: Insight Into Mechanisms of Overcoming Intracellular Barriers. *Front. Pharmacol.* **2018**, *9*. [CrossRef]
208. Lin, G.; Li, L.; Panwar, N.; Wang, J.; Tjin, S.C.; Wang, X.; Yong, K.-T. Non-Viral Gene Therapy Using Multifunctional Nanoparticles: Status, Challenges, and Opportunities. *Coord. Chem. Rev.* **2018**, *374*, 133–152. [CrossRef]
209. Uludag, H.; Ubeda, A.; Ansari, A. At the Intersection of Biomaterials and Gene Therapy: Progress in Non-Viral Delivery of Nucleic Acids. *Front. Bioeng. Biotechnol.* **2019**, *7*. [CrossRef]
210. Vermeulen, L.M.P.; Brans, T.; De Smedt, S.C.; Remaut, K.; Braeckmans, K. Methodologies to Investigate Intracellular Barriers for Nucleic Acid Delivery in Non-Viral Gene Therapy. *Nano Today* **2018**, *21*, 74–90. [CrossRef]
211. Shi, B.; Zheng, M.; Tao, W.; Chung, R.; Jin, D.; Ghaffari, D.; Farokhzad, O.C. Challenges in DNA Delivery and Recent Advances in Multifunctional Polymeric DNA Delivery Systems. *Biomacromolecules* **2017**, *18*, 2231–2246. [CrossRef] [PubMed]
212. Yao, J.; Fan, Y.; Li, Y.; Huang, L. Strategies on the Nuclear-Targeted Delivery of Genes. *J. Drug Target.* **2013**, *21*, 926–939. [CrossRef]
213. Zelmer, C.; Zweifel, L.P.; Kapinos, L.E.; Craciun, I.; Güven, Z.P.; Palivan, C.G.; Lim, R.Y.H. Organelle-Specific Targeting of Polymersomes into the Cell Nucleus. *Proc. Natl. Acad. Sci. USA* **2020**, *117*, 2770–2778. [CrossRef] [PubMed]
214. Kosugi, S.; Hasebe, M.; Matsumura, N.; Takashima, H.; Miyamoto-Sato, E.; Tomita, M.; Yanagawa, H. Six Classes of Nuclear Localization Signals Specific to Different Binding Grooves of Importin α. *J. Biol. Chem.* **2009**, *284*, 478–485. [CrossRef]

215. Pan, L.; He, Q.; Liu, J.; Chen, Y.; Ma, M.; Zhang, L.; Shi, J. Nuclear-Targeted Drug Delivery of TAT Peptide-Conjugated Monodisperse Mesoporous Silica Nanoparticles. *J. Am. Chem. Soc.* **2012**, *134*, 5722–5725. [CrossRef] [PubMed]
216. Van der Aa, M.A.E.M.; Mastrobattista, E.; Oosting, R.S.; Hennink, W.E.; Koning, G.A.; Crommelin, D.J.A. The Nuclear Pore Complex: The Gateway to Successful Nonviral Gene Delivery. *Pharm. Res.* **2006**, *23*, 447–459. [CrossRef] [PubMed]
217. Cartier, R.; Reszka, R. Utilization of Synthetic Peptides Containing Nuclear Localization Signals for Nonviral Gene Transfer Systems. *Gene Ther.* **2002**, *9*, 157–167. [CrossRef] [PubMed]
218. Escriou, V.; Carrière, M.; Scherman, D.; Wils, P. NLS Bioconjugates for Targeting Therapeutic Genes to the Nucleus. *Adv. Drug Deliv. Rev.* **2003**, *55*, 295–306. [CrossRef]
219. Zhao, M.; Li, J.; Ji, H.; Chen, D.; Hu, H. A Versatile Endosome Acidity-Induced Sheddable Gene Delivery System: Increased Tumor Targeting and Enhanced Transfection Efficiency. *Int. J. Nanomed.* **2019**, *14*, 6519–6538. [CrossRef] [PubMed]
220. Dean, D.; Strong, D.; Zimmer, W. Nuclear Entry of Nonviral Vectors. *Gene Ther.* **2005**, *12*, 881–890. [CrossRef]
221. Morille, M.; Passirani, C.; Vonarbourg, A.; Clavreul, A.; Benoit, J.-P. Progress in Developing Cationic Vectors for Non-Viral Systemic Gene Therapy against Cancer. *Biomaterials* **2008**, *29*, 3477–3496. [CrossRef] [PubMed]
222. Bogacheva, M.; Egorova, A.; Slita, A.; Maretina, M.; Baranov, V.; Kiselev, A. Arginine-Rich Cross-Linking Peptides with Different SV40 Nuclear Localization Signal Content as Vectors for Intranuclear DNA Delivery. *Bioorg. Med. Chem. Lett.* **2017**, *27*, 4781–4785. [CrossRef] [PubMed]
223. Siomi, H.; Dreyfuss, G. A Nuclear Localization Domain in the HnRNP A1 Protein. *J. Cell Biol.* **1995**, *129*, 551–560. [CrossRef]
224. Bremner, K.H.; Seymour, L.W.; Logan, A.; Read, M.L. Factors Influencing the Ability of Nuclear Localization Sequence Peptides To Enhance Nonviral Gene Delivery. *Bioconjug. Chem.* **2004**, *15*, 152–161. [CrossRef]
225. Nicolson, G.L. Mitochondrial Dysfunction and Chronic Disease: Treatment With Natural Supplements. *Integr. Med. Encinitas Calif* **2014**, *13*, 35–43.
226. Tiku, V.; Tan, M.-W.; Dikic, I. Mitochondrial Functions in Infection and Immunity. *Trends Cell Biol.* **2020**, *30*, 263–275. [CrossRef]
227. Gallay, P.; Stitt, V.; Mundy, C.; Oettinger, M.; Trono, D. Role of the Karyopherin Pathway in Human Immunodeficiency Virus Type 1 Nuclear Import. *J. Virol.* **1996**, *70*, 1027–1032. [CrossRef]
228. Robbins, J.; Dilworth, S.M.; Laskey, R.A.; Dingwall, C. Two Interdependent Basic Domains in Nucleoplasmin Nuclear Targeting Sequence: Identification of a Class of Bipartite Nuclear Targeting Sequence. *Cell* **1991**, *64*, 615–623. [CrossRef]
229. Donkuru, M.; Badea, I.; Wettig, S.; Verrall, R.; Elsabahy, M.; Foldvari, M. Advancing Nonviral Gene Delivery: Lipid- and Surfactant-Based Nanoparticle Design Strategies. *Nanomedicine* **2010**, *5*, 1103–1127. [CrossRef]
230. Preuss, M.; Tecle, M.; Shah, I.; Matthews, D.A.; D. Miller, A. Comparison between the Interactions of Adenovirus-Derived Peptides with Plasmid DNA and Their Role in Gene Delivery Mediated by Liposome–Peptide–DNA Virus-like Nanoparticles. *Org. Biomol. Chem.* **2003**, *1*, 2430–2438. [CrossRef]
231. Hébert, E. Improvement of Exogenous DNA Nuclear Importation by Nuclear Localization Signal-Bearing Vectors: A Promising Way for Non-Viral Gene Therapy? *Biol. Cell* **2003**, *95*, 59–68. [CrossRef]
232. Li, Q.; Huang, Y. Mitochondrial Targeted Strategies and Theirapplication for Cancer and Other Diseases Treatment. *J. Pharm. Investig.* **2020**, *50*, 271–293. [CrossRef]
233. Kim, S.; Nam, H.Y.; Lee, J.; Seo, J. Mitochondrion-Targeting Peptides and Peptidomimetics: Recent Progress and Design Principles. *Biochemistry* **2020**, *59*, 270–284. [CrossRef] [PubMed]
234. Jang, Y.; Lim, K. Recent Advances in Mitochondria-Targeted Gene Delivery. *Mol. J. Synth. Chem. Nat. Prod. Chem.* **2018**, *23*, 2316. [CrossRef] [PubMed]
235. Wang, H.; Fang, B.; Peng, B.; Wang, L.; Xue, Y.; Bai, H.; Lu, S.; Voelcker, N.H.; Li, L.; Fu, L.; et al. Recent Advances in Chemical Biology of Mitochondria Targeting. *Front. Chem.* **2021**, *9*, 321. [CrossRef]
236. Kang, Y.C.; Son, M.; Kang, S.; Im, S.; Piao, Y.; Lim, K.S.; Song, M.-Y.; Park, K.-S.; Kim, Y.-H.; Pak, Y.K. Cell-Penetrating Artificial Mitochondria-Targeting Peptide-Conjugated Metallothionein 1A Alleviates Mitochondrial Damage in Parkinson's Disease Models. *Exp. Mol. Med.* **2018**, *50*, 1–13. [CrossRef]
237. Horton, K.L.; Stewart, K.M.; Fonseca, S.B.; Guo, Q.; Kelley, S.O. Mitochondria-Penetrating Peptides. *Chem. Biol.* **2008**, *15*, 375–382. [CrossRef]
238. Horne, W.S.; Stout, C.D.; Ghadiri, M.R. A Heterocyclic Peptide Nanotube. *J. Am. Chem. Soc.* **2003**, *125*, 9372–9376. [CrossRef]
239. Wu, J.; Li, J.; Wang, H.; Liu, C.-B. Mitochondrial-Targeted Penetrating Peptide Delivery for Cancer Therapy. *Expert Opin. Drug Deliv.* **2018**, *15*, 951–964. [CrossRef]
240. Lin, R.; Zhang, P.; Cheetham, A.G.; Walston, J.; Abadir, P.; Cui, H. Dual Peptide Conjugation Strategy for Improved Cellular Uptake and Mitochondria Targeting. *Bioconjug. Chem.* **2015**, *26*, 71–77. [CrossRef]
241. Ezzat, K.; EL Andaloussi, S.; Zaghloul, E.M.; Lehto, T.; Lindberg, S.; Moreno, P.M.D.; Viola, J.R.; Magdy, T.; Abdo, R.; Guterstam, P.; et al. PepFect 14, a Novel Cell-Penetrating Peptide for Oligonucleotide Delivery in Solution and as Solid Formulation. *Nucleic Acids Res.* **2011**, *39*, 5284–5298. [CrossRef] [PubMed]
242. Cerrato, C.P.; Kivijärvi, T.; Tozzi, R.; Lehto, T.; Gestin, M.; Langel, Ü. Intracellular Delivery of Therapeutic Antisense Oligonucleotides Targeting MRNA Coding Mitochondrial Proteins by Cell-Penetrating Peptides. *J. Mater. Chem. B* **2020**, *8*, 10825–10836. [CrossRef] [PubMed]

243. Deshayes, S.; Konate, K.; Dussot, M.; Chavey, B.; Vaissière, A.; Van, T.N.N.; Aldrian, G.; Padari, K.; Pooga, M.; Vivès, E.; et al. Deciphering the Internalization Mechanism of WRAP:SiRNA Nanoparticles. *Biochim. Biophys. Acta BBA Biomembr.* **2020**, *1862*, 183252. [CrossRef] [PubMed]
244. Faria, R.; Vivés, E.; Boisguerin, P.; Sousa, A.; Costa, D. Development of Peptide-Based Nanoparticles for Mitochondrial Plasmid DNA Delivery. *Polymers* **2021**, *13*, 1836. [CrossRef]
245. Jain, A.; Chugh, A. Mitochondrial Transit Peptide Exhibits Cell Penetration Ability and Efficiently Delivers Macromolecules to Mitochondria. *FEBS Lett.* **2016**, *590*, 2896–2905. [CrossRef] [PubMed]
246. Fatouros, D.G.; Lamprou, D.A.; Urquhart, A.J.; Yannopoulos, S.N.; Vizirianakis, I.S.; Zhang, S.; Koutsopoulos, S. Lipid-like Self-Assembling Peptide Nanovesicles for Drug Delivery. *ACS Appl. Mater. Interfaces* **2014**, *6*, 8184–8189. [CrossRef]
247. Liang, J.; Wu, W.-L.; Xu, X.-D.; Zhuo, R.-X.; Zhang, X.-Z. PH Responsive Micelle Self-Assembled from a New Amphiphilic Peptide as Anti-Tumor Drug Carrier. *Colloids Surf. B Biointerfaces* **2014**, *114*, 398–403. [CrossRef]
248. Wang, Q.; Zhang, X.; Zheng, J.; Liu, D. Self-Assembled Peptide Nanotubes as Potential Nanocarriers for Drug Delivery. *RSC Adv.* **2014**, *4*, 25461. [CrossRef]
249. Zhang, C.; Xue, X.; Luo, Q.; Li, Y.; Yang, K.; Zhuang, X.; Jiang, Y.; Zhang, J.; Liu, J.; Zou, G.; et al. Self-Assembled Peptide Nanofibers Designed as Biological Enzymes for Catalyzing Ester Hydrolysis. *ACS Nano* **2014**, *8*, 11715–11723. [CrossRef] [PubMed]
250. Wang, M.; Wang, J.; Zhou, P.; Deng, J.; Zhao, Y.; Sun, Y.; Yang, W.; Wang, D.; Li, Z.; Hu, X.; et al. Nanoribbons Self-Assembled from Short Peptides Demonstrate the Formation of Polar Zippers between β-Sheets. *Nat. Commun.* **2018**, *9*, 5118. [CrossRef]
251. Pereira-Silva, M.; Jarak, I.; Alvarez-Lorenzo, C.; Concheiro, A.; Santos, A.C.; Veiga, F.; Figueiras, A. Micelleplexes as Nucleic Acid Delivery Systems for Cancer-Targeted Therapies. *J. Control. Release* **2020**, *323*, 442–462. [CrossRef]
252. Ibaraki, H.; Kanazawa, T.; Owada, M.; Iwaya, K.; Takashima, Y.; Seta, Y. Anti-Metastatic Effects on Melanoma via Intravenous Administration of Anti-NF-KB SiRNA Complexed with Functional Peptide-Modified Nano-Micelles. *Pharmaceutics* **2020**, *12*, 64. [CrossRef]
253. Zhang, Y.; Wang, Y.; Meng, L.; Huang, Q.; Zhu, Y.; Cui, W.; Cheng, Y.; Liu, R. Targeted Micelles with Chemotherapeutics and Gene Drugs to Inhibit the G1/S and G2/M Mitotic Cycle of Prostate Cancer. *J. Nanobiotechnol.* **2021**, *19*, 17. [CrossRef]
254. Mazza, M.; Hadjidemetriou, M.; de Lázaro, I.; Bussy, C.; Kostarelos, K. Peptide Nanofiber Complexes with SiRNA for Deep Brain Gene Silencing by Stereotactic Neurosurgery. *ACS Nano* **2015**, *9*, 1137–1149. [CrossRef]
255. Avila, L.A.; Aps, L.R.M.M.; Sukthankar, P.; Ploscariu, N.; Gudlur, S.; Šimo, L.; Szoszkiewicz, R.; Park, Y.; Lee, S.Y.; Iwamoto, T.; et al. Branched Amphiphilic Cationic Oligopeptides Form Peptiplexes with DNA: A Study of Their Biophysical Properties and Transfection Efficiency. *Mol. Pharm.* **2015**, *12*, 706–715. [CrossRef]
256. Abedi-Gaballu, F.; Dehghan, G.; Ghaffari, M.; Yekta, R.; Abbaspour-Ravasjani, S.; Baradaran, B.; Dolatabadi, J.E.N.; Hamblin, M.R. PAMAM Dendrimers as Efficient Drug and Gene Delivery Nanosystems for Cancer Therapy. *Appl. Mater. Today* **2018**, *12*, 177–190. [CrossRef]
257. Gorzkiewicz, M.; Konopka, M.; Janaszewska, A.; Tarasenko, I.I.; Sheveleva, N.N.; Gajek, A.; Neelov, I.M.; Klajnert-Maculewicz, B. Application of New Lysine-Based Peptide Dendrimers D3K2 and D3G2 for Gene Delivery: Specific Cytotoxicity to Cancer Cells and Transfection in Vitro. *Bioorg. Chem.* **2020**, *95*, 103504. [CrossRef] [PubMed]
258. Bahadoran, A.; Moeini, H.; Bejo, M.H.; Hussein, M.Z.; Omar, A.R. Development of Tat-Conjugated Dendrimer for Transdermal DNA Vaccine Delivery. *J. Pharm. Pharm. Sci.* **2016**, *19*, 325. [CrossRef] [PubMed]
259. Yang, L.; Tang, H.; Sun, H. Progress in Photo-Responsive Polypeptide Derived Nano-Assemblies. *Micromachines* **2018**, *9*, 296. [CrossRef]
260. Mura, S.; Nicolas, J.; Couvreur, P. Stimuli-Responsive Nanocarriers for Drug Delivery. *Nat. Mater.* **2013**, *12*, 991–1003. [CrossRef]
261. Lee, D.; Rejinold, N.S.; Jeong, S.D.; Kim, Y.-C. Stimuli-Responsive Polypeptides for Biomedical Applications. *Polymers* **2018**, *10*, 830. [CrossRef] [PubMed]
262. Han, K.; Chen, S.; Chen, W.-H.; Lei, Q.; Liu, Y.; Zhuo, R.-X.; Zhang, X.-Z. Synergistic Gene and Drug Tumor Therapy Using a Chimeric Peptide. *Biomaterials* **2013**, *34*, 4680–4689. [CrossRef] [PubMed]
263. Levine, M.N.; Raines, R.T. Trimethyl Lock: A Trigger for Molecular Release in Chemistry, Biology, and Pharmacology. *Chem. Sci.* **2012**, *3*, 2412. [CrossRef] [PubMed]
264. Kuppusamy, P.; Li, H.; Ilangovan, G.; Cardounel, A.J.; Zweier, J.L.; Yamada, K.; Krishna, M.C.; Mitchell, J.B. Noninvasive Imaging of Tumor Redox Status and Its Modification by Tissue Glutathione Levels. *Cancer Res.* **2002**, *62*, 307–312. [PubMed]
265. Cao, C.-Y.; Shen, Y.-Y.; Wang, J.-D.; Li, L.; Liang, G.-L. Controlled Intracellular Self-Assembly of Gadolinium Nanoparticles as Smart Molecular MR Contrast Agents. *Sci. Rep.* **2013**, *3*, 1024. [CrossRef]
266. Löwik, D.W.P.M.; Leunissen, E.H.P.; van den Heuvel, M.; Hansen, M.B.; van Hest, J.C.M. Stimulus Responsive Peptide Based Materials. *Chem. Soc. Rev.* **2010**, *39*, 3394–3412. [CrossRef]
267. Mackay, J.A.; Chilkoti, A. Temperature Sensitive Peptides: Engineering Hyperthermia-Directed Therapeutics. *Int. J. Hyperthermia* **2008**, *24*, 483–495. [CrossRef]
268. Sigg, S.J.; Postupalenko, V.; Duskey, J.T.; Palivan, C.G.; Meier, W. Stimuli-Responsive Codelivery of Oligonucleotides and Drugs by Self-Assembled Peptide Nanoparticles. *Biomacromolecules* **2016**, *17*, 935–945. [CrossRef]
269. Ouboter, D.d.B.; Schuster, T.; Shanker, V.; Heim, M.; Meier, W. Multicompartment Micelle-Structured Peptide Nanoparticles: A New Biocompatible Gene- and Drug-Delivery Tool. *J. Biomed. Mater. Res. A* **2014**, *102*, 1155–1163. [CrossRef] [PubMed]

270. Abbas, M.; Zou, Q.; Li, S.; Yan, X. Self-Assembled Peptide- and Protein-Based Nanomaterials for Antitumor Photodynamic and Photothermal Therapy. *Adv. Mater.* **2017**, *29*, 1605021. [CrossRef]
271. Ding, J.; Zhuang, X.; Xiao, C.; Cheng, Y.; Zhao, L.; He, C.; Tang, Z.; Chen, X. Preparation of Photo-Cross-Linked PH-Responsive Polypeptide Nanogels as Potential Carriers for Controlled Drug Delivery. *J. Mater. Chem.* **2011**, *21*, 11383. [CrossRef]
272. Liu, G.; Dong, C.-M. Photoresponsive Poly(S -(o -Nitrobenzyl)- L -Cysteine)- b -PEO from a L -Cysteine N -Carboxyanhydride Monomer: Synthesis, Self-Assembly, and Phototriggered Drug Release. *Biomacromolecules* **2012**, *13*, 1573–1583. [CrossRef] [PubMed]
273. Yang, Y.; Xia, X.; Dong, W.; Wang, H.; Li, L.; Ma, P.; Sheng, W.; Xu, X.; Liu, Y. Acid Sensitive Polymeric Micelles Combining Folate and Bioreducible Conjugate for Specific Intracellular SiRNA Delivery. *Macromol. Biosci.* **2016**, *16*, 759–773. [CrossRef] [PubMed]
274. Hu, Q.; Wang, K.; Sun, X.; Li, Y.; Fu, Q.; Liang, T.; Tang, G. A Redox-Sensitive, Oligopeptide-Guided, Self-Assembling, and Efficiency-Enhanced (ROSE) System for Functional Delivery of MicroRNA Therapeutics for Treatment of Hepatocellular Carcinoma. *Biomaterials* **2016**, *104*, 192–200. [CrossRef]
275. Wang, H.-X.; Song, Z.; Lao, Y.-H.; Xu, X.; Gong, J.; Cheng, D.; Chakraborty, S.; Park, J.S.; Li, M.; Huang, D.; et al. Nonviral Gene Editing via CRISPR/Cas9 Delivery by Membrane-Disruptive and Endosomolytic Helical Polypeptide. *Proc. Natl. Acad. Sci. USA* **2018**, *115*, 4903–4908. [CrossRef]
276. Reissmann, S.; Filatova, M.P. New Generation of Cell-Penetrating Peptides: Functionality and Potential Clinical Application. *J. Pept. Sci.* **2021**, *27*, e3300. [CrossRef]
277. Hallan, S.S.; Sguizzato, M.; Esposito, E.; Cortesi, R. Challenges in the Physical Characterization of Lipid Nanoparticles. *Pharmaceutics* **2021**, *13*, 549. [CrossRef]
278. Aldosari, B.N.; Alfagih, I.M.; Almurshedi, A.S. Lipid Nanoparticles as Delivery Systems for RNA-Based Vaccines. *Pharmaceutics* **2021**, *13*, 206. [CrossRef] [PubMed]
279. Reese, H.R.; Shanahan, C.C.; Proulx, C.; Menegatti, S. Peptide Science: A "Rule Model" for New Generations of Peptidomimetics. *Acta Biomater.* **2020**, *102*, 35–74. [CrossRef] [PubMed]
280. Qvit, N.; Rubin, S.J.S.; Urban, T.J.; Mochly-Rosen, D.; Gross, E.R. Peptidomimetic Therapeutics: Scientific Approaches and Opportunities. *Drug Discov. Today* **2017**, *22*, 454–462. [CrossRef] [PubMed]
281. Zhang, G.; Sun, H.J. Racemization in Reverse: Evidence That D-Amino Acid Toxicity on Earth Is Controlled by Bacteria with Racemases. *PLoS ONE* **2014**, *9*, e92101. [CrossRef] [PubMed]

International Journal of
Molecular Sciences

Article

Peptide Derivatives of the Zonulin Inhibitor Larazotide (AT1001) as Potential Anti SARS-CoV-2: Molecular Modelling, Synthesis and Bioactivity Evaluation

Simone Di Micco [1,*], Simona Musella [1], Marina Sala [2], Maria C. Scala [2], Graciela Andrei [3], Robert Snoeck [3], Giuseppe Bifulco [2], Pietro Campiglia [2] and Alessio Fasano [1,4]

1 European Biomedical Research Institute of Salerno (EBRIS), Via Salvatore de Renzi 50, 84125 Salerno, Italy; s.musella@ebris.eu (S.M.); AFASANO@mgh.harvard.edu (A.F.)
2 Dipartimento di Farmacia, Università degli Studi di Salerno, Via Giovanni Paolo II 132, 84084 Fisciano, Salerno, Italy; msala@unisa.it (M.S.); mscala@unisa.it (M.C.S.); bifulco@unisa.it (G.B.); pcampiglia@unisa.it (P.C.)
3 Department of Microbiology, Immunology and Transplantation, Rega Institute for Medical Research, KU Leuven, 3000 Leuven, Belgium; graciela.andrei@kuleuven.be (G.A.); robert.snoeck@kuleuven.be (R.S.)
4 Mucosal Immunology and Biology Research Center, Massachusetts General Hospital–Harvard Medical School, Boston, MA 02114, USA
* Correspondence: s.dimicco@ebris.eu

Citation: Di Micco, S.; Musella, S.; Sala, M.; Scala, M.C.; Andrei, G.; Snoeck, R.; Bifulco, G.; Campiglia, P.; Fasano, A. Peptide Derivatives of the Zonulin Inhibitor Larazotide (AT1001) as Potential Anti SARS-CoV-2: Molecular Modelling, Synthesis and Bioactivity Evaluation. *Int. J. Mol. Sci.* **2021**, 22, 9427. https://doi.org/10.3390/ijms22179427

Academic Editors: Nunzianna Doti and Menotti Ruvo

Received: 15 July 2021
Accepted: 26 August 2021
Published: 30 August 2021

Publisher's Note: MDPI stays neutral with regard to jurisdictional claims in published maps and institutional affiliations.

Abstract: A novel coronavirus, severe acute respiratory syndrome coronavirus 2 (SARS-CoV-2), has been identified as the pathogen responsible for the outbreak of a severe, rapidly developing pneumonia (Coronavirus disease 2019, COVID-19). The virus enzyme, called 3CLpro or main protease (M^{pro}), is essential for viral replication, making it a most promising target for antiviral drug development. Recently, we adopted the drug repurposing as appropriate strategy to give fast response to global COVID-19 epidemic, by demonstrating that the zonulin octapeptide inhibitor AT1001 (Larazotide acetate) binds M^{pro} catalytic domain. Thus, in the present study we tried to investigate the antiviral activity of AT1001, along with five derivatives, by cell-based assays. Our results provide with the identification of AT1001 peptide molecular framework for lead optimization step to develop new generations of antiviral agents of SARS-CoV-2 with an improved biological activity, expanding the chance for success in clinical trials.

Keywords: SARS-CoV-2; peptide; FRET; molecular docking; molecular dynamics; MM-GBSA; drug repurposing; antiviral

1. Introduction

The recent outbreak of COVID-19 pandemic, caused by severe acute respiratory syndrome-Coronavirus-2 (SARS-CoV-2), has raised serious global concern for public health. Due to the highly contagious nature of this life-threatening virus, new therapeutics are urgently required to counteract its transmission [1]. SARS-CoV-2 is a medium-sized, enveloped, positive-strand RNA virus (~30 kb) of genus *Betacoronavirus*, consisting of 29,903 nucleotides, flanked by two untranslated sequences of 254 and 229 nucleotides at the 5′-and 3′-ends, respectively [2,3]. The viral genome has 12 protein-coding regions, which deciphers two categories of proteins: several structural proteins conferring characteristic global shape to the virus and participating to viral entry in the host, and non-structural proteins which assist virion in the infection and replication by conserving a linear arrangement [4]. SARS-CoV-2 enters human cells through the binding to the angiotensin-converting enzyme 2 (ACE2) by its viral spike protein. The spike protein S forms the outer layer of the coronavirus, giving the characteristic crown-like aspect, and initiates host cell invasion [5]. The next step is the replication process, through the transcription of its RNA viral genome. A typical CoV genome contains at least six ORFs (*Open Reading Frame*) which usually encode four well-conserved and characterized structural proteins: S (spike), E (envelope), M (membrane) and N (nucleocapsid) proteins. Precisely, in all coronaviruses the replication phase is initiated by production of the replicase proteins with the translation of ORF1a and ORF1ab via a −1 ribosomal frame-shifting mechanism [6]. This mechanism produces two large viral polyproteins, pp1a and pp1ab (~450 and ~790 kDa, respectively), which are then cleaved into 16 non-structural proteins (Nsp1–Nsp16), required for correct viral replication and transcription [7,8]. The proteolytic process needed for the formation of the Nsps is mainly conducted by two cysteine proteases: papain-pike protease (PL^{pro}), which performs three cleavages, and main protease (M^{pro}), also called chymotrypsin-like protease (3CLpro), which is responsible for the remaining 11 cuts [8]. Therefore, processing of the viral polyproteins is required to generate the viral non-structural proteins involved in the formation of the replicase complex, which is responsible for properly structuring virions [9]. M^{pro} protease activity becomes thus crucial for viral life cycle and its inhibition can prevent the virus replication. Accordingly, the vital role of M^{pro} for SARS-CoV-2 replication and the lack of homologous human proteins make it the most promising target for antiviral drug development. M^{pro} digests the polyprotein 1ab at multiple cleavage sites by hydrolysis of the Gln-Ser peptide bond mainly in the Leu-Gln-Ser-Ala-Gly recognition sequence [10], leading to the formation of non-structural proteins (NSPs). This cleavage site in the substrate is distinct from the peptide sequence recognized by human homologs, decreasing the possibility to interfere with off-targets in the host.

X-ray crystallography investigations of M^{pro} reveal that the active form of the protein is a homodimer, containing two protomers. Each protomer is formed by a tertiary structure that consists of the catalytic domains I (residue 8–101) and II (residue 102–184) folded into a six-stranded β-barrel hosting the active site, and the helical domain III (residue 201–303) consisting of a cluster of five antiparallel α-helices functionals for the dimerization of the protease [11]. A flexible loop connects domain II to domain III. The M^{pro} active site contains a Cys-His catalytic dyad located in the cleft between domains I and II, in which the side chain of cysteine 145 functions as the nucleophile in the proteolytic process. The active site is composed by canonical sub-pockets that are denoted S1, S1′, S2, S3, and S4 (Figure 1) [12].

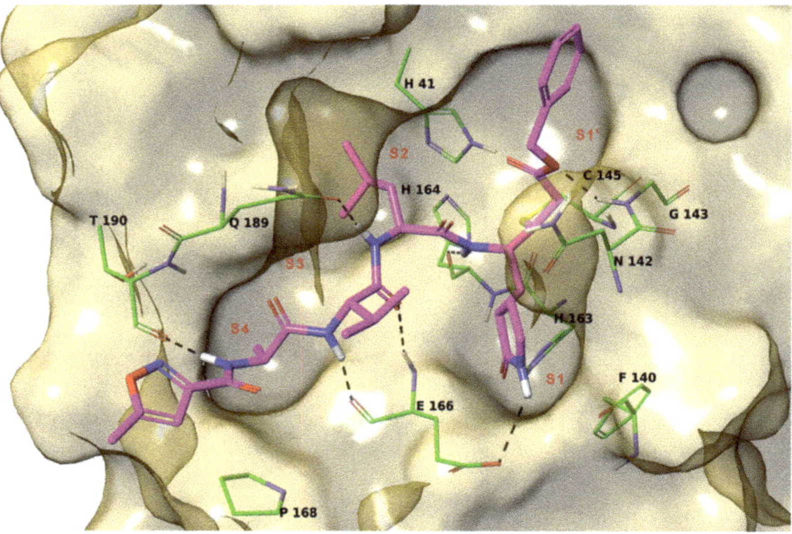

Figure 1. 3D model of the interaction between N3 and M^{pro}. The protein is represented by gold molecular surface and green tubes (atom color code: C, as tubes; polar H, sky blue; N, dark blue; O, red; S, yellow). N3 is depicted by magenta tubes (atom color code as for protein). The protein sub-pockets are indicated by red labels. The black dashed lines indicate ligand-protein H-bonds.

The amino acid sequence of the active site is highly conserved among coronaviruses. The catalytic dyad residues are H41 and C145, and residues involved in the binding of substrates include F140, H163, M165, E166, and Q189 [10]. These residues have been found to interact with the ligands co-crystallized with M^{pro} in different studies, like N3 compound that binds irreversibly M^{pro} pocket [13]. Crystallographic data also suggest that Ser1 of one protomer interacts with Phe140 and Glu166 of the other as the result of dimerization. These interactions stabilize the S1 binding pocket; thereby, dimerization of the main protease is likely responsible for its catalytic activity [10].

Recently, we adopted the drug repurposing approach [14] as suitable strategy to give fast response to global COVID-19 epidemic, by demonstrating that the zonulin octapeptide inhibitor AT1001 (Larazotide acetate) [15], currently in phase 3 trials in celiac disease, binds M^{pro} catalytic domain by means of an integrated approach of molecular modeling and fluorescence resonance energy transfer (FRET) assay [16]. Specifically, we observed that AT1001 shares a similar structural pattern to the peptidomimetic inhibitors of that enzyme, N3. These structural motifs, mainly represented by AVL residues in N3 and GVL in AT1001, provided the rationale to investigate AT1001 as a potential new inhibitor of M^{pro} enzyme. Our in silico analysis suggested that the octapeptide docks well in the catalytic domain of M^{pro}, presenting a global turn arrangement. In addition, there have been two in-vivo studies establishing the efficacy of AT1001 in mitigating ALI [15]. One of two shows the efficacy of AT1001 therapy during a lethal influenza, suggesting that the protective effect of the drug during influenza infection consists in acute lung injury (ALI) attenuation by diminishing pulmonary edema. To take advantage by previous studies that showed a strong safety profile in both AT1001 administration, systemically (IV) or locally (mucosal airways), we have proceeded in our investigation. The obtained results prompted us to consider AT1001 a new lead compound to a challenging development of potential protease inhibitor candidates. Thus, in the present study we have investigated the putative antiviral activity of AT1001 along with five derivatives, endowed with cap groups and different sequence length.

2. Results

Based on our previously reported results [16], we tried to investigate the antiviral activity of AT1001 (**1**, Table 1) by cell-based assays. Considering the intrinsic low membrane permeability of peptides, we designed AT1001 derivatives endowed of cap groups (**2–6**, Table 1) in order to mask the charged free N-and C-terminal positions of **1** at physiological pH to improve the cellular wall crossing as theoretically predicted (Table S1) [17]. In particular, we inserted small cap groups, such as acetyl and pivaloyl, at the N-terminus in order to preserve the main interactions given by the parent compound accordingly to the chemical modifications without any steric hindrances [16], obtaining: **2** (Ac-GGVLVQPG-NH_2), **3** (Ac-GGVLVQPG-$NHCH_3$), **4** (Piv-GGVLVQPG-$NHCH_3$). Furthermore, by maintaining a cap group on the N-terminus, a C-terminus methyl amide group was introduced to give peptide **3** and **4.** This kind of modification can increase lipophilicity and reduce the capability to form hydrogen bonds, thus facilitating the penetration of peptides across biological membranes and improving pharmacokinetic properties.

Finally, we also considered reducing the sequence length of peptide **2–4** to enhance the membrane passage designing the peptides **5** (Ac-GVLVQ-$NHCH_3$) and **6** (Ac-GVLV-$NHCH_3$). The rational design of shorter analogues was based on molecular dynamics investigation, integrated by MM-GBSA predictions, of AT1001 bound to M^{pro} showing that the N- and C-terminal residues (G1, Q6, P7 and G8) fluctuated largely than the remaining amino acids [16].

Table 1. Root mead square deviation (RMSD), predicted ΔG_{bind}, enzymatic inhibition, and activity against SARS-CoV-2 in Vero cells of **1–6**.

Peptide	RMSD (Å)	ΔG_{bind} (kcal/mol)	Inhibition [a] (%)	Antiviral Activity (EC_{50}, μM) [b]		Cytotoxicity (μM)	
				UC-1074	UC-1075	Cell Morphology (MCC) [c]	Cell Growth (CC_{50}) [d]
AT1001 (**1**)	0.000	−106.26	27.7 ± 0.2	>20	>20	100	82.5 ± 6.2
2	0.864	−112.34	27.1 ± 0.5	≥17.6 ± 2.4	>20	100	83.0 ± 17.0
3	0.664	−109.41	27.4 ± 1.1	>20	>20	100	81.8 ± 2.1
4	0.696	−105.79	26.2 ± 0.3	≥20	>20	100	74.9 ± 4.9
5	0.614	−95.86	27.0 ± 1.1	>20	>20	100	75.0 ± 2.2
6	1.081	−80.38	25.0 ± 0.5	≥20	>20	100	77.4 ± 0.7
Remdesivir	-	-	-	0.89 ± 0.44	1.06 ± 0.44	100	70.6 ± 4.9

[a] Relative to concentration range: 0.1–1 μM[16]. 50% of inhibition was observed for calpeptin. [b] Effective concentration required to reduce virus plaque formation by 50%. Virus input was 100 CID_{50}. [c] Minimum cytotoxic concentration that causes a microscopically detectable alteration of cell morphology. [d] Cytotoxic concentration required to reduce cell growth by 50%.

Following the same in silico investigation of AT1001 [16], we verified that the designed chemical modifications did not affect the global conformation of **2–6** in respect to the parent compound (Table 1 and Figure S1). Indeed, we observed small RMSD (Root Mean Square Deviation) values of **2–6** binding poses in respect to the parent peptide, ranging in 0.614–1.081 Å (Table 1). Moreover, the analysis of dihedral angles (Table S2) suggested that **2–6** preserve a global turn conformation as previously observed for **1** [16].

Furthermore, **2–6** kept most of the interactions observed for **1** with macromolecular counterparts (Figures 1, 2 and S1). In detail, the V3 is accommodated in S2 pocket delimited by H41, M49, M165, D187, Q189, and its backbone NH is H-bonded to Q189 side chain. The L4 accepts two H-bonds from main chain NH of G143 and C145 and gives van der Waals contacts with N142, H163, E166, F140, L141, H172 of S1 pocket. It is noteworthy that the van der Waals interactions identified for V3 and L4 of **1–6** are observed in the co-crystal structure with N3 (by its leucine and 3-methylpyrrolidin-2-one group) and calpeptin (by its leucine and butyl group) [18]. The G1 and G2 NH group are engaged in H-bond with a backbone of E166, and the CO group of G2 accepts an H-bond from NH of E166. The Q6 side chain is hydrogen bonded to the side chain of N119. The backbone CO of Q6 and P7 accept an H-bond from NH group of T26 and N142 side chain, respectively. The C-terminal carboxylate is H-bonded to side chains of S46 and Q189. The latter interactions were not hampered accordingly to the chemical substitution of carboxylic acid into amide group of **2** (Figures 2a and S1a). As expected, the $NHCH_3$ group of **3** and **4** preserves the hydrogen bond but at longer acceptor-donor distance (~0.6 Å) compared to the amide group of **2**. The acyl group at N-terminal also contribute with van der Waals contacts, and **3** and **4** are hydrogen bonded to side chain of Glu166 by their NH group of G1. For shorter peptides (**5** and **6**), the acetyl group is accommodated into S3 pocket formed by M165, L167, P168 and T190. Compared to N3 and calpeptin, **2–6** establish further van der Waals contacts with another deep crevice delimited by T25, T26, L27, H41, M49, and C145.

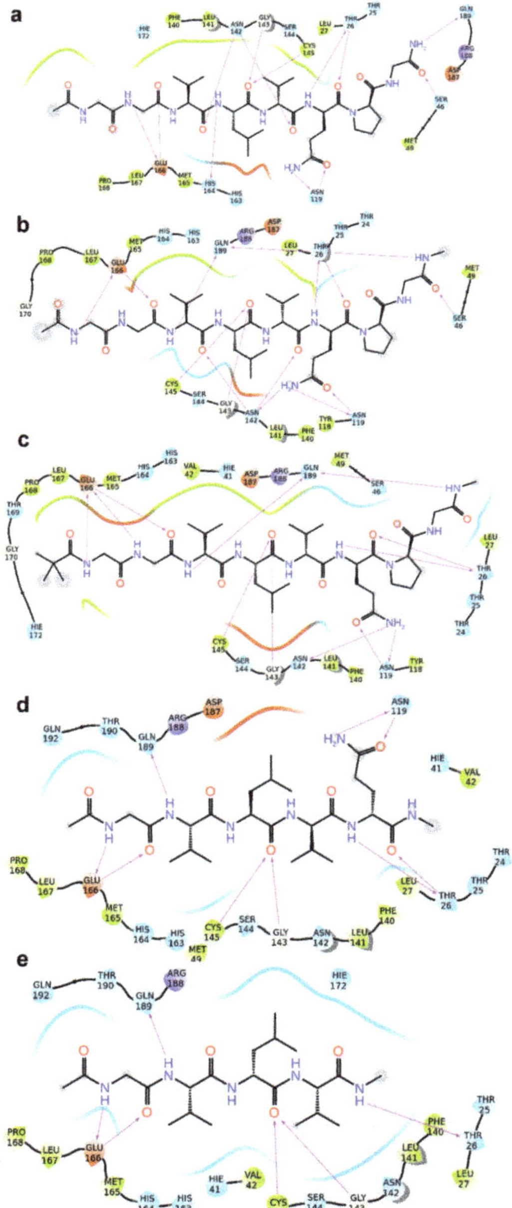

Figure 2. 2D panel representing interactions formed by fragments **2** (**a**), **3** (**b**), **4** (**c**), **5** (**d**) and **6** (**e**) with M^pro^. In each panel (**a**–**e**) the molecular structure of **2**–**6** is depicted in black, while the protein amino acids are indicated with three letter code, encircled with colored lines (negatively charged amino acids, red; positively charged amino acids, dark blue; polar amino acids, light blue; hydrophobic amino acids, green circles). The purple arrows indicate H-bonds. The arrows are directed from donor to acceptor of H-bond.

We monitored, over time, the crucial ligand-protein contacts identified for **2**–**6** docked poses into M^{pro} catalytic cavity by means of molecular dynamics simulations (100 ns,

310 K). The inspection of trajectories revealed that **2–6** keep most of the interactions with M^{pro} surrounding macromolecular residues for the duration of the whole simulations (>50%), especially with: T26, N119, N142, G143, C145, Hi164, E166, Q189 (Figure S2). The heavy-atom-positional RMSD (root mean square deviations) of **2–6**, referenced to protein main chain (Figures S3–S7) appears constant through the trajectory, and their atom-relative orientation is kept over time, confirming the RMSD values of **2–6** docked poses compared to AT1001. Moreover, unfavorable conformational rearrangement of the enzyme was not observed during trajectories (Figure S8). We evaluated the trend of predicted MM-GBSA ΔG_{bind} by calculations from molecular dynamics of each ligand-protein. The analysis of the averaged binding energies (Table S3) revealed the agreement with the values obtained from docked complexes, further corroborating their stability. Furthermore, the breakdown of the averaged MM-GBSA binding free energy of residues surrounding **2–6** at a distance of 5 Å (Table S4) agree with interactions observed from molecular docking and molecular dynamics investigations.

The compounds (**1–6**) were synthesized according to the microwave-assisted peptide synthesis using standard Fmoc methodology. In this work, the synthesis of this type of modified peptide, was performed on a Rink Amide-resin. The whole reaction was carried out following Scheme 1. The Rink amide Fmoc group was first removed using piperidine in DMF to give the free amine, which was then reacted with o-NBS-Cl and DIEA in NMP. The newly formed sulfonamide was deprotonated using DBU in NMP [19] and the resulting anion then reacted with methyl iodide. Next, the complete deprotection of o-NBS group was performed using DBU, as bases, and thiophenol, as thiols (Scheme 1, see also material and methods for further details).

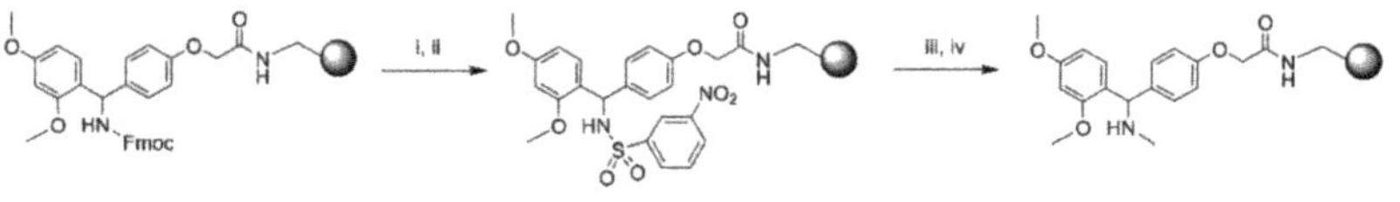

Reagents and conditions: (i) 30% piperidine/DMF; (ii) o-NBS-Cl (4 eq.), DIPEA (10 eq) in NMP (iii) DBU (10 eq), CH_3I (10 eq); (iv) PhSh (10 eq), DBU (5eq) in NMP.

Scheme 1. On resin N-methylation.

Before cell-based assay aimed to evaluate the antiviral activity, we checked the binding of AT1001 analogues by FRET assay. As previously kinetic measurements of enzymatic of AT1001 against M^{pro} showed a shape-bell dose-response profile with a maximum inhibitory activity at 27% at 0.1 µM, we checked if **2–6** maintained the inhibitory activity profile of the parent compound (Table 1). As expected, we observed a similar percentage of activity of AT1001 (Table 1).

Based on these results we proceeded with antiviral tests (Table 1). The different peptides were evaluated for their efficacy in inhibiting the replication of two SARS-CoV-2 clinical strains (UC-1074 = 1.58×10^4 $CCID_{50}$/mL and UC-1075 = 1.08×10^6 $CCID_{50}$/mL) in Vero cells with remdesivir included as reference compound. The peptides **1**, **3** and **5** a negligible activity against SARS-CoV-2; only peptide **2** against one of the SARS-CoV-2 strain (UC-1074) showed a micromolar range activity (EC_{50} = 17.6 ± 2.4 µM, Table 1), similar to the reported value of N3 (EC_{50} = 16.77 ± 1.70 µM) [12] but lower then calpeptin (EC_{50} = 0.072) [18] and remdesivir (Table 1). However, the remaining peptides (**4** and **6**) presented a comparable activity respect to **2**, even though they were lower. As expected, increasing the virus titer on Vero cells, there is a decrease in antiviral activity by tested compounds. All peptides altered cell morphology at a concentration of 100 µM and inhibited Vero cell growth with CC_{50} values in the range of 74–83 µM, comparable to the reference compound remdesivir.

We also evaluated the **1–6** antiviral activity against two unrelated viruses: cytomegalovirus (CMV, Table 2) and varicella-zoster virus (VZV, Table 3). In order to measure peptides effectiveness against cytomegalovirus, we used Ganciclovir and Cidofovir as known antiviral compounds; while we choose Aciclovir and Birivudine as specific reference compounds against VZV virus.

Table 2. Activity of the synthesized peptides against cytomegalovirus (HCMV) in human embryonic lung (HEL) cells.

Peptide	Antiviral Activity (EC_{50}, μM) [a]		Cytotoxicity (μM)	
	AD-169 Strain	Davis Strain	Cell Morphology (MCC) [b]	Cell Growth (CC_{50}) [c]
AT1001 (**1**)	>20	>20	≥100	ND [d]
2	>20	>20	≥20	ND [d]
3	>20	>100	≥20	ND [d]
4	>20	100	100	ND [d]
5	>20	>100	≥100	ND [d]
6	>4	>20	≥20	ND [d]
Ganciclovir	2.56 ± 0.23	1.18	≥394	350.23
Cidofovir	1.27 ± 0.18	0.67 ± 0.49	≥317	159.22 ± 76.5

[a] Effective concentration required to reduce virus cytopathic effect by 50%. Virus input was 100 $CCID_{50}$. [b] Minimum cytotoxic concentration that causes a microscopically detectable alteration of cell morphology. [c] Cytotoxic concentration required to reduce cell growth by 50%. [d] Not determined.

Table 3. Activity of the synthesized peptides against varicella-zoster virus (VZV) in human embryonic lung (HEL) cells.

Peptide	Antiviral Activity (EC_{50}, μM) [a]		Cytotoxicity (μM)	
	TK+ VZV Strain OKA	TK-VZV Strain 07–1	Cell Morphology (MCC) [b]	Cell Growth (CC_{50}) [c]
AT1001 (**1**)	44.14	59.06	≥100	ND [d]
2	50.05	48.42	>100	ND [d]
3	58.09	81.09	>100	ND [d]
4	78.20	>100	≥100	ND [d]
5	59.80	>100	>100	ND [d]
6	>20	>20	100	ND [d]
Aciclovir	8.39	116 ± 93	>444	>444
Birivudine	0.23 ± 0.04	2.13	>300.3	>300.3

[a] Effective concentration required to reduce virus plaque formation by 50%. Virus input was 20 plaque forming units (PFU). [b] Minimum cytotoxic concentration that causes a microscopically detectable alteration of cell morphology. [c] Cytotoxic concentration required to reduce cell growth by 50%. [d] Not determined.

Indeed, no activity for **1–6** was detected by treated the infected cells with cytomegalovirus, whereas against varicella-zoster virus was observed an activity in high micromolar range.

3. Discussion

Recently, we repurposed the zonulin octapeptide inhibitor AT1001 for the SARS-CoV-2 treatment. Our structural studies were limited to in silico analysis integrated by experiments demonstrating the inhibition of main protease, indicating a potential antiviral activity. Based on the previous results we investigated the anti-SARS-CoV-2 activity of AT1001, along with five derivatives. The latter are structurally featured with capped N-and C-terminals, considering the intrinsic low cell membrane permeability of peptides. AT1001 is currently in Phase 3 trials in celiac patients, showing a strong safety beyond a great efficacy for this indication. Thus, in order to remain consistent with structure of AT1001 and preserve its robust safety profile, we introduce small cap groups. The theoretical investigation integrated by FRET assay suggested that the chemical modifications did not affect the binding towards the virus enzyme. The biological activity investigation showed that **2**, followed by peptides **4** and **6**, gives a comparable antiviral activity respect to N3, and could be considered for the hit to lead optimization step. The better antiviral activity could be ascribed by an improved membrane permeability as theoretically predicted, beyond preserving the affinity towards the biological target. Moreover, the experimental results are in qualitative agreement with in silico analysis. Indeed, the peptide **2**, endowed with an amide at C-terminal, established more favorable H-bond with S46 and Q189 respect to **3** and **4** featured of $NHCH_3$ group in that position. We envisaged to increase the membrane crossing by shortening the sequence length of parent compound, by designing, synthesizing and testing the derivatives **5** and **6**. The shorter peptide **6** maintained the antiviral activity in the range of **2**, even though establishing less extended interactions with macromolecular counterparts gives rise to a lower biological activity than **2**. Compared to **5**, peptide **6** presents the right balance between affinity and cell membrane permeability. Indeed, the lower activity of **5** compared to **6**, could be ascribable to the lower cell permeability. It is worth of note that the peptides were incubated for five days in cells. Even though this long time could be detrimental for the activity due to possible peptidase hydrolysis, we observed an antiviral action against infected cells suggesting an improvement of antiviral profile in shorter incubation times.

It should be highlighted that for anti-cytomegalovirus and anti-varicella-zoster virus tests, we used the standard HEL cells, instead of Vero cells currently employed for anti-SARS-CoV-2 experiments. Thus, these data could not directly address a selectivity property by our tested compounds against SARS-CoV-2. However, the toxicology profiles obtained on Vero and human embryonic lung cells demonstrated strong safety by using the tested compounds.

Recently, the M^{pro} inhibitor PF-07321332 was announced by Pfizer (https://cen.acs.org/acs-news/acs-meeting-news/Pfizer-unveils-oral-SARS-CoV/99/i13/, accessed date: 15 July 2021) as promising Phase 1 clinical candidate. To date, in our opinion, our results could not be easily compared with PF-07321332 inhibitor. Firstly, peptides **1–6** are non-covalent M^{pro} inhibitors, despite the Pfizer candidate acting as covalent M^{pro} binder. Furthermore, our identified lead compounds require pharmacokinetics and pharmacodynamics improvements. However, the lack of an approved COVID-19 therapy still require drug discovery campaign and we do believe that by expanding the chemical diversity of the potential ligands, increases the possibility to reach more efficacious and safer treatment. To the best of our knowledge, our proposed leads represent the first class of peptide inhibitors of M^{pro}. Despite reported known M^{pro} binders so far, our structural investigation suggest that tested ligands interact with another deep subpocket (delimited by T25, T26, L27, H41, M49 and C145), giving new insights for drug design. In addition, our main goal remains the attempt to discover new drugs with a remarkable antiviral activity against SARS-CoV-2 maintaining a low cytotoxic profile that characterizes many drugs used, like Remdesivir. In fact, although remdesivir is currently approved by the USA-FDA to treat COVID-19 patients, its clinical efficacy remains debatable. Considering the strong safety profile of AT1001 administered to human subjects, it represents a good starting point to circumvent the adverse effect showed by most of repurposed drugs.

Overall, the present data has strongly suggested the identification of AT1001 peptide molecular framework for the hit to lead optimization step to develop new generations of antiviral agents for the treatment of SARS-CoV-2. Furthermore, the structural information and biological activities observed lay foundations for an ongoing rational design of both peptides and peptidomimetics with improved pharmacodynamics and pharmacokinetics properties. These outcomes could provide further chances of disclosing an interesting hit to be directed towards further investigations and for adding another piece to tackle the hard challenge to develop clinical candidates against SARS-CoV-2, which lacks any therapeutical treatment so far.

4. Materials and Methods

4.1. Molecular Docking

The Build Panel of Maestro (version 11, Schrödinger, LLC., New York, NY, USA) was used to construct the 3D structures of **2–6**, and successively optimizing their geometries through: OPLS3 force field [20], Polak-Ribière conjugate gradient algorithm (maximum derivative <0.001 kcal/mol), GB/SA (generalized Born/surface area) [21] solvent treatment of H_2O. Then, the peptides were processed by LigPrep [22], accounting for the protonation states at pH of 7.0 $\pm$ 1.0. Protein Preparation Wizard [23,24] was employed to process the X-ray structure of M^{pro} (PDB ID: 6LU7) [12]: bond order assignment and hydrogen addition; missing side chain and loop check; check of alternate positions of the residues, side chain charge assignment at pH 7.0 $\pm$ 1.0; H-bond network improvement through the optimize preference. The H_2O molecules were removed. Molecular docking predictions were carried out by Glide (v. 7.2, Schrödinger, LLC., New York, NY, USA), by using peptide specific protocol (SP-PEP) [25]. The docking protocol was validated by redocking the co-crystallized N3 with M^{pro} and overlapping the docked and experimental poses (Figure S9; RMSD = 1.076 Å). The receptor grid, proper for peptide docking, was sized as 10 Å inner and 22 Å outer boxes, with a center coordinates: −10.80 (x), 12.53 (y), 68.70 (z). A first set of SP-PEP was carried out by means of default parameters, with extended sampling option for conformer generation and expanded sampling for initial pose selection. 100 poses for **2–6** were generated, treating the ligands as flexible allowing only trans conformation for amide bonds. The following scoring contribution were considered: Epik state penalty; reward of intramolecular H-bonds; and aromatic hydrogen. The **2–6** docked poses obtained from the first calculation run were used as starting conformations for a second round by means of the same parameters generating 100 conformations for each input one. All 10,100 conformations from the first two calculation sets were collected and classified by docking score, and the best 100 ranked conformers were used as input for a third round of predictions generating further 10,000 docked poses. Predicted apparent Caco-2 cell permeability (QPPCaco) was calculated by QuikProp of Schrödinger suite [17], using default parameters and Caco-2 cells as model.

Maestro (version 11, Schrödinger, LLC., New York, NY, USA) was utilized for theoretic study and to generate all depictions.

4.2. MM-GBSA

The best 100 ranked conformers of **2–6** from the three rounds of molecular docking calculations were rescored by MM-GBSA predictions, by means of the Prime 3.1 [26,27] module of the Schrödinger suite (Schrödinger, LLC., New York, NY, USA) applying default parameters. A distance of 5 Å from each ligand was used to define flexible residues. Briefly, the binding energies of the protein and ligand are calculated by: Prime Energy + Implicit Solvent Energy in the free and bound states, accounting for the prediction of the energetic penalty due to strain between the ligand and protein in both states.

The φ, ψ and $\chi 1$ angles of best docked pose of each ligand were analyzed by PROMOTIF 3.0 (School of Animal and Microbial Sciences University of Reading, Whiteknights, UK; Biomolecular Structure and Modelling Unit, Department of Biochemistry and Molecular Biology, University College, Gower Street, London, UK) [28].

4.3. Molecular Dynamics

The docked complexes of **2–6** bound to M^{pro} were used for molecular dynamics simulation. These complexes were prepared by System Builder (Schrödinger, LLC., New York, NY, USA) [29] in Desmond v. 4.9 (DE Shaw Research, New York, NY, USA) [30,31], by suing: a cubic box with a 10 Å buffer distance, OPLS3 force field [20], the TIP3P [32] solvation model, Na^+ and Cl^- ions for electroneutrality, along with a NaCl solution (0.15 M). These systems were firstly optimized by the LBFGS methodology using default parameters and then underwent to the following relaxation protocol: (1) restrained solute heavy atom NVT simulation (2 ns, 10 K, small time steps); (2) restrained solute heavy atom NVT simulation (240 ps, 10 K with Berendsen thermostat, fast temperature relaxation constant) 1 ps of velocity resampling; (3) restrained solute heavy atom NPT simulation (240 ps, 10 K) with Berendsen thermostat and Berendsen barostat (1 atm), fast temperature relaxation constant, slow pressure relaxation constant, velocity resampling of 1 ps; (4) restrained solute heavy atom NPT ensemble simulation (240 ps) through Berendsen barostat (1 atm) and Berendsen thermostat (310 K), fast temperature relaxation constant, slow pressure relaxation constant, velocity resampling of 1 ps; (5) 480 ps NPT simulation employing Berendsen thermostat (310 K) and Berendsen barostat (1 atm), normal pressure relaxation constant and fast temperature relaxation constant. Unrestrained molecular dynamics of 100 ns (310 K) with NPT (1.01 bar) ensemble class were run, through 1.2 ps of recording time and 2.0 fs of integration time step. Each equilibration phase was evaluated by the Simulation Quality Analysis tool of Desmond, examining pressure, volume, temperature, total and potential energies.

4.4. *Synthesis*

4.4.1. Material

N^{α}-Fmoc-protected amino acids, coupling reagents (HOAt, HBTU), Fmoc-L-Gly-Wang resin, Rink Amide-resin, N, N-Diisopropylethylamine (DIEA), piperidine and trifluoroacetic acid (TFA) were purchased from Iris Biotech (Marktredwitz, Germany). Peptide synthesis solvents, reagents, as well as CH_3CN for High Performance Liquid Chromatography (HPLC) were reagent grade and were acquired from commercial sources and used without further purification unless otherwise noted.

4.4.2. Microwave Peptide Synthesis

The synthesis of peptides (**1–6**) was performed with a solid phase approach using a standard Fmoc methodology on a Biotage Initiator + Alstra automated microwave synthesizer (Biotage, Uppsala, Sweden).

Synthesis of **1**

Peptide was synthesized on an Fmoc-L-Gly-Wang resin (0.7 mmol/g, 150 mg), previously deprotected with 30% piperidine/DMF (1 × 3 min, 1 × 10 min) at room temperature. The resin was then washed with DMF (4 × 4.5 mL) and protected amino acids added on to the resin stepwise. Coupling reactions were performed using $N\alpha$-Fmoc amino acids (4.0 eq., 0.5 M), HBTU (3 eq, 0.6 M), HOAt (3 eq, 0.5 M), and DIEA (6 eq, 2 M) in N-methyl-2-pyrrolidone (NMP) for 10 min at 75 °C (2×). After each coupling step, the Fmoc protecting group was removed as described above. The resin was washed with DMF (4 × 4.5 mL) after each coupling and deprotection step. The N-terminal Fmoc group was removed, the resin was washed with DCM (7×), and the peptide released from the resin with TFA/TIS/H2O (ratio 90:5:5) for 3 h. The resin was removed by filtration and the crude peptide recovered by precipitation with cold anhydrous ethyl ether to give a white powder that was then lyophilized.

Synthesis of **2**

Peptide was synthesized onto a Rink Amide-resin (150 mg, loading 0.71 mmol/g), previously deprotected with 30% piperidine/DMF (1 × 3 min, 1 × 10 min) at room temperature. The synthesis was then performed using the conditions stated for **1**.

Synthesis of **3–6**

Peptides were synthesized onto a Rink Amide-resin (150 mg, loading 0.71 mmol/g), previously deprotected with 30% piperidine/DMF (1 × 3 min, 1 × 10 min) at room temperature. The resin was then washed with DMF (4 × 4.5 mL) and the primary amino group was protected and simultaneously activated by reacting it with o-NBS-Cl (4 eq) and DIEA (10 eq) in N-methyl-2-pyrrolidone (NMP) overnight. After this step, the methylation was carried out by treating the resin with DBU (10 eq, 30 min) and CH_3I (10 eq, 30 min) in NMP (2×). Then, the o-NBS protecting group was removed with thiophenol (10 eq) and DBU (5 eq) for 30 min. All steps were performed at room temperature. The synthesis was then performed as described above.

N-Terminal Capping

A capping step was performed: for peptides **2**, **3**, **5** and **6** by adding a solution of Ac_2O/DCM (1:3) shaking for 30 min and for **4** by adding a solution of Piv-Cl (4 eq.) e DIPEA (8 eq.) in DCM for 30 min.

4.4.3. Purification and Characterization

All crude peptides were purified by RP-HPLC on a preparative C18-bonded silica column (Phenomenex Kinetex AXIA 100 Å, 100 × 21.2 mm, 5 μm) using a Shimadzu SPD 20 A UV/VIS detector, with detection at 214 and 254 nm (Table S5). Mobile phase was: (A) H_2O and (B) ACN, both acidified with 0.1% TFA (v/v). Injection volume was 5000 μL; flow rate was set to 17 mL/min. The following gradient was employed: 0–18 min, 5–50% B, 18.01–20 min, 50–90% B, 20.01–21 min, returning to 5% B. Analytical purity and retention time (tr) of each peptide were determined using HPLC conditions in the above solvent system (solvents A and B) programmed at a flow rate of 0.500 mL/min, fitted with analytical C-18 column (Phenomenex, Aeris XB-C18 column, 100 mm × 2.1, 3.6 μm). LC gradient was the following: 0–7 min, 5–90% B, 7.01–8 min, returning to 5% B, 8–11 min, isocratic for 3 min. All analogues showed >99% purity when monitored at 220 nm. Homogeneous fractions, as established using analytical HPLC, were pooled and lyophilized.

Ultra high resolution mass spectra were obtained by positive ESI infusion on a LTQ Orbitrap XL mass spectrometer (Thermo Scientific, Dreieich, Germany), equipped with the Xcalibur software for processing the data acquired. The sample was dissolved in a mixture of water and methanol (50/50) and injected directly into the electrospray source, using a syringe pump, at constant flow (15 μL/min). Analytical data are shown in Supplementary Materials (Table S5 and Figures S10–S15).

4.5. Enzymatic Inhibition Assays

The recombinant SARS-CoV-2 M^{pro} (Proteros) (20 nM at a final concentration) was mixed with serial dilutions of AT1001 and Dabcyl-KTSAVLQSGFRKM-E(Edans)-NH_2 substrate (5 μM) in 20 μL (reaction volume) assay buffer solution (20 mM HEPES, pH 7.5, 1 mM DTT, 1 mM EDTA, 100 mM NaCl, 0.01% Tween20). The appropriate volume of substrate was added in reaction buffer along with 42.5 nL compound in 100% DMSO. Finally, the appropriate volume of target enzyme was added, and the reaction started with an incubation time of 10 min. The fluorescence signal of the Edans was monitored at an emission wavelength of 500 nm by exciting at 360 nm, by means of Pherastar FSX microplate Reader (BMG LABTECH GmbH, Ortenberg, Germany). Calpeptin was used as reference to set up the experiments.

4.6. Biological Activity

4.6.1. SARS-CoV-2

Vero cells (ATCC-CCL81^{TM}) were used to evaluate the activity of the peptides against SARS-CoV-2. Cells were grown in Dulbecco's Modified Eagle's Medium (DMEM, ThermoFisher, Merelbeke, Belgium) supplemented with 10% fetal calf serum (FCS), 2 mM L-glutamine, 0.1 mM non-essential amino acids, 1 mM sodium pyruvate and 10 mM HEPES at 37 °C in a 5% CO_2 humidified atmosphere. Two SARS-CoV-19 strains, denoted UC-1074 and UC-1075, were isolated in Vero cells from nasopharyngeal swabs of two COVID-19 patients who had, respectively, a Ct of 19 and 22, for detection of SARS-CoV-2 E protein by RT-qPCR real-time reverse transcription PCR (RT-qPCR). The infectious virus titer of the clinical isolates was determined in Vero cells and expressed as 50% cell culture infectious dose ($CCID_{50}$) per mL, being of 1.58×10^4 (UC-1074) and 1.08×10^6 (UC-1075) $CCID_{50}$/mL. For the antiviral assays, Vero cells were seeded in 96-well plates at a density of 1 × 104 cells per well in DMEM 10% FCS medium. After 24 h growth, the cell culture medium was removed and cells were treated with different compound concentrations in DMEM 2% FCS and mocked-infected or SARS-CoV-2-infected with 100 $CCID_{50}$/well (final volume 200 µL/well). After 5 days of incubation at 37 °C, viral CPE was recorded microscopically and the 50% effective concentration (EC_{50}) was calculated for each peptide (Figure S16) and remdesivir (reference anti-SARS-CoV-2 compound). In parallel, the cytotoxic effects of the derivatives were assessed by evaluating the MCC (minimum cytotoxic concentration that causes a microscopically detectable alteration of cell morphology). The effects of the compounds on cell growth were as well determined by counting the number of cells with a Coulter counter in mock-infected cultures and expressed as cytostatic concentration required to reduce cell growth by 50% (CC_{50}). All SARS-CoV-2-related work was conducted in the high-containment BSL3+ facilities of the KU Leuven Rega Institute (3CAPS) under licenses AMV 30,112,018 SBB 219 2018 0892 and AMV 23,102,017 SBB 219 2017 0589 according to institutional guidelines.

4.6.2. Cytomegalovirus and Varicella-Zoster Virus

The compounds were investigated against the following viruses: varicella-zoster virus (VZV) wild-type strain Oka (ATCC VR-795), thymidine kinase deficient (TK−) VZV strain 07−1 (kindly provided by Shiro Shigeta, Fukushima Medical Center, Fukushima, Japan), human cytomegalovirus (HCMV) strains AD-169 (ATCC VR-538) and Davis (VR-807). The antiviral assays are based on the inhibition of virus-induced cytopathic effect (HCMV) or plaque formation (VZV) in human embryonic lung (HEL) fibroblasts (HEL 299 (ATCC® CCL-137™). Confluent cell cultures in microtiter 96-well plates were inoculated with 100 $CCID_{50}$ of virus (1 $CCID_{50}$ being the virus dose to infect 50% of the cell cultures) or with 20 plaque forming units (PFU) (VZV). After adsorption for 2 h, the viral inoculum was removed and the cultures were further incubated in the presence of varying concentrations of the test compounds. Viral cytopathic effect or plaque formation was recorded after 5 (VZV) or 6−7 (CMV) days post-infection. Antiviral activity was expressed as the EC_{50} or compound concentration required inhibiting virus induced cytopathic effect or viral plaque formation by 50% (Figure S16). The cytostatic activity measurements were based on the inhibition of cell growth. HEL cells were S12 seeded into 96-well microtiter plates at a rate of 5×10^3 cells/well and incubated for 24 h. Then, medium containing the test compounds at different concentrations was added. After 3 days of incubation at 37 °C, the cell number was determined using a Coulter counter. The cytostatic concentration was calculated as the CC_{50}, or compound concentration required to reduce cell proliferation by 50% relative to the number of cells in the untreated controls. CC_{50} values were estimated from graphic plots of the number of cells (percentage of control) as a function of the concentration of the test compounds. Alternatively, cytotoxicity of the test compounds was expressed as the minimum cytotoxic concentration (MCC) or compound concentration that causes a microscopically detectable alteration of cell morphology.

Supplementary Materials: The following are available online at https://www.mdpi.com/article/10.3390/ijms22179427/s1, Table S1. docking scores and QPPCaco values of **1–6**; Table S2. Dihedral angle analysis of **2–6**; Table S3. average MM-GBSA ΔG_{bind} from molecular dynamics; Table S4. Averaged MM-GBSA ΔG_{bind} for residues from molecular dynamics; Table S5. Analytical data of peptides **1–6**; Figure S1. superimposition of AT1001 (purple) with **2–6** into M^{pro}; Figure S2. Protein-ligand contact histograms during the simulation; Figure S3–S7. RMSD of **2–6**; Figure S8. RMSD of protein Cα atoms bound to **2–6**; Figure S9. N3 docked and crystallized pose overlay; Figure S10–S15. HRMS spectra and HPLC chromatograms of peptide **1–6** title; Figure S16. Dose-response curves.

Author Contributions: Conceptualization, S.D.M. and A.F.; methodology, S.D.M.; validation, S.D.M.; formal analysis, S.D.M., M.S., S.M. and R.S.; investigation, S.D.M., M.C.S. and G.A.; resources, A.F., P.C. and G.B.; data curation S.D.M. and S.M.; writing—original draft preparation, S.D.M., S.M., M.S., G.A., R.S. and A.F.; writing—review and editing, S.D.M., P.C. and A.F.; visualization, S.D.M.; supervision, S.D.M.; project administration, S.D.M. and A.F.; funding acquisition, A.F. All authors have read and agreed to the published version of the manuscript.

Funding: This research was funded by the projects: Fase 2, studio multicentrico aperto per determinare la sicurezza, tollerabilità ed efficacia della larazotide acetato per l'uso urgente in pazienti anziani a rischio per la prevenzione di danno acuto polmonare (ali) e la sindrome da distress respiratorio acuto (ards) associate a infezione da covid-19—CUP G58D20000240002—SURF 20004BP000000011; Fighting Cancer Resistance: Multidisciplinary Integrated Platform for a Technological Innovative Approach to Oncotherapies (Campania Oncotherapies).

Institutional Review Board Statement: Not applicable.

Informed Consent Statement: Not applicable.

Acknowledgments: The authors are extremely grateful to Brecht Dirix for excellent technical assistance and dedication to evaluate the anti-SARS-CoV-2 activity of the derivatives.

Conflicts of Interest: The authors declare no conflict of interest.

References

1. Bontempia, E.; Vergalli, S.; Squazzoni, F. Understanding COVID-19 diffusion requires an interdisciplinary, multi-dimensional approach. *Environ. Res.* **2020**, *188*, 109814. [CrossRef]
2. Mousavizadeh, L.; Ghasemi, S. Genotype and phenotype of COVID-19: Their roles in pathogenesis. *J. Microbiol. Immunol. Infect.* **2021**, *54*, 159–163. [CrossRef]
3. Parlikar, A.; Kalia, K.; Sinha, S.; Patnaik, S.; Sharma, N.; Vemuri, S.G.; Sharma, G. Understanding genomic diversity, pan-genome, and evolution of SARS-CoV-2. *PeerJ* **2020**, *8*, e9576. [CrossRef]
4. Kirtipal, N.; Bharadwaj, S.; Kang, S.G. From SARS to SARS-CoV-2, insights on structure, pathogenicity and immunity aspects of pandemic human coronaviruses. *Infect. Genet. Evol.* **2020**, *85*, 104502. [CrossRef]
5. Sirois, S.; Zhang, R.; Gao, W.; Gao, H.; Li, Y.; Zheng, H.; Wei, D.-Q. Discovery of Potent Anti-SARS-CoV M^{pro} Inhibitors. *Curr. Comput. Aided Drug Des.* **2007**, *3*, 191–200. [CrossRef]
6. Bredenbeek, P.J.; Pachuk, C.J.; Noten, A.F.; Charité, J.; Luytjes, W.; Weiss, S.R.; Spaan, W.J. The primary structure and expression of the second open reading frame of the polymerase gene of the coronavirus MHV-A59; A highly conserved polymerase is expressed by an efficient ribosomal frameshifting mechanism. *Nucleic Acids Res.* **1990**, *18*, 1825–1832. [CrossRef]
7. Anand, K.; Palm, G.J.; Mesters, J.R.; Siddell, S.G.; Ziebuhr, J.; Hilgenfeld, R. Structure of coronavirus main proteinase reveals combination of a chymotrypsin fold with an extra a-helical domain. *EMBO J.* **2002**, *21*, 3213–3224. [CrossRef]
8. Yang, H.; Yang, M.; Ding, Y.; Liu, Y.; Lou, Z.; Zhou, Z.; Sun, L.; Mo, L.; Ye, S.; Pang, H.; et al. The crystal structures of severe acute respiratory syndrome virus main protease and its complex with an inhibitor. *Proc. Natl. Acad. Sci. USA* **2003**, *100*, 13190–13195. [CrossRef]
9. Báez-Santos, Y.M.; St John, S.E.; Mesecar, A.D. The SARS-coronavirus papain-like protease: Structure, function and inhibition by designed antiviral compounds. *Antivir. Res.* **2015**, *115*, 21–38. [CrossRef]
10. Ghahremanpour, M.M.; Tirado-Rives, J.; Deshmukh, M.; Ippolito, J.A.; Zhang, C.-H.; Cabeza de Vaca, I.; Liosi, M.-E.; Anderson, K.S.; Jorgensen, W.L. Identification of 14 Known Drugs as Inhibitors of the Main Protease of SARS-CoV-2. *ACS Med. Chem. Lett.* **2020**, *11*, 2526–2533. [CrossRef]
11. Kneller, D.W.; Galanie, S.; Phillips, G.; O'Neill, H.M.; Coates, L.; Kovalevsky, A. Malleability of the SARS-CoV-2 3CL M^{pro} active-site cavity facilitates binding of clinical antivirals. *Structure* **2020**, *28*, 1313–1320. [CrossRef] [PubMed]
12. Jin, Z.; Du, X.; Xu, Y.; Deng, Y.; Liu, M.; Zha, Y.; Zhang, B.; Li, X.; Zhang, L.; Peng, C.; et al. Structure of M^{pro} from 1 COVID-19 virus and discovery of its inhibitors. *Nature* **2020**, *582*, 289–293. [CrossRef]
13. Yang, H.; Xie, W.; Xue, X.; Yang, K.; Ma, J.; Liang, W.; Zhao, Q.; Zhou, Z.; Pei, D.; Ziebuhr, J.; et al. Design of wide-spectrum inhibitors targeting coronavirus main proteases. *PLoS Biol.* **2005**, *3*, e324.
14. Giordano, A.; Forte, G.; Massimo, L.; Riccio, R.; Bifulco, G.; Di Micco, S. Discovery of new erbB4 inhibitors: Repositioning an orphan chemical library by inverse virtual screening. *Eur. J. Med. Chem.* **2018**, *152*, 253–263. [CrossRef]
15. Troisi, J.; Venutolo, G.; Terracciano, C.; Delli Carri, M.; Di Micco, S.; Landolfi, A.; Fasano, A. The therapeutic use of the zonulin inhibitor AT-1001 (Larazotide) for a variety of acute and chronic inflammatory diseases. *Curr. Med. Chem.* **2021**, *28*. [CrossRef] [PubMed]
16. Di Micco, S.; Musella, S.; Scala, M.C.; Sala, M.; Campiglia, P.; Bifulco, G.; Fasano, A. In silico Analysis Revealed Potential Anti-SARS-CoV-2 Main Protease Activity by the Zonulin Inhibitor Larazotide Acetate. *Front. Chem.* **2021**, *8*, 628609. [CrossRef]
17. *Schrödinger Release 2017-1, QikProp*; Schrödinger, LLC.: New York, NY, USA, 2017.
18. Günther, S.; Reinke, P.Y.A.; Fernández-García, Y.; Lieske, J.; Lane, T.J.; Ginn, H.M.; Koua, F.H.M.; Ehrt, C.; Ewert, W.; Oberthuer, D.; et al. X-ray screening identifies active site and allosteric inhibitors of SARS-CoV-2 main protease. *Science* **2021**, *372*, 642–646. [CrossRef] [PubMed]
19. Biron, E.; Kessler, H. Convenient synthesis of N-methylamino acids compatible with Fmoc solid-phase peptide synthesis. *J. Org. Chem.* **2005**, *70*, 5183–5189. [CrossRef] [PubMed]
20. Harder, E.; Damm, W.; Maple, J.; Wu, C.J.; Reboul, M.; Xiang, J.Y.; Wang, L.; Lupyan, D.; Dahlgren, M.K.; Knight, J.L.; et al. OPLS3: A force field providing broad coverage of drug-like small molecules and proteins. *J. Chem. Theory Comput.* **2016**, *12*, 281–296. [CrossRef] [PubMed]

21. Still, W.C.; Tempczyk, A.; Hawley, R.C.; Hendrickson, T. Semianalytical treatment of solvation for molecular mechanics and dynamics. *J. Am. Chem. Soc.* **1990**, *112*, 6127–6129. [CrossRef]
22. *Schrödinger Release 2017-1, LigPrep*; Schrödinger, LLC.: New York, NY, USA, 2017.
23. *Protein Preparation Wizard Schrödinger LLC*; Schrödinger LLC.: New York, NY, USA, 2017.
24. Sastry, G.M.; Adzhigirey, M.; Day, T.; Annabhimoju, R.; Sherman, W. Protein and ligand preparation: Parameters, protocols, and influence on virtual screening enrichments. *J. Comput. Aided Mol. Des.* **2013**, *27*, 221–234. [CrossRef] [PubMed]
25. Tubert-Brohman, I.; Sherman, W.; Repasky, M.; Beuming, T. Improved docking of polypeptides with glide. *J. Chem. Inf. Model.* **2013**, *53*, 1689–1699. [CrossRef] [PubMed]
26. *Prime, Schrödinger, LLC*; Prime, Version 3.1; Schrödinger, LLC.: New York, NY, USA, 2012.
27. Knight, J.L.; Krilov, G.; Borrelli, K.W.; Williams, J.; Gunn, J.R.; Clowes, A.; Cheng, L.; Friesner, R.A.; Abel, R. Leveraging data fusion strategies in multireceptor lead optimization MM/GBSA end-point methods. *J. Chem. Theory Comput.* **2014**, *10*, 3207–3220. [CrossRef] [PubMed]
28. Hutchinson, E.G.; Thornton, J.M. PROMOTIF-a program to identify and analyze structural motifs in proteins. *Protein Sci.* **1996**, *5*, 212–220. [CrossRef]
29. *System Builder, Schrödinger LLC*; Schrödinger LLC.: New York, NY, USA, 2015.
30. *Desmond*; DE Shaw Research: New York, NY, USA, 2017.
31. Bowers, K.J.; Chow, D.E.; Xu, H.; Dror, R.O.; Eastwood, M.P.; Gregersen, B.A.; Klepeis, J.L.; Kolossvary, I.; Moraes, M.A.; Sacerdoti, F.D.; et al. Scalable algorithms for molecular dynamics simulations on commodity clusters. In Proceedings of the ACM/IEEE Conference on Supercomputing (SC06), Tampa, FL, USA, 11–17 November 2006.
32. Jorgensen, W.L.; Chandrasekhar, J.; Madura, J.D.; Impey, R.W.; Klein, M.L. Comparison of simple potential functions for simulating liquid water. *J. Chem. Phys.* **1983**, *79*, 926–935. [CrossRef]

International Journal of
Molecular Sciences

Article

Anti-Cancer Effects of Cyclic Peptide ALOS4 in a Human Melanoma Mouse Model

Bar Levi [1,†], Shiri Yacobovich [1,†], Michael Kirby [1], Maria Becker [2], Oryan Agranyoni [1], Boris Redko [3], Gary Gellerman [3], Albert Pinhasov [1,2], Igor Koman [4] and Elimelech Nesher [1,4,*]

1 Department of Molecular Biology, Faculty of Natural Sciences, Ariel University, Ariel 4070000, Israel; barsh@ariel.ac.il (B.L.); shiriya@ariel.ac.il (S.Y.); michael.kirby566@gmail.com (M.K.); oryanag@ariel.ac.il (O.A.); albertpi@ariel.ac.il (A.P.)
2 Adelson School of Medicine, Ariel University, Ariel 4070000, Israel; mariabe@ariel.ac.il
3 Department of Chemical Sciences, Faculty of Natural Sciences, Ariel University, Ariel 4070000, Israel; borisr@ariel.ac.il (B.R.); garyg@ariel.ac.il (G.G.)
4 Institute for Personalized and Translational Medicine, Ariel University, Ariel 4070000, Israel; igorko@ariel.ac.il
* Correspondence: elimelechn@ariel.ac.il
† Equal contributors.

Abstract: We examined the effects of ALOS4, a cyclic peptide discovered previously by phage library selection against integrin $\alpha_V\beta_3$, on a human melanoma (A375) xenograft model to determine its abilities as a potential anti-cancer agent. We found that ALOS4 promoted healthy weight gain in A375-engrafted nude mice and reduced melanoma tumor mass and volume. Despite these positive changes, examination of the tumor tissue did not indicate any significant effects on proliferation, mitotic index, tissue vascularization, or reduction of αSMA or Ki-67 tumor markers. Modulation in overall expression of critical downstream $\alpha_V\beta_3$ integrin factors, such as FAK and Src, as well as reductions in gene expression of *c-Fos* and *c-Jun* transcription factors, indirectly confirmed our suspicions that ALOS4 is likely acting through an integrin-mediated pathway. Further, we found no overt formulation issues with ALOS4 regarding interaction with standard inert laboratory materials (polypropylene, borosilicate glass) or with pH and temperature stability under prolonged storage. Collectively, ALOS4 appears to be safe, chemically stable, and produces anti-cancer effects in a human xenograft model of melanoma. We believe these results suggest a role for ALOS4 in an integrin-mediated pathway in exerting its anti-cancer effects possibly through immune response modulation.

Keywords: cancer; cyclic peptide; integrin; $\alpha_V\beta_3$; ALOS4; melanoma

Citation: Levi, B.; Yacobovich, S.; Kirby, M.; Becker, M.; Agranyoni, O.; Redko, B.; Gellerman, G.; Pinhasov, A.; Koman, I.; Nesher, E. Anti-Cancer Effects of Cyclic Peptide ALOS4 in a Human Melanoma Mouse Model. *Int. J. Mol. Sci.* **2021**, *22*, 9579. https://doi.org/10.3390/ijms22179579

Academic Editors: Menotti Ruvo and Nunzianna Doti

Received: 17 August 2021
Accepted: 30 August 2021
Published: 3 September 2021

Publisher's Note: MDPI stays neutral with regard to jurisdictional claims in published maps and institutional affiliations.

1. Introduction

Integrin $\alpha_V\beta_3$ has been shown to play an essential role in different stages of cancer progression [1], metastasis [2], invasion [3,4], and angiogenesis [5]. Structurally, integrin $\alpha_V\beta_3$ possesses a common integrin-binding motif and an Arg-Gly-Asp (RGD) recognition sequence [6] shared with several extra-cellular matrix (ECM) proteins including vitronectin, fibronectin, and fibrinogen [7]. Due to high expression in activated proliferating and angiogenetic [5] endothelial cells, integrin $\alpha_V\beta_3$ has become a cancer theraputic target [8] and is considered a cancer prognostic biomarker [9] that correlates well with tumor progression [10,11] and invasion in such cancers as glioma [12], prostate carcinoma [13,14], osteosarcoma [2], breast cancer [9,15], and melanoma [16]. Melanoma is known to be one of the most fatal types of skin cancer, with a five-year relative survival rate of less than 20% for patients diagnosed with active metastasis [17,18]. Current therapeutic approaches to treatment of malignant melanoma include surgical resection of the tumor, immunotherapy, biological therapy, chemotherapy, radiation therapy, and combination targeted therapy [19]. The search for new therapeutic targets for a melanoma cure has revealed that overexpressed integrin $\alpha_V\beta_3$ in transformed melanocytes [16] mediates tumor angiogenesis and is associated with organ-specific metastasis of human malignant melanoma [16], which has suggested a number of therapeutic approach possibilities for targeting $\alpha_v\beta_3$. Among the approaches [20] used to inhibit integrin signal transduction, tumor growth, angiogenesis, and metastasis are blocking $\alpha_v\beta_3$ with monoclonal antibodies [21], cyclic RGD antagonist peptides [22], or other antagonists [8]. Unfortunately, despite demonstrated anti-cancer activity in nude mice, previous attempts for developing $\alpha_v\beta_3$ inhibitors such as the cyclic peptide Cilengitide [23] and functional anti-$\alpha_v\beta_3$ antibodies such as Abegrin [24] have failed in clinical trials.

ALOS4, a synthetic 9-amino acid cyclic non-RGD peptide (NH_2-CSSAGSLFC-COOH (MW = 871.98)) was previously discovered using a phage–display technique targeted to integrin $\alpha_v\beta_3$ binding [25,26]. Using a murine melanoma model, we previously demonstrated anti-cancer properties of ALOS4 [25]. In this study, we investigated the effects of ALOS4 on a subcutaneous xenograft model of A375 human melanoma for effects on tumor growth, tumor tissue development, and expression of downstream targets of $\alpha_v\beta_3$. In addition, we also characterized the physiochemical aspects of ALOS4 formulated stability and toxicity issues such as alterations in mouse behavior, blood cell profile, and blood chemistry in healthy (nominally cancer-free) mice. Our findings suggest that ALOS4 is stable in chemical formulation and poses no overt toxicity risks, yet is effective in melanoma tumor reduction by an $\alpha_v\beta_3$-related mechanism and perhaps other mechanisms.

2. Results

2.1. ALOS4 Selectively Affects Tumor Development in the A375 Xenograft Model

In our previous research, we have shown that ALOS4 treatment leads to tumor growth inhibition and increased survival of C57BL/6J mice inoculated with murine B16F10 melanoma cells. In this study, we used a xenograft model to further confirm ALOS4 anti-cancer properties using immunodeficient nude mice, which were SC inoculated with human A375 melanoma cells followed by administration with 0.3, 3, or 30 mg/kg ALOS4. We found that 3 and 30 mg/kg of ALOS4 preserved normal weight gain of nude mice compared with untreated control animals, whose weight was significantly decreased during tumor development (two-way ANOVA followed by a Bonferroni means separation test: Interaction between weight and time $F[39,490] = 0.5072$, $p = 0.9947$; time $F[30,490] = 37.52$, $p < 0.0001$; treatment $F[3,490] = 17.47$, $p < 0.0001$; Figure 1A). A ROC analysis of tumor mass in mice treated with 30 mg/kg ALOS4 (Figure 1B) yielded a Youden's index cut-off value of 0.22 ($p = 0.047$), which differentiated between responder and non-responder individuals, excluding two animals from analysis (Figure 1C, circled). Tumor mass data from lower doses of ALOS4 when analyzed by ROC did not yield significant results. Comparison of tumor mass collected at termination point at day 18 (not including two non-responders) demonstrated a dose-dependent inhibition of tumor growth by ALOS4 treatment (Figure 1C; Kruskal-Wallis ANOVA followed by a Dunn's test, $p = 0.0239$).

We also observed that ALOS4 in a dose-dependent manner inhibited tumor growth (by estimated volume) in all examined concentrations showing maximal two-fold changes in growth inhibition with 30 mg/kg on day 17 (Figure 1D; two-way ANOVA followed by a Bonferroni means separation test: Interaction: $F[33,310] = 3.590$, $p < 0.0001$; Day: $F[11,310] = 13.71$, $p < 0.0001$; Treatment: $F[3,310] = 42.13$, $p < 0.0001$). We similarly conducted an ROC analysis of the results to distinguish responders from non-responders (Figure 1G–I). Responders (Figure 1E) and non-responders (Figure 1F) for each ALOS4 dose both yielded significant reductions in tumor volume compared with saline-injected control mice (two-way ANOVA followed by Bonferroni means separation test: Figure 1E: Interaction: $F[33,156] = 2.860$, $p < 0.0001$; Day: $F[11,156] = 1.898$, $p = 0.0433$; Treatment: $F[3,156] = 36.04$, $p < 0.0001$; Figure 1F: Interaction: $F[33,202] = 1.809$, $p = 0.0072$; Day: $F[11,202] = 9.847$, $p < 0.0001$; Treatment: $F[3,202] = 20.43$, $p < 0.0001$). Youden's index values for ROC analysis of each ALOS4 dose were as follows: 231.6, $p = 0.0056$ (Figure 1G), 220.5, $p = 0.0111$ (Figure 1H), 226.9, $p = 0.0210$ (Figure 1I). TGI% values for each ALOS4 treatment group were similar and were as follows (ALOS4 mg/kg): 0.3, 61.1; 3, 66.3; 30, 61.5.

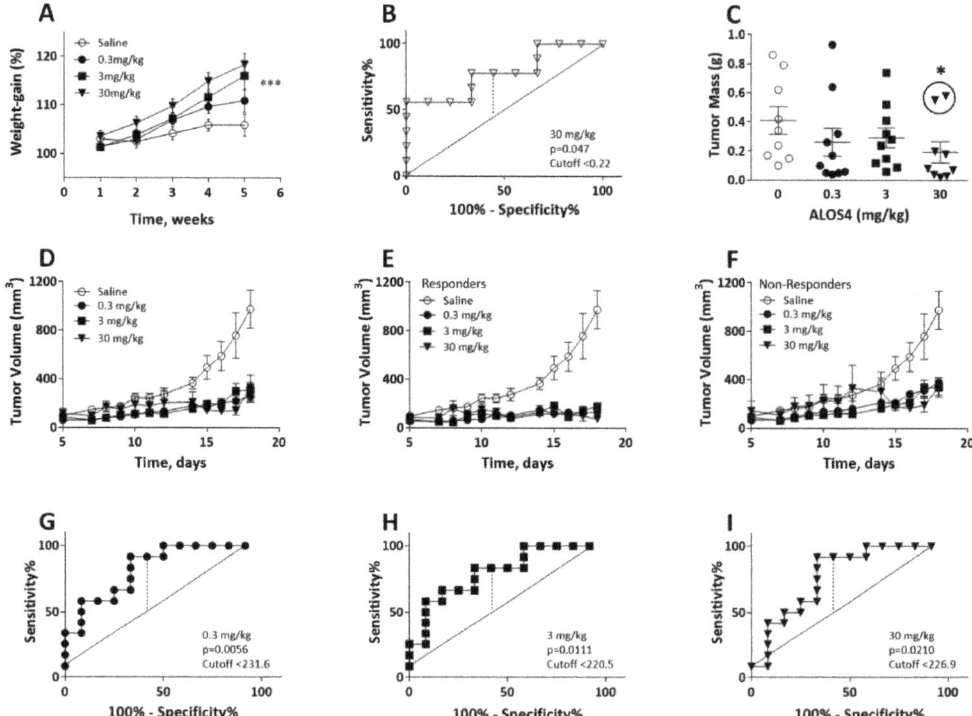

Figure 1. Effect of ALOS4 on body weight and tumor volume in SC A375 human melanoma mouse model. (**A**) Differences in the body weight gain of the nude mice inoculated with xenograft A375 SC tumor after 5 weeks administration with 0.3, 3, or 30 mg/kg of ALOS4. (**B**) ROC analyses of 30 mg/kg ALOS4-treated mice to determine threshold for positive drug response (Youden's Index) at day 18 ($n = 10$). (**C**) Tumor mass (g) with two excluded (circled) non-responder mice based on ROC cutoff value. *, Dunn's test $p < 0.05$. (**D–I**) ROC Analysis of responders and non-responders to ALOS4 treatment in SC A375 model. (**D**) Tumor volume growth in all treated nude mice. (**E**) Saline and ALOS4-treated responders only. (**F**) Saline and ALOS4-treated non-responders only. (**G–I**) ROC analyses of ALOS4-treated mice to determine threshold for positive drug response (Youden's Index). * $p < 0.05$; *** $p < 0.001$. ($n = 8$).

We further performed immunohistochemistry staining in ex vivo tumors obtained from SC xenografts to identify the effects of ALOS4 treatment on common hallmarks of cancer development and progression. Pleomorphism grades did not differ among control and ALOS4 treatments (all were rated at 2) and all examined tissue sections, regardless of treatment, had evidence of vascular invasion of the tumor mass (except tumor samples from one individual treated with 30 mg/kg ALOS4). Mitotic indices were also similar between controls and ALOS4 treatments (ALOS4 mg/kg, mean ± SD: 0, 6.2 ± 0.51; 0.3, 5.3 ± 1.13; 3, 6.68 ± 0.67; 30, 5.72 ± 0.98).

Analysis of the effect of ALOS4 on the expression of alpha smooth muscle actin (αSMA), a marker of vascular smooth muscle cells, was used to assess the number of blood vessels in the tissue sections to indicate the vascular invasion (Figure 2A–D). Controls treated with saline showed relatively low to moderate vascular density around and within the tumor tissue (Figure 2A). ALOS4 treatments of 0.3 and 30 mg/kg similarly showed moderate vascular density around and within the tumor tissue (Figure 2B,D), whereas ALOS4 treatment of 3 mg/kg showed relative moderate to high vascular density around and within the tumor tissue (Figure 2C). Overall, ALOS4 did not appear to produce any significant effects on tumor vascularization in the xenograft model of human melanoma (Figure 2I).

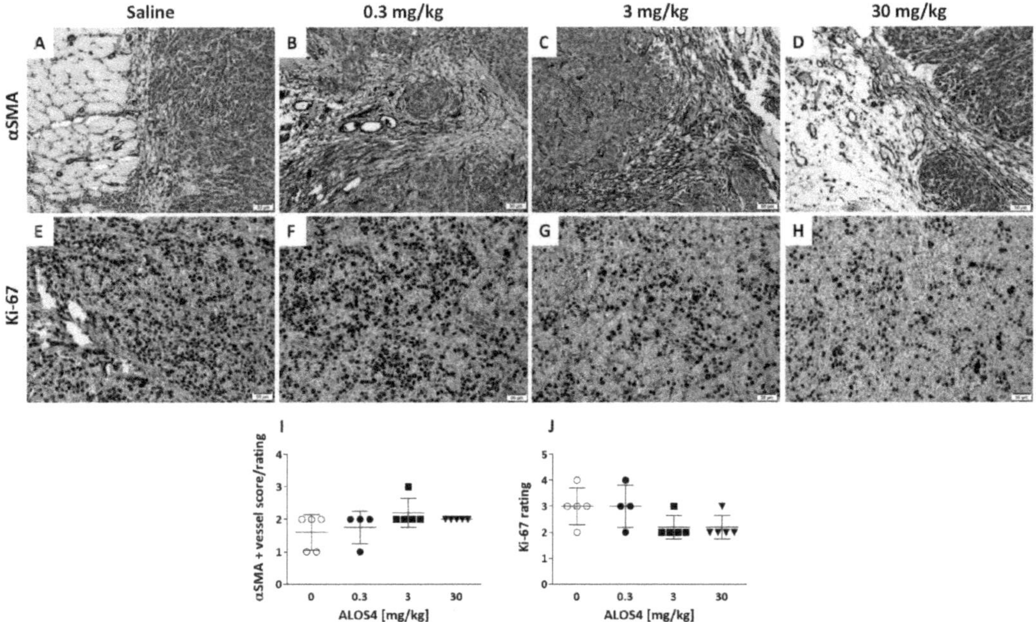

Figure 2. Effect of ALOS4 on carcinogenesis markers presentation in human melanoma A375 SC tumors from nude mice. (**A–D**) Representative photographs of slides stained for αSMA marker showing the number of blood vessels in the tissue sections (10×). Arrows demonstrate representative vessels in the tumor tissues. Scale: 50 µm. (**A**) Tumor of a saline-treated mouse shows low to moderate αSMA expression. (**B**) Tumor of an ALOS4 0.3 mg/kg-treated mouse demonstrates relative moderate vascular density around and within the tumor tissue. (**C**) Tumor of an ALOS4 3 mg/kg-treated mouse demonstrates relative moderate to high vascular density around and within the tumor tissue. (**D**) Tumor of an ALOS4 30 mg/kg-treated mouse demonstrates relative moderate vascular density around and within the tumor tissue. (**E–H**) Representative photographs of slides stained for Ki-67 marker. (**E**) Tumor of a saline-treated mouse shows a high number of positive cells within the neoplastic cell population. (**F**) Tumor of an ALOS4 0.3 mg/kg-treated mouse demonstrates a high number of positive cells within the neoplastic cell population. (**G**) Tumor of an ALOS4 3 mg/kg-treated mouse demonstrates a moderate to low number of positive cells within the neoplastic cell population. (**H**) Tumor of ALOS4 30 mg/kg-treated mouse demonstrates a moderate to low number of positive cells within the neoplastic cell population. Scale: 50 µm. (**I,J**) Quantification of histopathological evaluation scoring grades for αSMA (**I**) and Ki-67 (**J**) markers.

Non-parametric Kruskal–Wallis ANOVA analysis of Ki-67 proliferation marker showed a tendency toward dose-dependent reduction of expression in tumors treated with ALOS4 ($p = 0.089$). Thus, ALOS4 0.3 mg/kg dose and saline-treated controls both appeared to have a higher score in Ki-67-positive cells within the neoplastic cell population (Figure 2E,F). ALOS4 treatment with 3 mg/kg demonstrated a moderate to low number of Ki-67-positive cells (Figure 2G), whereas ALOS4 treatment with 30 mg/kg demonstrated a relatively low number of Ki-67-positive cells (Figure 2H). Comparisons of the Ki-67 results were performed using a tumor pathology scoring index for clinical relevance; however, and despite the appearance of dose-dependent reductions in Ki-67 expression, these reductions are not considered clinically meaningful.

2.2. The Effect of ALOS4 on Integrin-Related Signal Transduction

Since ALOS4 was discovered based on $\alpha_v\beta_3$ integrin binding, we analyzed the effect of ALOS4 on integrin-related signal transduction. Integrin mediated "outside-in" signals, activate growth factor receptors and cytoplasmic kinases, which regulate gene expression of immediate early genes [27]. Activation of $\alpha_v\beta_3$ integrin is known to induce the Fyn/Ras/Raf/MEK/ERK cascade, also called the MAPK pathway [28]. This pathway is highly or constantly activated in most cancer types and contributes to cancer proliferation, survival and migration [29]. Since we showed previously that ALOS4 treatment in B16F10 cells reduced migration [25], we hypothesized that ALOS4 may affect the MAPK pathway.

A375 cells were treated with concentrations of 0.01, 0.1, or 1.0 μM for 48 h and protein extracts were prepared for Western blots. We observed that 1.0 μM of ALOS4 significantly upregulated focal adhesion kinase (FAK), as well as proto-oncogene tyrosine protein kinase (Src) and pSrc levels (Figure 3A,B; One-way ANOVA: FAK, $F[3,8] = 12.86$, $p = 0.0020$; Src, $F[3,8] = 12.1$, $p = 0.0024$; pFAK, $F[3,8] = 5.28$, $p = 0.0267$; pSrc, $F[3,8] = 14.44$, $p = 0.0019$), while not affecting levels of extracellular signal-regulated kinase (ERK) and pERK (ERK, $F[3,8] = 1.288$, $p = 0.3429$; pERK, $F[3,8] = 0.3928$, $p = 0.7617$).

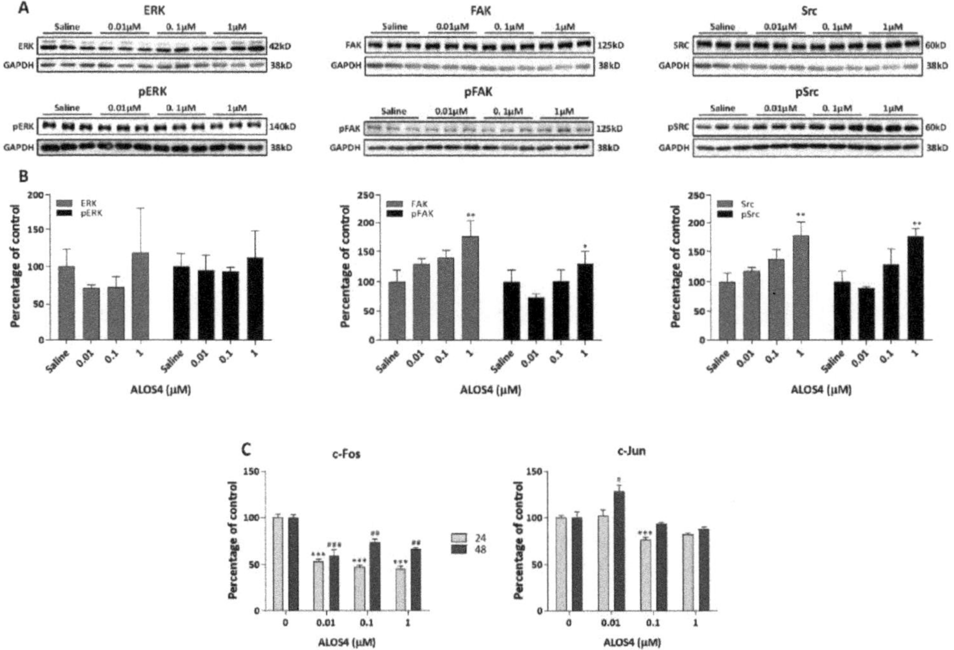

Figure 3. Effect of ALOS4 on $\alpha_v\beta_3$ integrin signaling. Representative gel bands (**A**) and Western blot densitometry results (**B**) performed for A375 cells treated for 48 h with ALOS4 at 0.01, 0.1, or 1.0 μM and analyzed for ERK/pERK, FAK/pFAK, and Src/pSrc protein expression. Data are presented as percentage of control normalized to GAPDH, $n = 3$ for each treated group. (**C**) A375 cells treated for 24 ($n = 6$) and 48 ($n = 3$) h with ALOS4 at 0.01, 0.1, or 1.0 μM were analyzed for *c-Fos* and *c-Jun* mRNA expression using qRT-PCR. Data presented as percentage of control. */# at $p < 0.05$, **/## at $p < 0.01$, and ***/### at $p < 0.0001$.

We also examined the expression of the immediate early genes *c-Fos* and *c-Jun* in ALOS4-treated A375 human melanoma cells. A375 cells were treated with ALOS4 at concentrations of 0.01, 0.1, or 1.0 μM for 24 h or 48 h and RNA was extracted for qRNA analysis. We found that ALOS4 treatment significantly decreased *c-Fos* gene expression after 24 and 48 h (Figure 3C, left panel; One-way ANOVA: 24 h, $F[3,20] = 76.99$, $p < 0.0001$; 48 h, $F[3,8] = 19.19$, $p = 0.0005$). A similar phenomenon was observed in *c-Jun* transcription levels, which showed significant decrease after 24 h at higher doses (Figure 3C, right panel; One-way ANOVA: $F[3,15] = 10.69$, $p = 0.0005$), whereas *c-Jun* was increased by 0.01 μM ALOS4 at 48 h (Figure 3C, right panel; One-way ANOVA: $F[3,5] = 9.331$, $p = 0.0172$).

2.3. ALOS4 Does Not Adhere to Inert Materials and Is Stable over a Range of Acid/Base and Temperature Conditions

We chose to perform a series of chemical stability and recoverability tests on ALOS4 to determine its practical applicability as a drug in formulation. To ensure that ALOS4 was stable and did not adhere to standard laboratory materials, we incubated ALOS4 formulated in 0.9% NaCl solution at a range of concentrations from 1–100 μM in polypropylene microtubes for 60 min at room temperature, then transferred solutions to either polypropylene or borosilicate glass liquid chromatography (LC) vials. LC-MS analysis showed that ALOS4 recoverability was near 100% in both borosilicate LC glass vials and standard polypropylene LC vials (Figure 4A).

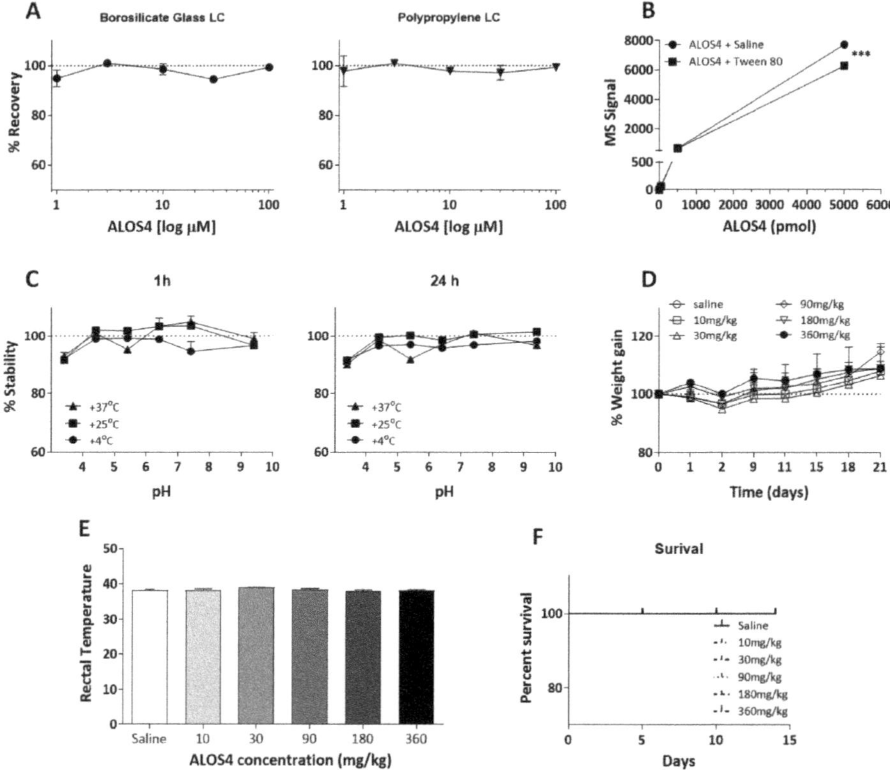

Figure 4. ALOS4 inert materials adherence, pH stability, temperature stability, safety, and toxicity. (**A–C**) LC-MS analysis of ALOS4 recoverability following inert materials exposure and storage under different pH, temperature, and formulation time conditions. (**A**) ALOS4 does not adhere to borosilicate LC glass vials or polypropylene LC vials at concentrations up to 100 μM. Data represent mean ± SEM (n = 3). (**B**) ALOS4 showed optimal stability in standard saline solution (0.9% NaCl) in comparison with saline containing 0.1% Tween-80 at concentrations of 0.5, 5, 50, 500, or 5000 pmol/μL ALOS4 incubated for 60 min at 25 °C in standard polypropylene tubes. Data represent mean ± SEM (n = 3), *** at $p < 0.001$. (**C**) ALOS4 demonstrates stability in a variety of pH and temperature conditions. ALOS4 (10 μM) was incubated in standard polypropylene tubes under different pH conditions at and stored at 4, 25, or 37 °C for 1 or 24 h. Data represent mean ± SEM (n = 3). (**D**,**E**) ALOS4 shows no effect on ICR mouse (**D**) body weight or (**E**) body-temperature following repeated IV administration at 10, 30, 90, 180, or 360 mg/kg ALOS4 (n = 5). Weight-gain of mice was measured three times per week, one hour prior to ALOS4 injections. Percent of weight change was calculated according to baseline weight prior to treatment. Body temperature was measured 30 min after ALOS4 administration. (**F**) ALOS4 repeated doses (10, 30, 90, 180, or 360 mg/kg) administrated intravenously for 14 days did not affect survival of ICR mice (n = 5).

Solubilizing agents are commonly used to stabilize peptides in solution and to reduce inert substrate interaction. Therefore, we compared two solvent options for ALOS4, standard saline solution (0.9% NaCl) and saline solution containing 0.1% Tween-80 (polysorbate). Solutions of ALOS4 ranging from 0.5–5000 pmol/μL formulated in both solvents were incubated for 60 min at room temperature in standard polypropylene tubes. LC-MS analysis showed that ALOS4 formulated in the saline solution containing Tween-80 had significantly reduced peptide stability by 1.2-fold in comparison with ALOS4 formulated in 0.9% NaCl only (Figure 4B; Two-way ANOVA, $F(1,10) = 3217.54$, $p < 0.0001$).

Evaluation of ALOS4 (10 μM) stability at different ranges of acid/base and temperature conditions was conducted in saline after 1 or 24 h in different pH solutions above and below the physiological pH (7.4): pH 3.4, 4.4, 5.4, 6.4, 7.4, and 9.4. Stability analysis was also performed under three regimes: at 4 °C (storage temperature), 25 °C (room temperature), and 37 °C (body temperature). LC-MS analysis indicated that ALOS4 was highly stable (90–100% recoverability) at all measured temperatures and all analyzed acid/basic conditions at both time points (Figure 4C).

2.4. ALOS4 Shows High Safety and No Toxicity In Vivo

To evaluate ALOS4 safety and toxicity, uninoculated (nominally cancer-free) ICR mice were IV administrated ALOS4 and monitored for clinical signs of toxicity including body weight changes, body temperature, alopecia, nasal bleeding, and mortality. ICR mice were injected with doses of ALOS4 every other day over 21 days with 10, 30, 90, 180, or 360 mg/kg (for a total of ten doses). Mouse weights were taken daily and rectal temperatures were recorded 30 min following ALOS4 injection. We observed no alterations in weight-gain (Figure 4D) or body temperature changes (Figure 4E) of treated mice in comparison with control mice during the course of the trial. Further, to determine the maximal tolerant dose (MTD) of ALOS4, ICR mice were ALOS4 IV-injected at a range of doses from 10–360 mg/kg and monitored for survival and clinical symptoms. We found that even at the maximum tested repeated dose of 360 mg/kg ALOS4, 21-day survival was 100% (Figure 4F). No adverse overt clinical signs were observed during the trial. Necropsy assessment for organ damage (histology of liver, spleen, kidney, lungs, and brain) did not reveal any overt signs of tissue damage or inflammation.

To evaluate potential effect of ALOS4 on mouse locomotory activity and anxiety-like behaviors, we used two standard behavioral paradigms: the open-field ambulation and elevated plus maze (EPM) tests.

We found that an IV acute single dose of ALOS4 of 30, 90, or 180 mg/kg administered to ICR mice did not affect locomotor activity in general (Figure 5A: One-way ANOVA analysis, $F[3,16] = 0.9406$, $p = 0.4441$) with only an exception for total traveled distance at the dose of 180 mg/kg (Figure 5B: One-way ANOVA, $F[3,16] = 3.308$, $p = 0.0471$, followed by Bonferroni's means separation test [180 mg/kg, $p = 0.0329$]), and did not produce any anxiety-like behaviors (Figure 5C: One-way ANOVA, $F[3,16] = 0.1064$, $p = 0.9551$; Figure 5D: One-way ANOVA, $F[3,16] = 0.04374$, $p = 0.9874$).

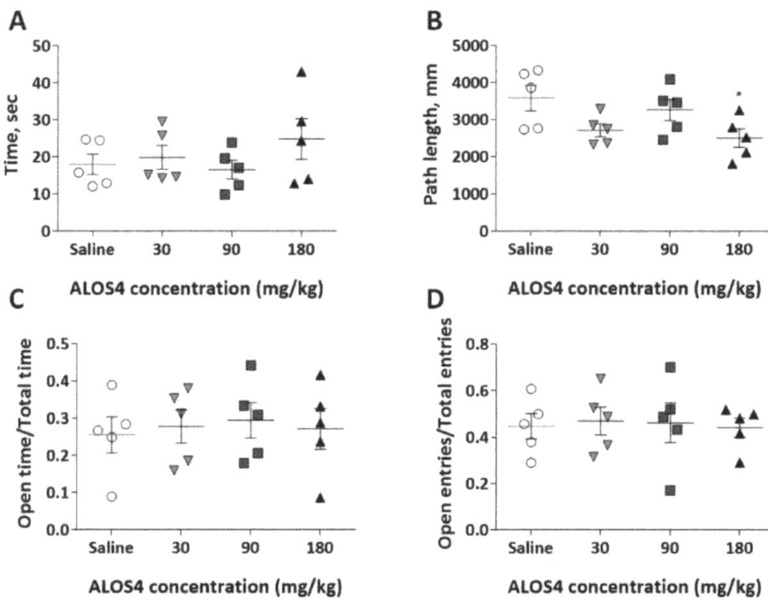

Figure 5. ALOS4 does not affect mouse locomotion or produce anxiety-like behaviors. (**A**,**B**) Behavior of ICR mice ($n = 5$) was not affected by intravenous administration with three acute single doses of ALOS4 (30, 90, or 180 mg/kg). Mice tested in the open-field arena for locomotory activity showed no changes in either cumulative central arena area dwell time (**A**) or total traveled distance, * at $p < 0.05$ (**B**). Mice tested in the EPM test for anxiety-like behavior showed no changes in two analyzed parameters (with exception to 180 mg/kg ALOS4 for OE/TE ratio): Open arm time (OT):/Total time (TT) ratio (**C**); Open entry (OE)/Total entry (TE) ratio (**D**).

2.5. ALOS4 Does Not Affect Blood Cell Counts or Blood Chemistry

Blood of ICR mice collected 24 h after ALOS4 IV treatment with acute single doses of 30, 90, or 180 mg/kg and was evaluated for complete blood count (CBC) and basic blood chemistry profile. Analysis of blood compared with the normal rage of ICR mice blood scores [30] showed that ALOS4 generally does not affect blood counts of ICR mice when compared with the control group injected with saline. There were several cell count values that differed from the established laboratory normal range with some doses of ALOS4 treatment; however, control mice treated with saline also deviated from the laboratory normal values as well. Specifically, in the CBC (Table 1), lower values were observed in white blood cells (WBC), mean corpuscular hemoglobin (MCHC), and platelet counts (except 90 mg/kg dose) of ALOS4-treated animals. However, to attribute this decrease to an ALOS4-specific effect may not be correct, since in most of these cases the saline-treated mice also had scores laying out of normal range and may simply be an injection response (Table 1). We observed a 40% decrease in WBC count with 180 mg/kg ALOS4, which was far outside the normal range. Blood biochemical results of ALOS4 and control-treated mice showed higher than normal or normal values in cholesterol, TP, and alkaline phosphate concentrations and lower than normal range in both total bilirubin and chlorides in all injected groups (Table 2). Thus, ALOS4-dependent alterations of blood parameters were minor and clinically non-significant in comparison with controls.

Table 1. Complete blood count of ALOS4-treated ICR mice.

	Normal Range	Saline	ALOS4 30 mg/kg	ALOS4 90 mg/kg	ALOS4 180 mg/kg
WBC $10^3/\mu L$	6.5–24.5	5.49 ± 1.8	4.51 ± 1.3	6.07 ± 1.2	3.2 ± 0.06
RBC $10^6/\mu L$	7.31–10.03	9.21 ± 0.8	9.16 ± 0.3	9.78 ± 0.3	8.02 ± 0.6
HGB g/dL	13.1–16.2	14.68 ± 1.02	14.3 ± 0.6	15.42 ± 0.5	13.2 ± 1.04
Hematocrit %	36.8–48.7	44.8 ± 3.2	43.96 ± 1.9	47.88 ± 1.1	40.12 ± 3.15
MCV fL	46.0–50.9	48.94 ± 1.2	47.96 ± 0.6	48.96 ± 0.24	49.96 ± 0.32
MCV pg	15–18	16.06 ± 0.3	15.6 ± 0.2	15.76 ± 0.15	16.42 ± 0.2
MCHC g/dL	33.7–36.4	32.82 ± 0.5	32.54 ± 0.23	32.2 ± 0.4	32.88 ± 0.4
Platelets $10^3/\mu L$	674–1675	535.4 ± 139.7	582.4 ± 132	770.2 ± 189	661.8 ± 160.5

n = 5, mean counts are presented.

Table 2. Plasma Biochemistry of ALOS4 treated ICR mice.

	Normal Range	Saline	ALOS4 30 mg/kg	ALOS4 90 mg/kg	ALOS4 180 mg/kg
Creatinine mg/dL	0.2–0.4	0.31 ± 0.03	0.26 ± 0.014	0.25 ± 0.06	0.27 ± 0.02
Calcium mg/dL	9.8–10.8	11.34 ± 0.25	10.29 ± 0.11	10.38 ± 0.29	10.42 ± 0.1
Phosphorus mg/dL	6.4–11.3	10.43 ± 0.9	8.6 ± 0.56	9.53 ± 0.4	8.44 ± 0.75
Glucose mg/dL	169–282	176.2 ± 7.4	169.8 ± 6.9	170.75 ± 10.7	183.2 ± 14.3
Urea mg/dL	39–62	55.48 ± 2.9	46.08 ± 1.7	48.23 ± 3.2	46.02 ± 3.9
Cholesterol mg/dL	56–133	140.4 ± 7.9	108.2 ± 9.5	142.25 ± 13.7	141.8 ± 9.1
TP g/dL	4.7–5.8	6.23 ± 0.11	6.17 ± 0.18	6.32 ± 0.13	6.28 ± 0.09
Alb g/dL	3.3–4.0	4.3 ± 0.09	4.3 ± 0.14	4.3 ± 0.06	4.36 ± 0.024
Globulin g/dL	1.4–2.0	1.93 ± 0.12	1.87 ± 0.07	2.02 ± 0.07	1.59 ± 0.4
Total Bilirubin mg/dL	0.16–0.32	0.1 ± 0.02	0.11 ± 0.02	0.14 ± 0.02	0.11 ± 0.03
Alkaline Phos IU/L	43–125	0.31 ± 0.03	0.26 ± 0.014	0.25 ± 0.06	0.27 ± 0.02
SGOT IU/L	69–191	11.34 ± 0.25	10.29 ± 0.11	10.38 ± 0.29	10.42 ± 0.1
SGTP IU/L	26–120	10.43 ± 0.9	8.6 ± 0.56	9.53 ± 0.4	8.44 ± 0.75
Sodium mmol/L	151–156	176.2 ± 7.4	169.8 ± 6.9	170.75 ± 10.7	183.2 ± 14.3
Potassium mmol/L	7.3–10.2	55.48 ± 2.9	46.08 ± 1.7	48.23 ± 3.2	46.02 ± 3.9
Chloride mmol/L	110–119	140.4 ± 7.9	108.2 ± 9.5	142.25 ± 13.7	141.8 ± 9.1

n = 5, mean counts are presented.

3. Discussion

$\alpha_v\beta_3$ integrin is an important cell adhesion receptor involved in various biological activities [18,31] acting through cell signal transduction from the cell membrane to several cytosolic pathways [27]. Due to overexpression of $\alpha_v\beta_3$ in many cancers [21], this integrin is a desirable therapeutic target for cancer treatment. Several anti-cancer peptides [23,32] were developed for $\alpha_v\beta_3$ inhibition targeting the Arginine-Glycine-Aspartate (RGD) motif. Despite their promising potential, these peptides failed in late clinical trials [33], possibly due to their competitive binding to the ECM proteins, which also have RGD sites [34]. In this work, using A375 human melanoma cells we demonstrated anti-cancer properties of ALOS4, a non-RGD peptide thought to target the $\alpha_v\beta_3$ integrin signaling pathway. This peptide was discovered in our laboratory using a phage display technique and previously demonstrated an anti-cancer efficacy in a murine melanoma model [25]. The potential of ALOS4 as a formulated drug is further demonstrated in its physical and chemical stability, as well as appearing to have a good safety profile.

Using a subcutaneous model of A375 human melanoma, we demonstrated high efficacy of ALOS4 in tumor growth inhibition during an 18-day trial. Since there were limitations regarding the number of mice permitted for study, thus also restricting the number of dosing groups, we applied receiver operating characteristic (ROC) analyses [35] and calculated Youden's indices to account for result variability. This enabled us to distinguish between treatment responder and non-responder mice at examined doses of ALOS4 and helped to explain the observed group variability (i.e., segregated responder and non-responder mice were highly internally consistent in their responses to treatment). The existence of "non-responders" is a phenomena extensively discussed in the medical literature describing cancer patients who do not respond to conventional therapies [36,37] and can be explained by variability in the patient microbiome [38] and differential expression of cancer cell surface proteins [39]. These explanations may also be applicable to ALOS4 treatment non-responders observed in our experiments. However, such extrapolation requires further characterization to account for the mechanistic basis of differential treatment responses to ALOS4. Determining the underlying mechanism of ALOS4 has allowed us to eliminate a few possibilities. For example, ex vivo analysis of xenograft tumors did not indicate any significant effect of ALOS4 on angiogenesis or proliferation rates, despite the ALOS4 dose-dependent decrease in tumor size observed. Finally, it is interesting to mention that mice treated with ALOS4 did not lose body weight in contrast with untreated animals. In fact, mice treated with ALOS4 even gained weight leading us to believe that ALOS4 may be used for treatment of cancer patients at different stages of disease suffering from cachexia [40,41], which is considered in 20–40% of cases as an immediate cause of death [42,43].

Since ALOS4 was developed as an $\alpha_v\beta_3$ integrin-targeted molecule, it is likely that observed anti-cancer effects were achieved through modulation of $\alpha_v\beta_3$ integrin signaling. To confirm this suggestion, we analyzed changes in expression of selected candidates from the $\alpha_v\beta_3$ integrin signaling pathway, including extracellular signal-regulated kinases (ERK; also known as mitogen-activated protein kinases or MAPK) initiated by activation of focal adhesion kinases (FAK) and Src kinases, which in complex or individually further activate downstream ERK signaling [44,45]. We found that ALOS4 does not alter total ERK, despite significant upregulation of FAK, or alter Src protein expression at high doses in human melanoma cells in vitro. Since ALOS4 alters FAK and Src, but not ERK which acts as the last messenger of MAPK/ERK pathway prior to entering the nucleus and activating transcription factors of genes involved in proliferation and metastasis [46], we suggest that additional modulation occurs interrupting downstream signals. Furthermore, the final products of ERK signaling, the oncogenes *c-Fos* and *c-Jun*, were both affected by ALOS4 treatment in vitro. ALOS4 significantly reduced *c-Fos* mRNA levels at all doses, while downregulation of the *c-Jun* gene was significant at 0.1 μM dosage after 24 h and showed a non-significant tendency to decrease after 48 h. We speculate that the differences in levels of significance in *c-Fos* expression at 24 and 48 h of treatment may indicate the attempt of the cancer cells to stabilize expression of this oncogene, whereas its downregulation by ALOS4 remains to be explained. These results indirectly confirm that ALOS4 is able to modulate $\alpha_v\beta_3$ integrin signaling and differences in the effect of ALOS4 on *c-Fos* and *c-Jun* expression may be explained by additional ALOS4-independent processes involved in *c-Fos* and *c-Jun* transcription. The *c-Fos* results are not without precedent considering the actions of other peptide-based integrin antagonists (flavoridin) in melanoma cell lines, which increase activation of downstream integrin pathway elements (such as increased FAK phosphorylation) while also effecting downregulation *c-Fos* expression [47]. Thus, since ALOS4 was developed targeting $\alpha_v\beta_3$ integrin and its ability to bind $\alpha_v\beta_3$ leading to metastatic arrest was previously demonstrated [25], we believe that our new results indirectly confirm the involvement of ALOS4 in the modulation of selected components of integrin signaling. However, unaffected ERK in the presence of upregulated FAK and Src suggests that an additional intervening pathway modulating ERK-related signaling, possibly through integrin-initiated RAS-RAF activating cascade [48,49] or integrin independent signaling pathway [50], is present and a further study of molecular mechanisms of action at a higher-resolution with ALOS4 is required.

We also examined ALOS4 safety and stability as a potential drug candidate. Due to a known tendency of peptides to adhere to standard inert laboratory materials [51], we demonstrated that regardless of ALOS4 concentration, we achieved nearly 100% peptide recovery in both analyzed materials (borosilicate LC glass vials and standard polypropylene LC vials). We posit that the allosterically-constrained, cyclical structure of the peptide may be the reason ALOS4 does not significantly interact with typically problematic laboratory materials as do other peptides. Furthermore, despite the fact that most peptides in solution undergo degradation by hydrolysis or oxidation [52], ALOS4 was highly stable in saline solution over a wide range of acid/base conditions at different temperatures, features which are beneficial for long-term storage and ease-of-use for therapeutic applications [53]. We also examined the toxicity of ALOS4 in nominally cancer-free mice, which is considered an essential factor for pharmaceutical safety [54,55]. ALOS4 demonstrated no toxicity in vivo with repeated treatments over a range of doses from 10 to 360 mg/kg and no mortality or serious adverse events were observed. We also elected to examine whether repeated ALOS4 dosing would produce any unfavorable behavioral features, such as sedation, hyperactivity, or anxiety-like behaviors. No adverse behavioral effects were observed. Hence, when comparing with the therapeutic doses of other anticancer peptides, which range from 2.5 mg/kg (Cilengitide [56]) to 60 mg/kg (HM-3 [57]), ALOS4 stands out as a remarkably non-toxic compound. Moreover, whereas most conventional chemotherapies are accompanied by severe side effects that require medical intervention [58], the safety profile of ALOS4 shows potential as an anti-cancer drug that may be tolerable for patients.

Blood chemistry was also examined during toxicity studies and revealed reduction of white blood cell counts (40%) with ALOS4 treatment at 180 mg/kg. This reduction could indicate higher levels of tumor-infiltrating lymphocytes (TILs), suggesting that the immune system may be involved in the ALOS4 activity. Unfortunately, potential effects and mechanistic outcomes of immune interactions of ALOS4 could not be determined in this study due to the immunodeficient nature of the mice required for the xenograft model. Nevertheless, cumulative results from this work and prior studies suggest an interaction of ALOS4 with immune system elements, which needs to be further evaluated.

In summary, ALOS4 appears to be completely non-toxic, remarkably prolongs lifespan, and increases weight of treated mice. The latter feature makes ALOS4 beneficial to counteract cachexia experienced by cancer patients during the process of disease progression. We believe that demonstrating the anti-cancer activity through modulation of components of integrin signaling together with its safety profile suggests that ALOS4 peptide is a promising patient-tolerable prospective anti-cancer drug candidate.

4. Materials and Methods

4.1. ALOS4

ALOS4 was developed based on $\alpha_v\beta_3$ binding using phage display technology. This synthetic cyclic peptide is composed of the following nine-amino-acid sequence: H-cycl(Cys-Ser-Ser-Ala-Gly-Ser-Leu-Phe-Cys)-OH. ALOS4 was custom-synthesized by Shanghai Hanhong Scientific Co. (Cat#P120301-LG221431, Shanghai, China). Stock solutions of ALOS4 at 10 mM, were prepared in sterile physiological saline solution (0.9%; Sigma-Aldrich, Cat#7647-14-5, Darmstadt, Germany) with the addition of 0.02% BSA (Biological Industries, Cat#1522089, Kibbutz Beit-Haemek, Israel) and maintained at either −20 °C for short term use, or −80 °C for long-term storage. For each experiment, ALOS4 was thawed and freshly diluted to working concentrations in physiological saline.

4.2. Cell Cultures

A375 human melanoma cells (ATCC; Cat#CRL-1619, Manassas, VA, USA) were grown in Dulbecco's Modified Eagle Medium (DMEM; Fisher Scientific [Gibco], Cat#41965-039, Hampton, NH, USA) with 4.5 g/L glucose and L-glutamine, supplemented with 10% fetal bovine serum (FBS; Fisher Scientific [Gibco], Cat#16000-036, Hampton, NH, USA) and 1% penicillin-streptomycin (Fisher Scientific, Cat#10378-016, Hampton, NH, USA). Cells were maintained on uncoated dishes in atmosphere of 5% CO_2 at 370 °C.

4.3. Chemical Properties Assays

Adhesiveness to inert materials was measured for ALOS4 0.9% NaCl(aq) solution in concentrations of 1, 3 10, 30, or 100 μM and incubated in standard laboratory polypropylene microtubes for 60 min at room temperature. Solutions were transferred to either polypropylene LC vials or borosilicate glass LC vials. Optimal formulation stability of ALOS4 was analyzed in either saline solution (0.9% NaCl) or saline solution containing 0.1% tween 80 (polysorbate; Sigma-Aldrich, Cat#P1754, Darmstadt, Germany) at concentrations of 0.5, 5, 50, 500, or 5000 pmol/µL. Solutions were incubated in standard polypropylene tubes for 60 min at room temperature. Acid/base stability was measured for ALOS4 formulated in saline at a concentration of 10 µM and incubated for 1 or 24 h at 3.4, 4.4, 5.4, 6.4, 7.4, or 9.4 pH at temperatures of 4, 25, or 37 °C. Recovery of ALOS4 following these materials assays was assessed by LC-MS.

4.4. Animals

To investigate the effect of ALOS4 on human melanoma cancer cells, nude Fox nu/nu mice were used for SC- or IV-injected inoculations. Additionally, uninoculated (nominally cancer-free) ICR mice were used for safety and toxicity studies. Mice were obtained from Envigo, Israel, and arrived at the age of 4–5 weeks old. Upon arrival, mice were habituated to vivarium conditions for one week before initiation of experiments. All mice were maintained under a 12:12 light–dark cycle and provided Purina rodent chow (Envigo, Ness-Ziona, Israel) and water ad libitum. Animals were housed five to a cage in a room maintained at 22 ± 0.5 °C (nude mice cages were held in a laminar-flow cabinet).

4.5. Behavioral Models

4.5.1. Open Field

To evaluate the effect of ALOS4 on mouse locomotor activity, we used the open field (OF) behavioral test [59,60]. This assay consists of an arena (30 × 40 cm) with no grid markings and uses an infrared imaging system. The number of entries into the arena center zone was recorded using EthoVision 7.1 software (Noldus Information Technology, Wageningen, The Netherlands). Each mouse was placed individually in the center of the arena and evaluated for 6 min. Arena center dwell time versus arena border dwell time, as well as total traveled distance, were recorded. To provide a less stressful environment, the test was performed in a semi-dark room. One hour prior to the test, all mice were placed in the behavioral experiment room for acclimation. Between subjects, the apparatus was thoroughly washed with 70% ethanol and dried.

4.5.2. Elevated plus Maze

To evaluate the effect of ALOS4 on mouse anxiety-like behaviors, we used the elevated plus maze test (EPM) [60]. The EPM consists of a plus-shaped arena with two open (10 × 45 × 40 cm) and two enclosed (10 × 45 × 40 cm) open-roof arms, elevated 70 cm from the floor. Each mouse was placed in the center of the maze and was free to move in the arena for 5 min. The number of entries into open and closed arms, as well as time spent in the open and closed arms (dwell time), was recorded using EthoVision 7.1 software (Noldus Information Technology, Wageningen, The Netherlands). To provide a less stressful environment, the test was performed in a semi-dark room. One hour prior to the test, all mice were placed in the behavioral experiment room for acclimation. Between subjects, the apparatus was thoroughly washed with 70% ethanol and dried.

4.5.3. Toxicity Assessment

Acute single or repeated doses of ALOS4 at 10, 30, 90, 180, or 360 mg/kg were administered intravenously to uninoculated (nominally cancer-free) ICR mice. Body weight, rectal temperature, and survival were evaluated for 14 days following injections. Mice were also evaluated for locomotory and anxiety-like behaviors (open-field and EPM) on treatment day 14.

4.5.4. Subcutaneous Model of Melanoma

We used a subcutaneous (SC) melanoma model to study the effect of ALOS4 on localized solid tumor growth. Nude mice (Fox nu/nu) were SC injected with A375 cells at 2×10^6 cells in 100 μL in serum-free DMEM medium/mouse. Following inoculation, all mice were randomly divided into experimental groups, then treated IP with either ALOS4 or saline (negative control) at day one post-inoculation. Mouse body weights were monitored during the course of the experiment. The base-weight of mice was determined by the weight on the second day to account for acclimation-related changes, then mice were weighed twice a week until tumor appearance and thereafter daily until the experiment was terminated. Termination resulted from mouse death, when tumor diameter reached or exceeded 1500 mm^3, or when 30 days of treatment had elapsed, whereupon mice were CO_2 euthanized. Mice were IP-injected with ALOS4 (0.1, 0.3, or 30 mg/kg; assumed therapeutic range) or saline (control) at identical fixed volumes. SC inoculation of cells usually formed a palpable tumor in 7–14 days. Tumor volumes were estimated by digital caliper and calculated with the following equation: $V(\text{tumor}, mm^3) = \pi/6 \times \text{width} \times \text{length} \times \text{height}$. Survival rate of mice was documented at the end point of experiments. Tumor growth rates were calculated by the following formula: TGI% = (relative tumor volume ALOS4-treated)/(relative tumor volume saline-treated).

4.5.5. Immunohistochemistry of A375 Tumor

Nude mice (Fox nu/nu) were SC-inoculated with A375 human melanoma cells (2×10^6 cells in 100 μL normal saline/mouse) and treated for 18 days with ALOS4 (0, 0.3, 3, 30 mg/kg) injected IP with daily monitoring for clinical signs. Mice were euthanized by CO_2 asphyxiation when the first mouse reached the ethical protocol limit of 1500 mm^3 tumor size. Tumors were harvested, weighed, and fixed in 4% formalin. After 24 h fixation, samples were rinsed with PBS and transferred to 70% ethanol for transport to the pathology laboratory. Embedding, 5 μm sectioning, and slide preparation were performed for the 5 tumors from each experimental group (n = 4 for the 0.3 mg/kg due to the technical issues within processing) according to routine procedure.

Tumor pathology was rated by a certified veterinary pathologist (Patho-Logica, Rehovot, Israel) using the following scales when examined at 40× magnification: Pleomorphism (0, none; 1, mild; 2, moderate; 3, severe), mitotic index (mitotic indicators were counted in 10 different 40× fields and averaged), degree of vascular invasion (−, none; +, invasion). Prepared tumor tissue slides were also evaluated for the presence of tumor-related-markers by monoclonal antibody staining for the α-smooth muscle actin (αSMA), the marker of vascular smooth muscle cells, as well as the nuclear protein cell proliferation marker Ki-67 using the following rating scales: αSMA (0, not present; 1, mild [10–20 positive vessels]; 2, moderate [20–50 positive vessels]; 3, severe [>50 positive vessels]), Ki-67 (0, not present; 1, <10%; 2, 10–50%; 3, 50–75%; 4, >75%).

4.5.6. RNA Extraction and qRT-PCR

RNA from A375 cells 24 and 48 h after the treatment was purified from cells using a quick RNA miniprep kit (Zymo Research, Cat#R1018, Irvine, CA, USA). DNase treatment was performed using on-column DNase digestion. RNA concentration was measured at 260 nm using NanoDrop spectrophotometer (Thermo Scientific, Wilmington, DE, USA, Cat#DE19810) and 260/280 ratio method was used to verify that the samples met proper purification standards around 2. A total of 1 μg of total RNA was reverse-transcribed using a reverse transcription system (Promega, Cat#A3500, Madison, WI, USA). The master mix for cDNA synthesis insisted of 10× Reverse Transcription buffer, dNTP mix, oligo (dT) (18T) primers, and AMV enzyme. The reverse transcription reaction was performed in a thermocycler (Bio-Rad Laboratories, T100, Hercules, CA, USA) using a two-step program: 42 °C for 60 min followed by heating to 70 °C for 15 min to terminate the reaction, and maintained at 4 °C. The quantitative RT-PCR for *c-Fos* and *c-Jun* was performed using 2× PCR SYBR Green Master Mix (Applied Biosystems, Cat#4344463, Warrington, UK), with a 100 nM mixture of forward and reverse primers (*c-Fos* forward: ctggcgttgtgaagaccat and reverse: tcccttcggattctcctttt; *c-Jun* forward: atcaaggcggagaggaagc and reverse: tgagcatgttggccgtggac; as well as HPRT used as an endogenous normalization factor, forward: cctggcgtcgtgattagtgat and reverse: tcgagcaagacgttcagtcc), 4 μg of cDNA and RNase/DNase free water. Samples were placed in Real-Time PCR (AriaMx; Cat#G88230A, Santa Clara, CA, USA,) and reactions were performed in a thermocycler: 180 s at 95 °C, followed by 40 cycles of 3 s at 95 °C and 30 s at 60 °C.

4.5.7. Protein Extraction and Western Blot Analysis

Proteins from A375 cells were extracted in RIPA lysis buffer solution (150 mM NaCl, 50 mM Tris-HCl pH 8.0, 1% Triton X-100, 0.5% sodium deoxycholate, 0.1% SDS) with freshly added 1 mM sodium orthovanadate (Na_3VO_4; Sigma-Aldrich, Cat#S6508, Darmstadt, Germany), 5 mM sodium fluoride (NaF; Sigma-Aldrich, Cat#S7920, Darmstadt, Germany), protease inhibitor cocktail (Millipore, Cat#539134, Burlington, MA, USA), and phosphatase inhibitor (Sigma-Aldrich, Cat#4906845001, Darmstadt, Germany). Protein concentrations were assessed using a Bradford assay. Proteins were separated by gel electrophoresis in an 8% polyacrylamide gel. Blots were incubated with blocking solution (5% BSA in TBST: Bio-Lab, Cat#208923, Jerusalem, Israel) for 1 h with gentle shaking at room temperature. After blocking, separate membranes were each probed with one of target-specific antibodies (ERK1/2, Millipore, Cat#MABS827, Burlington, MA, USA; pERK, Cell Signaling, Cat#C33E10, Danvers, MA, USA; FAK, Cell Signaling, Cat#3285; pFAK, Santa Cruz, Cat#sc-374668, Dallas TX, USA; c-Src, Novus Biologicals, Cat#5A18, Littleton CO, USA; p-c-Src, Santa Cruz, Cat#sc-166860), then hybridized with horseradish peroxidase-conjugated streptavidin secondary antibodies (Abcam, Cat#ab6802 and ab205719, Cambridge, UK) and developed using ECL solution (Immobilon Crescendo Western HRP substrate, Millipore, Cat#ELLUR0100, Burlington, MA, USA) according to the manufacturer protocol. After the antibody of a specific protein on each membrane was evaluated, we performed a GAPDH (Millipore, Cat#MABS819, Burlington, MA, USA) re-probe for all of membranes to quantify the target proteins. FAK and Src were performed after stripping on the same membrane and that is why they share their common GAPDH. Blots were visualized using ChemiDoc™ Imaging System (Bio-Rad Laboratories, Hercules, CA, USA) apparatus and densitometry analysis was performed using Image Lab Software (Bio-Rad Laboratories, Hercules, CA, USA). The grouped data sets representing phosphorylated and total proteins demonstrates percentage of each treated group normalized to untreated control.

4.5.8. Complete Blood Cell Count and Blood Chemistry

ICR mice were treated with single IV injections of ALOS4 (30, 90, or 180 mg/kg) and blood samples were collected at 24 h post-injection to EDTA and serum tubes. Complete blood cell count (CBC) and blood chemistry analyses were performed for saline and ALOS4 treatment groups at a certified animal laboratory (Herzliya Medical Center, Herzliya, Israel).

4.5.9. Statistical Analysis

All data are expressed as means ± SE (±SD in a few measures). Threshold for significance was set to $\alpha = 0.05$. Multiple treatments were Bonferroni-corrected and compared by unmatched one-way ANOVA for single time point results or by two-way ANOVA for multiple treatment outcomes over time. ANOVA tests were followed with a Bonferroni means separation test to identify specific differences between treatments. For ordinal or nominal data of tumor immunohistochemistry scoring, groups were compared by Kruskal–Wallis ANOVA followed by a Dunn's post-test for intergroup comparisons [61]. To differentiate between responders and non-responders for tumor growth effects, we performed responder operator curve analyses (ROC) to separate groups. ROC analyses and ANOVAs including post-tests were performed with GraphPad Prism 7.0.

5. Patents

Pinhasov A. has an ALOS4 patent (62/127,854).

Author Contributions: Conceptualization, A.P., I.K., and E.N.; Methodology, E.N., G.G., and A.P.; Validation, B.L., S.Y., B.R., O.A., and M.B.; Analysis, M.K., M.B., and B.R.; Data curation, M.B., M.K., G.G., and E.N.; Writing—original draft preparation, E.N. and M.K.; Writing—review and editing, M.K., E.N., I.K., and A.P. Funding acquisition, I.K. All authors have read and agreed to the published version of the manuscript.

Funding: This research was funded by The Institute for Personalized and Translational Medicine, Ariel University, Israel. Grant Number: RA1600000120.

Institutional Review Board Statement: All procedures with animals were conducted under supervision of The Institutional Animal Care and Use Committee of Ariel University and the Israel Ministry of Health (IL-108-06-16, approved on June 2016; IL-123-02-17, approved on February 2017).

Informed Consent Statement: Not applicable.

Data Availability Statement: All data are provided as figures and tables and included in this paper.

Conflicts of Interest: Pinhasov A. has an ALOS4 patent (62/127,854). Pinhasov A., Koman I., and Nesher E. have an ALOS4-related pending patent (European Patent Application No. 16758561.1).

References

1. Cooper, C.R.; Chay, C.H.; Pienta, K.J. The role of alpha(v)beta(3) in prostate cancer progression. *Neoplasia* **2002**, *4*, 191–194. [CrossRef]
2. Takayama, S.; Ishii, S.; Ikeda, T.; Masamura, S.; Doi, M.; Kitajima, M. The relationship between bone metastasis from human breast cancer and integrin alpha(v)beta3 expression. *Anticancer Res.* **2005**, *25*, 79–83.
3. Koistinen, P.; Heino, J. Integrins in Cancer Cell Invasion. In *Madame Curie Bioscience Database [Internet]*; Landes Bioscience: Austin, TX, USA, 2013.
4. Attieh, Y.; Clark, A.G.; Grass, C.; Richon, S.; Pocard, M.; Mariani, P.; Elkhatib, N.; Betz, T.; Gurchenkov, B.; Vignjevic, D.M. Cancer-associated fibroblasts lead tumor invasion through integrin-beta3-dependent fibronectin assembly. *J. Cell Biol.* **2017**, *216*, 3509–3520. [CrossRef] [PubMed]
5. Weis, S.M.; Cheresh, D.A. alphaV integrins in angiogenesis and cancer. *Cold Spring Harb. Perspect. Med.* **2011**, *1*, a006478. [CrossRef] [PubMed]
6. Xiong, J.P.; Stehle, T.; Zhang, R.; Joachimiak, A.; Frech, M.; Goodman, S.L.; Arnaout, M.A. Crystal structure of the extracellular segment of integrin alpha Vbeta3 in complex with an Arg-Gly-Asp ligand. *Science* **2002**, *296*, 151–155. [CrossRef] [PubMed]
7. Bellis, S.L. Advantages of RGD peptides for directing cell association with biomaterials. *Biomaterials* **2011**, *32*, 4205–4210. [CrossRef] [PubMed]
8. Rocha, L.A.; Learmonth, D.A.; Sousa, R.A.; Salgado, A.J. Alphavbeta3 and alpha5beta1 integrin-specific ligands: From tumor angiogenesis inhibitors to vascularization promoters in regenerative medicine? *Biotechnol. Adv.* **2018**, *36*, 208–227. [CrossRef]
9. Sloan, E.K.; Pouliot, N.; Stanley, K.L.; Chia, J.; Moseley, J.M.; Hards, D.K.; Anderson, R.L. Tumor-specific expression of alphavbeta3 integrin promotes spontaneous metastasis of breast cancer to bone. *Breast Cancer Res. BCR* **2006**, *8*, R20. [CrossRef]
10. Guo, W.; Giancotti, F.G. Integrin signalling during tumour progression. *Nat. Reviews. Mol. Cell Biol.* **2004**, *5*, 816–826. [CrossRef]
11. Rathinam, R.; Alahari, S.K. Important role of integrins in the cancer biology. *Cancer Metastasis Rev.* **2010**, *29*, 223–237. [CrossRef] [PubMed]
12. Bello, L.; Francolini, M.; Marthyn, P.; Zhang, J.; Carroll, R.S.; Nikas, D.C.; Strasser, J.F.; Villani, R.; Cheresh, D.A.; Black, P.M. Alpha(v)beta3 and alpha(v)beta5 integrin expression in glioma periphery. *Neurosurgery* **2001**, *49*, 380–389. [CrossRef]
13. van den Hoogen, C.; van der Horst, G.; Cheung, H.; Buijs, J.T.; Pelger, R.C.; van der Pluijm, G. Integrin alphav expression is required for the acquisition of a metastatic stem/progenitor cell phenotype in human prostate cancer. *Am. J. Pathol.* **2011**, *179*, 2559–2568. [CrossRef]
14. McCabe, N.P.; De, S.; Vasanji, A.; Brainard, J.; Byzova, T.V. Prostate cancer specific integrin alphavbeta3 modulates bone metastatic growth and tissue remodeling. *Oncogene* **2007**, *26*, 6238–6243. [CrossRef]
15. Felding-Habermann, B.; O'Toole, T.E.; Smith, J.W.; Fransvea, E.; Ruggeri, Z.M.; Ginsberg, M.H.; Hughes, P.E.; Pampori, N.; Shattil, S.J.; Saven, A.; et al. Integrin activation controls metastasis in human breast cancer. *Proc. Natl. Acad. Sci. USA* **2001**, *98*, 1853–1858. [CrossRef]
16. Huang, R.; Rofstad, E.K. Integrins as therapeutic targets in the organ-specific metastasis of human malignant melanoma. *J. Exp. Clin. Cancer Res. CR* **2018**, *37*, 92. [CrossRef]
17. Heistein, J.B.; Acharya, U. *Malignant Melanoma*; StatPearls: Treasure Island, FL, USA, 2021.
18. Siegel, R.L.; Miller, K.D.; Jemal, A. Cancer statistics, 2019. *CA A Cancer J. Clin.* **2019**, *69*, 7–34. [CrossRef]

19. Domingues, B.; Lopes, J.M.; Soares, P.; Populo, H. Melanoma treatment in review. *ImmunoTargets Ther.* **2018**, *7*, 35–49. [CrossRef] [PubMed]
20. Meerovitch, K.; Bergeron, F.; Leblond, L.; Grouix, B.; Poirier, C.; Bubenik, M.; Chan, L.; Gourdeau, H.; Bowlin, T.; Attardo, G. A novel RGD antagonist that targets both alphavbeta3 and alpha5beta1 induces apoptosis of angiogenic endothelial cells on type I collagen. *Vasc. Pharmacol.* **2003**, *40*, 77–89. [CrossRef]
21. McNeel, D.G.; Eickhoff, J.; Lee, F.T.; King, D.M.; Alberti, D.; Thomas, J.P.; Friedl, A.; Kolesar, J.; Marnocha, R.; Volkman, J.; et al. Phase I trial of a monoclonal antibody specific for alphavbeta3 integrin (MEDI-522) in patients with advanced malignancies, including an assessment of effect on tumor perfusion. *Clin. Cancer Res. Off. J. Am. Assoc. Cancer Res.* **2005**, *11*, 7851–7860. [CrossRef] [PubMed]
22. Russo, M.A.; Paolillo, M.; Sanchez-Hernandez, Y.; Curti, D.; Ciusani, E.; Serra, M.; Colombo, L.; Schinelli, S. A small-molecule RGD-integrin antagonist inhibits cell adhesion, cell migration and induces anoikis in glioblastoma cells. *Int. J. Oncol.* **2013**, *42*, 83–92. [CrossRef]
23. Carter, A. Integrins as target: First phase III trial launches, but questions remain. *J. Natl. Cancer Inst.* **2010**, *102*, 675–677. [CrossRef] [PubMed]
24. Hersey, P.; Sosman, J.; O'Day, S.; Richards, J.; Bedikian, A.; Gonzalez, R.; Sharfman, W.; Weber, R.; Logan, T.; Buzoianu, M.; et al. A randomized phase 2 study of etaracizumab, a monoclonal antibody against integrin alpha(v)beta(3), + or − dacarbazine in patients with stage IV metastatic melanoma. *Cancer* **2010**, *116*, 1526–1534. [CrossRef] [PubMed]
25. Yacobovich, S.; Tuchinsky, L.; Kirby, M.; Kardash, T.; Agranyoni, O.; Nesher, E.; Redko, B.; Gellerman, G.; Tobi, D.; Gurova, K.; et al. Novel synthetic cyclic integrin alphavbeta3 binding peptide ALOS4: Antitumor activity in mouse melanoma models. *Oncotarget* **2016**, *7*, 63549–63560. [CrossRef]
26. Redko, B.; Tuchinsky, H.; Segal, T.; Tobi, D.; Luboshits, G.; Ashur-Fabian, O.; Pinhasov, A.; Gerlitz, G.; Gellerman, G. Toward the development of a novel non-RGD cyclic peptide drug conjugate for treatment of human metastatic melanoma. *Oncotarget* **2017**, *8*, 757–768. [CrossRef]
27. Harburger, D.S.; Calderwood, D.A. Integrin signalling at a glance. *J. Cell Sci.* **2009**, *122*, 159–163. [CrossRef]
28. Vellon, L.; Menendez, J.A.; Lupu, R. A bidirectional "alpha(v)beta(3) integrin-ERK1/ERK2 MAPK" connection regulates the proliferation of breast cancer cells. *Mol. Carcinog.* **2006**, *45*, 795–804. [CrossRef]
29. Dhillon, A.S.; Hagan, S.; Rath, O.; Kolch, W. MAP kinase signalling pathways in cancer. *Oncogene* **2007**, *26*, 3279–3290. [CrossRef] [PubMed]
30. Serfilippi, L.M.; Pallman, D.R.; Russell, B. Serum clinical chemistry and hematology reference values in outbred stocks of albino mice from three commonly used vendors and two inbred strains of albino mice. *Contemp. Top. Lab. Anim. Sci.* **2003**, *42*, 46–52. [PubMed]
31. Duffy, M.J.; McGowan, P.M.; Gallagher, W.M. Cancer invasion and metastasis: Changing views. *J. Pathol.* **2008**, *214*, 283–293. [CrossRef]
32. Zuo, H. iRGD: A Promising Peptide for Cancer Imaging and a Potential Therapeutic Agent for Various Cancers. *J. Oncol.* **2019**, *2019*, 9367845. [CrossRef]
33. Chinot, O.L. Cilengitide in glioblastoma: When did it fail? *Lancet. Oncol.* **2014**, *15*, 1044–1045. [CrossRef]
34. Alberts, S.R.; Fishkin, P.A.; Burgart, L.J.; Cera, P.J.; Mahoney, M.R.; Morton, R.F.; Johnson, P.A.; Nair, S.; Goldberg, R.M.; North Central Cancer Treatment, G. CPT-11 for bile-duct and gallbladder carcinoma: A phase II North Central Cancer Treatment Group (NCCTG) study. *Int. J. Gastrointest. Cancer* **2002**, *32*, 107–114. [CrossRef]
35. Zhang, J.; Mueller, S.T. A note on ROC analysis and non-parametric estimate of sensitivity. *Psychometrika* **2005**, *70*, 203–212. [CrossRef]
36. Vasan, N.; Baselga, J.; Hyman, D.M. A view on drug resistance in cancer. *Nature* **2019**, *575*, 299–309. [CrossRef] [PubMed]
37. Balmativola, D.; Marchio, C.; Maule, M.; Chiusa, L.; Annaratone, L.; Maletta, F.; Montemurro, F.; Kulka, J.; Figueiredo, P.; Varga, Z.; et al. Pathological non-response to chemotherapy in a neoadjuvant setting of breast cancer: An inter-institutional study. *Breast Cancer Res. Treat.* **2014**, *148*, 511–523. [CrossRef]
38. Gopalakrishnan, V.; Spencer, C.N.; Nezi, L.; Reuben, A.; Andrews, M.C.; Karpinets, T.V.; Prieto, P.A.; Vicente, D.; Hoffman, K.; Wei, S.C.; et al. Gut microbiome modulates response to anti-PD-1 immunotherapy in melanoma patients. *Science* **2018**, *359*, 97–103. [CrossRef]
39. Sherlach, K.S.; Roepe, P.D. Drug resistance associated membrane proteins. *Front. Physiol.* **2014**, *5*, 108. [CrossRef]
40. Dhanapal, R.; Saraswathi, T.; Govind, R.N. Cancer cachexia. *J. Oral Maxillofac. Pathol. JOMFP* **2011**, *15*, 257–260. [CrossRef]
41. Argiles, J.M.; Lopez-Soriano, F.J.; Stemmler, B.; Busquets, S. Therapeutic strategies against cancer cachexia. *Eur. J. Transl. Myol.* **2019**, *29*, 7960. [CrossRef]
42. Fox, K.M.; Brooks, J.M.; Gandra, S.R.; Markus, R.; Chiou, C.F. Estimation of Cachexia among Cancer Patients Based on Four Definitions. *J. Oncol.* **2009**, *2009*, 693458. [CrossRef]
43. Tisdale, M.J. Cachexia in cancer patients. *Nat. Reviews. Cancer* **2002**, *2*, 862–871. [CrossRef] [PubMed]
44. Bolos, V.; Gasent, J.M.; Lopez-Tarruella, S.; Grande, E. The dual kinase complex FAK-Src as a promising therapeutic target in cancer. *OncoTargets Ther.* **2010**, *3*, 83–97. [CrossRef] [PubMed]

45. Sawai, H.; Okada, Y.; Funahashi, H.; Matsuo, Y.; Takahashi, H.; Takeyama, H.; Manabe, T. Activation of focal adhesion kinase enhances the adhesion and invasion of pancreatic cancer cells via extracellular signal-regulated kinase-1/2 signaling pathway activation. *Mol. Cancer* **2005**, *4*, 37. [CrossRef] [PubMed]
46. Guo, Y.J.; Pan, W.W.; Liu, S.B.; Shen, Z.F.; Xu, Y.; Hu, L.L. ERK/MAPK signalling pathway and tumorigenesis. *Exp. Ther. Med.* **2020**, *19*, 1997–2007. [CrossRef]
47. Oliva, I.B.; Coelho, R.M.; Barcellos, G.G.; Saldanha-Gama, R.; Wermelinger, L.S.; Marcinkiewicz, C.; Benedeta Zingali, R.; Barja-Fidalgo, C. Effect of RGD-disintegrins on melanoma cell growth and metastasis: Involvement of the actin cytoskeleton, FAK and c-Fos. *Toxicon Off. J. Int. Soc. Toxinology* **2007**, *50*, 1053–1063. [CrossRef] [PubMed]
48. Wortzel, I.; Seger, R. The ERK Cascade: Distinct Functions within Various Subcellular Organelles. *Genes Cancer* **2011**, *2*, 195–209. [CrossRef]
49. Yee, K.L.; Weaver, V.M.; Hammer, D.A. Integrin-mediated signalling through the MAP-kinase pathway. *IET Syst. Biol.* **2008**, *2*, 8–15. [CrossRef]
50. Sundaram, M.V. Canonical RTK-Ras-ERK signaling and related alternative pathways. In *WormBook: The Online Review of C. elegans Biology [Internet]*; WormBook: Pasadena, CA, USA, 2018. [CrossRef]
51. Goebel-Stengel, M.; Stengel, A.; Tache, Y.; Reeve, J.R., Jr. The importance of using the optimal plasticware and glassware in studies involving peptides. *Anal. Biochem.* **2011**, *414*, 38–46. [CrossRef]
52. Furman, J.L.; Chiu, M.; Hunter, M.J. Early engineering approaches to improve peptide developability and manufacturability. *AAPS J.* **2015**, *17*, 111–120. [CrossRef] [PubMed]
53. Bottger, R.; Hoffmann, R.; Knappe, D. Differential stability of therapeutic peptides with different proteolytic cleavage sites in blood, plasma and serum. *PLoS ONE* **2017**, *12*, e0178943. [CrossRef] [PubMed]
54. Gad, S.C. *Preclinical Development Handbook: ADME and Biopharmaceutical Properties*; Wiley: Hoboken, NJ, USA, 2008; p. 1352.
55. Allen, D.D.; Caviedes, R.; Cardenas, A.M.; Shimahara, T.; Segura-Aguilar, J.; Caviedes, P.A. Cell lines as in vitro models for drug screening and toxicity studies. *Drug Dev. Ind. Pharm.* **2005**, *31*, 757–768. [CrossRef] [PubMed]
56. Dolgos, H.; Freisleben, A.; Wimmer, E.; Scheible, H.; Kratzer, F.; Yamagata, T.; Gallemann, D.; Fluck, M. In vitro and in vivo drug disposition of cilengitide in animals and human. *Pharmacol. Res. Perspect.* **2016**, *4*, e00217. [CrossRef] [PubMed]
57. Yassin, S.; Hu, J.; Xu, H.; Li, C.; Setrerrahmane, S. In vitro and in vivo activities of an antitumor peptide HM-3: A special dose-efficacy relationship on an HCT116 xenograft model in nude mice. *Oncol. Rep.* **2016**, *36*, 2951–2959. [CrossRef]
58. Nurgali, K.; Jagoe, R.T.; Abalo, R. Editorial: Adverse Effects of Cancer Chemotherapy: Anything New to Improve Tolerance and Reduce Sequelae? *Front. Pharmacol.* **2018**, *9*, 245. [CrossRef] [PubMed]
59. Gross, M.; Sheinin, A.; Nesher, E.; Tikhonov, T.; Baranes, D.; Pinhasov, A.; Michaelevski, I. Early onset of cognitive impairment is associated with altered synaptic plasticity and enhanced hippocampal GluA1 expression in a mouse model of depression. *Neurobiol. Aging* **2015**, *36*, 1938–1952. [CrossRef]
60. Nesher, E.; Gross, M.; Lisson, S.; Tikhonov, T.; Yadid, G.; Pinhasov, A. Differential responses to distinct psychotropic agents of selectively bred dominant and submissive animals. *Behav. Brain Res.* **2013**, *236*, 225–235. [CrossRef]
61. Gibson-Corley, K.N.; Olivier, A.K.; Meyerholz, D.K. Principles for valid histopathologic scoring in research. *Vet. Pathol.* **2013**, *50*, 1007–1015. [CrossRef]

International Journal of
Molecular Sciences

MDPI

Article

Environment-Sensitive Fluorescent Labelling of Peptides by Luciferin Analogues

Marialuisa Siepi [1,†], Rosario Oliva [2,†], Antonio Masino [1,†], Rosa Gaglione [2], Angela Arciello [2], Rosita Russo [3], Antimo Di Maro [3], Anna Zanfardino [1], Mario Varcamonti [1], Luigi Petraccone [2], Pompea Del Vecchio [2], Marcello Merola [1], Elio Pizzo [1], Eugenio Notomista [1,*,‡] and Valeria Cafaro [1,‡]

1 Department of Biology, University of Naples Federico II, 80126 Naples, Italy; marialuisa.siepi@unina.it (M.S.); antonio.masino@unina.it (A.M.); anna.zanfardino@unina.it (A.Z.); varcamon@unina.it (M.V.); m.merola@unina.it (M.M.); elipizzo@unina.it (E.P.); vcafaro@unina.it (V.C.)

2 Department of Chemical Sciences, University of Naples Federico II, 80126 Naples, Italy; rosario.oliva2@unina.it (R.O.); rosa.gaglione@unina.it (R.G.); anarciel@unina.it (A.A.); luigi.petraccone@unina.it (L.P.); pompea.delvecchio@unina.it (P.D.V.)

3 Department of Environmental, Biological and Pharmaceutical Sciences and Technologies, University of Campania "Luigi Vanvitelli", 81100 Caserta, Italy; rosita.russo@unicampania.it (R.R.); antimo.dimaro@unicampania.it (A.D.M.)

* Correspondence: notomist@unina.it

† Equally contributing authors.

‡ Also these authors contributed equally.

Citation: Siepi, M.; Oliva, R.; Masino, A.; Gaglione, R.; Arciello, A.; Russo, R.; Di Maro, A.; Zanfardino, A.; Varcamonti, M.; Petraccone, L.; et al. Environment-Sensitive Fluorescent Labelling of Peptides by Luciferin Analogues. *Int. J. Mol. Sci.* **2021**, *22*, 13312. https://doi.org/10.3390/ijms222413312

Academic Editor: Herbert Schneckenburger

Received: 12 November 2021
Accepted: 7 December 2021
Published: 10 December 2021

Abstract: Environment-sensitive fluorophores are very valuable tools in the study of molecular and cellular processes. When used to label proteins and peptides, they allow for the monitoring of even small variations in the local microenvironment, thus acting as reporters of conformational variations and binding events. Luciferin and aminoluciferin, well known substrates of firefly luciferase, are environment-sensitive fluorophores with unusual and still-unexploited properties. Both fluorophores show strong solvatochromism. Moreover, luciferin fluorescence is influenced by pH and water abundance. These features allow to detect local variations of pH, solvent polarity and local water concentration, even when they occur simultaneously, by analyzing excitation and emission spectra. Here, we describe the characterization of (amino)luciferin-labeled derivatives of four bioactive peptides: the antimicrobial peptides GKY20 and $ApoB_L$, the antitumor peptide p53pAnt and the integrin-binding peptide RGD. The two probes allowed for the study of the interaction of the peptides with model membranes, SDS micelles, lipopolysaccharide micelles and *Escherichia coli* cells. K_d values and binding stoichiometries for lipopolysaccharide were also determined. Aminoluciferin also proved to be very well-suited to confocal laser scanning microscopy. Overall, the characterization of the labeled peptides demonstrates that luciferin and aminoluciferin are previously neglected environment-sensitive labels with widespread potential applications in the study of proteins and peptides.

Keywords: fluorescent peptide; environment-sensitive fluorophore; peptide labeling; luciferin; membrane-binding peptide; antimicrobial peptide; antitumor peptide; RGD peptide

1. Introduction

Environment-sensitive fluorophores are fluorophores, the excitation spectra, emission spectra and/or quantum yields (QY) of which depend on variables such as pH, solvent polarity, viscosity and even molecular crowding/aggregation state. Several different molecular mechanisms can contribute to modulation of the fluorescence of environment-sensitive fluorophores [1]. Solvatochromic fluorophores, the λ_{max} of which changes with solvent polarity, are often push-pull molecules with an electron-donating and an electron-withdrawing group bound to an aromatic moiety. Solvent polarity strongly affects the intramolecular charge transfer of these molecules, thus modulating their fluorescence [1]. In a less common type of solvatochromic fluorophores, an excited-state intramolecular proton-transfer event generates tautomeric forms with different absorption and emission spectra [1]. In other cases, conformational changes or variations in the aggregation state strongly influence the fluorescence of the probe [1]. Finally, pH probes have two or more ionizations states with different fluorescence properties [2]. Often, only one of the pH-dependent species is fluorescent.

These fluorophores are very valuable tools in the study of a wide variety of molecular and cellular processes [1,2]. As protein and peptide labels, they find applications in the study of conformational variations of proteins, of protein/protein, protein/ligand and protein/membrane interactions but also in the design of biosensors [1,3,4]. For these reasons, the search for new environment-sensitive fluorescent probes and, in particular, for protein and peptide labels, is a very active research field.

Firefly luciferin (Luc), the substrate of firefly luciferase, is the most popular bioluminescent compound [5]. It is less known that Luc is also an environment-sensitive fluorophore with unusual properties, summarized in Figure 1. Even if Luc is typical push-pull molecule, the hydroxyl group at position 6 of the benzothiazole moiety behaves as a weak acid with a $pK_a \approx 8.7$ (Figure 1A) [6]. Both the phenol and the phenolate forms are strongly fluorescent but with very different excitation and emission wavelengths (Figure 1B) [7]. However, as Luc is a "photoacid" [7–9], it undergoes a light-induced dissociation with a $pK_a < 1$ (Figure 1B). Thus, in the presence of water or other proton acceptors, Luc fluorescence is exclusively the result of the magenta and green pathways shown in Figure 1B. Even in neutral and acidic aqueous media, only the emission of the phenolate at 530 nm is generally observed, regardless of the excitation wavelength. Only in anhydrous organic solvents, e.g., acetonitrile or DMSO, does Luc show the blue fluorescence emitted by the phenol form (blue pathway in Figure 1B) [8,9]. However, very low amounts of water cause the appearance of the green emission. For example, in acetonitrile containing 14% water, the intensity of the blue and green emissions is comparable [8].

Two synthetic Luc analogues (Figure 1B), methoxyluciferin (mLuc) and aminoluciferin (aLuc), bearing non-dissociable groups at position 6 of the benzothiazole moiety show a single emission peak, with emission maxima similar to those of Luc phenol and phenolate forms, respectively [10,11]. mLuc is a very weak fluorophore [10], whereas aLuc is brighter than Luc itself; very interestingly, it is a strongly solvatochromic fluorophore with shifts up to 40 nm in the λ_{max} values [11].

In spite of these intriguing features, there are very few reports of probes exploiting (amino)luciferin fluorescence to detect cell thiols [12–14] or Furin activity [15]. To our knowledge, no example of direct fluorescent labeling of proteins or peptides based on Luc exists. This is even more surprising, considering that a Luc moiety can be incorporated very easily at the N-terminus of proteins and peptides. The final step of biological synthesis of Luc is a spontaneous condensation between D-cysteine and 6-hydroxy-2-cyanobenzothiazole (Figure 1C). This reaction is included among the so-called "click reactions" that are quantitative, irreversible and very fast in physiological conditions (phosphate buffer, pH 7–7.4, 25–37 °C) [16,17].

L-cysteine at the N-terminus of a peptide reacts with 2-cyanobenzothiazole (CBT) similarly to free cysteine, as the carboxyl group is not involved in the reaction (Figure 1C). Moreover, the nature of the substituent at position 6 of CBT also has a limited impact on the reaction. Therefore, peptides with an N-terminal cysteine residue are efficiently labeled trough a Luc-like spacer by using 6-substituted 2-cyanobenzothiazoles carrying any desired molecule (e.g., biotin, metal chelates, fluorophores, etc.) conjugated at position 6 (Figure 1C) [16,17]. In these bioconjugates, the label is usually attached through acylation of the amino group of aLuc or, less commonly, through alkylation of the hydroxyl group of Luc (Figure 1C), thus strongly reducing the fluorescence of (amino)luciferin [10,14,18]. Very interestingly, even if CBT reacts quantitatively and irreversibly with N-terminal cysteines, it generally forms a reversible adduct with the thiol group of internal cysteine residues, which can be removed by adding free cysteine or other aminothiols [16,17]. Therefore, an N-terminal cysteine can also be selectively labeled in the presence of cysteine residues at other positions on the peptide.

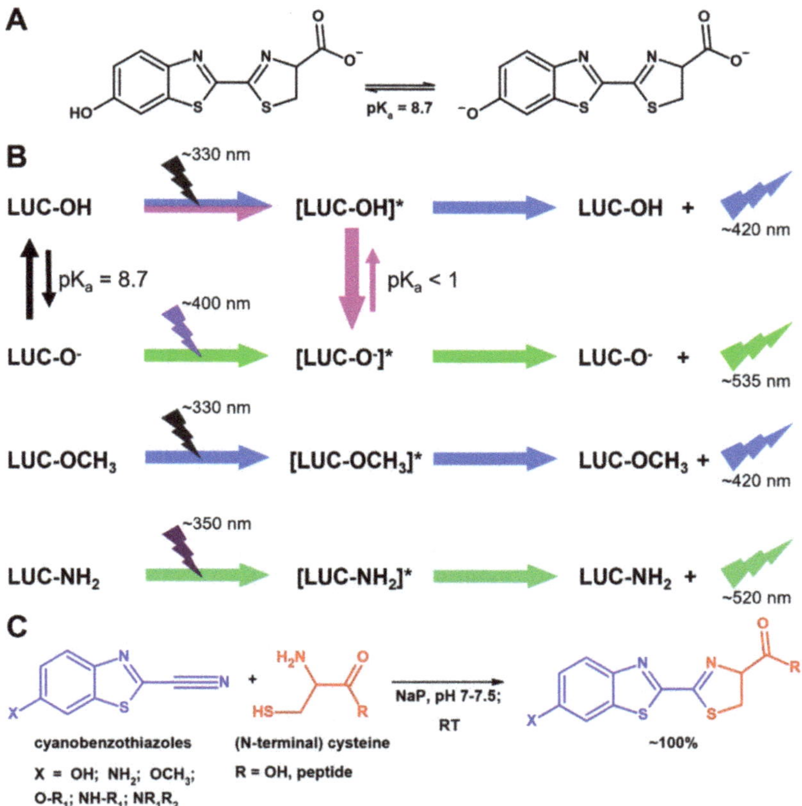

Figure 1. Structure, fluorescence and synthesis of luciferin and its analogues. (**A**) Structure of luciferin in phenol and phenolate forms. (**B**) Fluorescence of luciferin (Luc-OH), luciferin phenolate (Luc-O$^-$), methoxyluciferin (Luc-OCH3) and aminoluciferin (Luc-NH2). Excited states are indicated by an asterisk. (**C**) Condensation reaction between 6-substituted 2-cyanobenzothiazoles and cysteine (free or as N-terminal residue of a peptide). R_1 and R_2 can be a wide variety of alkyl and acyl groups. The nature of R_1 and R_2 has little or no impact on the reaction but can strongly influence fluorescence.

In order to study the possible exploitation of (amino)luciferin as an environment-sensitive fluorescent label for peptides, we synthesized (amino)luciferin conjugates of four peptides of different lengths, compositions and biological properties. Characterization of the fluorescence properties of the labeled peptides demonstrates that Luc and its analogues are previously neglected bright and photostable fluorescent labels suited for in vitro studies like peptide/membrane or peptide/cell interactions and microscopy/imaging studies.

2. Results

2.1. Preparation and Labeling of the Peptides

To evaluate the suitability of Luc and aLuc as environment-sensitive fluorescent labels, we selected four previously described bioactive peptides: GKY20 (20 residues), ApoB$_L$ (37 residues), p53pAnt (37 residues) and RGD (6 residues) (Figure S1, Supplementary Materials). GKY20 [19] and ApoB$_L$ [20] are cryptic cationic antimicrobial peptides (CAMP), peptides unusually rich in cationic and hydrophobic residues, which kill bacteria by damaging structural integrity and functions of bacterial membranes [21–23]. Moreover, CAMPs often show additional pharmacologically relevant biological activities, like antifungal, antitumor and anti-inflammatory activity [23,24]. p53pAnt (also spelled p53p-Ant) is a designed antitumor peptide containing a sequence at the N-terminus that induces apoptosis by interacting with p53, as well as a "cell-penetrating peptide" (CPP) at the C-terminus that derived from the Antennapedia protein homeodomain, which can drive peptides inside eukaryotic cells both by directly crossing the cytoplasmic membrane and by mediating endocytosis [25–27]. RGD is a natural ligand of integrins, cell surface proteins involved in adhesion to the extracellular matrix [28,29]. It is very different from the other three peptides; it is very short, hydrophilic and uncharged. RGD interaction with eukaryotic cell membranes depends on binding to a protein receptor rather than on a direct interaction with membrane lipids, as in the case of CAMPs and CPPs. Immobilized RGD-like peptides are frequently used to promote the adhesion of eukaryotic cells to surfaces, polymers, hydrogels, etc. Moreover, they may find applications in diagnostic and imaging fields [30,31]. Peptides GKY20, $ApoB_L$ and p53pAnt with an additional cysteine residue at the N-terminus—herein named (C)GKY20, (C)$ApoB_L$ and (C)p53pAnt—were prepared using a recently described strategy for the preparation of recombinant toxic peptides in *E. coli* based on the selective cleavage of a carrier/peptide fusion bearing the acid sensitive sequence Asp-Cys [32] (details in Supplementary Materials, Sections S2–S4). Peptide RGD with an additional cysteine residue at the N-terminus—(C)RGD—was obtained by chemical synthesis. The four peptides were labeled with either a Luc or an aLuc moiety at the N-terminus by incubation with a slight molar excess of 6-hydroxy-2-cyanobenzothiazole or 6-amino-2-cyanobenzothiazole, respectively, in aqueous buffer at pH 7.4 at 25 °C (Supplementary Materials, Section S5). As expected, based on the well-known high efficiency of this reaction, Luc-labeled peptides (Luc-GKY20, Luc-$ApoB_L$, Luc-p53pAnt and Luc-RGD) and aLuc-labeled peptides (aLuc-GKY20, aLuc-$ApoB_L$, aLuc-p53pAnt and aLuc-RGD) were obtained with yields close to 100% after very short incubation times (30–60 min). (C)GKY20 was also treated with 6-methoxy-2-cyanobenzothiazole to obtain an mLuc-labeled peptide (mLuc-GKY20) and with PyMPO maleimide, the thiol-reactive version of PyMPO, a widely used photostable solvatochromic fluorophore [33–35], to obtain the peptide PyMPO-(C)GKY20. After purification (Section 4 and Supplementary Materials), the mass of all labeled peptides was confirmed by MALDI-MS (Table S1). SDS-PAGE analysis of labeled GKY20, $ApoB_L$ and p53pAnt shows their bright fluorescence under UV light (365 nm) and their solvatochromic nature appreciable even by the naked eye (Figure S2).

2.2. Stability and Biological Activity of Luc-Labeled Peptides

When choosing the best fluorophore to label a peptide, in addition to excitation and emission wavelengths, other relevant parameters are (photo)stability and molecular features, which might influence the biological activity of the peptide to be labeled (molecular weight, charge, hydrophobicity, etc.).

Although Luc is a fairly stable molecule, in order to evaluate the photostability of Luc-labeled peptides, we monitored the fluorescence of Luc-GKY20, aLuc-GKY20 and mLuc-GKY20 for 60 min, constantly exciting the three labeled peptides at their respective maximum absorption. As comparative control, we used the very photostable fluorophore PyMPO [33] by treating PyMPO-(C)GKY20 in the same conditions. Luc-GKY20 lost only 30% of its initial fluorescence (Figure S3), thus proving to be even more photostable than PyMPO-(C)GKY20, which lost 43% of its initial fluorescence. aLuc-GKY20 and mLuc-GKY20 proved to be slightly less photostable than Luc-GKY20, with a loss of initial fluorescence of about 55% and 60%, respectively. Therefore, Luc can be considered a very stable label, though aLuc is also a suitable tool.

An ideal fluorescent label or probe should not significantly change the properties (size, hydrophobicity, charge, etc.) of the peptide to be labeled in order to minimize the risk of altering its chemical-physical behavior and/or biological activity. Therefore, we compared molecular weight, accessible surface area (ASA), polar ASA, ratio of polar ASA/ASA and number of charged groups (Table S2) of Luc, aLuc and a wide panel of commonly used solvatochromic and non-solvatochromic fluorescent labels that can be attached either at the N-terminus of a peptide (generally through an amide or a sulfonamide bond) or at the side-chain sulfur of a cysteine residue (generally through a thioether bond). Only NBD and some coumarin dyes among fluorescent labels that can be attached at the N-terminus and only bimane among fluorescent labels that can be attached to the side chain of a cysteine residue have an ASA smaller than that of Luc and aLuc. Moreover, several solvatochromic and not-solvatochromic fluorophores are either quite hydrophobic (e.g., pyrene-1-butirrate, Atto 495, most BODIPY labels, BADAN etc.) or very hydrophilic and with several charged groups (e.g., Alexa Fluor™ 405 and Alexa Fluor 488-C5-maleimide), whereas Luc and aLuc have an intermediate ratio of polar ASA/ASA (0.31–0.34) and no (aLuc) or low charge (Luc at neutral pH). Therefore, Luc and aLuc can be confidently considered fluorescent labels with a low impact on the molecular properties of the labeled peptide.

In order to determine the influence of Luc labels on the biological activity of the peptides considered in this study, we compared the antimicrobial activity of labeled GKY20 and $ApoB_L$ with that of the corresponding unlabeled recombinant peptides, (P)GKY20 [36] and (P)$ApoB_L$ [20] (Supplementary Materials, Section S6). The cytotoxic activity of labeled and unlabeled p53pAnt was also investigated (Supplementary Materials, Section S7). The minimum inhibitory concentrations (MIC) of Luc-GKY20 and aLuc-GKY20 were found to be identical to those measured for (P)GKY20 on Gram-negative and Gram-positive strains (Table S3). The MIC values of Luc-$ApoB_L$ and aLuc-$ApoB_L$ were found to be very similar to those measured for (P)$ApoB_L$ (the observed differences, not higher than one dilution, are generally considered not significant in the microdilution assay).

Luc-p53pAnt and (C)p53pAnt showed similar toxicity when assayed on two human cell lines, namely HaCaT noncancerous immortalized keratinocytes and HeLa cervical cancer cells (Figure S4). It is worth noting that labeled and unlabeled p53pAnt showed significantly higher toxicity for HeLa cancer cells than for HaCaT cell, as expected of an anticancer peptide.

Finally, we tested the effect Luc-RGD and unlabeled RGD on HaCaT and HeLa cells. As expected, we did not observe any significant toxic effect.

2.3. Response of the Labeled Peptides to Solvent Polarity Changes

Most solvatochromic fluorophores show an increase in fluorescence emission and a blue shift of the λ_{max} as solvent polarity decreases. Therefore, we measured emission spectra and QY of the labeled peptides in sodium phosphate 10 mM, pH 7.4 (NaP) and in the same buffer containing 50% (*v:v*) either isopropanol or methanol.

The emission spectra in NaP of the Luc- and aLuc-labeled peptides showed maximum emission at 539 and 526 nm, respectively (Figure S5 and Table 1). As expected, all the labeled peptides showed a blue shift of the λ_{max} in the presence of the organic solvents (Table 1). The blue shift was less in the case of Luc-labeled peptides (6–7 nm) and greater in the case of aLuc-labeled peptides (9–15 nm). Furthermore, aLuc-labeled peptides showed a considerably greater blue shift in isopropanol than in methanol, thus confirming that aLuc is a probe more sensitive to solvent polarity compared to Luc. mLuc-GKY20 showed blue-shift values very similar to those of aLuc-GKY20, whereas PyMPO-(C)GKY20 showed a 5 nm blue shift only in the presence of isopropanol.

Intriguingly, emission intensities (Figure S5) and QY values (Table 2) showed complex and partially unexpected variation characteristic of each peptide. For example, all the variants of GKY20 showed very large increases in emission intensities and QY values in the presence of the organic solvent (2.5–6 times higher in 50% isopropanol than in water). On the contrary, Luc-p53pAnt and aLuc-p53pAnt showed a 30–35% reduction in QY. Both labeled variants of ApoB$_L$ and RGD showed little or no variation in emission intensities (Figure S5) and QY (Table 2). These puzzling variations are likely the result of a very well-known phenomenon, i.e., the quenching of fluorophores bound to protein/peptides. Even if several mechanisms can contribute to quenching, the most common is photoinduced electron transfer (PET), a reversible light-triggered transfer of electrons from amino-acid residues to the fluorophore [37]. The most efficient donors are tryptophan and tyrosine, although to a lesser extent, histidine and methionine can also contribute significantly to quenching [38]. Obviously, the redox potential of a fluorophore and therefore its propensity to accept electrons, influences its sensitivity to quenching [37].

Table 1. λ_{max} and blue-shift values of the labeled peptides.

Peptide	λ_{max} (nm) [a]							
	NaP pH 7.4	MeOH 50%	IPA 50%	SDS (25 mM)	POPC +POPG	POPC	LPS (200 μg/mL)	*E. coli* Cells
Luc-GKY20	538	533 (5)	531 (7)	526 (12)	520 (19)	520 (19)	516 (23)	522 (16)
Luc-ApoB$_L$	539	533 (6)	533 (6)	526 (13)	522 (17)	539 (0)	508 (31)	534 (5)
Luc-p53pAnt	539	533 (6)	533 (6)	526 (13)	522 (17)	539 (0)	nd [b]	nd
Luc-RGD	539	533 (6)	533 (6)	533 (6)	539 (0)	539 (0)	nd	nd
aLuc-GKY20	525	515 (10)	510 (15)	506 (19)	499 (26)	499 (26)	503 (22)	509 (16)
aLuc-ApoB$_L$	527	516 (11)	512 (15)	505 (22)	497 (27)	527 (0)	486 (41)	518 (9)
aLuc-p53pAnt	526	516 (10)	515 (11)	504 (22)	499 (25)	526 (0)	nd	nd
aLuc-RGD	526	517 (9)	513 (13)	510 (16)	526 (0)	526 (0)	nd	nd
mLuc-GKY20	439	429 (10)	424 (15)	422 (17)	nd	nd	nd	nd
PyMPO-(C)GKY20	563	563 (0)	558 (5)	556 (7)	535 (28)	541 (22)	nd	nd

[a] Blue-shift values with respect to the λ_{max} in NaP are shown in parenthesis. [b] nd = not determined.

Table 2. Relative quantum yield of the labeled peptides.

Peptide	QY [a] (Variation Relative to NaP pH 7.4)		
	NaP pH 7.4	IPA 50% [b]	NaAc [c] pH 5.0
PyMPO-GKY20	0.080	0.373 (4.66)	nd[d]
mLuc-GKY20	0.008	0.021 (2.63)	nd
Luc-GKY20	0.111	0.639 (5.76)	0.269 (2.42)
aLuc-GKY20	0.202	0.969 (4.80)	nd
Luc-ApoB$_L$	0.365	0.389 (1.07)	0.450 (1.23)
aLuc-ApoB$_L$	0.574	0.757 (1.32)	nd
Luc-p53pAnt	0.261	0.184 (0.70)	0.418 (1.60)
aLuc-p53pAnt	0.432	0.274 (0.63)	nd
Luc-RGD	0.519	0.443 (0.85)	0.510 (0.98)
aLuc-RGD	0.937	1.040 (1.11)	nd

[a] Errors $\leq$ 5% of the reported values. [b] Isopropanol:NaP pH 7.4 (1:1, v/v). [c] Sodium acetate 20 mM. [d] nd = not determined.

PET is also very sensitive to distance and orientation of the donor/acceptor couple, so the quenching efficiency can be influenced even by minor variations in the conformation of the labeled protein/peptide. In fact, PET-mediated quenching is a powerful tool for detection of conformational variations in proteins and peptides [37].

Very interestingly, the peptides that show the largest variations, namely GKY20 and p53pAnt, contain several residues with strong quenching ability (tryptophan, histidine and two tyrosines in GKY20; two tryptophans, three histidines and a methionine in p53pAnt; Figure S1).

The QY values of the Luc-labeled peptides were also measured in sodium acetate at pH 5.0 to induce the protonation of the histidine residues. Protonation of histidine residue can influence PET quenching both directly, by reducing the donor ability of the imidazole ring, and indirectly, by inducing conformational changes. As expected, significant increases in QY values were observed only for the three peptides containing histidine residues: GKY20, p53pAnt and $ApoB_L$ (Table 2). To date, it has not been possible to determine the relative contribution of the direct and indirect effect of histidine protonation on quenching.

2.4. Response of Luc-Labeled Peptides to pH

Luc has a phenolic hydroxyl, which behaves as a weak acid, with a pK_a of about 8.7 [6]. As Luc and its phenolate have quite different excitation spectra, with maxima at 330 and 395 nm, respectively, the excitation spectra of the Luc-labeled peptides recorded at pH 7.4 show a characteristic shoulder at 390–400 nm originating from the small amount of phenolate ion present at this pH (Figure S6). It should be noted that the shoulder is also visible in the presence of organic solvents (Figure S5). On the other hand, the shoulder disappears in buffers with pH values below 6, as Luc becomes protonated, whereas the peak at 400 nm prevails at pH values greater than 8.5 (Figure S6). By recording the emission at 539 nm of the Luc-labeled peptides after excitation at 400 nm in buffers with different pH values, we determined the actual pK_a values of the Luc moieties (Table 3). All pK_a values were found to be slightly lower than those of free Luc, with small differences among the peptides, likely due to differences in the net charge and distribution of positively charged residues. In particular, the highest pK_a value was found for Luc-RGD, the only peptide with a negative net charge at pH values close to the pK_a of the Luc moiety (Table 3). The remaining three peptides have a high positive net charge, which could stabilize the phenolate anion, thus lowering the pK_a value. The possibility of selectively exciting the phenolic or phenolate form of Luc not only makes Luc a useful pH probe for the pH range of 7–8 but also allows for monitoring of the formation of peptide/ligand complexes in which the ionization of the Luc moiety is suppressed or altered. The next sections show some applications of this peculiar feature of Luc.

Table 3. pK_a values of the phenolic group in Luc and Luc-labeled peptides.

Peptide	Net Charge of the Peptidyl Moiety [a]	pK_a
Free Luc	not applicable	8.70 [b]
Luc-RGD	−1	8.40 ± 0.03
Luc-GKY20	+4	8.13 ± 0.04
Luc-$ApoB_L$	+6	8.11 ± 0.04
Luc-p53pAnt	+11	7.97 ± 0.03

[a] Theoretical net charge at pH 7.5–9.0 calculated attributing a charge = +1 to each lysine/arginine residue and a charge = −1 to each aspartate/glutamate residue and to the free C-terminus. [b] From reference [6].

2.5. Interaction of Labeled Peptides with Liposomes

Both CAMPs and CPPs are able to interact with biological membranes, and this interaction is essential to their properties. Therefore, in order to verify whether Luc-labeling can be used to investigate peptide/membrane interaction, we studied the behavior of the labeled peptides in the presence of liposomes composed either of pure palmitoyl-oleoyl-phosphatidylcholine (POPC) or of a mixture of POPC and palmitoyl-oleoyl-phosphatidylglycerol (POPG) at a molar ratio of 4:1. As the POPC head group is zwitterionic, liposomes composed only of this lipid are neutral and are usually considered mimetic of eukaryotic cell membranes. On the contrary, POPG-containing liposomes are negatively charged and are considered a simplified model of bacterial cell membranes.

The short, hydrophilic and negatively charged labeled RGD was used as a negative control. As expected, the excitation and emission spectra of Luc-RGD were essentially identical in NaP and in the presence of the two types of liposomes (Figure 2).

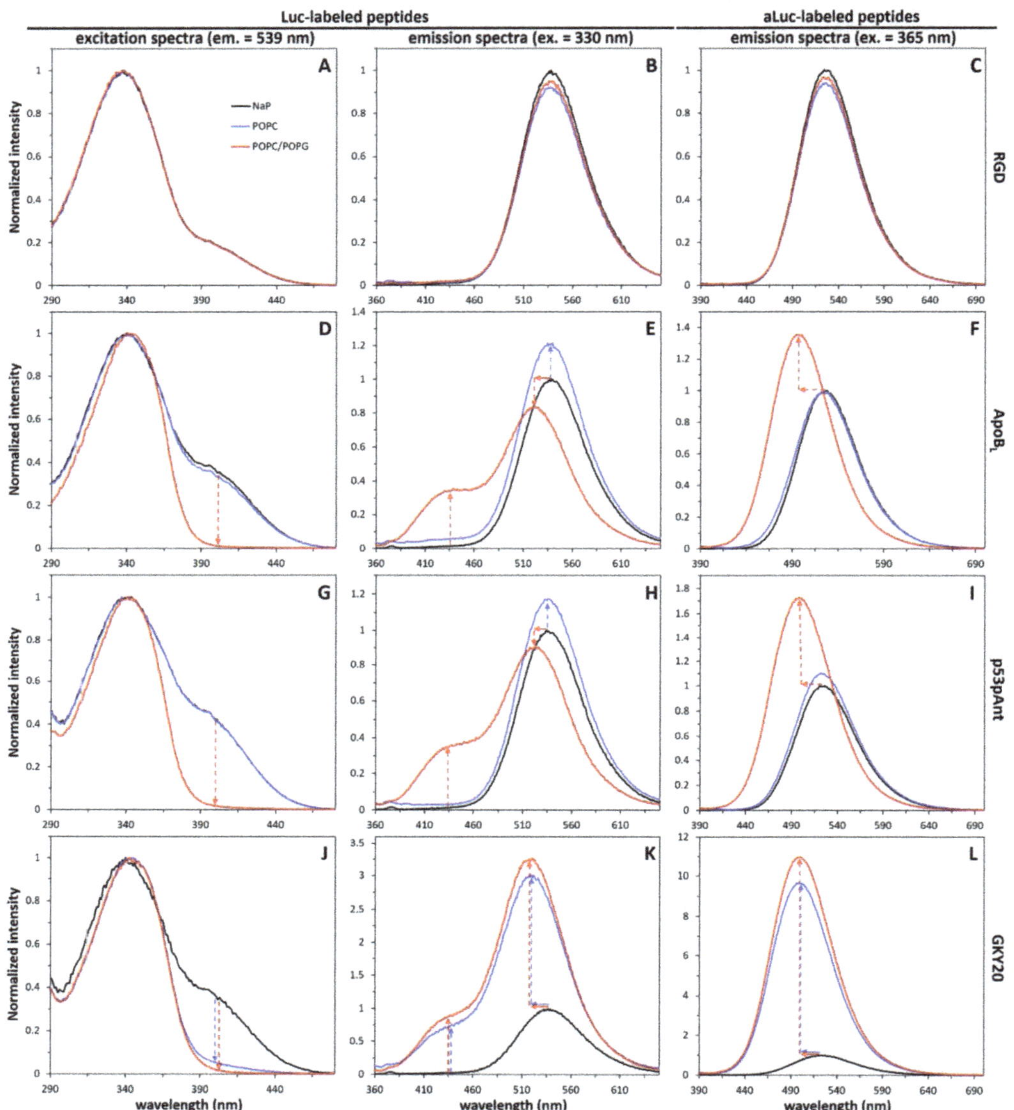

Figure 2. Fluorescence of labeled peptides in the presence of liposomes of POPC or POPC/POPG (5:1). Excitation spectra of luciferin (left), emission spectra of luciferin (center) and emission spectra of aminoluciferin (right) were recorded for peptides RGD (**A–C**), ApoB$_L$ (**D–F**), p53pAnt (**G–I**) and GKY20 (**J–L**). Spectra recorded in the presence of liposomes were normalized to the corresponding spectra in NaP. Arrows highlight the main changes with respect to NaP.

The same result was obtained in the case of the emission spectrum of aLuc-RGD (Figure 2). On the contrary, in the case of Luc-ApoB$_L$, the excitation spectrum in the presence of POPC/POPG liposomes was clearly different from the spectra in NaP and in the presence of POPC liposomes, completely lacking a shoulder at 400 nm (Figure 2). This suggests that in the presence of POPC/POPG liposomes, the deprotonation of the Luc phenolic group is inhibited. Two mechanisms could explain this finding: (i) embedding of the phenolic group of Luc among the lipids would directly prevent deprotonation; or (ii) binding of the Luc probe to the surface of the negatively charged POPC/POPG liposomes could prevent deprotonation as a consequence of the more acidic local environment. In fact, it is well-known that polyanionic surfaces and polymers determine the formation of acidic local environments by attracting protons from the bulk solution [39,40]. Further information was obtained from analysis of the emission spectra. Once more, the emission spectrum in the presence of POPC/POPG liposomes was very different from the other two spectra, showing the presence of a large shoulder at 425–430 nm (Figure 2). As mentioned above, this blue emission is characteristic of the phenolic form of Luc and can only be observed when Luc is in an environment with very low water content, a condition able to suppress the photoinduced dissociation. Therefore, the presence of a blue shoulder in the emission spectrum of Luc-ApoB$_L$ is a clear indication that the Luc moiety is deeply embedded into the POPC/POPG bilayer. This is further confirmed by the 16–18 nm blue shift (from 539 to about 522 nm) of the peak in the green region (Table 1). The analysis of the emission spectra of aLuc-ApoB$_L$ is simpler but not less informative. This peptide shows a considerable increase in the fluorescence emission and a 27–28 nm blue shift (from 525 to about 497 nm) only in the presence of POPC/POPG liposomes (Figure 2 and Table 1). Given the solvatochromic nature of aLuc, this is an indication that, only in the case of negatively charged liposomes, the probe is embedded in a hydrophobic environment. Therefore, three different and independent phenomena confirm binding and embedding of labeled ApoB$_L$ into the POPC/POPG bilayer: (i) inhibition of the deprotonation of Luc; (ii) inhibition of the photoinduced dissociation of Luc; (iii) the blue shift of Luc and aLuc emission peaks. In this regard, it is worth noting that the interaction of ApoB$_L$ with anionic liposomes of phosphatidylcholine and phosphatidylglycerol has been recently demonstrated by using differential scanning calorimetry [41].

Labeled p53pAnt showed a behavior essentially similar to labeled ApoB$_L$ (Figure 2), whereas GKY20 showed that it can interact with both types of liposomes. Indeed, the emission spectra of Luc and aLuc-GKY20 and the excitation spectra of Luc-GKY20 in the presence of POPC and POPC/POPG liposomes are very similar (Figure 2). All the emission spectra show a large blue shift compared to those recorded in NaP (Table 1). Furthermore, the emission spectra of Luc-GKY20 show a blue shoulder, thus indicating that the (a)Luc moiety is embedded in a less polar environment. The spectra of PyMPO-(C)GKY20 in the presence of liposomes were found to be very similar to those of aLuc-GKY20 (Figure S7), thus demonstrating that the binding to both liposome types is not an artifact due to (a)Luc. These findings agree with previous studies performed using unlabeled GKY20 and the same model membranes [42]. We do not have a straightforward explanation for the observed varying behavior of the two CAMPs. However, very interestingly, further analyses conducted on labeled GKY20 and ApoB$_L$ evidence several additional differences, as described in the next sections.

2.6. Interaction of Labeled CAMPs with SDS and LPS Micelles

In order to further confirm the ability of Luc probes to reveal the interaction of peptides with lipidic structures, we also recorded fluorescence spectra in the presence of 25 mM SDS (Figure 3A–C). At this concentration, well above the critical micelle concentration (CMC) of about 8 mM, SDS forms micelles containing an average of 60 molecules [43]. For this reason, SDS has been frequently used as a membrane mimetic, such as for determination of the NMR structure of membrane-binding peptides. However, it should be remembered that SDS is a strong detergent able to interact unspecifically with peptides. Accordingly, in the presence of SDS, not only labeled GKY20, ApoB$_L$ and p53pAnt but also labeled RGD showed spectral behavior, suggesting interaction with micelles (Figure 3A–C). Nonetheless, the spectra of labeled RGD were not completely superimposable to those of the other three peptides. For example, in the excitation spectra of Luc-RGD in the presence of SDS, a tail at 400–420 nm is still visible, indicating that not all of the peptide is strongly associated to the micelles (Figure 3A–C). In the emission spectrum of Luc-RGD, the blue peak is less evident than in the corresponding spectra of the other three peptides (Figure 3A–C), and the ratio between the area of the peaks in the blue (380–465 nm) and green (466–640 nm) regions was

0.16, whereas the same ratio for Luc-GKY20, Luc-ApoB$_L$ and Luc-p53pAnt was 0.27, 0.36 and 0.33, respectively. It is also evident that the peak in the green region is less blue-shifted than in the case of Luc-GKY20, Luc-ApoB$_L$ and Luc-p53pAnt (Figure 3A–C and Table 1).

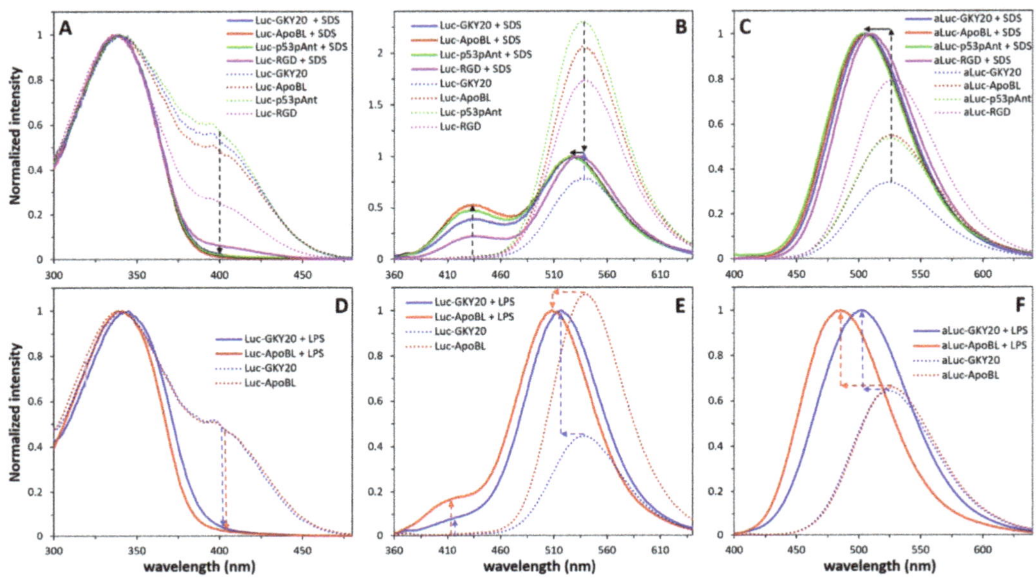

Figure 3. Fluorescence of labeled peptides in the presence of SDS or LPS micelles. (**A**,**D**) Excitation spectra of Luc-labeled peptides (em. = 539; Luc-GKY20 + LPS, em. = 516 nm; Luc-ApoBL + LPS, em. = 508 nm). (**B**,**E**) Emission spectra of Luc-labeled peptides (ex. = 330 nm). (**C**,**F**) Emission spectra of aLuc-labeled peptides (ex. = 363 nm). Solid line, spectra recorded in the presence of SDS or LPS; dotted line, spectra recorded in NaP. In (**B**,**C**,**E**,**F**), spectra recorded in NaP were normalized to the corresponding spectra recorded in the presence of SDS or LPS. Arrows highlight the main changes with respect to NaP.

As in the case of the emission spectra recorded in the presence of liposomes, the spectra of aLuc-labeled peptides recorded in the presence of SDS also showed a significant increase in emission intensity, as well as a blue shift. However, again, the blue shift was smaller in the case of aLuc-RGD (Figure 3A–C and Table 1). These results indicate that the hydrophilic RGD peptide interacts less tightly with the SDS micelles than Luc-GKY20, Luc-ApoB$_L$ and Luc-p53pAnt.

GKY20 and ApoB$_L$ bind to and neutralize lipopolysaccharides (LPS), the main components of Gram-negative bacteria outer membrane, with strong proinflammatory effects in higher eukaryotes [19,20,44]. These very complex molecules have three portions with different compositions: (i) lipid A, a phosphorylated disaccharide bearing 4–6 fatty acid residues; (ii) the core, an oligosaccharide with a variable number of phosphate groups; (iii) the O-antigen, a species and strain-specific polysaccharide [45]. It should be noted that pure LPS, like SDS, does not form regular bilayers but negatively charged micelles [46]. However, the CMC of LPS is usually in the low micromolar range [46]. The ability to bind and neutralize the strong pro-inflammatory and sometimes life-threatening effects of LPS is one of the most pharmacologically relevant properties of CAMPs [45]. Therefore, the study of the CAMP/LPS interaction is of outstanding importance.

Accordingly, we studied the interaction of labeled GKY20 and ApoB$_L$ by using a commercially available LPS from *E. coli* strain 0111:B4. Firstly, we recorded the excitation and the emission spectra of Luc-GKY20 and Luc-ApoB$_L$ in the presence of 200 μg/mL LPS, i.e., a concentration well above the CMC of this LPS (about 1.3–1.6 μM corresponding to 13–16 μg/mL) [46]. As expected, the binding of the two peptides to LPS caused the disappearance of the shoulder at 400 nm in the excitation spectra (Figure 3D–F). The emission spectra of Luc-GKY20 and Luc-ApoB$_L$ show the expected blue shift of the peak in the green region—even larger than those observed in the presence of SDS and liposomes (Table 1) However, surprisingly, only a very low shoulder was observed in the blue region, especially in the case of Luc-GKY20 (Figure 3D–F). The ratio between the area of the peaks in the blue (365–435 nm) and green (436–640 nm) regions is 0.046, and 0.095 for Luc-GKY20 and Luc-ApoB$_L$, respectively, i.e., even lower than the ratio observed for the Luc-RGD peptide in the presence of SDS. Very interestingly, the peak in the blue region is also blue-shifted (about 20 nm) with respect to the same peak observed in the case of liposomes and SDS. Finally, the emission spectra of aLuc-GKY20 and aLuc-ApoB$_L$ show, as expected, an increase in emission intensity and a large blue shift, particularly in the case of aLuc-ApoB$_L$ (41 nm). These findings suggest that the orientation of the Luc moiety when the peptides are bound to LPS is very different from that adopted when the peptides are bound to liposomes and SDS micelles. The high blue-shift values suggest that the (a)Luc moiety is in a very non-polar environment, while the very weak emission in the blue region indicate that the phenolic OH group of Luc points toward the solvent or a proton acceptor in the LPS (e.g., a basic group in the lipid A of LPS). Possible orientations of the Luc moiety in LPS and SDS or liposomes explaining the observed variations in excitation and emission spectra are schematically drawn in Figure S8.

2.7. Interaction of Labeled CAMPs with Non-Micellar LPS

Next, we recorded the emission spectra of Luc-GKY20, Luc-ApoB$_L$, aLuc-GKY20 and aLuc-ApoB$_L$ at a constant peptide concentration in the presence of increasing concentrations of LPS (Figure 4). All peptides showed a turn-off of the fluorescence associated with a considerable blue shift of λ_{max} values for LPS concentrations up to 10–20 μg/mL, followed by a turn-on phase with smaller changes in λ_{max} values. The biphasic nature of the process is clearly visible by plotting the area beneath the spectra and the λ_{max} values as a function of the LPS concentration (Figure 4B,D,F,H). Only in the case of (a)Luc-ApoB$_L$, a turn-on phase with no shift in λ_{max} values was visible at very low LPS concentrations (0–2 μg/mL).

A similar behavior has been previously described for GKY25, HVF18 and VFR12, three CAMPs derived, like GKY20, from the C-terminus of human thrombin. In particular, GKY25 is a variant of GKY20, with five additional residues at the C-terminus [47]. In that case, the authors, exploiting the intrinsic fluorescence of the single tryptophan residue present in all the thrombin-derived CAMPs, observed a biphasic process with a turn-off phase for LPS concentrations below 10 μg/mL and a turn-on phase at higher concentrations. Therefore, the turn-off/turn-on switch seems to be independent of the nature and position of the fluorophore. As the CMC of *E. coli* LPS is about 16 μg/mL, it can be speculated that the turn-off and turn-on phases might be the result of the binding to free and micellar LPS, respectively. The additional turn-on phase observed at very low LPS concentrations only in the case of (a)Luc-ApoB$_L$ might be due to a conformational change in this peptide induced by the presence of small amounts of LPS. In this regard, it is worth noting that the CD spectra of unlabeled GKY20 and ApoB$_L$ in the presence of LPS are quite different [19,20]. In the case of ApoB$_L$, circular dichroism studies suggest that this peptide adopts a β-sheet conformation upon interaction with LPS [20]. On the other hand, the CD spectrum of GKY20 in the presence of LPS is not similar to any of the CD spectra of model conformations [19]. It could also be hypothesized that LPS might form small aggregates even below the CMC, which could be responsible for the turn-off phase, whereas the first turn-on phase observed at very low LPS concentrations would be due to the association of (a)Luc-ApoB$_L$ with truly monomeric LPS molecules. This aspect would require further investigation, which lies outside the scope of this work.

We also recorded the emission spectra of Luc-GKY20 and Luc-ApoB$_L$ after excitation at 400 nm in order to follow the binding process by monitoring the disappearance of the phenolate form in the solution (Figure 5). As expected for a saturable binding process, we observed a progressive reduction in the fluorescence, which reached a minimum at about 50 μg/mL LPS for both peptides.

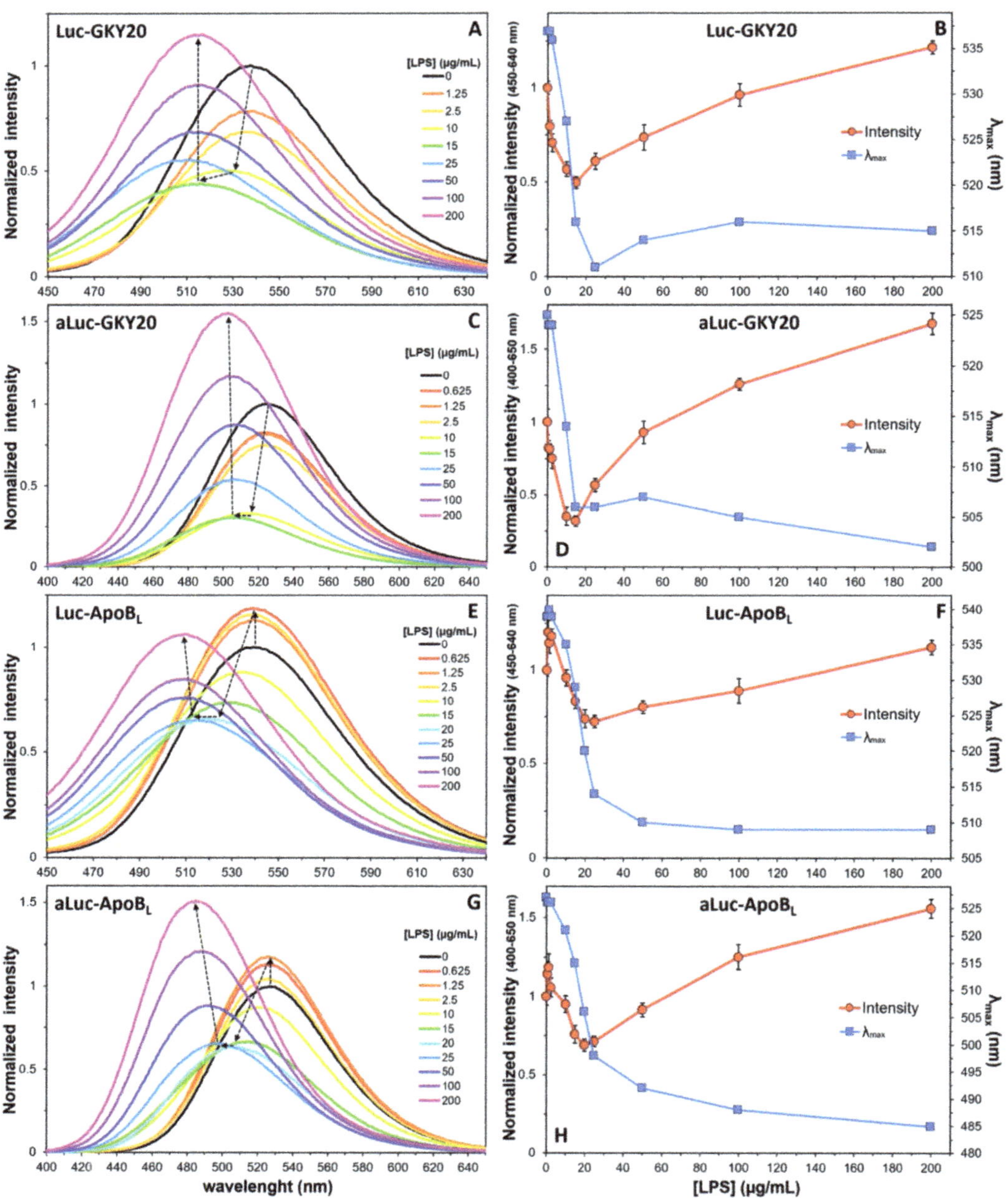

Figure 4. Fluorescence of the labeled peptides in the presence of increasing concentrations of LPS (0–200 μg/mL). (**A**,**C**,**E**,**G**) Emission spectra of the peptides recorded after excitation at 330 nm (Luc-labelled peptides) and 363 nm (aLuc-labelled peptides). Spectra recorded in the presence of LPS were normalized to the corresponding spectra recorded in NaP (black lines). Arrows highlight the main changes with respect to NaP. (**B**,**D**,**F**,**H**) Variation of total fluorescence (in the indicated ranges) and of the λ_{max} values as a function of LPS concentration.

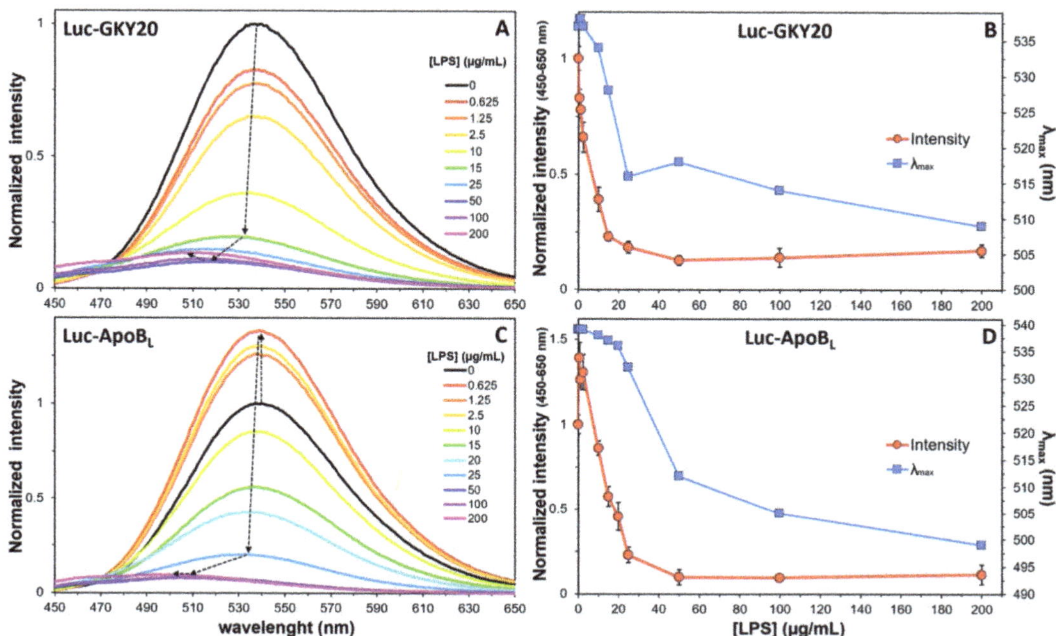

Figure 5. Fluorescence of the Luc-labeled peptides in the presence of increasing concentrations of LPS (0–200 µg/mL). (**A**,**C**) Emission spectra of Luc-labeled peptides recorded after excitation at 400 nm. Spectra recorded in the presence of LPS were normalized to the corresponding spectra recorded in NaP. Arrows highlight the main changes with respect to NaP. (**B**,**D**) Variation of total fluorescence (in the indicated ranges) and of the λ_{max} values as a function of LPS concentration.

2.8. Quantitative Analysis of the Peptide/LPS Interaction

The curves measured at constant peptide concentration and variable LPS concentration could be used to determine K_d values (Supplementary Materials, Section S9). However, as such K_d values would be the result of measurements obtained using concentrations below and above the CMC of LPS, their meaning would be questionable. In order to determine the K_d value of Luc-labeled peptides for micellar LPS, we repeated the experiment at variable Luc-labeled peptide concentration and constant LPS concentration (40 µg/mL corresponding to ~4 µM). The experimental data were fitted to the model described in Section S9, Supplementary Materials. The model allows for the estimation not only the K_d value but also the number of binding sites and hence the stoichiometry of binding. The spectra and the fittings are shown in Figure 6. Luc-GKY20, Luc-ApoB$_L$ and Luc-p53pAnt show K_d values in the range 50–400 nM. As expected, Luc-RGD did not bind to LPS, and the resulting plot of the fluorescence emission as function of the peptide concentration was a straight line (Figure 6). The fact that the anticancer peptide Luc-p53pAnt binds to LPS with affinity comparable to those of the two CAMPs is not surprising, considering that it has an amino-acid composition similar to that of the CAMPs (Figure S1) and that it is the most cationic of the three peptides (Table 3).

On the other hand, the three peptides show very different binding stoichiometries (Figure 6). Only Luc-ApoB$_L$ binds to LPS in a 1:1 ratio. In the case of Luc-GKY20, three molecules of the peptide bind to two molecules of LPS, whereas in the case of Luc-p53pAnt, two molecules of the peptide bind to three molecules of LPS. The higher number of binding sites found for Luc-GKY20 might be due to the fact that this peptide is considerably shorter than the other two. However, Luc-ApoB$_L$ and Luc-p53pAnt have exactly the same length; therefore, the different stoichiometry might be due to a different mode of binding or a different fold adopted by the peptides upon binding.

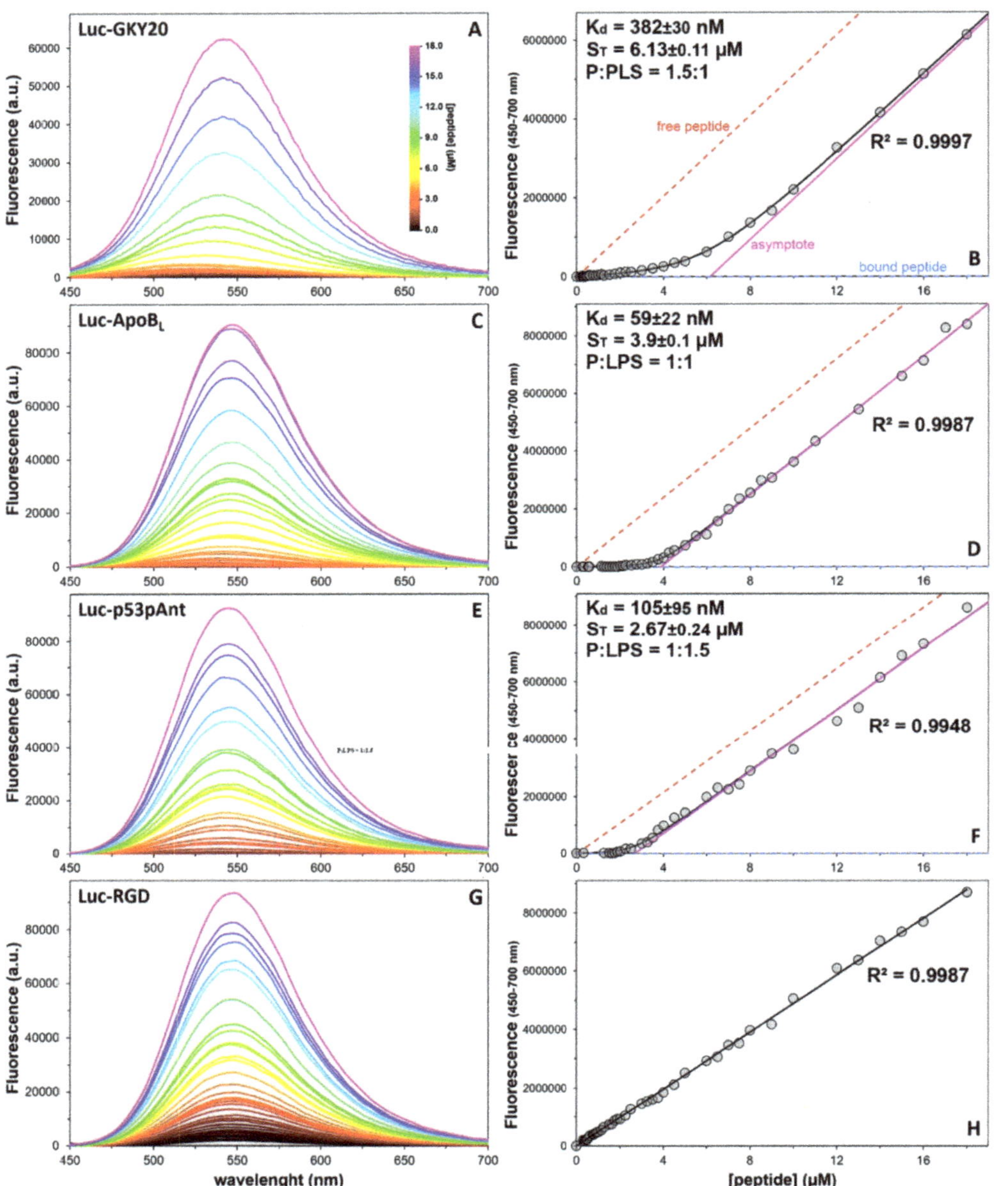

Figure 6. Determination of the K_d values and stoichiometry for the peptide/micellar LPS interaction. (**A**,**C**,**E**,**G**) Emission spectra of Luc-labeled peptides (0–18 µM) in the presence of 40 µg/mL LPS (ex. = 400, 415, 425 and 410 nm, respectively). (**B**,**D**,**F**,**H**) Variation of total fluorescence (450–700 nm) as a function of peptide concentration. The dashed lines are the expected fluorescence of the free and bound peptide, respectively. Black lines, K_d and stoichiometry (S_T) values were obtained using the equation described in Supplementary Materials. The ratio P:LPS was calculated from S_T, assuming that *E. coli* LPS has an average molecular weight of 10 kDa. In (**H**), data were fitted to a straight line.

2.9. Interaction of Labeled CAMPs with E. coli Cells

In order to study the interaction of GKY20 and $ApoB_L$ with whole bacterial cells, we recorded the emission spectra of the labeled peptides in the presence of *E. coli* cells at an optical density of 0.1 OD_{600} (corresponding to about 0.63×10^9 CFU/mL). The emission spectra of Luc-GKY20 and Luc-$ApoB_L$ obtained after excitation at 330 nm show that in both cases, binding to *E. coli* cells causes a moderate decrease in the fluorescence emission (Figure S9A), accompanied by a blue shift of 16 nm in the case of Luc-GKY20 (a value slightly lower than those observed in the case of liposomes and micellar LPS) and of only 5 nm in the case of Luc-$ApoB_L$ (Table 1). In the blue region, the emission of Luc-GKY20 was higher than that of Luc-$ApoB_L$, which is the opposite of what was observed in the case of LPS (Figure S9B). The differences between the spectra of the labeled peptides in the presence of whole bacterial cells and those obtained in the presence of liposomes and purified LPS might be due to the fact that the outer membrane of Gram-negative bacteria is a very complex mixture of LPS, phospholipids and proteins.

The emission spectra of Luc-GKY20 and Luc-$ApoB_L$ obtained after excitation at 400 nm show a strong decrease in fluorescence emission (Figure S9C), likely due to a reduced hydrolysis of the Luc hydroxyl group of the cell-bound peptides.

Finally, *E. coli* cells caused a significant increase in the fluorescence emission of aLuc-GKY20 and a slight decrease in the fluorescence emission of aLuc-$ApoB_L$ (Figure S9D). In the case of aLuc-labeled peptides, we also observed a blue shift lower than that observed in the case of liposomes and micellar LPS (Table 1).

The spectra shown in Figure S9 were recorded after an incubation time of 20 min in the case of labeled GKY20 and of 120 min in the case of labeled $ApoB_L$. The different incubation times were necessary for an unexpected difference in the binding kinetic of the two peptides, as shown in Figure S9E. In the case of Luc-GKY20, the slope of the curve obtained by plotting fluorescence intensity (ex. = 400 nm; em. = 539 nm) as a function of time was about 6.7 times higher than that observed in the case of Luc-$ApoB_L$. Intriguingly, in the case of purified LPS, the binding process was complete within the preparation time of the samples (about 60 s) for both peptides. The reasons for such differences were not further investigated. Nonetheless, these results highlight another useful application of Luc labeling.

We also observed *E. coli* cells treated with the labeled peptides (3 μM) using a fluorescence microscope equipped with a mercury arc lamp (Figures S10 and S11). In the case of GKY20-treated cells, in addition to homogeneously labeled cells, we observed several cells with a heterogeneous labeling pattern (Figures S10K,N and S11E). The same pattern was observed in *E. coli* cells treated with PYMPO-(C)GKY20, indicating that heterogeneous labeling is not an artifact of Luc labeling (Figure S10A–D). For incubation times longer than 30 min, we observed an increased amount of highly fluorescent and large bodies (Figure S11), likely aggregates of cell debris and/or dead cells. This is not surprising, as GKY20 and $ApoB_L$, like many CAMPs, cause cell lysis [19,44].

2.10. Confocal Laser Scanning Microscopy

Confocal Laser Scanning Microscopy (CLSM) allows for the attainment of high resolution tridimensional images of biological samples by stacking several bidimensional images taken at different depths [48]. As in CLSM, fluorophore excitation is obtained through high-power but very narrow laser beams. It is mandatory that the excitation peak of the chosen fluorescent label overlaps the wavelength of one of the available laser lines [48]. Moreover, the fluorescent label should be photostable enough to avoid a quick bleaching of the sample. This is especially important in the case of live-imaging applications. We have already shown that Luc and aLuc are very stable fluorophores. As regards the excitation wavelength, it is interesting to note that the excitation spectra of neutral Luc and its phenolate show an isosbestic point at about 355 nm, a wavelength very close to that of the argon-ion laser (351 nm). Therefore, this laser could be used to simultaneously excite both forms of Luc. A blue diode laser (405 nm) could be used to excite aLuc, the broad excitation peak of which is centered at 360–370 nm but has also a remarkable tail in the violet region. Differently from an argon-ion laser, which is not common, the blue diode laser is present on most CLSM devices used to excite blue fluorophores. In order to evaluate the suitability of aLuc as label for CLSM, we incubated HaCaT and HeLa cells with aLuc-p53pAnt or aLuc-RGD for 60–150 min; hence, samples were either observed or incubated for 15 additional min with LysoTracker™ Red DND-99, a red probe that is actively taken up by acidic organelles [49], and/or NucRed™ Live 647, a far-red cell-permeant vital stain for nucleic acids [50,51]. Samples were observed without any further wash or treatment in order to minimize artifacts and to mimic the conditions of live imaging.

Non-cell-specific uptake of p53pAnt has been previously demonstrated by Western blotting of cell lysates and immunostaining by an antibody specific to the p53-derived portion of p53pAnt [25,52], whereas confocal microscopy with a rhodamine B-labeled peptide showed that it accumulates both in the cytosol and the nuclei of two prostate cancer cell lines [52]. Our CLSM analysis of HaCaT and HeLa cells treated with aLuc-p53pAnt confirm the previous findings. In the case of HaCaT cells treated with aLuc-p53pAnt, the peptide was localized at the cell periphery, mainly in the form of circular spots, thus suggesting the presence of the peptide in an endosomal compartment (Figure S12). A very small fraction of the cells, however, displayed a very strong signal, partly diffused into the cell and partly associated to the nucleus (Figure S13). The signal was very strong at the nuclear periphery but also present inside the nucleus in the form of one or more patches of different dimensions (Figure S13). As discussed below, colocalization studies with NucRed Live confirm that the peptide binds to nucleic acids. Very interestingly, in the case of HeLa cells, the proportion between the two types of staining pattern was reversed, with the highly stained cells being predominant (Figure S13). Considering that p53pAnt is much more toxic for HeLa cancer cells than for HaCaT cells (Figure S4) and that it has been suggested that p53pAnt induces apoptosis, it can be speculated that highly stained cells are apoptotic or pre-apoptotic cells. It is well known that apoptosis causes relevant alterations of cell membranes, e.g., externalization of phosphatidylserine [53], a negatively charged lipid, which could determine an increased afflux into the cytosol of p53pAnt. Once in the cytosol, p53pAnt could migrate into the nucleus and bind nucleic acids due to its high positive charge.

HaCaT cells treated with aLuc-p53pAnt and LysoTracker Red showed, as expected, numerous red spots (Figures 7 and S15). Very interestingly, we observed partial colocalization between LysoTracker Red and aLuc-p53pAnt (Figures 7E,F and S14A–D). Cells showing a strong aLuc-p53pAnt fluorescence did not show the presence of LysoTracker Red (Figure 7A–D). The same pattern was observed in the case of HeLa cells, except that, again, the frequency of cells showing only the strong fluorescence of aLuc-p53pAnt was higher than in the case of HaCaT cells (Figure S15A–D). Accumulation of LysoTracker Red, requiring acidification of endosomes, is expected only in metabolically active cells; therefore, these findings are in agreement with the hypothesis that cells strongly stained with aLuc-p53pAnt are apoptotic or pre-apoptotic cells.

Unexpectedly, our attempts to perform three-color imaging by staining aLuc-p53pAnt-treated HaCaT cells with LysoTracker Red and NucRed Live for 15 min revealed an alteration of the localization both of LysoTracker Red and of aLuc-p53pAnt (Figure S14E–H). The change was particularly evident in the case of LysoTracker Red, which appeared more homogeneously diffused inside the cell than in the absence of NucRed. On the contrary, the signal of aLuc-p53pAnt appeared less diffused and with more defined spots (Figure S14E). The alteration in the distribution of aLuc-p53pAnt was more pronounced by co-incubating HaCaT cells with NucRed Live and aLuc-p53pAnt for one hour (Figure S14I–L). In that case, in the majority of the cells, aLuc-p53pAnt appeared as numerous large and well-defined spots at the cell periphery. In the case of aLuc-p53pAnt-treated HeLa cells, staining with NucRed Live for 15 min had minor effects on the appearance of the putative apoptotic/pre-apoptotic cells (Figure S15E–H). Interestingly, inside the nuclei of these cells, aLuc-p53pAnt and NucRed Live were essentially colocalized (Figure S15M–O), thus suggesting that aLuc-p53pAnt is bound to nucleic acids. Similarly to what was observed in the case of HaCaT cells, LysoTracker Red was more diffused in the presence of NucRed Live (Figure S15). When HeLa cells were treated with NucRed Live for one hour, we no longer observed the putative apoptotic/pre-apoptotic cells (Figure S15I–L), and the staining pattern was very similar to that of the HaCaT cells treated for one hour with NucRed Live (Figure S14I–L). Very interestingly, NucRed Live also proved to change the behavior of aLuc-RGD, as described below.

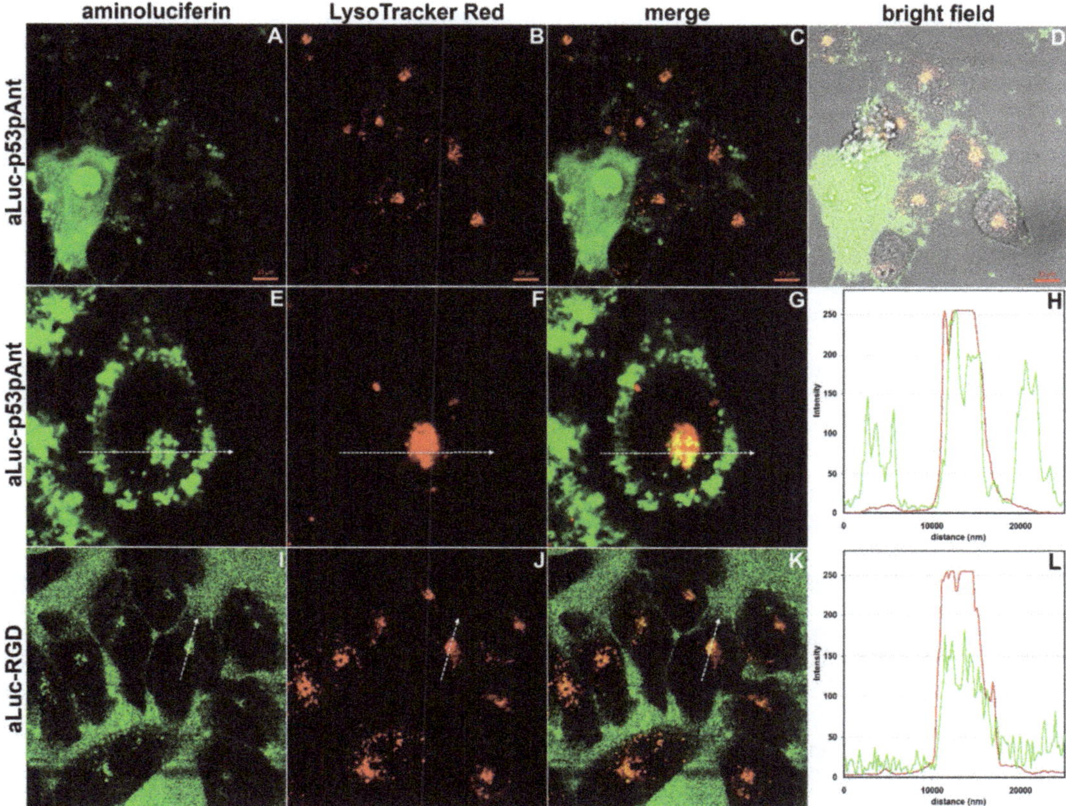

Figure 7. CLSM images of HaCaT cells treated with aLuc-p53pAnt or aLuc-RGD. (**A–D**) Cells were incubated with aLuc-p53pAnt for 120 min and with LysoTracker™ Red DND-99 for 15 additional minutes. Bar = 10 μm. (**E**,**F**) Single representative HaCaT cell treated with aLuc-p53pAnt (from Figure S14A–D). White arrow = 25 μm. (**I–K**) Cells were incubated with aLuc-RGD for 90 min and with LysoTracker™ Red DND-99 for 15 additional minutes (bright field in Figure S16E). White arrow = 25 μm. (**H**,**L**) Fluorescence intensity across the white arrows in panels (**E–G**) and (**I–K**), respectively (green curve, aLuc; red curve, LysoTracker Red).

RGD is a known integrin ligand derived from fibronectin; therefore, it is expected to undergo receptor-mediated endocytosis, as demonstrated for other similar peptides [54]. In particular, it has been demonstrated that peptide GRGDNP, which differs by one amino acid (N at position 5) from RGD, is endocytosed in Molt-4 cells (leukaemic T-cell line) [55]. Differently from aLuc-p53pAnt, aLuc-RGD was homogeneously dispersed in the culture medium in the case of both HaCaT and HeLa cells, which appeared as non-fluorescent regions (Figures 7I–L and S16, respectively). Nonetheless, as expected, an intracellular signal was visible and colocalized with LysoTracker Red (Figures 7I–L and S16). In the presence of NucRed Live, aLuc-RGD was no longer visible inside the cells, whereas the signal of LysoTracker Red appeared homogeneously diffused inside the cytosol (Figure S16), as observed in the case of the analysis of aLuc-p53pAnt.

Overall, NucRed Live seems to be able to interfere with endocytosis and intracellular trafficking. NucRed Live is a relatively recent stain, and we did not find other reports regarding its potential biological effects. On the other hand, it has been demonstrated that another far-red, membrane-permeable nuclear stain, DRAQ5™, alters membrane fluidity and inhibits the internalization of bacterial toxins [56]. Reduction in membrane fluidity and/or alteration of membrane potential by NucRed Live might prevent translocation of p53pAnt from the endosomes to cytosol, thus causing accumulation of the peptide in the lumen of endosomes. This, in turn, would prevent its interaction with p53, thus inhibiting the proapoptotic effect of the peptide. The effect on LysoTracker Red could be explained, at least in part, by assuming that NucRed Live is also able to inhibit endosome acidification. We did not further study the effects of NucRed Live, as this is beyond the scope of the current work; nonetheless, the analysis reported herein clearly shows that aLuc is very well suited as a probe for CLSM.

3. Discussion

We have shown that Luc and aLuc are very well suited as environment-sensitive fluorescent labels for peptides. Labeling is fast, quantitative, very specific and can be performed in very mild conditions (RT, buffer phosphate at pH 7–7.5) using commercially available reagents and peptides with an N-terminal cysteine that can be prepared either by chemical synthesis or by several recombinant strategies ([32] and references therein). Luc and aLuc are photostable fluorophores with a large Stoke shift (about 210 and 145 nm, respectively) and a high quantum yield. Moreover, they are small molecules (low molecular weight and low solvent-accessible surface), uncharged (aLuc) or with a low percentage of ionized form at pH 7 (Luc), neither highly polar nor particularly hydrophobic, thus minimizing the impact of labeling on the structure of the peptides and presumably on their properties. Furthermore, N-terminal labeling is very well suited for most peptides, and indeed, it is a very common choice. On the other hand, Luc and aLuc possess complementary properties. Luc has a very uncommon dual emission with a main emission in the green region (~539 nm), commonly observed in aqueous buffers and a blue emission (~450 nm), which can only be observed in an environment with a very low water content. Moreover, both peaks undergo significant blue shifts (20–30 nm) in hydrophobic environments. This behavior makes Luc a very useful probe for the study of the interactions of peptides with membranes, liposomes, micelles of detergents and LPS. The very large difference between the excitation maxima of the neutral and the phenolate form (330 and 400 nm, respectively) makes Luc an intriguing probe for the pH range of 6.5–9.5 (hypothesizing pK_a values in the range 7.5–8.5 for peptide-bound Luc). Moreover, it can be exploited to detect binding events that influence its ionization state, as shown in the case of binding to liposomes, SDS, LPS and *E. coli* cells, and to determine K_d values and stoichiometries. aLuc is not sensitive to pH and shows less pronounced variations in fluorescence emission. However, it is, in turn, a strongly solvatochromic fluorophore, showing blue shifts up to 40 nm. This makes aLuc an alternative probe for the study of the interactions of peptides with their targets. In addition, aLuc is very well suited as a label for CLSM, and it does not require special equipment. It can be efficiently excited by the common 404 nm laser, and emitting at 500–520 nm, it allows for colocalization studies with the many commercially available and widely used orange, red and far-red probes, as was shown by our analysis of the colocalization of p53pAnt and RGD with LysoTracker™ Red DND-99 and NucRed™ Live 647.

It is worth noting that we have only explored a minimal part of the potentialities of Luc and aLuc as fluorescent labels for peptides. For example, when N-terminal labeling is not suitable or when large proteins have to be labeled, a 1,2-aminothiol functionality could be introduced at internal positions, e.g., as N^{ε}-cysteinyl-L-lysine [57], or by modifying a cysteine residue [58]. Furthermore, Luc and aLuc might be bound to internal cysteine or lysine residues via the carboxyl group of luciferins, exploiting conventional chemical strategies to crosslink carboxyl groups to amines and thiols [59]. Even more interestingly, in the attempt to find new substrates for firefly luciferase, an astonishing number of Luc and, in particular, aLuc derivatives and analogues have been published [60,61]. Many of these compounds are fluorescent and show intriguing properties; for example, many aLuc derivatives show red-shifted λ_{max} values (up to 576 nm) and/or altered solvatochromic behavior (increased or decreased, depending on the nature of the substituents bound to the N6 nitrogen atom) [11,61,62]. Moreover, halogenated luciferins show decreased pK_a values—e.g., 7-F- and 7-Cl-luciferin have pK_a values of 7.1 and 6.7, respectively [63]—thus expanding the useful pH range in applications based on pH-dependent fluorescence. Most of the cited derivatives and analogues were synthetized by reacting the corresponding 2-cyano-benzothiazole with cysteine. Thus, they could be directly generated at the N-terminus of peptides with a terminal cysteine residue. The others might be linked to peptides through the activation of their carboxylate group, which, being essential for the catalytic activity of firefly luciferase, is present in all the analogues.

Therefore, Luc and aLuc can reasonably be regarded as the prototypes of a huge new and variegated family of fluorescent labels for proteins and peptides.

4. Materials and Methods

4.1. Materials and General Methods

Materials and general methods can be found in Section S1 (Supplementary Materials). The sequences of the peptides GKY20, ApoBL, p53pAnt and RGD are shown in Figure S1, and their preparation, purification and labeling are described in Sections S2–S6 (Supplementary Materials).

4.2. Steady-State Fluorescence Spectroscopy in Water/Organic Solvent Mixtures and SDS

Fluorescence spectra were recorded on a Fluoromax-4 fluorometer (Horiba, Edison, NJ, USA) using a 1 cm path length quartz cuvette at a temperature of 25 °C, using a peltier that can ensure an accuracy of ±0.1 °C. All experiments were carried out at a fixed peptide concentration (2 μM) in 10 mM sodium phosphate (NaP), pH 7.4, at 25 °C, unless otherwise stated. The excitation wavelengths were set to 330 nm (Luc-labeled peptides, phenol form), 400 nm (Luc-labeled peptides, phenolate form), 363 nm (aLuc-labeled peptides), 330 nm (mLuc-GKY20) and 408 nm for 1-[2-(maleimido)ethyl]-4-[5-(4-methoxyphenyl)-2-oxazolyl]pyridinium-labeled GKY20 [PyMPO-(C)GKY20]. The excitation spectra were recorded by varying the wavelength of excitation between 200 nm and 500 nm (em. = 539 nm). To evaluate the solvatochromic properties of labeled peptides, fluorescence spectra were recorded in the presence of NaP:methanol (50% *v:v*), NaP:isopropanol (50% *v:v*) and SDS (25 mM) under the experimental condition described above.

4.3. Steady-State Fluorescence Spectroscopy in the Presence of Liposomes

Liposome preparation is described in Section S8, Supplementary Materials. Fluorescence spectra were recorded on a FluoroMax-4 fluorometer (Horiba, Kyoto, Japan). The emission spectra of Luc-labeled peptides were acquired, upon excitation at 330 nm, in the range 350–650 nm. For the aLuc-labeled peptides, the emission spectra were acquired in the range 380–700 nm, upon excitation at 363 nm. In addition, for Luc-labeled peptides, excitation spectra were also recorded. The excitation spectra were recorded by varying the wavelength of excitation between 275 nm and 480 nm and monitoring the emission at 539 nm. All the spectra were recorded at a lipid-to-peptide ratio of 200 in 10 mM NaP, pH 7.4. The concentration of peptides was in the range of 2.2–3.6 μM.

4.4. Quantum-Yield Determination

Fluorescence quantum yields were determined for all the labeled peptides in 10 mM phosphate buffer, pH 7.4, in a mixture composed of phosphate buffer and isopropanol in the ratio 1:1 (*v:v*) and in acetate buffer, pH 5. Determination of the quantum yields was performed by comparing the fluorescence of samples to that of a standard, as previously described [64]. Fluorescein (in 0.1 M NaOH) was used as standard in the case of (a)Luc-

and PyMPO-labeled peptides, whereas coumarin-6 (in pure ethanol) was used in the case of mLuc-labeled GKY20.

4.5. pK$_a$ Determination of Luc-Labeled Peptides

In order to measure pK$_a$ values, phenolate concentration was evaluated by titrating a solution of Luc-labeled peptide (2 μM) as a function of pH (0.2 M sodium acetate, pH 4–6; 0.2 M sodium phosphate, pH 6–7.4; 0.2 M Tris/HCl, pH 7–9; 0.2 M Glycine/NaOH, pH 8–11). The excitation wavelength was set to 400 nm. Excitation spectra were recorded at 539 nm. pK$_a$ values were determined by GraphPad Prism software (version 6, San Diego, CA, USA) by plotting variation of total fluorescence (450–700 nm) as a function of pH values.

4.6. Interaction of Labeled CAMPs with LPS

Binding of CAMPs (2 μM) to LPS (200 μg/mL) from *E. coli* 0111:B4 (MW 10,000) [46] was performed in 10 mM NaP buffer, pH 7.4. Mixtures were equilibrated at 25 °C for 10 min before recording emission (Luc-peptides, ex. = 330; aLuc-peptides, ex. = 363 nm) and excitation spectra (Luc-GKY20, em. = 516 nm; Luc-ApoB$_L$, em. = 508 nm) by means of a FluoroMax-4 fluorimeter. To test the influence of micellar and sub-micellar LPS concentration on CAMP fluorescence (2 μM), emission spectra (Luc-peptides, ex. = 330 or 400 nm; aLuc-peptides, ex. = 363 nm) were also recorded in the presence of increasing concentrations of LPS (0.62 e 200 μg/mL). Variation of total fluorescence (450–700 nm) was reported as a function of LPS concentration. The assays to determine K$_d$ and binding stoichiometry of Luc-peptides toward LPS were carried out in 96-well polystyrene microtiter plates containing 100 μL of peptide/LPS mixtures. Spectra were recorded using a Synergy^{TM} H4 microplate reader (BioTek Instruments Inc., Winooski, VT, USA) in 10 mM NaP buffer, pH 7.4, in the presence of 40 μg/mL (≈4 μM; MW 10,000) LPS and Luc-peptides (0.25–18 μM). Mixtures were incubated 15 min before emission spectra were recorded by excitation between 400 and 425 nm (phenolate form). Variation of total fluorescence (450–700 nm) was reported as a function of peptide concentration, and data were fitted to the model using Graphpad Prism (Supplementary Materials, Section S9).

4.7. Interaction of Labeled CAMPs with E. coli Cells

Bacterial *E. coli* ATCC 25922 strain was cultured in LB medium at 37 °C overnight. Culture was diluted 1:100 in fresh LB medium, and bacteria were grown until 1 OD$_{600}$ optical density. Cells were collected by centrifugation at 8000× *g* for 5 min at 4 °C, washed three times in 10 mM NaP buffer, pH 7.4, and suspended at 1 OD$_{600}$ concentration (10x cell stock solution) in the same buffer. The bacteria mixture was stored on ice until use. Binding of labeled CAMPs to *E. coli* cells was performed in 10 mM NaP buffer, pH 7.4, in the presence of 0.1 OD$_{600}$ bacterial cells and 2 μM peptides. Mixtures were incubated at 25 °C for 20 min [(a)Luc-GKY20] and 120 min [(a)Luc-ApoB$_L$] before recording emission spectra (Luc-peptides, ex. = 330 and 400 nm; aLuc-peptides, ex. = 363 nm).

4.8. Kinetic Analysis

Binding kinetic to LPS and *E. coli* cells was carried out in 10 mM NaP buffer, pH 7.4, in the presence of either 50 μg/mL LPS or 0.1 OD$_{600}$ bacterial cells prepared as described above. Binding reactions were started by adding peptides (2 μM) and manually mixing the samples for 40 s. Binding was monitored, exciting at 400 nm and reading at 539 nm. One reading per minute was performed over 16 min observation time. Samples were not irradiated in the period between two readings in order to minimize peptide photobleaching. Photobleaching of peptides was also verified by control experiments carried out in the absence of LPS and cells.

4.9. Microscopy Analysis of E. coli Cells Treated with the Labeled Peptides

Binding of labeled CAMPs (3 μM) to *E. coli* cells was performed in 10 mM NaP buffer, pH 7.4, in the presence of 0.1 OD_{600} bacterial cells. Fluorescence microscopy images of treated and untreated *E. coli* cells were taken over 50 min incubation at 25 °C. For this purpose, 10 μL of each sample was observed with an Olympus BX51 fluorescence microscope (Olympus, Tokyo, Japan) using DAPI (aLuc-labeled peptides) and FITC (Luc-labeled peptides) filters. Standard acquisition times were 1000 ms. Images were captured using an Olympus DP70 digital camera. The experiments were performed at least three times.

4.10. Interaction of aLuc-p53pAnt and aLuc-RGD with HeLa and HaCaT Cells

Normal human keratinocytes (HaCaT) and human cancer epithelial cells (HeLa cells) were cultured in Dulbecco's Modified Eagle's Medium (DMEM), supplemented with 10% fetal bovine serum (FBS), 2 mM L-glutamine and 1% penicillin–streptomycin in a 5% CO_2 humidified atmosphere at 37°C. HaCaT and HeLa cells were seeded in chambered well plates (500 μL/well; Nunc™ Lab-Tek™ Chambered Coverglass systems, Thermo Fisher Scientific, Waltham, MA, USA) with a density of 4.5×10^4 and 2.5×10^4/well, respectively, and then grown at 37 °C for 48 h. Cells were washed three times with PBS and then incubated with aLuc-peptides (10 μM) for 1 h in medium without FBS, supplemented with 2 mM L-glutamine and 1% penicillin–streptomycin in a 5% CO_2 humidified atmosphere at 37 °C. Lysotracker™ Red DND-99 and NucRed™ Live 647 (Thermo Fisher Scientific, Waltham, MA, USA) were added to the cells at the concentrations recommended by the producer and incubated for 15 min in a 5% CO_2 humidified atmosphere at 37 °C. The samples were then analyzed using a confocal laser scanning microscope (Zeiss LSM 710, Zeiss, Germany) and a 63X objective oil-immersion system. Acquired images were analyzed using the Zen Lite 2.3 software package. In the case of the images shown in Figure 7A–D the intensity of the green channel was increased in order to show both the apoptotic and non-apoptotic cells. Each experiment was performed in triplicate.

Supplementary Materials: The following are available online at https://www.mdpi.com/article/10.3390/ijms222413312/s1.

Author Contributions: E.N., E.P. and V.C.: conceptualization, methodology, supervision, writing—original draft preparation; R.O., M.S., A.M., R.G., A.A., R.R. and A.Z.: investigation, writing—review and editing; A.D.M., P.D.V., L.P., M.V. and M.M.: formal analysis, writing—review and editing. All authors have read and agreed to the published version of the manuscript.

Funding: This research was funded by the Italian Cystic Fibrosis Research Foundation, grant numbers FFC#16/2017 and FFC#18/2018.

Data Availability Statement: The data presented in this study are available in the article or Supplementary Materials. The raw datasets are available from the corresponding authors upon reasonable request.

Conflicts of Interest: The authors declare no conflict of interest. The funders had no role in the design of the study; in the collection, analyses or interpretation of data; in the writing of the manuscript or in the decision to publish the results.

References

1. Klymchenko, A.S. Solvatochromic and Fluorogenic Dyes as Environment-Sensitive Probes: Design and Biological Applications. *Acc. Chem. Res.* **2017**, *50*, 366–375. [CrossRef] [PubMed]
2. Han, J.; Burgess, K. Fluorescent Indicators for Intracellular pH. *Chem. Rev.* **2010**, *110*, 2709–2728. [CrossRef] [PubMed]
3. Loving, G.S.; Sainlos, M.; Imperiali, B. Monitoring protein interactions and dynamics with solvatochromic fluorophores. *Trends Biotechnol.* **2010**, *28*, 73–83. [CrossRef] [PubMed]
4. Donadio, G.; Di Martino, R.; Oliva, R.; Petraccone, L.; Del Vecchio, P.; Di Luccia, B.; Ricca, E.; Isticato, R.; Di Donato, A.; Notomista, E. A new peptide-based fluorescent probe selective for zinc(II) and copper(II). *J. Mater. Chem. B* **2016**, *4*, 6979–6988. [CrossRef]
5. Li, S.; Ruan, Z.; Zhang, H.; Xu, H. Recent achievements of bioluminescence imaging based on firefly luciferin-luciferase system. *Eur. J. Med. Chem.* **2021**, *211*, 113111. [CrossRef]

6. Morton, R.A.; Hopkins, T.A.; Seliger, H.H. Spectroscopic properties of firefly luciferin and related compounds; an approach to product emission. *Biochemistry* **1969**, *8*, 1598–1607. [CrossRef]
7. Ando, Y.; Akiyama, H. PH-dependent fluorescence spectra, lifetimes, and quantum yields of firefly-luciferin aqueous solutions studied by selective-excitation fluorescence spectroscopy. *Jpn. J. Appl. Phys.* **2010**, *49*, 117002. [CrossRef]
8. Presiado, I.; Erez, Y.; Huppert, D. Excited-state intermolecular proton transfer of the firefly's chromophore d-luciferin. 2. water-methanol mixtures. *J. Phys. Chem. A* **2010**, *114*, 9471–9479. [CrossRef]
9. Kuchlyan, J.; Banik, D.; Roy, A.; Kundu, N.; Sarkar, N. Excited-state proton transfer dynamics of fireflys chromophore d -luciferin in DMSO-water binary mixture. *J. Phys. Chem. B* **2014**, *118*, 13946–13953. [CrossRef]
10. Vieira, J.; Da Silva, L.P.; Da Silva, J.C.G.E. Advances in the knowledge of light emission by firefly luciferin and oxyluciferin. *J. Photochem. Photobiol. B Biol.* **2012**, *117*, 33–39. [CrossRef]
11. Kakiuchi, M.; Ito, S.; Yamaji, M.; Viviani, V.R.; Maki, S.; Hirano, T. Spectroscopic Properties of Amine-substituted Analogues of Firefly Luciferin and Oxyluciferin. *Photochem. Photobiol.* **2017**, *93*, 486–494. [CrossRef]
12. Zheng, M.; Huang, H.; Zhou, M.; Wang, Y.; Zhang, Y.; Ye, D.; Chen, H.Y. Cysteine-Mediated Intracellular Building of Luciferin to Enhance Probe Retention and Fluorescence Turn-On. *Chem. A Eur. J.* **2015**, *21*, 10506–10512. [CrossRef]
13. Miao, Q.; Li, Q.; Yuan, Q.; Li, L.; Hai, Z.; Liu, S.; Liang, G. Discriminative Fluorescence Sensing of Biothiols in Vitro and in Living Cells. *Anal. Chem.* **2015**, *87*, 3460–3466. [CrossRef]
14. Zheng, M.; Wang, Y.; Shi, H.; Hu, Y.; Feng, L.; Luo, Z.; Zhou, M.; He, J.; Zhou, Z.; Zhang, Y.; et al. Redox-Mediated Disassembly to Build Activatable Trimodal Probe for Molecular Imaging of Biothiols. *ACS Nano* **2016**, *10*, 10075–10085. [CrossRef]
15. Zhao, X.; Lv, G.; Peng, Y.; Liu, Q.; Li, X.; Wang, S.; Li, K.; Qiu, L.; Lin, J. Targeted Delivery of an Activatable Fluorescent Probe for the Detection of Furin Activity in Living Cells. *ChemBioChem* **2018**, *19*, 1060–1065. [CrossRef]
16. Ren, H.; Xiao, F.; Zhan, K.; Kim, Y.P.; Xie, H.; Xia, Z.; Rao, J. A biocompatible condensation reaction for the labeling of terminal cysteine residues on proteins. *Angew. Chem. Int. Ed.* **2009**, *48*, 9658–9662. [CrossRef]
17. Chen, K.T.; Ieritano, C.; Seimbille, Y. Early-Stage Incorporation Strategy for Regioselective Labeling of Peptides using the 2-Cyanobenzothiazole/1,2-Aminothiol Bioorthogonal Click Reaction. *ChemistryOpen* **2018**, *7*, 256–261. [CrossRef]
18. Shinde, R.; Perkins, J.; Contag, C.H. Luciferin derivatives for enhanced in vitro and in vivo bioluminescence assays. *Biochemistry* **2006**, *45*, 11103–11112. [CrossRef]
19. Kasetty, G.; Papareddy, P.; Kalle, M.; Rydengård, V.; Mörgelin, M.; Albiger, B.; Malmsten, M.; Schmidtchen, A. Structure-activity studies and therapeutic potential of host defense peptides of human thrombin. *Antimicrob. Agents Chemother.* **2011**, *55*, 2880–2890. [CrossRef]
20. Gaglione, R.; Dell'Olmo, E.; Bosso, A.; Chino, M.; Pane, K.; Ascione, F.; Itri, F.; Caserta, S.; Amoresano, A.; Lombardi, A.; et al. Novel human bioactive peptides identified in Apolipoprotein B: Evaluation of their therapeutic potential. *Biochem. Pharmacol.* **2017**, *130*, 34–50. [CrossRef]
21. Wiesner, J.; Vilcinskas, A. Antimicrobial peptides: The ancient arm of the human immune system. *Virulence* **2010**, *1*, 440–464. [CrossRef]
22. Pane, K.; Durante, L.; Crescenzi, O.; Cafaro, V.; Pizzo, E.; Varcamonti, M.; Zanfardino, A.; Izzo, V.; Di Donato, A.; Notomista, E. Antimicrobial potency of cationic antimicrobial peptides can be predicted from their amino acid composition: Application to the detection of "cryptic" antimicrobial peptides. *J. Theor. Biol.* **2017**, *419*, 254–265. [CrossRef]
23. Pizzo, E.; Cafaro, V.; Di Donato, A.; Notomista, E. Cryptic Antimicrobial Peptides: Identification Methods and Current Knowledge of their Immunomodulatory Properties. *Curr. Pharm. Des.* **2018**, *24*, 1054–1066. [CrossRef]
24. Dell'Olmo, E.; Gaglione, R.; Cesaro, A.; Cafaro, V.; Teertstra, W.R.; de Cock, H.; Notomista, E.; Haagsman, H.P.; Veldhuizen, E.J.A.; Arciello, A. Host defence peptides identified in human apolipoprotein B as promising antifungal agents. *Appl. Microbiol. Biotechnol.* **2021**, *105*, 1953–1964. [CrossRef]
25. Selivanova, G.; Iotsova, V.; Okan, I.; Fritsche, M.; Ström, M.; Groner, B.; Grafström, R.C.; Wiman, K.G. Restoration of the growth suppression function of mutant p53 by a synthetic peptide derived from the p53 C-terminal domain. *Nat. Med.* **1997**, *3*, 632–638. [CrossRef]
26. Li, Y.; Mao, Y.; Rosal, R.V.; Dinnen, R.D.; Williams, A.C.; Brandt-Rauf, P.W.; Fine, R.L. Selective induction of apoptosis through the FADD/Caspase-8 pathway by a p53 C-terminal peptide in human pre-malignant and malignant cells. *Int. J. Cancer* **2005**, *115*, 55–64. [CrossRef]
27. Li, Y.; Rosal, R.V.; Brandt-Rauf, P.W.; Fine, R.L. Correlation between hydrophobic properties and efficiency of carrier-mediated membrane transduction and apoptosis of a p53 C-terminal peptide. *Biochem. Biophys. Res. Commun.* **2002**, *298*, 439–449. [CrossRef]
28. Pierschbacher, M.D.; Ruoslahti, E. Cell attachment activity of fibronectin can be duplicated by small synthetic fragments of the molecule. *Nature* **1984**, *309*, 30–33. [CrossRef]
29. Ruoslahti, E. RGD and other recognition sequences for integrins. *Annu. Rev. Cell Dev. Biol.* **1996**, *12*, 697–715. [CrossRef]
30. Knetsch, P.A.; Zhai, C.; Rangger, C.; Blatzer, M.; Haas, H.; Kaeopookum, P.; Haubner, R.; Decristoforo, C. [68Ga]FSC-(RGD)3 a trimeric RGD peptide for imaging αvβ3 integrin expression based on a novel siderophore derived chelating scaffold-synthesis and evaluation. *Nucl. Med. Biol.* **2015**, *42*, 115–122. [CrossRef]
31. Karimi, F.; O'Connor, A.J.; Qiao, G.G.; Heath, D.E. Integrin Clustering Matters: A Review of Biomaterials Functionalized with Multivalent Integrin-Binding Ligands to Improve Cell Adhesion, Migration, Differentiation, Angiogenesis, and Biomedical Device Integration. *Adv. Healthc. Mater.* **2018**, *7*, e1701324. [CrossRef] [PubMed]

32. Pane, K.; Verrillo, M.; Avitabile, A.; Pizzo, E.; Varcamonti, M.; Zanfardino, A.; Di Maro, A.; Rega, C.; Amoresano, A.; Izzo, V.; et al. Chemical Cleavage of an Asp-Cys Sequence Allows Efficient Production of Recombinant Peptides with an N-Terminal Cysteine Residue. *Bioconjug. Chem.* **2018**, *29*, 1373–1383. [CrossRef] [PubMed]
33. Litak, P.T.; Kauffman, J.M. Syntheses of reactive fluorescent stains derived from 5(2)-aryl-2(5)-(4-pyridyl)oxazoles and bifunctionally reactive linkers. *J. Heterocycl. Chem.* **1994**, *31*, 457–479. [CrossRef]
34. Dou, Y.; Goodchild, S.J.; Velde, R.V.; Wu, Y.; Fedida, D. The neutral, hydrophobic isoleucine at position I521 in the extracellular S4 domain of hERG contributes to channel gating equilibrium. *Am. J. Physiol. Cell Physiol.* **2013**, *305*, 468–478. [CrossRef]
35. Wakabayashi, H.; Fay, P.J. Molecular orientation of Factor VIIIa on the phospholipid membrane surface determined by fluorescence resonance energy transfer. *Biochem. J.* **2013**, *452*, 293–301. [CrossRef]
36. Pane, K.; Durante, L.; Pizzo, E.; Varcamonti, M.; Zanfardino, A.; Sgambati, V.; Di Maro, A.; Carpentieri, A.; Izzo, V.; Di Donato, A.; et al. Rational design of a carrier protein for the production of recombinant toxic peptides in Escherichia coli. *PLoS ONE* **2016**, *11*, e0146552. [CrossRef]
37. Doose, S.; Neuweiler, H.; Sauer, M. Fluorescence quenching by photoinduced electron transfer: A reporter for conformational dynamics of macromolecules. *ChemPhysChem* **2009**, *10*, 1389–1398. [CrossRef]
38. Chen, H.; Ahsan, S.S.; Santiago-Berrios, M.B.; Abruña, H.D.; Webb, W.W. Mechanisms of quenching of alexa fluorophores by natural amino acids. *J. Am. Chem. Soc.* **2010**, *132*, 7244–7245. [CrossRef]
39. Goldstein, L.; Levin, Y.; Katchalski, E. A Water-insoluble Polyanionic Derivative of Trypsin. II. Effect of the Polyelectrolyte Carrier on the Kinetic Behavior of the Bound Trypsin. *Biochemistry* **1964**, *3*, 1913–1919. [CrossRef]
40. Maurel, P.; Douzou, P. Catalytic implications of electrostatic potentials: The lytic activity of lysozyme as a model. *J. Mol. Biol.* **1976**, *102*, 253–264. [CrossRef]
41. Gaglione, R.; Smaldone, G.; Cesaro, A.; Rumolo, M.; De Luca, M.; Di Girolamo, R.; Petraccone, L.; Del Vecchio, P.; Oliva, R.; Notomista, E.; et al. Impact of a Single Point Mutation on the Antimicrobial and Fibrillogenic Properties of Cryptides from Human Apolipoprotein B. *Pharmaceuticals* **2021**, *14*, 631. [CrossRef]
42. Oliva, R.; Del Vecchio, P.; Grimaldi, A.; Notomista, E.; Cafaro, V.; Pane, K.; Schuabb, V.; Winter, R.; Petraccone, L. Membrane disintegration by the antimicrobial peptide (P)GKY20: Lipid segregation and domain formation. *Phys. Chem. Chem. Phys.* **2019**, *21*, 3989–3998. [CrossRef]
43. Aniansson, E.A.G.; Wall, S.N.; Almgren, M.; Hoffmann, H.; Kielmann, I.; Ulbricht, W.; Zana, R.; Lang, J.; Tondre, C. Theory of the kinetics of micellar equilibria and quantitative interpretation of chemical relaxation studies of micellar solutions of ionic surfactants. *J. Phys. Chem.* **1976**, *80*, 905–922. [CrossRef]
44. Gaglione, R.; Cesaro, A.; Dell'Olmo, E.; Della Ventura, B.; Casillo, A.; Di Girolamo, R.; Velotta, R.; Notomista, E.; Veldhuizen, E.J.A.; Corsaro, M.M.; et al. Effects of human antimicrobial cryptides identified in apolipoprotein B depend on specific features of bacterial strains. *Sci. Rep.* **2019**, *9*, 6728. [CrossRef]
45. Rosenfeld, Y.; Shai, Y. Lipopolysaccharide (Endotoxin)-host defense antibacterial peptides interactions: Role in bacterial resistance and prevention of sepsis. *Biochim. Biophys. Acta Biomembr.* **2006**, *1758*, 1513–1522. [CrossRef]
46. Yu, L.; Tan, M.; Ho, B.; Ding, J.L.; Wohland, T. Determination of critical micelle concentrations and aggregation numbers by fluorescence correlation spectroscopy: Aggregation of a lipopolysaccharide. *Anal. Chim. Acta* **2006**, *556*, 216–225. [CrossRef]
47. Saravanan, R.; Holdbrook, D.A.; Petrlova, J.; Singh, S.; Berglund, N.A.; Choong, Y.K.; Kjellström, S.; Bond, P.J.; Malmsten, M.; Schmidtchen, A. Structural basis for endotoxin neutralisation and anti-inflammatory activity of thrombin-derived C-terminal peptides. *Nat. Commun.* **2018**, *9*, 2762. [CrossRef]
48. Claxton, N.S.; Fellers, T.J.; Davidson, M.W. Microscopy, Confocal. In *Encyclopedia of Medical Devices and Instrumentation*; Webster, J.G., Ed.; John Wiley & Sons, Inc.: Hoboken, NJ, USA, 2016; pp. 449–477. [CrossRef]
49. Dolman, N.J.; Kilgore, J.A.; Davidson, M.W. A review of reagents for fluorescence microscopy of cellular compartments and structures, part I: BacMam labeling and reagents for vesicular structures. *Curr. Protoc. Cytom.* **2013**, *65*, 1–27. [CrossRef]
50. Nogueira, E.; Cruz, C.F.; Loureiro, A.; Nogueira, P.; Freitas, J.; Moreira, A.; Carmo, A.M.; Gomes, A.C.; Preto, A.; Cavaco-Paulo, A. Assessment of liposome disruption to quantify drug delivery in vitro. *Biochim. Biophys. Acta Biomembr.* **2016**, *1858*, 163–167. [CrossRef]
51. Gargotti, M.; Lopez-Gonzalez, U.; Byrne, H.J.; Casey, A. Comparative studies of cellular viability levels on 2D and 3D in vitro culture matrices. *Cytotechnology* **2018**, *70*, 261–273. [CrossRef]
52. Dinnen, R.D.; Drew, L.; Petrylak, D.P.; Mao, Y.; Cassai, N.; Szmulewicz, J.; Brandt-Rauf, P.; Fine, R.L. Activation of Targeted Necrosis by a p53 Peptide: A Novel Death Pathway That Circumvents Apoptotic Resistance. *J. Biol. Chem.* **2007**, *282*, 26675–26686. [CrossRef]
53. Elmore, S. Apoptosis: A Review of Programmed Cell Death. *Toxicol. Pathol.* **2007**, *35*, 495–516. [CrossRef]
54. Mana, G.; Valdembri, D.; Serini, G. Conformationally active integrin endocytosis and traffic: Why, where, when and how? *Biochem. Soc. Trans.* **2020**, *48*, 83–93. [CrossRef]
55. Buckley, C.D.; Pilling, D.; Henriquez, N.V.; Parsonage, G.; Threlfall, K.; Scheel-Toellner, D.; Simmons, D.L.; Akbar, A.N.; Lord, J.M.; Salmon, M. RGD peptides induce apoptosis by direct caspase-3 activation. *Nature* **1999**, *397*, 534–539. [CrossRef]
56. Webb, J.N.; Koufos, E.; Brown, A.C. Inhibition of Bacterial Toxin Activity by the Nuclear Stain, DRAQ5^{TM}. *J. Membr. Biol.* **2016**, *249*, 503–511. [CrossRef]

57. Nguyen, D.P.; Elliott, T.; Holt, M.; Muir, T.W.; Chin, J.W. Genetically encoded 1,2-aminothiols facilitate rapid and site-specific protein labeling via a bio-orthogonal cyanobenzothiazole condensation. *J. Am. Chem. Soc.* **2011**, *133*, 11418–11421. [CrossRef]
58. Yuan, Y.; Wang, X.; Mei, B.; Zhang, D.; Tang, A.; An, L.; He, X.; Jiang, J.; Liang, G. Labeling thiols on proteins, living cells, and tissues with enhanced emission induced by FRET. *Sci. Rep.* **2013**, *3*, 3523. [CrossRef]
59. Koniev, O.; Wagner, A. Developments and recent advancements in the field of endogenous amino acid selective bond forming reactions for bioconjugation. *Chem. Soc. Rev.* **2015**, *44*, 5495–5551. [CrossRef]
60. Takakura, H. Molecular Design of d-Luciferin-Based Bioluminescence and 1,2-Dioxetane-Based Chemiluminescence Substrates for Altered Output Wavelength and Detecting Various Molecules. *Molecules* **2021**, *26*, 1618. [CrossRef]
61. Sharma, D.K.; Adams, S.T.; Liebmann, K.L.; Miller, S.C. Rapid Access to a Broad Range of 6′-Substituted Firefly Luciferin Analogues Reveals Surprising Emitters and Inhibitors. *Org. Lett.* **2017**, *19*, 5836–5839. [CrossRef]
62. Mofford, D.M.; Reddy, G.R.; Miller, S.C. Aminoluciferins extend firefly luciferase bioluminescence into the near-infrared and can be preferred substrates over d-luciferin. *J. Am. Chem. Soc.* **2014**, *136*, 13277–13282. [CrossRef] [PubMed]
63. Takakura, H.; Kojima, R.; Ozawa, T.; Nagano, T.; Urano, Y. Development of 5′- and 7′-Substituted Luciferin Analogues as Acid-Tolerant Substrates of Firefly Luciferase. *ChemBioChem* **2012**, *13*, 1424–1427. [CrossRef] [PubMed]
64. Fery-Forgues, S.; Lavabre, D. Are Fluorescence Quantum Yields So Tricky to Measure? A Demonstration Using Familiar Stationery Products. *J. Chem. Educ.* **1999**, *76*, 1260. [CrossRef]

International Journal of
Molecular Sciences

Article

Conformational Preferences and Antiproliferative Activity of Peptidomimetics Containing Methyl 1′-Aminoferrocene-1-carboxylate and Turn-Forming Homo- and Heterochiral Pro-Ala Motifs

Monika Kovačević [1], Mojca Čakić Semenčić [1], Kristina Radošević [2], Krešimir Molčanov [3], Sunčica Roca [4], Lucija Šimunović [1], Ivan Kodrin [5,*] and Lidija Barišić [1,*]

1. Department of Chemistry and Biochemistry, Faculty of Food Technology and Biotechnology, University of Zagreb, 10000 Zagreb, Croatia; monika.kovacevic@pbf.hr (M.K.); mcakic@pbf.hr (M.Č.S.); lsimunovic@pbf.hr (L.Š.)
2. Department of Biochemical Engineering, Faculty of Food Technology and Biotechnology, University of Zagreb, 10000 Zagreb, Croatia; kradosev@pbf.hr
3. Division of Physical Chemistry, Ruđer Bošković Institute, 10000 Zagreb, Croatia; Kresimir.Molcanov@irb.hr
4. NMR Centre, Ruđer Bošković Institute, 10000 Zagreb, Croatia; sroca@irb.hr
5. Department of Organic Chemistry, Faculty of Science, University of Zagreb, 10000 Zagreb, Croatia

* Correspondence: ikodrin@chem.pmf.hr (I.K.); lidija.barisic@pbf.hr (L.B.); Tel.: +385-1-4606-403 (I.K.); +385-1-4605-069 (L.B.)

Citation: Kovačević, M.; Čakić Semenčić, M.; Radošević, K.; Molčanov, K.; Roca, S.; Šimunović, L.; Kodrin, I.; Barišić, L. Conformational Preferences and Antiproliferative Activity of Peptidomimetics Containing Methyl 1′-Aminoferrocene-1-carboxylate and Turn-Forming Homo- and Heterochiral Pro-Ala Motifs. *Int. J. Mol. Sci.* **2021**, *22*, 13532. https://doi.org/10.3390/ijms222413532

Academic Editors: Menotti Ruvo and Nunzianna Doti

Received: 24 November 2021
Accepted: 14 December 2021
Published: 16 December 2021

Abstract: The concept of peptidomimetics is based on structural modifications of natural peptides that aim not only to mimic their 3D shape and biological function, but also to reduce their limitations. The peptidomimetic approach is used in medicinal chemistry to develop drug-like compounds that are more active and selective than natural peptides and have fewer side effects. One of the synthetic strategies for obtaining peptidomimetics involves mimicking peptide α-helices, β-sheets or turns. Turns are usually located on the protein surface where they interact with various receptors and are therefore involved in numerous biological events. Among the various synthetic tools for turn mimetic design reported so far, our group uses an approach based on the insertion of different ferrocene templates into the peptide backbone that both induce turn formation and reduce conformational flexibility. Here, we conjugated methyl 1′-aminoferrocene-carboxylate with homo- and heterochiral Pro-Ala dipeptides to investigate the turn formation potential and antiproliferative properties of the resulting peptidomimetics **2–5**. Detailed spectroscopic (IR, NMR, CD), X-ray and DFT studies showed that the heterochiral conjugates **2** and **3** were more suitable for the formation of β-turns. Cell viability study, clonogenic assay and cell death analysis showed the highest biological potential of homochiral peptide **4**.

Keywords: antiproliferative activity; chirality; conformational analysis; density functional theory (DFT); ferrocene; hydrogen bonds; peptidomimetic; X-ray

1. Introduction

Despite their enormous biological importance and drug-like properties, the medical use of peptides is still limited by their poor proteolytic stability, poor absorption, and low selectivity. Peptidomimetics, that is *"compounds whose essential elements (pharmacophore) mimic a natural peptide or protein in 3D space and which retain the ability to interact with the biological target and produce the same biological effect"* [1] are an efficient answer to these drawbacks. In the last four decades, the concept of peptidomimetics, i.e., the art of transforming peptides into drugs, has emerged as the powerful tool in medicinal chemistry [2].

One of the synthetic approaches in the development and optimization of peptidomimetics is based on mimicking the peptide secondary structures (α-helices, β-sheets or turns)

involved in protein-protein interactions (PPIs) [3]. Within a cell, PPIs form an "interactome", an intricate network involved in physiological and pathological processes such as signal transduction, cell proliferation, growth, differentiation, apoptosis, etc. Protein-protein interfaces have bioactive "hotspots" consisting of four to eight amino acid segments, and half of them are arranged in turns [4]. The development of mimetics of PPIs "hotspot" regions that act as modulators or inhibitors of PPIs is a promising strategy for drug discovery [5].

Turns are the protein sites mainly composed of Asn, Gly and Pro where the polypeptide chain folds back on itself, making the proteins compact and globular. Since the turns are usually located at the protein surface, they are exposed to cell receptors and therefore involved in biological interactions [6]. Depending on their length and hydrogen bonding pattern, turns are classified as α-(13-membered hydrogen bonded (HB) ring), β-(10-membered HB ring) and γ-turns (7-membered HB ring). Recently, Trabocchi and Lenci [2] reviewed several conceptually different synthetic toolboxes for the design of β-turns that are involved in numerous biological recognition processes, such as peptide-antibody interactions, and recognition between peptide ligands and proteins. The turn inducing elements approach is based on the replacement of the amino acid at $i + 1$ and/or $i + 2$ with an element that both induces the formation of the turn and reduces the conformational flexibility, while the small molecular scaffolds as structural mimetics approach involves the replacement of the entire peptide backbone with the rigid scaffold that allows the alignment of the side chains in a spatial arrangement corresponding to the peptide turn residues.

Using the first approach, our group has made serious efforts in the synthesis and conformational analysis of ferrocene-containing peptidomimetics [7–14]. Due to the distance between the cyclopentadienyl (Cp) rings of 3.3 Å, the peptide chains when attached to the 1,1′-disubstituted ferrocene templates, i.e., -NH-Fn-CO- and -NH-Fn-NH- (Fn = ferrocenylene), come close enough to form 12- (**I**) [7–12] and 14-membered interstrand hydrogen-bonded rings (**II**) [13,14], respectively, in symmetrically disubstituted ferrocene peptides (Figure 1). We have therefore shown that 1,1′-disubstituted ferrocenes are capable to nucleate β-turns and β-sheet-like structures upon conjugation with amino acids and short peptides.

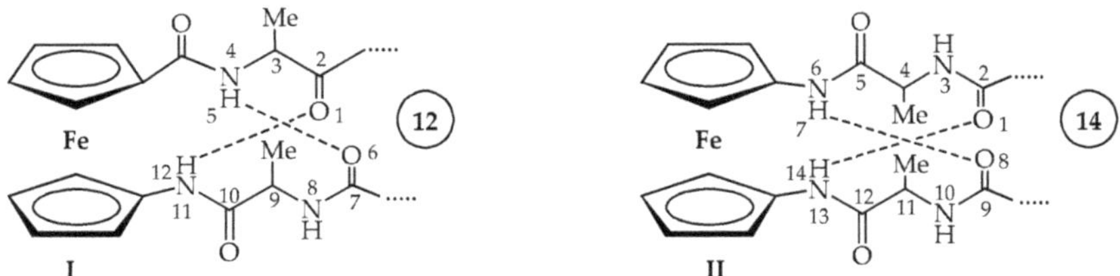

Figure 1. The symmetrically disubstituted peptidomimetics containing turn-inducing ferrocene templates -NH-Fn-CO- (**I**) and -NH-Fn-NH- (**II**).

In 2018, Moriuchi et al. gave a review of previously synthesized symmetrically disubstituted ferrocene-dipeptide conjugates that adopt β- and γ-turn-like structures [15].

To investigate whether the asymmetrically disubstituted ferrocene conjugates with amino acids are involved in hydrogen-bonded turns, we first prepared the conjugates of methyl 1′-aminoferrocene-carboxylate with Ala (**III**) [16] and Pro (**IV**) [17] (Figure 2). Detailed spectroscopic analysis revealed two different conformational patterns consisting of seven-membered intra- (**A**, γ-turn) and nine-membered interstrand hydrogen bonded rings (**B**) in the Ala-dipeptides **III**, whereas the Pro-dipeptides **IV** adopted pattern **A** only. It was found that the different bulkiness and basicity of the Boc and Ac group affected the hydrogen-bonding patterns to some extent.

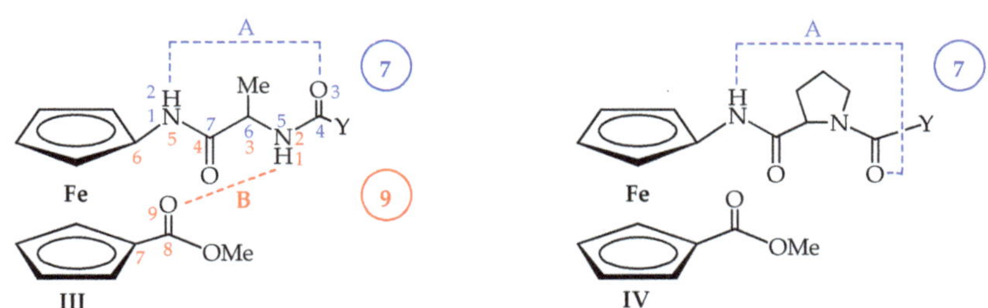

Figure 2. The asymmetrically disubstituted peptidomimetics (**III**) and (**IV**) containing turn-inducing ferrocene template -NH-Fn-CO- (Y = Boc or Ac).

Our previous studies have shown that even the monosubstituted conjugates between the ferrocene and chiral amino acids or short peptides may induce a different sign in the circular dichroism spectra (CD) near the absorption maximum of a ferrocene chromophore (around 470 nm) [18–20]. We have pointed out a strong correlation between the sign of the Cotton effect and the sign of the dihedral angle between two planes, one containing the cyclopentadienyl ring and the other containing amide bond [18–20]. Predominance of a specific conformer arises from the amino acid sequence, which triggers a different helicity of the folded peptide through intramolecular interactions, primarily hydrogen bonds. The second substituent connected to the opposite cyclopentadienyl ring adds additional hydrogen bond donor and acceptor sites, resulting with more rigid structure through the subsequent formation of intra-or interstrand hydrogen bonds (HBs) inducing the helical chirality of the ferrocene moiety by restricting the torsional twist about the Cp(centroid)-Fe-Cp(centroid) axis. The helical arrangement (*M*- or *P*-) of the ferrocene moiety depended on the chirality of the bound amino acids or peptides. In addition, the peptide sequence, backbone homo-or heterochirality, hydrogen bond-acceptor potential of the *N*-terminal groups, and hydrogen-bonding patterns were found to regulate the size of the hydrogen-bonded ring, i.e., the type of turn.

Therefore, we decided to investigate the conformational consequences of introduction of additional hydrogen bond donor and acceptor via Ac/Boc-L-Pro and Ac/Boc-D-Pro sequence at the *N*-terminus of peptide 1. Considering that (i) the most favourable conformation of Pro is in a tight turn [21] and (ii) Pro-Xaa sequence is recognised as a β-turn sequence [22–24], we naturally expected an altered conformational space with more complex intramolecular hydrogen bond (IHB) patterns of the new conjugates **2**–**5** in comparison with previously synthesized compounds with only one amino acid.

The antitumor activity of ferrocenes was first reported in 1978 [25]. Since then, various ferrocene compounds have been investigated as candidates for anticancer, antibacterial, antifungal and antiparasitic drugs [26,27]. A review of the anticancer activity of ferrocene hybrids with amino acids/peptides, azoles, chalcones, coumarins, indoles, steroids, sugars, etc. was recently given by Wang et al. Due to the ability of drug-amino acid/peptide hybrids to overcome multi-drug resistance in chemotherapy and bind to specific receptors expressed on cancer cells, ferrocene-amino acid/peptide hybrids could be used to identify new anticancer agents [28]. With this in mind, we decided to investigate conjugates **2**–**5** for their antitumor activity.

2. Results and Discussion

2.1. *Synthesis of Peptides* 2–5

The previously established simple and efficient synthetic route to ferrocene-containing peptides [13,14,16,17] was applied here to obtain peptides **2**–**5** (Scheme 1). Boc-deprotection of Boc-L-Ala-NH-Fn-COOMe **1** [16] in the presence of gaseous HCl gave a hydrochloride salt, which was processed with an excess of NEt_3 to give the free amine required for the coupling step. Then, C-activated Boc-L-Pro-OH and Boc-D-Pro-OH were added to the unstable amine to obtain diastereomeric Boc-peptides **2** and **4**, respectively. Conversion of carbamates **2** and **4** to acetamides **3** and **5** was accomplished by (i) acidic Boc-deprotection and (ii) Ac-protection in the presence of acetyl chloride [14]. The characterization data with IR, NMR, and MS spectra of conjugates **2–5** can be found in the Supplementary Material, Figures S2–S51.

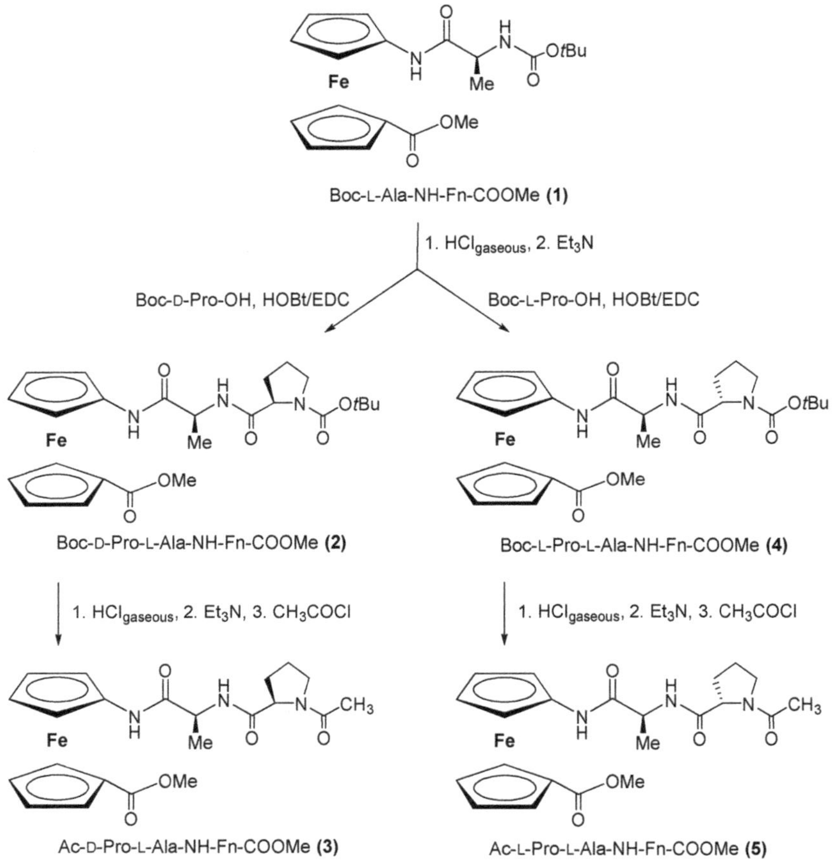

Boc-L-Ala-NH-Fn-COOMe **(1)**

1. $HCl_{gaseous}$, 2. Et_3N

Boc-D-Pro-OH, HOBt/EDC

Boc-L-Pro-OH, HOBt/EDC

Boc-D-Pro-L-Ala-NH-Fn-COOMe **(2)**

Boc-L-Pro-L-Ala-NH-Fn-COOMe **(4)**

1. $HCl_{gaseous}$, 2. Et_3N, 3. CH_3COCl

1. $HCl_{gaseous}$, 2. Et_3N, 3. CH_3COCl

Ac-D-Pro-L-Ala-NH-Fn-COOMe **(3)**

Ac-L-Pro-L-Ala-NH-Fn-COOMe **(5)**

Scheme 1. Synthesis of Boc-(**2**,**4**) and Ac-protected peptides (**3**,**5**) containing homo-and heterochiral Pro-Ala sequences.

2.2. Computational Study

As we have already mentioned, previously investigated conjugates **III** [16] and **IV** [17] show two different HB patterns, the one including 7-membered intrastrand and 9-membered interstrand HB rings in Ala amino acid derivatives **III**, and other with only one 7-membered ring in Pro amino acid derivatives **IV** (Figure 2). Without changing the ester substituent on one Cp ring, we can simply modify the type and the number of hydrogen bond donor and acceptor groups by incorporating an additional amino acid, e.g., Pro, to test the robustness of the existing hydrogen bond patterns. We expected formation of 10-membered rings because Pro-Xaa sequence is known as good β-turn-inducer.

A detailed conformational analysis of four newly synthesized compounds was performed hierarchically, starting from molecular mechanics and finishing with optimization of the most stable conformers in implicitly modelled solvent (SMD) with the B3LYP-D3 functional and $6-311+G(d,p)$ basis set, LanL2DZ for iron. Additionally, all hydrogen bonds were confirmed based on the Quantum theory of atoms in molecules (QTAIM) analysis of the bond critical points between hydrogen bond acceptors and hydrogens (more details in Materials and Methods). The results of the computational study are displayed in Figures 3–5 and in Table S1 in Supplementary Materials.

The heterochiral D-Pro-L-Ala sequence in compounds **2** and **3** promotes formation of two hydrogen bonds (pattern **A**). The one is an intrastrand $NH_{Fc}\cdots O{=}C_{Boc/Ac}$ HB, which is exactly a 10-membered ring (β-turn) as expected from Pro-Ala sequence. The other is interstrand $NH_{Ala}\cdots O{=}C_{COOMe}$ HB that forms a 9-membered ring. The same IHB pattern labelled as pattern **A** is formed no matter of a relative orientation of the second Cp ring because ester group can freely rotate to accommodate best position on both sides to establish the same type of the interstrand hydrogen bond, $NH_{Ala}\cdots O{=}C_{COOMe}$, in conformers with both type of helicity (*P*- in **2–1**, **2–2**, **3–1** and **3–4**; while *M*- in **2–3**, **3–2** and **3–3**). Many energetically close conformers differ only by puckering modes of pyrrolidine ring.

In comparison, homochiral L-Pro-L-Ala sequence favours formation of *M*-helical peptides and it is more dependent on the type of the protecting group attached to *N*-terminus. The most stable conformer of both **4** and **5** has an appropriate relative arrangement of both substituents to form interstrand 9-membered ring connected through $NH_{Ala}\cdots O{=}C_{COOMe}$ hydrogen bond (IHB pattern **B**). However, bulkier *tert*-butyl (Boc-protection) group in comparison with smaller methyl (Ac-protection) prevents formation of an additional 10-membered ring, described as IHB pattern A that is observed in **4–1**, but not in other conformers of **4** and **5**. The other, less stable conformers of **4** and **5** are folded under the influence of the same types of $NH_{Ala}\cdots O{=}C_{COOMe}$ (IHB pattern B) and $NH_{Fc}\cdots O{=}C_{Boc/Ac}$ (IHB pattern C) hydrogen bonds, but only when acting individually.

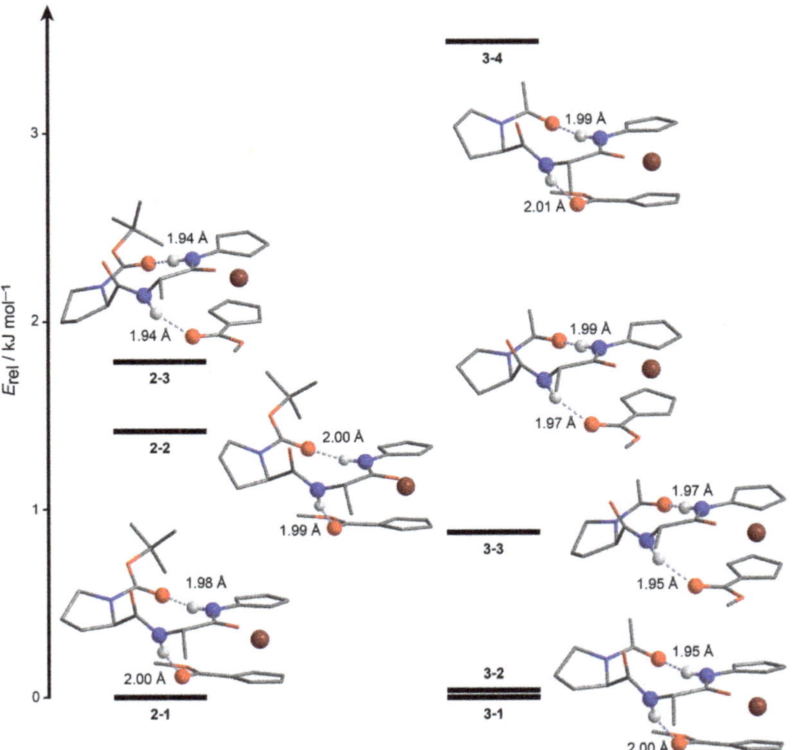

Figure 3. DFT optimized geometries of the most stable conformers of heterochiral peptides **2** and **3**. Dashed lines represent hydrogen bonds confirmed by the QTAIM analysis. Nonpolar hydrogen atoms are omitted for clarity.

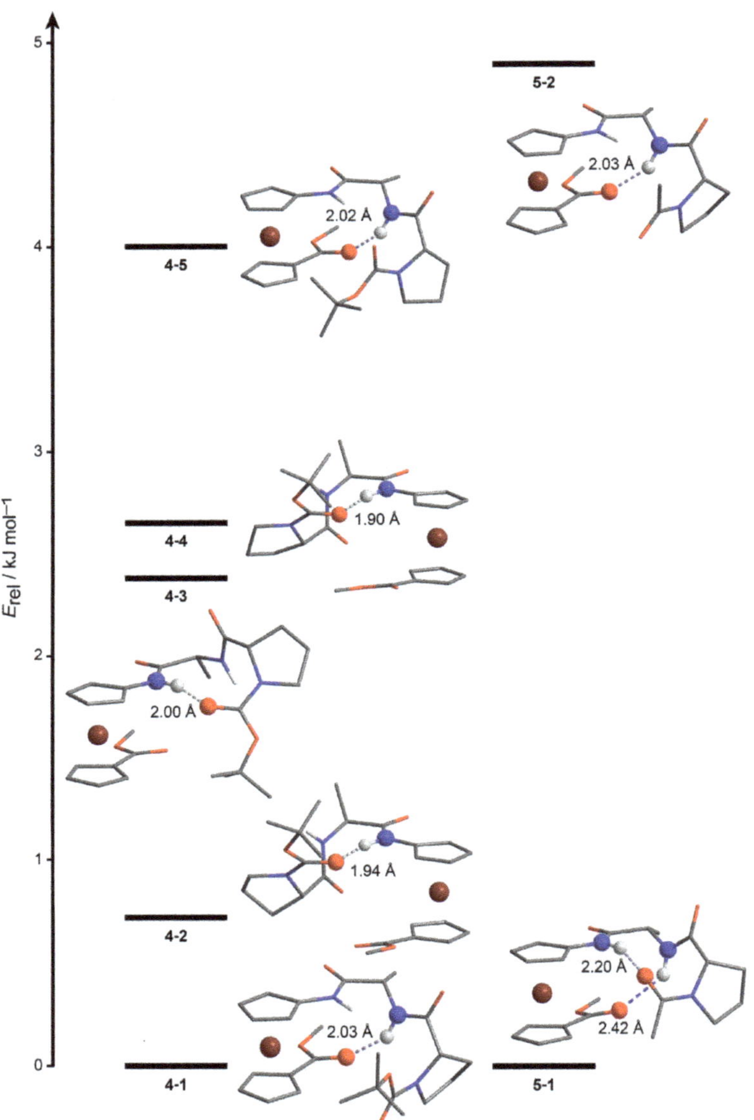

Figure 4. DFT optimized geometries of the most stable conformers of homochiral peptides **4** and **5**. Dashed lines represent hydrogen bonds confirmed by the QTAIM analysis. Nonpolar hydrogen atoms are omitted for clarity.

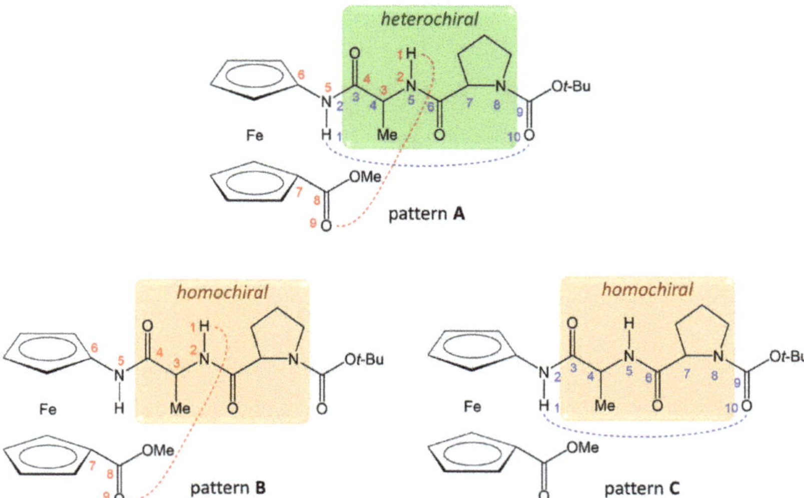

Figure 5. Intramolecular hydrogen bond (IHB) patterns (labelled as **A**, **B** and **C**) determined in DFT optimized geometries of the most stable conformers of **2–5**. Atom-numbering scheme for interstrand 9-(red colour) and intrastrand 10-membered (blue colour) rings. IHB **A** consists of two hydrogen-bonded rings, while IHBs **B** and **C** consist of one hydrogen-bonded ring.

2.3. IR Spectroscopy

Our next goal was to connect the results of the computational analysis with experimental data, especially with the determined HB patterns predicted by DFT in peptides **2–5**. HBs have a significant effect on the IR spectrum causing a red shift and increase in the intensity of the X–H stretching frequency when X–H···Y hydrogen bonds are formed [29,30]. Therefore, a closer look at the amide A region in the IR spectra of the studied compounds revealed the presence of free (~3420 cm^{-1}) and associated (~3300–3325 cm^{-1}) NH groups, and the red-shifted stretching frequencies of the $CO_{Ac/Boc}$ groups strongly suggest their participation in HBs [31] (Figure 6a,b and Figures S12, S24, S36 and S48 in Supplementary Materials). The ratios of free and associated NH bands in homochiral peptides **4** and **5** (~1:1.5) (Figure 6b) and heterochiral peptides **2** and **3** (~1:2.4) (Figure 6a) suggest a higher extent of hydrogen bonding in heterochiral conjugates. The intramolecular nature of the hydrogen bonds was proved by the concentration-independent IR spectra (Figure 6a,b). Otherwise, the concentration dependence of the IR spectra, i.e., the decrease in the intensities of intermolecularly engaged NH groups would be observed upon dilution. The domination of associated NH groups was also observed in the solid state of compounds **2–5** (Figure 6c and Figures S13, S25, S37 and S49 in Supplementary Materials).

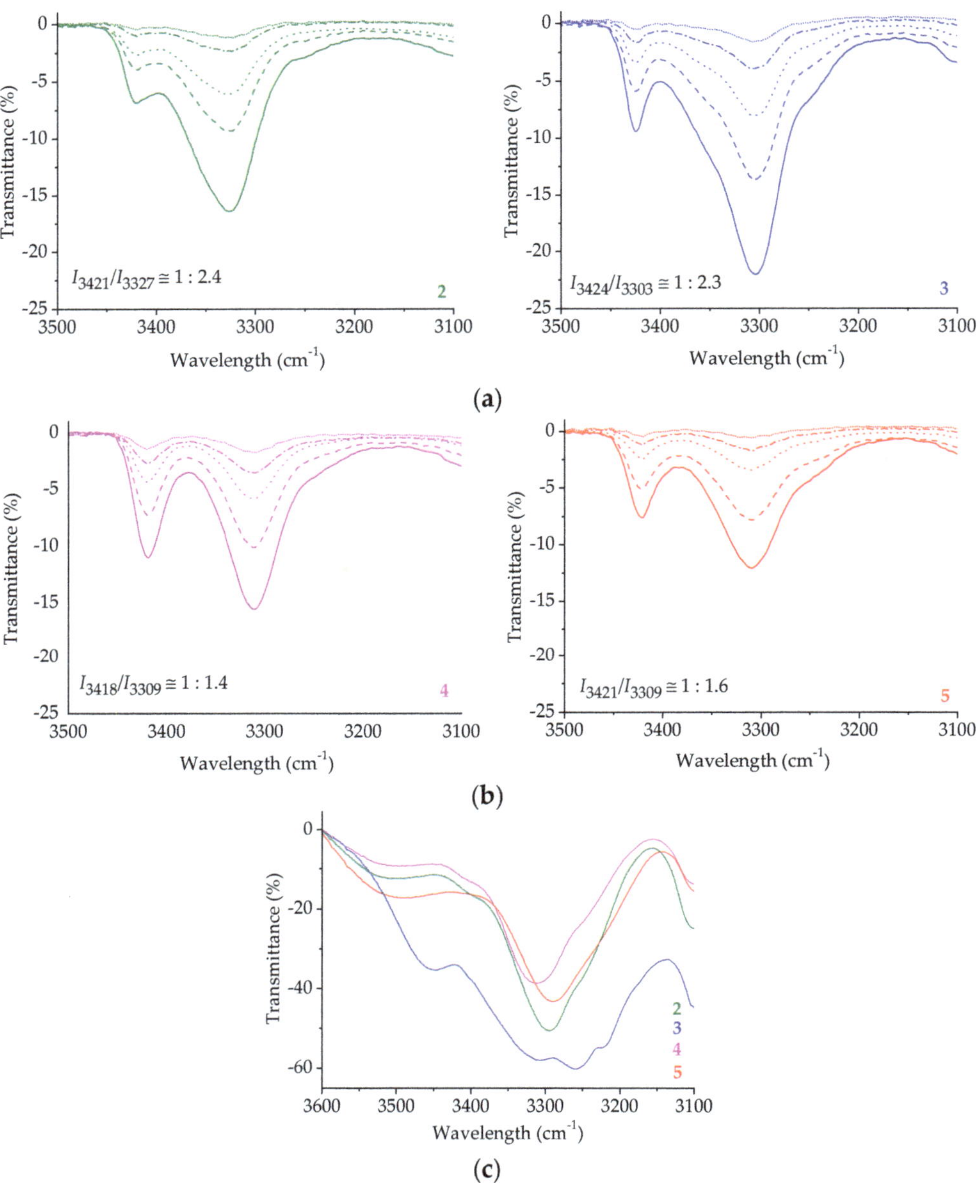

Figure 6. The NH stretching vibrations in concentration-dependent IR spectra of (**a**) hetero-(**2**, **3**) and (**b**) homochiral peptides (**4**, **5**) in CH_2Cl_2 ((—) $c = 5 \times 10^{-2}$ M, (——) $c = 2.5 \times 10^{-2}$ M, (· · ·) $c = 1.25 \times 10^{-2}$ M, (——), $c = 6.13 \times 10^{-3}$ M, (·····), $c = 3 \times 10^{-3}$ M) and ratios of free and associated NH bands, (**c**) the NH stretching vibrations of compounds **2–5** (2 mg) in KBr (200 mg).

2.4. NMR Spectroscopy

NMR spectroscopy is a powerful tool for studying the structure and interactions of peptides and proteins, allowing the identification of bioactive conformations responsible for their drug-like properties. An overview of NMR methods for obtaining the 3D structure of small, unlabelled peptides was recently provided by Vincenzi et al. [32]. Here, we performed detailed 1D (^{1}H, ^{13}C) and 2D NMR studies (^{1}H-^{1}H COSY, ^{1}H-^{1}H NOESY,

^{1}H-^{13}C HMQC, and ^{1}H-^{13}C HMBC) to assign the proton resonances and determine the individual hydrogen bonds and their strength.

Due to the hydrogen bonding deshielding, the resonances of the involved amide protons are downfield shifted (δ > 7 ppm), and the higher values of the chemical shifts indicate stronger hydrogen bonding [33]. Since the NH_{Fn} resonances of the tested peptides are significantly downfield shifted (δ ~ 8–8.6 ppm) compared to NH_{Ala} (δ ~ 6.9–7.2 ppm), their participation in stronger HBs is suggested (Figure 7, Figures S4, S16, S28 and S40 in Supplementary Materials).

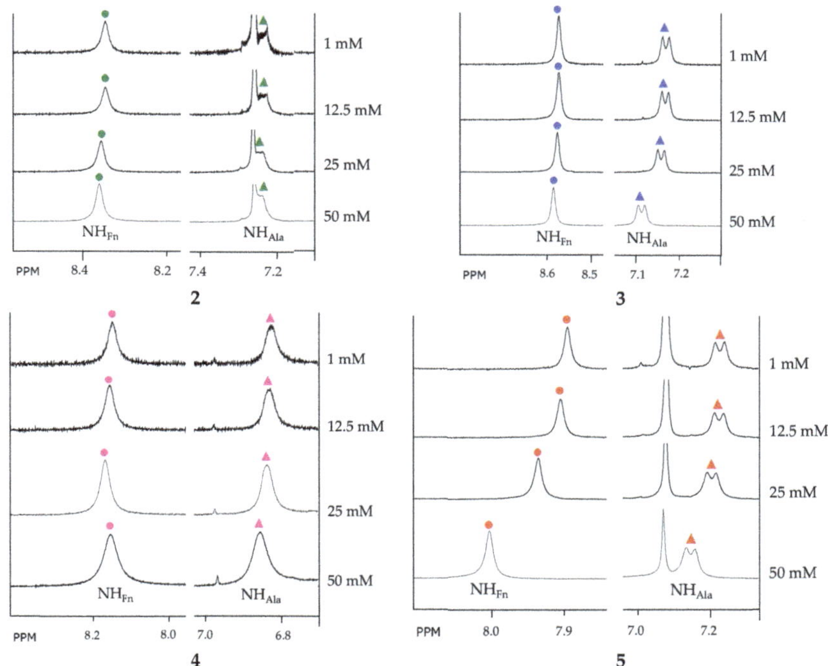

Figure 7. Concentration-dependent NH chemical shifts of hetero-(**2**, **3**) and homochiral peptides (**4**, **5**).

Next, concentration-, temperature- and solvent-dependent NMR spectroscopy was performed to obtain further details about the conformational space of the tested conjugates. NH_{Fn} and NH_{Ala} showed no significant upfield shifts ($\delta < 0.1$ ppm) at high (50 mM) vs. low concentrations (1 mM) (Figure 7), supporting their involvement in IHBs, as suggested by the concentration-independent IR data (Figure 6a,b).

We have further estimated the stability of the intramolecularly hydrogen-bonded structures in peptides **2**–**5** by examining the temperature dependence of the shifts of the amide protons. A smaller shift corresponds to a more stable conformation [34]. The signals of NH_{Fn} and NH_{Ala} of heterochiral peptide **3** showed the smallest upfield shifts compared to those of peptides **2**, **4**, and **5** (with the exception of NH_{Ala} from **4** which was slightly downfield shifted) (Figure 8). Therefore, the heterochiral Ac-peptide **3** is expected to adopt the most stable conformations compared to its counterparts, where both NHs are involved in relatively strong hydrogen bonds.

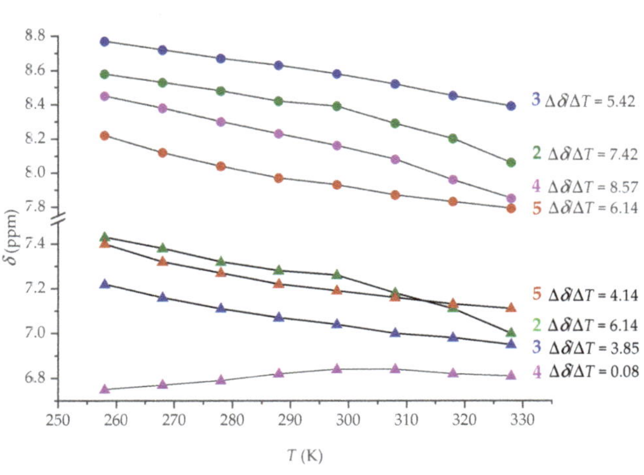

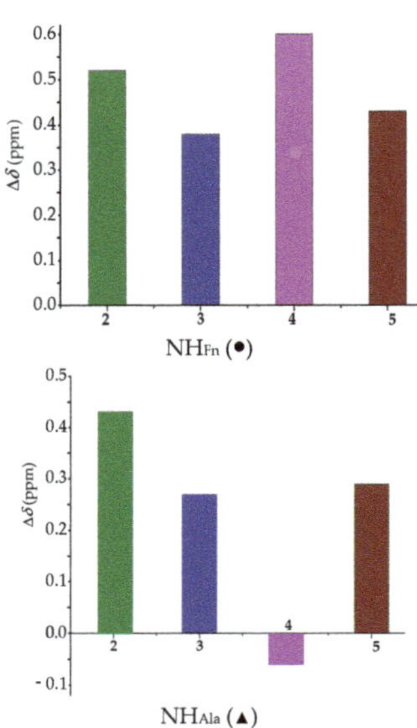

Figure 8. Changes in chemical shifts (Δδ) of NH_{Fn} (•) and NH_{Ala} (▲) in peptides **2–5** in $CDCl_3$ from 258–328 K ($c = 1 \times 10^{-3}$ M in $CDCl_3$).

Temperature coefficients ($\Delta\delta/\Delta T$), i.e., the variations in the chemical shifts of the amide protons with temperature, imply if the NH groups are exposed to or shielded (hydrogen bonded) from the solvent, and therefore provide information about hydrogen bonding. While low $\Delta\delta/\Delta T$ values (-2.4 ± 0.5 ppb K^{-1}) correspond to either shielded protons which were initially downfield shifted, or protons exposed to $CDCl_3$, larger $\Delta\delta/\Delta T$ values always reflect initially shielded NH protons exposed during unfolding of intramolecularly hydrogen-bonded structures or dissociation of aggregates upon heating. [13,14,35–43] Since the concentration-independent IR and NMR spectra excluded the self-assembly, the observed larger temperature coefficients are an additional confirmation of the intramolecularly folded conformations in peptides **2–5**.

The multiple resonances of the amide protons at lower temperatures indicate a slow *cis/trans* isomerization of the proline imide bond [37,44,45]. At higher temperatures, the rate of isomerization increases, and the signals of the amide protons involved in weak hydrogen bonds coalesce. Conversely, the slow proline isomerization and decreased coalescence occur at high temperatures when the amide protons are involved in strong HBs that can induce isomer locking [46].

The multiple resonances observed for NH_{Fn} and NH_{Ala} of peptides **2**, **4** and **5** at 258 K and the coalescence that occurred upon subsequent heating to 328 K are consistent with their involvement in weak HBs. Also, the absence of multiple resonances of the amide protons of the heterochiral Ac-peptide **3** at lower temperatures is an additional confirmation of its involvement in strong IHBs and hydrogen-bond-induced folding into a stable turn structure (see Supplementary Materials Figures S11, S23, S35, S47 and S50).

The chemical shifts of the four NH_{Fn} and four NH_{Ala} showed different variations upon titration of $CDCl_3$ solution with DMSO-d_6 (Figure 9). Although these protons are involved in IHBs, they showed downfield shift with increasing DMSO content, which could be due to exposure to the hydrogen-bond-accepting solvent. The NH_{Ala} of the tested peptides showed a high degree of solvent sensitivity (Δδ ~ 0.94–1.46 ppm), which is probably due to their involvement in weak HBs. However, increasing the DMSO content from 0–56% affected the NH_{Fn} of heterochiral peptides **2** and **3** much less (Δδ ~ 0.3–0.45 ppm) than those of homochiral peptides **4** and **5** (Δδ ~ 1 ppm), indicating their involvement in stronger IHBs and folding into more stable turn structures.

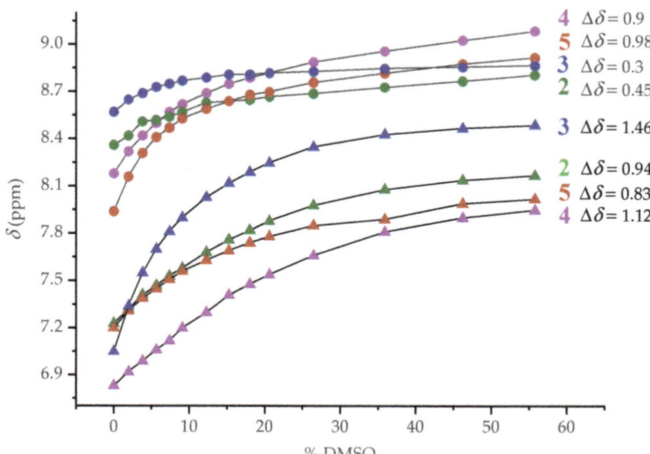

Figure 9. Solvent dependence of of NH_{Fn} (•) and NH_{Ala} (▲) chemical shifts of peptides **2**–**5** at increasing concentrations of d_6-DMSO in $CDCl_3$ (c = 25 mM, 298 K) to probe exposed vs. hydrogen-bonded amides.

As for the *cis/trans* proline isomerization, it is expected that when a nonpolar or less polar solvent is used, the *trans* isomer is more pronounced [14,17,47,48]. Moreover, the *trans* isomer will dominate when the IHBs are formed [14,17,49]. When $CDCl_3$ solutions of Boc-peptides **2** and **4** were titrated with DMSO, the population of the *cis* isomer almost reached the amount of the *trans* fraction, which could be indicative of their participation in weaker HBs. However, the addition of DMSO had a much smaller effect on heterochiral Ac-peptide **3**, which retained almost exclusively *trans* fraction in the presence of DMSO due to its involvement in strong IHBs (Supplementary Materials Figures S10, S22, S34, S46 and S51).

Two-dimensional NOESY spectroscopy was performed to further investigate the folded conformations of peptides **2**–**5** (Figure 10). We were focused on intrastrand NOE interactions between NH_{Fn} and NH_{Ala} with the *N*-terminal Boc or Ac group and on interstrand NOE interactions between NH_{Fn} and NH_{Ala} and the ester methyl group.

The observed NOE contacts between NH_{Fn} and Ac or Boc methyl protons of heterochiral peptides **2** and **3** clearly indicate the presence of intrastrand hydrogen bonds $NH_{Fn}\cdots O{=}C_{Boc/Ac}$ corresponding to β-turns. NOE contact between NH_{Ala} and the ester methyl group of peptides **3** is observed, indicating the presence of interstrand hydrogen bond $NH_{Ala}\cdots O{=}C_{COOMe}$ as in IHB pattern A with two hydrogen bonds (Figure 5).

As can be seen from the above 1D NMR data, the homochirality of the peptide backbone affects the conformational behaviour and makes the hydrogen bonding more sensitive and weaker compared to heterochiral analogues. The absence of NOE contact between NH_{Fn} and Boc methyl protons for homochiral peptide **4** suggests the presence of weaker HBs and lower degree of chiral organization compared to heterochiral peptides **2** and **3**. The NOEs between NH_{Fn}/NH_{Ala} and Ac-methyl protons of homochiral peptide **5** were not observed, indicating that the intramolecular hydrogen bonding in **5** is very weak and its structure is not helically ordered. These observations confirm the results of the computational study. Heterochiral peptides **4** and **5** equilibrate mostly between the conformations having either IHB pattern **B** or **C**, both with one hydrogen bond, that makes them more flexible for different orientation of the Boc/Ac protecting groups relative to NH. Based on NMR study, hydrogen bonding pattern **A** with two simultaneous intra- and interstrand HBs (Figure 5) is found only in heterochiral conjugate **3** which is therefore expected to have the most ordered chiral surrounding compared to its homologues.

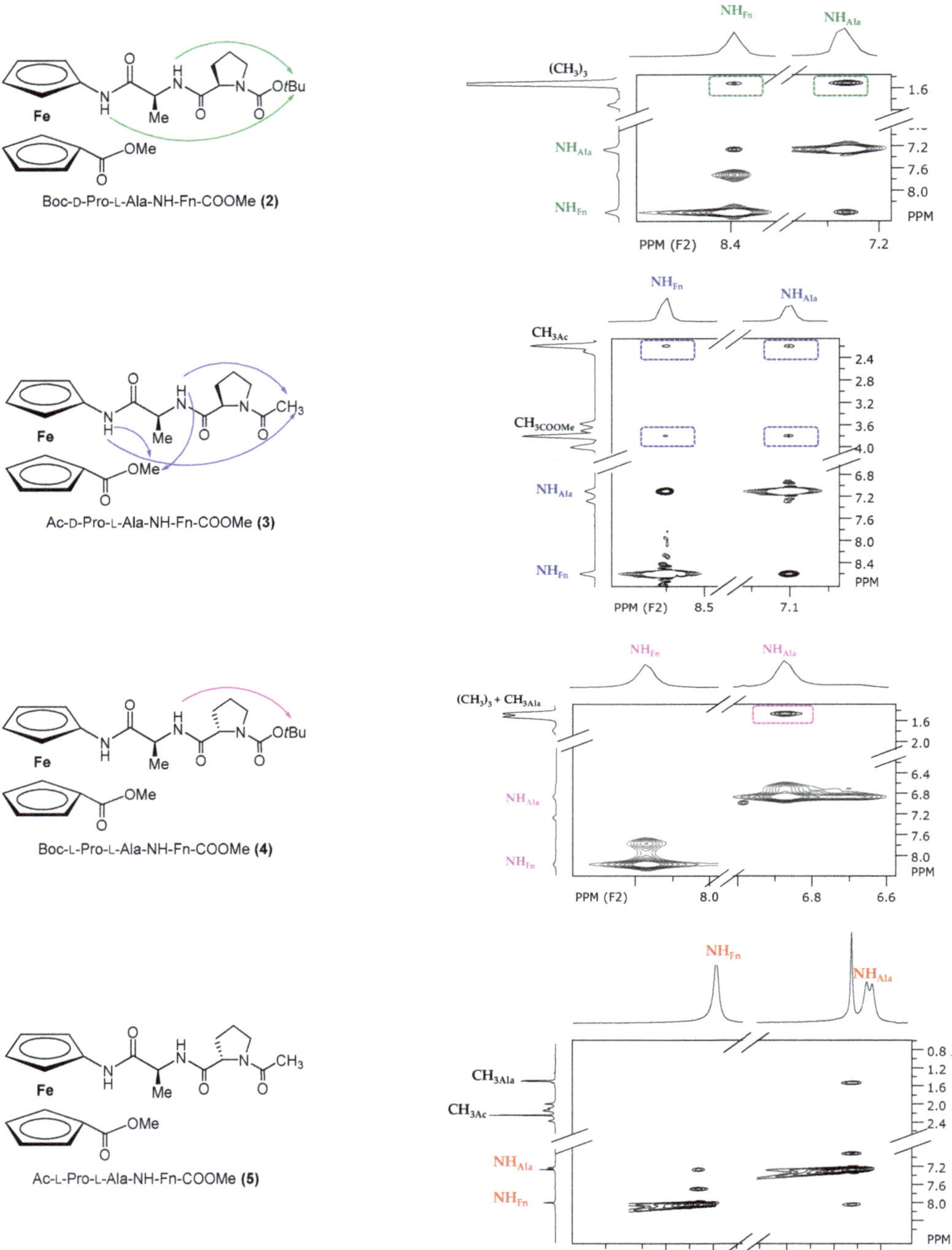

Figure 10. The intrastrand NOE contacts in the spectra of conjugates **2–4** and interstrand NOE contacts in the spectrum of conjugate **3** are depicted with arrows. The NOE contacts attributed to intra- and/or intrastrand HBs are not observed in spectrum of conjugate **5**.

2.5. CD Spectroscopy

Recent reviews describe CD (circular dichroism) as one of the most useful techniques for measuring conformational changes in the secondary and tertiary structures of peptides and proteins during aggregation, thermal or chemical unfolding, and ligand binding interactions [50,51].

When the ferrocene scaffold is inserted into a chiral peptide chain, the formation of turns stabilized by hydrogen bonds occurs, and β-sheet-like structure is formed. Consequently, restriction of the free rotation of the ferrocene rings gives rise to helical chirality of the ferrocene core and Cotton effects in the region of ferrocene-based transitions around 480 nm. The positive Cotton effects correspond to the *P*-helicity of the ferrocene unit, while *M*-conformers induce negative Cotton effects. The most pronounced CD activity (M_θ ~ 700,000 deg cm^2 dmol^{-1}) was measured for symmetrically disubstituted β-sheet-like mimetics **II** [13,14] composed of homo- and heterochiral Ala-Pro dipeptides bound to an -NH-Fn-NH- template and was attributed to a highly ordered chiral environment. Their conformational stability, realized by two strong interstrand hydrogen bonds, was confirmed by the preservation of more than 70% of CD activity upon titration with DMSO [14]. However, the noticeable loss of CD activity for asymmetrically disubstituted ferrocene peptides **III** [16] and **IV** [17] (M_θ ~ 500–800 deg cm^2 dmol^{-1}) is attributed to the reduction and weakening of hydrogen bonds. We have shown that even monosubstituted aminoferrocene incorporated at the *C*-terminus of di- and tripeptide sequences can sense the chiral environment (M_θ ~ 1000–2000 deg cm^2 dmol^{-1}) resulting from the turn structures established in the attached peptide fragment [19,20].

The CD silent ferrocene region in the spectrum of homochiral peptide **5** (Figure 11) confirms the absence of chiral order predicted by NMR. Since conformational analysis indicated the presence of the folded structures in peptides **2–4,** they were expected to show CD activity, but with different sign depending on their homo-or heterochirality. The negative Cotton effect (M_θ ~ 2500 deg cm^2 dmol^{-1}) for homochiral peptide **4** indicated *M*-helicity, whereas heterochiral peptides **2** and **3** showed almost 2-fold stronger positive Cotton effects (M_θ ~ 4200–4800 deg cm^2 dmol^{-1}) related to *P*-helicity of the ferrocene core (Figure 11a). These findings are in the agreement with the computational study. The most stable conformers of **2** and **3** have a significant contribution of the *P*-1,2′ helical conformations (Table S1 in Supplementary Materials) in the total population, thus resulting with a positive Cotton effect. However, *M*-1,1′ helical structures were determined to be the most abundant in derivatives **4** and **5**, especially in derivative **5** where all of the most stable conformers adopt this type of helicity. Smaller values of the pseudotorsion angles (due to 1,1′ relative orientation of the substituents on the opposite Cp rings) will result with a smaller intensity in CD spectrum. However, the conformers *M*-1,2′ that coexist in **4** will additionally enhance the negative Cotton effect.

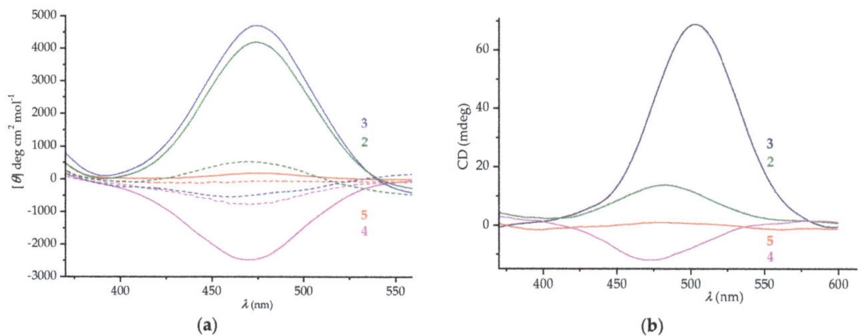

Figure 11. The Cotton effects in chirality-organized ferrocene peptides **2–4** (**a**) in solution (CH_2Cl_2, $c = 1 \times 10^{-3}$ M, (—) and CH_2Cl_2, $c = 1 \times 10^{-3}$ M containing 20% of DMSO (------)) and (**b**) in solid state (2 mg in 200 mg KBr).

As shown in our previous studies [14,18–20], the increase and strengthening of hydrogen bonding in peptides **2** and **3** leads to a higher degree of chiral organization compared to their counterpart **4** and analogues **III** [16] and **IV** [17]. It was expected that heterochiral peptides **2** and **3**, which showed less DMSO-induced shifts of NH_{Fn} compared to homochiral peptides **4** and **5**, would adopt a more stable turn conformations realized by intrastrand hydrogen bonds. However, it was found that all the tested peptides lost their CD activity in the presence of 20% DMSO, confirming our previous finding that the less stable conformations were generally formed by hydrogen bonding within the same strand [16–18]. Considering the hydrogen bonding potential of water (both as a donor and acceptor of hydrogen bonds), the changes in the hydrogen-bonding pattern of the peptides studied here are also expected in the water environment.

The CD activity of the peptides **2**–**5** in the solid state resembles their behaviour in the solution in terms of the sign of the Cotton effects. Here, the intensity of CD activity of heterochiral peptide **3** is much more pronounced compared to those of peptides **2**, **4**, and **5** than in the solution state, indicating the presence of a more stable turn conformation in the solid state (Figure 11b).

2.6. X-ray Crystal Structure Analysis

We applied the same crystallization procedure for all goal compounds, i.e., recrystallization from a solution of dichloromethane, chloroform, and ethyl acetate, but only compounds **2** and **5** gave single crystals of suitable quality for X-ray structural analysis.

The conformation in the solid state is usually affected by the anisotropic environment (i.e., the crystal field) and specific intermolecular interactions forming in the crystals, and often differs from the conformation in solution. The formation of medium-strong N–H···O hydrogen bonds (energies in the range 5–10 kcal mol^{-1}) is particularly favorable, and their energies (in the range of 5–10 kcal mol^{-1}) are high enough to affect the molecular conformation.

Compound **2** adopts the bent conformation with intermolecular hydrogen bond N1–H1···O3 (Figure 12), which is consistent with pattern A shown in Figure 5. It is probably supported by a weaker C5–H5···O3 hydrogen bond (Table S55 in Supplementary Materials). The other proton donor, N2, is oriented outward and participates in intermolecular hydrogen bonding with the carbonyl oxygen O1 of a neighbouring molecule related by translation (Figure 12, Table S55 in Supplementary Materials); thus, chains parallel to the crystallographic direction [100] (i.e., parallel to axis a, Figure S56 in Supplementary Materials) are formed.

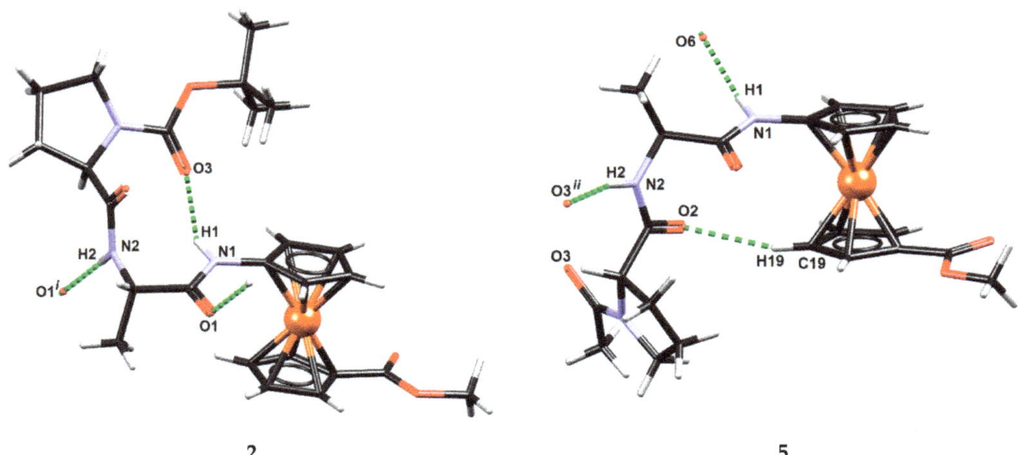

Figure 12. Conformations of peptides **2** and **5** determined by X-ray structure analysis. Hydrogen bonds are shown as green dashed lines. Symmetry operators: (i) $-1 + x, y, z$; (ii) $\frac{1}{2} - x, -\frac{1}{2} + y, -z$.

The conformation of compound **5** is also bent, but without intramolecular N–H···O hydrogen bonding (Figure 12). It crystallises as a monohydrate, so the water molecule O6 interferes with the hydrogen bonding: it acts as a proton acceptor for the group N1-H1 (Supplementary Materials Table S55). This gives rise to zig-zag chains extending in the direction [10] (i.e., crystallographic axis *b*, Figure S57 in Supplementary Materials). The bent conformation of **5** is stabilized by a single weak hydrogen bond C19–H19···O2 (Figure 12, Table S55 in Supplementary Materials).

The reported conformations differ from those determined by the computational study, although β-turn prevails in heterochiral derivative **2**. However, the energy penalty connected with a reorganization of the individual molecule from the most stable conformer (as determined by computational study) to less stable conformation (as determined by X-ray analysis) is usually overcome by favourable intermolecular interactions that additionally stabilize molecules in crystal [16].

2.7. Biological Evaluation

A further step after the successful synthesis and characterization of the investigated peptidomimetics is the determination of their biological activity. Based on the literature data and our previously published work [13,17], peptidomimetics **2–5** are expected to possess antiproliferative and/or antitumor activity. Therefore, the biological activity of **2–5** was evaluated based on their ability to inhibit the growth of MCF-7 and HeLa carcinoma cells. The cytotoxicity of the synthesized compounds was measured using the CellTiter 96® AQ_{ueous} One Solution Cell Proliferation Assay. Results are expressed as cell viability (%) of treated cells versus control, non-treated cells and shown in Figure 13a,b.

All tested compounds **2–5** have inhibitory effects on HeLa and MCF-7 cell lines at concentrations of 100 μM and higher, as shown in Figure 13a,b, respectively. The effect of the tested peptidomimetics on cell viability is dose-dependent, i.e., the growth inhibition is proportional to the increase in the concentration of the tested compound. Boc-protected peptides **2** and **4** have stronger inhibitory effect compared to Ac-peptides **3** and **5**. Viability of cells treated with the highest concentration (500 μM) was from 35.4377% (HeLa) to 54.3296% (MCF-7) for **2** and 37.2897% (HeLa) to 48.3693% (MCF-7) for **4**. In vitro cytototoxicity results are quantified as the IC_{50} value (the half-maximal inhibitory concentration), which is defined as the concentration of the test compound that results in 50% inhibition of cell growth. These values for both cell lines and the four compounds tested were calculated from the best-fitted equations of dose-response curves and are shown in Table 1.

Table 1. IC_{50} values calculated from dose-response curves of cell viability on HeLa and MCF7 cells.

Compound	HeLa Cells	MCF-7 Cells
2	436.1959 μM	n.d. [1]
3	n.d. [1]	n.d. [1]
4	370.3969 μM	270.6925 μM
5	n.d. [1]	n.d. [1]

[1] n.d. = not detected.

According to the IC_{50} values, compound **4** has the strongest inhibitory effect on MCF-7 cells, with slightly less pronounced impact on HeLa cells. For compound **2**, an IC_{50} value was calculated only for HeLa cells, while no IC_{50} values were calculated for other compounds from the experimental data. In the range of tested concentrations (10 μM−500 μM), no 50% inhibition of cell growth was observed, so for compounds **2**, **3** and **5** on MCF-7 cells, and for compounds **3** and **5** on HeLa cells, the IC_{50} value can be considered higher than 500 μM. The cytotoxicity assay revealed that HeLa cells are somewhat more sensitive to the effect of the tested peptidomimetics **2–5**, therefore the remaining experiments were performed on HeLa cells.

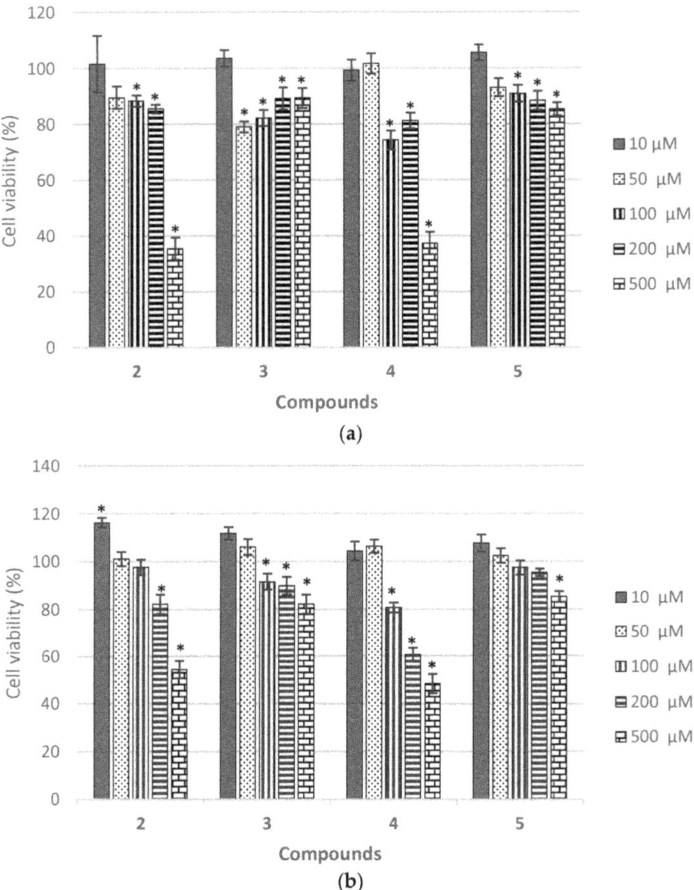

Figure 13. Cell viability of HeLa (**a**) and MCF-7 (**b**) cells treated with peptides **2–5** for 72 h in the range of concentration from 10 μM to 500 μM and assessed by the CellTiter 96® AQueous. One Solution Cell Proliferation Assay. Cell viability (%) was expressed as percentage of treated cells versus control cells. and the data from three individual experiments were expressed as the means (n = 5) ± S.D. * significant difference was considered at a p value < 0.05.

Another way to assess cell survival and determine the efficacy of cytotoxic agents, is a colony formation test or clonogenic assay, *in vitro* method based on the ability of a single cell to grow into a colony. After treatment with the tested compounds, the surviving cells take about 1–3 weeks to form colonies, but only a small fraction of exposed cells retain the ability to form colonies [52]. Since clone formation is in some respects a property of unlimited growth, which is a special feature of tumor cells, the clonogenic assay may serve as a good indicator of the antitumor potential of the test compounds. Therefore, peptides **2–5** were also analyzed by a clonogenic assay on HeLa cells treated with 100 μM and 500 μM concentrations of **2–5**. After 17 days of in vitro cultivation, colonies became visible and were then coloured with 0.5 % crystal-violet, counted, and photographed (see Supplementary Materials Figure S58).

Based on the number of colonies counted, the plating efficiency (PE) and surviving fraction (SF) were calculated for compounds **2–5** (Table 2) according to the equations in the protocol of Franken et al. [53]. A higher SF value means that a higher colony forming ability is maintained after treatment with the test substance, which could be related to less pronounced cytotoxic efficacy of that compound.

Table 2. Plating efficiency (PE) and surviving fraction (SF) for peptides **2–5**.

PE (%)	Concentration	SF (2)	SF (3)	SF (4)	SF (5)
31.75	100 μM	0.1575	1.0394	0	0.7874
	500 μM	0	0	0	0.0315

When HeLa cells were treated with 100 μM of the tested compounds, higher values of PE were calculated for Ac-peptides **3** and **5** than for Boc-peptides **2** and **4**, while survival of cells treated with 500 μM was seen only after treatment with compound **5**. The results of the clonogenic analysis are consistent with their cytotoxicity, as there is a significant difference in the number of visible colonies grown after treatment with 100 μM of Boc-peptides **2** and **4**, in contrast to Ac-peptides **3** and **5**, which showed a weaker effect on the growth and survival of HeLa cells.

The observed cytotoxicity of the tested peptidomimetics could be the result of their impact on two basic cell processes, cell division and/or cell death. Programmed cell death through the process of apoptosis was originally defined based on morphological characteristics, so the initial identification of apoptotic cells is often observed under the microscope. Control and treated HeLa cells were stained with the fluorescent dyes fluorescein diacetate (FDA) and propidium iodide (PI), examined, and photographed under the EVOS FLoid Cell Imaging Station fluorescence microscope, as shown in Figure 14.

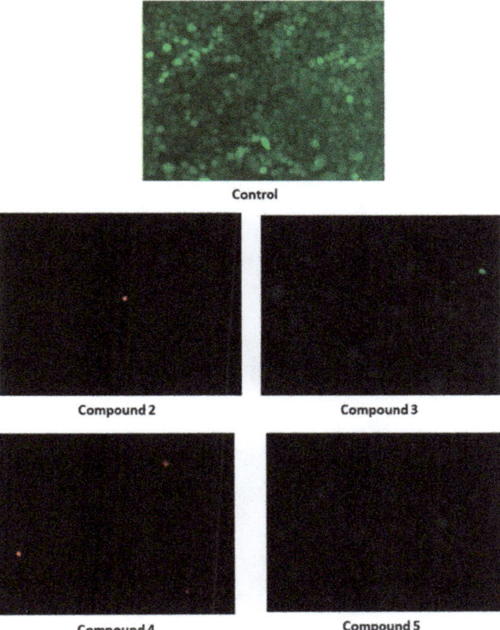

Figure 14. Morphological appearance of control, non-treated and HeLa cells treated with peptides **2–5** (500 μM) photographed after staining with FDA and PI under the fluorescence microscopy.

The use of FDA and PI fluorescent stain allows differentiation between living, necrotic, and apoptotic cells in the sample of treated cells. In a photograph of control HeLa cells, there are plenty of live, green-stained cells, whereas in the case of HeLa cells treated with compounds **2** and **4**, respectively, only a few dead cells stained red can be seen (Figure 14).

To further confirm and quantify the observed morphological changes indicative of the induction of cell death during treatment with the tested peptides **2–5,** flow cytometry analysis was conducted using the Muse® Cell Analyser and the Muse™ Annexin V & Dead Cell Kit. The results of this analysis for HeLa cells are shown in Figure 15.

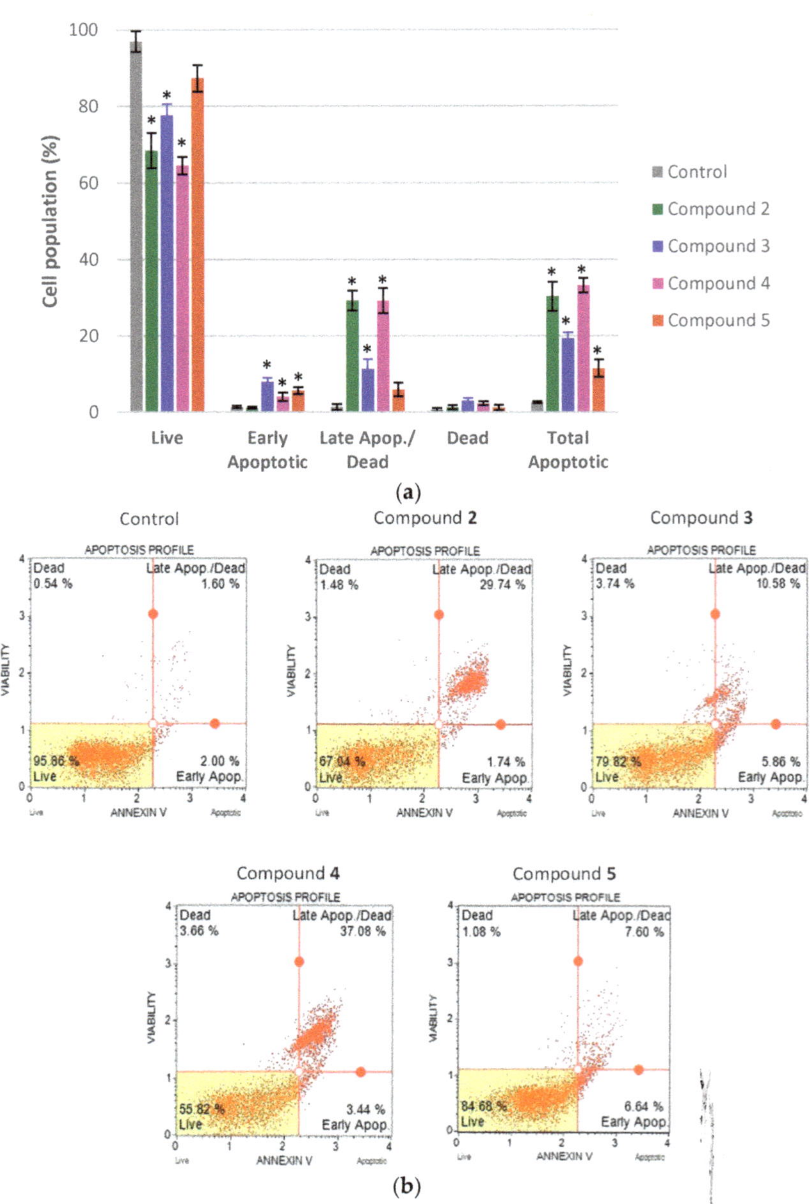

Figure 15. Cell death analysis of HeLa cells treated with the highest tested concentration of Boc-(**2**,**4**) and Ac-peptides (**3**,**5**) (500 μM) for 72 h. Distribution of four distinct cell populations (**a**). Values were represented as the mean (n = 2) ± SD from two independent experiments. * significant difference of treated cells versus control cells was considered at a p value < 0.05. Representative histograms for control and treated cells were given from the Muse Annexin V & Dead Cell assay (**b**).

Double staining with 7-aminoactinomycin D (7-AAD) and Annexin-V allows differentiation between populations of live, dead, and early/late apoptotic cells in the sample by flow cytometry. The results of the analysis show that HeLa cells treated with compound **4** (500 μM) have the highest percentage of the total apoptotic cells (33.19% ± 19,123), followed by compound **2** with a total apoptosis of 30.32% ± 3,8201 (Figure 15). The lowest percentage of total apoptotic cells was determined in a sample treated with 500 μM of compound **5** (11.45%), which had the weakest effect on HeLa cells according to all other methods used for the biological evaluation of peptides **2–5**. The induction of apoptosis by the action of peptidomimetics has been reported previously [53], whereby peptidomimetics with anticancer properties bind to target proteins and mimic interactions that activate specific death pathways in cancer cells, which then die by apoptotic cell death. A recent review on peptidomimetics highlights their crucial role as inhibitors of protein-protein interactions in cancer cells. Based on the traditional classification of peptidomimetics into types I to III, which are divided into distinct classes A–D [54], peptidomimetics **2–5** belong to class A, like type I mimetics, which are involved in apoptosis regulation through inhibitory effects on the p53–MDM2 and p53–MDMX complexes, among other modes of action [54].

Our previous work on ferrocene peptidomimetics has shown that conformational patterns do not have a decisive influence on biological activity, and that lipophilicity contributes to biological activity, i.e., ferrocene peptidomimetics with larger retention factors (R_f) had better antiproliferative capacity [13,17]. The results obtained here on the increased antiproliferative activity of Boc-peptides **4** (R_f = 0.55) and **2** (R_f = 0.52) compared to the more polar Ac-peptides **3** (R_f = 0.33) and **5** (R_f = 0.35) are consistent with the above findings. Moreover, structural modification of the inactive Boc-L-Pro-NH-Fn-COOMe (**IV**) [17] by insertion of L-Ala between ferrocene core and Pro resulted with the promising outcome on Boc-peptide **4**, which could serve as a leading compound for further research and development. Considering the results obtained so far, we are encouraged to continue this study with the aim of developing an antitumor drug.

3. Materials and Methods

3.1. General Procedure and Methods

The synthesis of peptides **2–5** was carried out under argon atmosphere and the chemicals used for the reactions were analytically pure. CH_2Cl_2 used for synthesis, CD measurements and FTIR was dried (P_2O_5), distilled over CaH_2, and stored over molecular sieves (4 Å). EDC (Acros Organics, Geel, Belgium), HOBt (Aldrich, Santa Clara, CA, USA) and acetyl chloride (Aldrich), were used as received. The synthesis of Boc-L-Ala-NH-Fn-COOMe (**1**) has been described previously [16]. Its *N*-terminus was deprotected by exposure to gaseous HCl. The *N*-termini of L- and D-proline were protected in the presence of sodium hydroxide, aqueous dioxane and di-*tert*-butyldicarbonate to give Boc-L-Pro-OH and Boc-D-Pro-OH, respectively. Boc-L-Pro-OH and Boc-D-Pro-OH were activated with the coupling reagent HOBt for 1 h in CH_2Cl_2. The products were purified by preparative thin layer chromatography on silica gel (Kieselgel 60 HF254, KGaA, Darmstadt, Germany) using EtOAc/CH_2Cl_2 mixture or pure EtOAc as eluent. Infrared spectra were recorded as CH_2Cl_2 solutions between NaCl windows or in KBr using a MB 100 mid-FTIR spectrometer (Bomem, Saint-Jean-Baptiste, Canada) ((s) = strong, (m) = medium, (w) = weak, (br) = broad, (sh) = shoulder). The ^{1}H- and ^{13}C-NMR spectra were recorded at 600 MHz using an Avance spectrometer (Bruker, Rheinstetten, Germany) with a 5 mm TBI probe at the Ruđer Bošković Institute and were referenced to the peak of the residual solvent ($CDCl_3$-*d*, ^{1}H: δ = 7.24 ppm, ^{13}C: δ = 77.23 ppm). In the case of the $CDCl_3$-*d*/DMSO-d_6 mixture, calibration was performed using Me_4Si as an internal standard (^{1}H: δ = 0.0 ppm). Double resonance experiments (COSY, NOESY, HMQC and HMBC) were performed to facilitate the assignment of signals; ((s) = singlet, (d) = doublet, (m) = multiplet, (dd) = doublet of doublets, (td) = triplet of doublets, (dq) = doublet of quartets). Unless otherwise stated, all spectra were recorded at 298 K. NMR titrations were performed by adding 10 μL portions of DMSO-d_6 to NMR tubes containing $CDCl_3$-*d* solutions of the peptides under study ($c = 2.5 \times 10^{-2}$ M). Spectra were recorded after each addition, and DMSO-d_6 was added until no change in the chemical shift of the amide protons was observed. CD spectra were recorded using a model 810 spectropolarimeter (Jasco, Tokyo, Japan) in CH_2Cl_2 or KBr. Molar ellipticity coefficients $[\theta]$ are given in degrees, concentration c in $molL^{-1}$, and path length l in cm, so that the unit for $[\theta]$ is deg cm^2 $dmol^{-1}$. Mass spectra were recorded using HPLC-MS system coupled to a triple-quadrupole mass spectrometer, operating in a negative ESI mode (Agilent, Palo Alto, CA, USA). High-resolution mass spectra were recorded using a MALDI-TOF/TOF 4800+ analyser (SCIEX Headquarters, Framingham, MA, USA). Melting points were determined using Reichert Thermovar apparatus (Reichert, Vienna, Austria).. Single crystal measurements were performed with an Xcalibur Nova R system (Oxford Diffraction, Wroclaw, Poland).

3.1.1. Synthesis of Boc-D-Pro-L-Ala-NH-Fn-COOMe (**2**) and Boc-L-Pro-L-Ala-NH-Fn-COOMe (**4**)

The HCl_{gas} was purged through the suspension of Boc-L-Ala-NH-Fn-COOMe (**1**) (1000 mg, 2.32 mmol) in dry CH_2Cl_2 (5 mL) at 0 °C. After 30 min, the solvent was evaporated in vacuo, leaving a dark yellow hydrochloride salt, which was then suspended in CH_2Cl_2 and treated with NEt_3 (pH ~ 8) to afford an unstable free amine suitable for coupling to Boc-L-Pro-OH or Boc-D-Pro-OH (998 mg, 4.64 mmol) using the standard EDC/HOBt method; EDC (1007 mg, 5.57 mmol), HOBt (753 mg, 5.57 mmol). The reaction mixtures were then stirred at room temperature until the ferrocene amine was completely consumed, which was monitored by TLC (~1 h). Standard work-up (washing with a saturated aqueous solution of $NaHCO_3$, a 10% aqueous solution of citric acid and brine, drying over Na_2SO_4 and evaporation in vacuo) including TLC purification of the crude products (EtOAc: CH_2Cl_2 = 1: 5; R_f = 0.52 (**2**), R_f = 0.55 (**4**)) gave orange solids of **2** (1107 mg, 89%) and **4** (1213 mg, 92%).

Boc-D-Pro-L-Ala-NH-Fn-COOMe (**2**): m.p. = 119.2 °C. IR (CH_2Cl_2) $\tilde{\upsilon}_{max}/cm^{-1}$: 3418 w ($NH_{free}$), 3325 m ($NH_{assoc.}$), 1705 s ($C = O_{COOMe}$), 1684 s, 1671 s ($C = O_{CONH}$), 1557 s, 1531 s (amide II). IR (KBr) $\tilde{\upsilon}_{max}/cm^{-1}$: 3509 w($NH_{free}$), 3308 m ($NH_{assoc.}$), 1714 s ($C = O_{COOMe}$), 1695 s, 1671 s ($C = O_{CONH}$). ^{1}H-NMR (600 MHz, $CDCl_3$) δ/ppm: 8.37 (s, 0.89H, $NH_{Fn\ trans}$), 7.70 (s, 0.11H, $NH_{Fn\ cis}$), 7.24 (d, J = 6.3 Hz, 0.89H, $NH_{Ala\ trans}$), 6.66 (d, 0.11H, $NH_{Ala\ cis}$), 5.09 (s, 1H, H-3), 4.81 (s, 2H, H-8), 4.72–4.69 (m, 2H, H-9, CH_{Ala}) 4.58 (s, 1H, H-4), 4.40 (s, 1H, H-7), 4.35 (s, 1H, H-10), 4.16 (s, 1H, CH-α (Pro)), 4.01 (s, 1H, H-2), 3.95 (s, 1H, H-5), 3.79 (s, 3H, COOMe), 3.53 (td, J = 10.3 Hz, 6,7 Hz, 1H, CH_2-δ (Pro)), 3.46 (s, 1H, CH_2-δ′ (Pro)), 2.22–2.07 (m, 2H, CH_2-β, CH_2-β′ (Pro)), 1.89–1.86 (m, 2H, CH_2-γ, CH_2-γ′ (Pro)), 1.50 (s, 9H, $(CH_3)_{3\ Boc}$), 1.48 (d, J = 7.30 Hz, 3H, $CH_{3\ Ala}$). ^{13}C-NMR (150 MHz, $CDCl_3$) δ/ppm: 172.47 (CO_{Fn}), 172.10 (CO_{COOMe}), 170.89 (CO_{Ala}), 155.26 (CO_{Boc}), 96.31 (C-1, Fn), 80.37 ($C_{q\ Boc}$), 72.79 (C-7), 72.45 (C-10), 71.88 (C-6), 71.75 (C-8), 71.43 (C-9), 66.59 (C-2), 65.40 (C-5), 63.94 (C-4), 62.54 (C-3), 61.10 (C-α, Pro), 51.87 ($CH_{3\ COOMe}$), 48.59 (CH_{Ala}), 47.53 (CH_2-δ, Pro), 29.87 (CH_2-β, Pro), 28.58 ($(CH_3)_{3\ Boc}$), 24.96 (CH_2-γ, Pro), 17.56 ($CH_{3\ Ala}$). ESI-MS (H_2O:MeOH = 50:50): m/z 526.1 ($(M - H)^-$). MALDI-HRMS m/z = 527.1708 (calculated for $C_{25}H_{33}N_3O_6Fe$ = 527.1718).

Boc-D-Pro-L-Ala-NH-Fn-COOMe (**4**): m.p. = 66.9 °C. IR (CH_2Cl_2) $\tilde{\upsilon}_{max}/cm^{-1}$: 3418 w ($NH_{free}$), 3310 m ($NH_{assoc.}$), 1705 s ($C = O_{COOMe}$), 1674 s ($C = O_{CONH}$), 1555 s, 1503 s (amide II). IR (KBr) $\tilde{\upsilon}_{max}/cm^{-1}$: 3505 w ($NH_{free}$), 3296 m ($NH_{assoc.}$), 1715 s ($C = O_{COOMe}$), 1669 s, 1660 s ($C = O_{CONH}$). ^{1}H-NMR (600 MHz, $CDCl_3$) δ/ppm: 8.15 (s, 0.9H, $NH_{Fn\ trans}$), 7.74 (s, 0.1H, $NH_{Fn\ cis}$), 6.85 (d, J = 6.3 Hz, 0.9H, $NH_{Ala\ trans}$), 6.67 (d, 0.1H, $NH_{Ala\ trans}$), 4.81 (s, 1H, H-3), 4.76 (s, 2H, H-8, H-9), 4.60 (s, 1H, H-4), 4.48 (m, 1H, CH_{Ala}), 4.38 (s, 1H, H-7), 4.37 (s, 1H, H-10), 4.34 (s, 1H, CH-α (Pro)), 4.00 (s, 2H, H-2, H-5), 3.78 (s, 3H, COOMe), 3.50–3.45 (m, 2H, CH_2-δ, CH_2-δ′ (Pro)), 2.22–2.16 (m, 2H, CH_2-β, CH_2-β′ (Pro)), 1.92–1.91 (m, 2H, CH_2-γ, CH_2-γ′ (Pro)), 1.49 (s, 9H, $(CH_3)_{3\ Boc}$), 1.42 (d, J = 6.98 Hz, 3H, $CH_{3\ Ala}$). ^{13}C-NMR (150 MHz, $CDCl_3$) δ/ppm: 172,35 (CO_{Fn}), 171.95 (CO_{COOMe}), 170.45 (CO_{Ala}), 156.34 (CO_{Boc}), 95.79 (C-1, Fn), 81.24 ($C_{q\ Boc}$), 73.08 (C-7), 72.97 (C-10), 71.90 (C-6), 71.41 (C-8), 71.08 (C-9), 66.56 (C-2), 66.39 (C-5),

63.15 (C-4), 62.71 (C-3), 60.97 (C-α, Pro), 51.70 ($CH_{3\,COOMe}$), 49.74 (CH_{Ala}), 47.57 (CH_2-δ, Pro), 29.14 (CH_2-β, Pro), 28.54 ($(CH_3)_{3\,Boc}$), 24.84 (CH_2-γ, Pro), 17.54 ($CH_{3\,Ala}$). ESI-MS (H_2O:MeOH = 50:50): m/z 526.1 ($(M-H)^-$). MALDI-HRMS m/z = 527.1729 (calculated for $C_{25}H_{33}N_3O_6Fe$ = 5.271.718).

3.1.2. Synthesis of Ac-D-Pro-L-Ala-NH-Fn-COOMe (**3**) and Ac-L-Pro-L-Ala-NH-Fn-COOMe (**5**)

The transformation of carbamates **2** and **4** (1000 mg, 1.89 mmol) to acetamides **3** and **5** began with the acidic Boc-deprotection described above. Their free amines, obtained by treating the hydrochloride salt with NEt_3 (2.07 mL, 23.7 mmol), were cooled to 0 °C and acetyl chloride (807 µL, 11.34 mmol) was added dropwise, stirring in an ice bath. After TLC monitoring showed complete conversion of the starting materials, the reaction mixtures were poured into water and extracted with CH_2Cl_2. The combined organic phases were washed with a brine, dried over Na_2SO_4 and evaporated to dryness in vacuo. The resulting crude products were purified by TLC on silica gel (EtOAc; R_f = 0.33 (**3**), R_f = 0.35 (**5**)) to give orange solids of **3** (1132 mg, 60%) and **5** (1213 mg, 64%).

Ac-D-Pro-L-Ala-NH-Fn-COOMe (**3**): m.p. = 132.3 °C. IR (CH_2Cl_2) $\tilde{\upsilon}_{max}/cm^{-1}$: 3424 w ($NH_{free}$), 3303 m ($NH_{assoc.}$), 1706 s ($C=O_{COOMe}$), 1688 s, 1630 s ($C=O_{CONH}$), 1557 s, 1541 s, 1521 m (amide II). IR (KBr) $\tilde{\upsilon}_{max}/cm^{-1}$: 3542 w ($NH_{free}$), 3308 s, 3259 s, 3222 m ($NH_{assoc.}$), 1714 s ($C=O_{COOMe}$), 1690 s, 1679 s, 1688 s ($C=O_{CONH}$). ^{1}H-NMR (600 MHz, $CDCl_3$) δ/ppm: 8.58 (s, 1H, NH_{Fn}), 7.08 (d, J = 9.1 Hz, 1H, NH_{Ala}), 5.05 (s, 1H, H-3), 4.78 (s, 1H, H-8), 4.76 (s, 1H, H-9), 4.70 (s, 1H, H-4), 4.62 (dq, J = 8.5 Hz, 7.1 Hz, 1H, CH_{Ala}), 4.40 (s, 1H, H-7), 4.35 (s, 1H, H-10), 4.26 (dd, J = 7.7 Hz, 5,6 Hz, 1H, CH-α (Pro)), 3.98 (s, 1H, H-2), 3.95 (s, 1H, H-5), 3.77 (s, 3H, COOMe), 3.68 (td, J = 9.8 Hz, 6,9 Hz, 1H, CH_2-δ (Pro)), 3.55 (td, J = 9.8 Hz, 6,5 Hz, 1H, CH_2-δ′ (Pro)), 2.28–2.23 (m, 1H, CH_2-γ Pro)), 2.21–2.18 (m, 1H, CH_2-β (Pro)), 2.16 (s, 3H, $CH_{3\,Ac}$), 2.15–2.11 (m, 1H, CH_2-β′ (Pro)), 2.00–1.95 (s, 1H, CH_2-γ′ (Pro)), 1.49 (d, J = 7.2 Hz, 3H, $CH_{3\,Ala}$). ^{13}C-NMR (150 MHz, $CDCl_3$) δ/ppm: 172.08 (CO_{Ac}), 172.03 (CO_{COOMe}), 170.63 (CO_{Ala}), 170.31 (CO_{Fn}), 96.56 (C-1, Fn), 72.80 (C-7), 72.64 (C-10), 71.91 (C-6), 71.46 (C-8), 71.20 (C-9), 66.45 (C-2), 65.74 (C-5), 63.25 (C-4), 62.69 (C-3), 61.19 (C-α, Pro), 51.78 ($CH_{3\,COOMe}$), 49.11 (CH_{Ala}), 48.66 (CH_2-δ, Pro), 29.23 (CH_2-β, Pro), 25.52 (CH_2-γ, Pro), 22.87 ($CH_{3\,Ac}$), 17.55 ($CH_{3\,Ala}$). ESI-MS (H_2O:MeOH = 50:50): m/z 468.1 ($(M-H)^-$). MALDI-HRMS m/z = 469.1280 (calculated for $C_{22}H_{27}N_3O_5Fe$ = 4.691.300).

Ac-L-Pro-L-Ala-NH-Fn-COOMe (**5**): m.p. = 125.1 °C. IR (CH_2Cl_2) $\tilde{\upsilon}_{max}/cm^{-1}$: 3420 w ($NH_{free}$), 3309 m ($NH_{assoc.}$), 1705 s ($C=O_{COOMe}$), 1696 s, 1680 s, 1636 s ($C=O_{CONH}$), 1555 s, 1540 s, 1507 m (amide II). IR (KBr) $\tilde{\upsilon}_{max}/cm^{-1}$: 3499 w ($NH_{free}$), 3288 m ($NH_{assoc.}$), 1714 s ($C=O_{COOMe}$), 1673 s, 1630 s ($C=O_{CONH}$). ^{1}H-NMR (600 MHz, $CDCl_3$) δ/ppm: 7.99 (s, 0.95H, $NH_{Fn\,trans}$), 7.65 (s, 0.05H, $NH_{Fn\,cis}$), 7.22 (d, J = 7.1 Hz, 0.94H, $NH_{Ala\,trans}$), 6.86 (d, 0.06H, $NH_{Ala\,cis}$), 4.78 (s, 1H, H-8), 4.76 (s, 1H, H-3), 4.74 (s, 1H, H-9), 4.67 (s, 1H, H-4), 4.57 (m, 1H, CH-α (Pro)), 4.48 (dq, J = 8.4 Hz, 7,1 Hz, 1H, CH_{Ala}), 4.39 (s, 1H, H-7), 4.37 (s, 1H, H-10), 4.02 (s, 1H, H-2), 3.96 (s, 1H, H-5), 3.77 (s, 3H, COOMe), 3.67–3.64 (m, 1H, CH_2-δ, (Pro)), 3.52–3.50 (m, 1H, CH_2-δ′ (Pro)), 2.32 (s, 1H, CH_2-γ, Pro)), 2.18 (s, 3H, $CH_{3\,Ac}$), 2.08–2.06 (m, 2H, CH_2-β, CH_2-β′ (Pro)), 1.93 (s, 1H, CH_2-γ′ (Pro)), 1.42 (d, J = 7.0 Hz, 3H, $CH_{3.Ala}$). ^{13}C-NMR (150 MHz, $CDCl_3$) δ/ppm: 171.82 (CO_{Ac}), 171.81 (CO_{COOMe}), 171.66 (CO_{Ala}), 170.28 (CO_{Fn}), 95.76 (C-1, Fn), 72.67 (C-7), 72.65 (C-10), 72.21 (C-6), 71.43 (C-8), 71.08 (C-9), 66.62 (C-2), 66.01 (C-5), 63.04 (C-4), 62.87 (C-3), 60.65 (C-α, Pro), 51.76 ($CH_{3\,COOMe}$), 49.58 (CH_{Ala}), 48.67 (CH_2-δ, Pro), 28.68 (CH_2-β, Pro), 25.21 (CH_2-γ, Pro), 22.93 ($CH_{3\,Ac}$), 17.22 ($CH_{3\,Ala}$). ESI-MS (H_2O:MeOH = 50:50): m/z 468.1 ($(M-H)^-$). MALDI-HRMS m/z = 469.1280 (calculated for $C_{22}H_{27}N_3O_5Fe$ = 469.1300).

3.1.3. Computational Details

Conformational analyses of compounds **2–5** were done in three stages. First, a series of low-level optimizations with molecular mechanics, OPLS2005 force field, were performed in MacroModel v10.3 [55–57]. The most stable conformers were selected for further optimizations at a high level of theory and run in Gaussian16 [58] with a default grid and convergence criteria B3LYP/Lanl2DZ. The last stage included optimization of the most stable conformers at the B3LYP/6-311+G(d,p) (LanL2DZ basis set on Fe) level of theory while surrounding solvent (chloroform) were described as polarizable continuum (SMD) [59]. Vibrational analysis was performed to verify each structure as a minimum on the potential energy surface and the reported energies refer to standard Gibbs free energies at 298 K. QTAIM theory were used to characterize hydrogen bonds in AIMAll package [60]. Topological parameters of the displayed bond critical points between hydrogen bond acceptors and hydrogen atoms were calculated and verified according to the Koch and Popelier criteria [61,62].

3.1.4. Crystallographic Study

X-ray diffraction: single crystal measurements were performed on a Rigaku Oxford Diffraction Xtalab Synergy S (2) and an Oxford Diffraction Xcalibur Nova R (5), using mirror-monochromated CuKα radiation. Program package CrysAlis PRO [63] was used for data reduction and numerical absorption correction.

The structures were solved using SHELXS97 [64] and refined with SHELXL-2017 [65]. Models were refined using the full-matrix least squares refinement; all non-hydrogen atoms were refined anisotropically. Rigid-body restraints were applied to ADPs of C atoms of cyclopentadienyl ring C1→C5 in **5**. Hydrogen atoms were located in a difference Fourier map and refined either as riding entities. Hydrogen atoms of water molecule O6 in 5 could not be located from the difference map and were therefore not modelled. Molecular geometry calculations were performed by PLATON [66] and molecular graphics were prepared using ORTEP-3 [67], and Mercury [68]. Crystallographic and refinement data for the structures reported in this paper are shown in in Table S52 in Supplementary Material. The crystallographic data have been deposited in the Cambridge Structural Database as entries No. 2122149 and 2122150.

3.1.5. Biological Activity

Materials: Trypsin-EDTA (0.25%), FBS (fetal bovine serum) and PBS (phosphate buffer saline) were purchased from Sigma-Aldrich while DMEM (Dulbecco's Modified Eagle Medium) was purchased from Capricorn Scientific GmbH (Ebsdorfergrund, Germany). The CellTiter 96® AQ_{ueous} One Solution Cell Proliferation Assay was purchased from Promega (Madison, WI, USA). Fluorescein diacetate (FDA) and propidium iodide (PI) were purchased from Sigma-Aldrich. The Muse Annexin V Dead Cell kit was purchased from EMD Milipore Corporation (Merck KGaA).

Cell culture and cultivation conditions: Two adherent human cell lines used in this work were obtained from the Ruđer Bošković Institute (Zagreb, Croatia). The HeLa cell line derived from the cervical adenocarcinoma (ATCC No. CCL-2) and the MCF-7 cell line derived from breast adenocarcinoma (ATCC No. HTB-22) were cultured in DMEM supplemented with 5% FBS and maintained in BioLite petri dishes for cell culture (Thermo Fisher Scientific, Waltham, MA, USA) in an incubator under a humidified atmosphere and 5% CO_2 at 37 °C. Cells in the exponential growth phase were trypsinized, counted by the trypan blue method using an improved Neubauer hemocytometer, and used to set up individual experiments. BioLite 6-well and 96-well plates were used for individual experiments to test compounds of interest (Thermo Fisher Scientific).

Evaluation of cytotoxicity: The effect of peptides **2–5** on cell viability was examined using the CellTiter 96® AQ_{ueous} One Solution Cell Proliferation Assay, which was performed according to the manufacturer's instructions with minor modifications and as described [69]. In brief, HeLa and MCF-7 cells were seeded in 96-well plates at a density of 3×10^4 cells per well in 100 µL of media. Stock solutions of peptides 2–5 were prepared as 10 mM solutions of compounds in ethanol, sterilized by filtration through a 0.22 µM filters and then, prior to each experiment, diluted in culture medium. After overnight incubation, HeLa and MCF-7 cells were treated with peptides 2–5 at nominal concentrations ranging from 10 μM to 500 μM. After the 72 h treatment, 10 μL of CellTiter 96® AQ_{ueous} One Solution Cell Proliferation reagent was added to each well, and the cells were incubated for an additional 3 h. Subsequently, absorbance was measured at 490 nm on the microplate reader (Tecan, Mannedorf,, Switzerland). Cell viability was expressed as the percentage of treated cells versus control cells. Experiments were performed three times with five parallels for each concentration of compound tested and data were expressed as mean ± SD. The corresponding IC_{50} values were calculated from the dose-response curves using equations of best-fitted trend lines.

Clonogenic assay: The clonogenic analysis began by seeding pre-cultured HeLa cells in 6-well plates at an initial concentration of 200 cells in 2 mL of culture medium per well. The cells were incubated under optimal conditions and treated with peptides **2–5** at a concentration of 100 μM and 500 μM after 24 h. There was also a control cells that were not treated with peptidomimetics. Three days after the cells were treated, the growth medium containing the test compounds was removed and replaced with fresh growth medium, after which the plate with the HeLa cells was returned to the incubator for further cultivation. After treatment with the test substances, the surviving cells need about 1–3 weeks to form colonies. In this work, the colonies formed were visible 17 days after initial seeding of the cells. Staining the grown colonies with crystal-violet begins by removing the growth medium and washing the cells with 1 mL of PBS buffer. Then 2.5 mL of methanol was added to fix the cells, which was removed after 10 min. The plates are then allowed to air dry completely. A 0.5% solution of crystal-violet is then added and incubated for 10 min. In the final step, the dye is removed and the colonies in the wells are rinsed with 1 mL of PBS buffer and deionized water. The number of colonies grown was then counted and the plating efficiency (PE) and survival fraction (SF) were calculated according to the equations in the protocol of Franken et al. [52]. PE is the ratio of the number of colonies to the number of seeded cells, while SF is the number of colonies formed after treatment of the cells, expressed as PE.

Analysis of cell death by fluorescence microscopy and flow cytometry: for fluorescein diacetate and propidium iodide staining, HeLa cells were seeded in 6-well plates at a concentration approximately about 1×10^5 cells mL^{-1} and treated with 500 μM of peptides **2–5** after 24 h. After the 72 h treatment, cells were washed with PBS, trypsinized, centrifuged, and resuspended in 0.2 mL of PBS. Cell were stained with FDA and PI according to the method described by us [70] and immediately examined with the fluorescent microscope EVOS FLoid Cell Imaging Station (Thermo Fisher Scientific).

Quantitative analysis of live, apoptotic, and dead cells treated with peptides **2–5** was performed with the Muse Cell Analyzer (EMD Millipore Corporation, Burlington, MA, USA) using the Muse Annexin V & Dead Cell Kit according to the manufacturer's specifications. In brief, HeLa cells were plated into a 6-well culture at a density of 5×10^4 cells mL^{-1} (2 mL per well) and treated with the 500 μM concentration of conjugates **2–5** for 72 h. After treatment with the test compounds, both floating and adherent cells were collected, centrifuged (600 g min^{-1}), and suspended in cell culture medium to adjust the cell concentration according to the manufacturer's protocol. Then, 100 μL aliquots of the cell suspension were added to 100 μL of Muse Annexin V & Dead Cell Reagent and incubated for 20 min in the dark at RT. Cells were then analyzed using the Muse Cell Analyzer. Each compound was tested in duplicate, and each experiment was performed twice.

The data in the graphs are expressed as mean ± standard deviation (±SD), and the error bars in the figures indicate the SD. Differences between means were analyzed using the ANOVA test, followed by post-hoc Tukey's test. A significant difference was considered at a p value < 0.05.

4. Conclusions

New insight is provided into the effects of the constituent homo- and heterochiral Pro-Ala sequences on the conformational properties of the asymmetrically disubstituted ferrocene peptidomimetics.

The results of the DFT study agree quite well with the spectroscopic (IR, NMR) data, also confirming the theoretically predicted differences in heterochiral vs. homochiral analogs. According to the complementary experimental and computational study, the heterochiral Pro-Ala sequence initiated more complex hydrogen-bonding patterns consisting of intrastrand 10-membered (β-turn) and interstrand 9-membered rings. In comparison, the homochiral Pro-Ala sequence resulted in more flexible conformations in which mostly one of these two hydrogen bonds occur.

Of the four peptides tested, Boc-peptide **4** has the strongest inhibitory effect on tumor cells HeLa and MCF-7, the greatest potential to promote apoptotic cell death in HeLa cells and the highest ability to reduce survival of treated HeLa cells. All in all, we can conclude that peptidomimetic **4**, although its IC_{50} value is quite high compared to referent antitumour drug such as cisplatin, has the potential to serve for development of new antitumor drugs. Moreover, the obtained results are certainly a valuable guide for future research on the synthesis of new peptidomimetics that are structurally similar to compound **4** and hopefully will have improved biological properties.

Supplementary Materials: The following are available online at https://www.mdpi.com/article/10.3390/ijms222413532/s1.

Author Contributions: Conceptualization, L.B. and I.K.; methodology, L.B.; software, I.K.; validation, S.R., I.K., K.R. and K.M.; investigation, M.K., M.Č.S., K.R., K.M., S.R., L.Š., I.K. and L.B.; writing—original draft preparation, L.B., I.K., K.M. and K.R.; writing—review and editing, L.B. and I.K.; supervision, L.B.; project administration, L.B.; funding acquisition, L.B. All authors have read and agreed to the published version of the manuscript.

Funding: This work has been fully supported by Croatian Science Foundation under the projects IP-2014-09-7899 and IP-2020-02-9162.

Institutional Review Board Statement: Not applicable.

Informed Consent Statement: Not applicable.

Data Availability Statement: The crystallographic data have been deposited in the Cambridge Structural Database as entries No. 2122149 and 2122150. The DFT, spectroscopic and biological evaluation data are provided as figures and tables and are included in this paper.

Acknowledgments: Computational resources were provided by the Isabella cluster at Zagreb University Computing Centre (SRCE).

Conflicts of Interest: The authors declare no conflict of interest.

References

1. Vagner, J.; Qu, H.; Hruby, V.J. Peptidomimetics, a synthetic tool of drug discovery. *Curr. Opin. Chem. Biol.* **2008**, *12*, 292–296. [CrossRef]
2. Lenci, E.; Trabocchi, A. Peptidomimetic toolbox for drug discovery. *Chem. Soc. Rev.* **2020**, *49*, 3262–3277. [CrossRef]
3. Lu, H.; Zhou, Q.; He, J.; Jiang, Z.; Peng, C.; Tong, R.; Shi, J. Recent advances in the development of protein-protein interactions modulators: Mechanisms and clinical trials. *Signal. Transduct. Target. Ther.* **2020**, *5*, 213–235. [CrossRef]
4. Hoang, H.N.; Hill, T.A.; Ruiz-Gómez, G.; Diness, F.; Mason, J.M.; Wu, C.; Abbenante, G.; Shepherd, N.E.; Fairlie, D.P. Twists or turns: Stabilising alpha vs. beta turns in tetrapeptides. *Chem. Sci.* **2019**, *10*, 10595–10600. [CrossRef]
5. Laxio Arenas, J.; Kaffy, J.; Ongeri, S. Peptides and peptidomimetics as inhibitors of protein-protein interactions involving β-sheet secondary structures. *Curr. Opin. Chem. Biol.* **2019**, *52*, 157–167. [CrossRef]
6. Nair, R.V.; Baravkar, S.B.; Ingole, T.S.; Sanjayan, G.J. Synthetic turn mimetics and hairpin nucleators. *Chem. Commun.* **2014**, *50*, 13874–13884. [CrossRef] [PubMed]
7. Barišić, L.; Dropučić, M.; Rapić, V.; Pritzkow, H.; Kirin, S.I.; Metzler-Nolte, N. The first oligopeptide derivative of 1′-aminoferrocene-1-carboxylic acid shows helical chirality with antiparallel strands. *Chem. Commun.* **2004**, *17*, 2004–2005. [CrossRef] [PubMed]
8. Barišić, L.; Čakić, M.; Mahmoud, K.A.; Liu, Y.-N.; Kraatz, H.-B.; Pritzkow, H.; Kirin, S.I.; Metzler-Nolte, N.; Rapić, V. Helically Chiral Ferrocene Peptides Containing 1′-Aminoferrocene-1-Carboxylic Acid Subunits as Turn Inducers. *Chem. Eur. J.* **2006**, *12*, 4965–4980. [CrossRef]

9. Barišić, L.; Rapić, V.; Metzler-Nolte, N. Incorporation of the Unnatural Organometallic Amino Acid 1′-Aminoferrocene-1-carboxylic Acid (Fca) into Oligopeptides by a Combination of Fmoc and Boc Solid-Phase Synthetic Methods. *Eur. J. Inorg. Chem.* **2006**, 4019–4021. [CrossRef]
10. Čakić Semenčić, M.; Siebler, D.; Heinze, K.; Rapić, V. Bis- and Trisamides Derived From 1′-Aminoferrocene-1-carboxylic Acid and α-Amino Acids: Synthesis and Conformational Analysis. *Organometallics* **2009**, *28*, 2028–2037. [CrossRef]
11. Čakić Semenčić, M.; Heinze, K.; Förster, C.; Rapić, V. Bioconjugates of 1′-Aminoferrocene-1-carboxylic Acid with (S)-3-Amino-2-methylpropanoic Acid and L-Alanine. *Eur. J. Inorg. Chem.* **2010**, 1089–1097. [CrossRef]
12. Čakić Semenčić, M.; Barišić, L. Ferrocene Bioconjugates. *Croat. Chem. Acta* **2017**, *90*, 537–569. [CrossRef]
13. Kovačević, M.; Kodrin, I.; Cetina, M.; Kmetič, I.; Murati, T.; Semenčić, M.Č.; Roca, S.; Barišić, L. The conjugates of ferrocene-1,1′-diamine and amino acids. A novel synthetic approach and conformational analysis. *Dalton Trans.* **2015**, *44*, 16405–16420. [CrossRef]
14. Kovačević, M.; Kodrin, I.; Roca, S.; Molčanov, K.; Shen, Y.; Adhikari, B.; Kraatz, H.B.; Barišić, L. Helically Chiral Peptides That Contain Ferrocene-1,1′-diamine Scaffolds as a Turn Inducer. *Chem. Eur. J.* **2017**, *23*, 10372–10395. [CrossRef]
15. Moriuchi, T.; Ohmura, S.D.; Moriuchi-Kawakami, T. Chirality Induction in Bioorganometallic Conjugates. *Inorganics* **2018**, *6*, 111. [CrossRef]
16. Barišić, L.; Kovačević, M.; Mamić, M.; Kodrin, I.; Mihalić, Z.; Rapić, V. Synthesis and Conformational Analysis of Methyl *N*-Alanyl-1-aminoferrocene-1-carboxylate. *Eur. J. Inorg. Chem.* **2012**, *11*, 1810–1822. [CrossRef]
17. Kovačević, M.; Molčanov, K.; Radošević, K.; Srček, G.V.; Roca, S.; Čače, A.; Barišić, L. Conjugates of 1′-aminoferrocene-1-carboxylic acid and proline: Synthesis, conformational analysis and biological evaluation. *Molecules* **2014**, *21*, 12852–12880. [CrossRef]
18. Kovač, V.; Čakić Semenčić, M.; Kodrin, I.; Roca, S.; Rapić, V. Ferrocene-dipeptide conjugates derived from aminoferrocene and 1-acetyl-1′-aminoferrocene: Synthesis and conformational studies. *Tetrahedron* **2013**, *69*, 10497–10506. [CrossRef]
19. Nuskol, M.; Studen, B.; Meden, A.; Kodrin, I.; Čakić Semenčić, M. Tight turn in dipeptide bridged ferrocenes: Synthesis, X-ray structural, theoretical and spectroscopic studies. *Polyhedron* **2019**, *161*, 137–144. [CrossRef]
20. Nuskol, M.; Šutalo, P.; Đaković, M.; Kovačević, M.; Kodrin, I.; Čakić, S.M. Testing the Potential of the Ferrocene Chromophore as a Circular Dichroism Probe for the Assignment of the Screw-Sense Preference of Tripeptides. *Organometallics* **2021**, *40*, 1351–1362. [CrossRef]
21. Ganguly, H.K.; Basu, G. Conformational landscape of substituted prolines. *Biophys. Rev.* **2020**, *12*, 25–39. [CrossRef] [PubMed]
22. Byun, B.J.; Song, I.K.; Chung, Y.J.; Ryu, K.H.; Kang, Y.K. Conformational preferences of X-Pro sequences: Ala-Pro and Aib-Pro motifs. *J. Phys. Chem. B.* **2010**, *114*, 14077–14086. [CrossRef]
23. Martin, V.; Legrand, B.; Vezenkov, L.L.; Berthet, M.; Subra, G.; Calmès, M.; Bantignies, J.-L.; Martinez, J.; Amblard, M. Turning Peptide Sequences into Ribbon Foldamers by a Straightforward Multicyclization Reaction. *Angew. Chem. Int. Ed.* **2015**, *54*, 13966–13970. [CrossRef]
24. Metrano, A.J.; Abascal, N.C.; Mercado, B.Q.; Paulson, E.K.; Hurtley, A.E.; Miller, S.J. Diversity of Secondary Structure in Catalytic Peptides with β-Turn-Biased Sequences. *J. Am. Chem. Soc.* **2017**, *139*, 492–516. [CrossRef]
25. Fiorina, V.J.; Dubois, R.J.; Brynes, S. Ferrocenyl polyamines as agents for the chemoimmunotherapy of cancer. *J. Med. Chem.* **1978**, *21*, 393–395. [CrossRef]
26. Astruc, D. Why is Ferrocene so Exceptional? *Eur. J. Inorg. Chem.* **2017**, 6–29. [CrossRef]
27. Patra, M.; Gasser, G. The medicinal chemistry of ferrocene and its derivatives. *Nat. Rev. Chem.* **2017**, *1*, 66. [CrossRef]
28. Wang, R.; Chen, H.; Yan, W.; Zheng, M.; Zhang, T.; Zhang, Y. Ferrocene-containing hybrids as potential anticancer agents: Current developments, mechanisms of action and structure-activity relationships. *Eur. J. Med. Chem.* **2020**, *190*, 112109–112129. [CrossRef] [PubMed]
29. Arunan, E.; Desiraju, G.R.; Klein, R.A.; Sadlej, J.; Scheiner, S.; Alkorta, I.; Clary, D.C.; Crabtree, R.H.; Dannenberg, J.J.; Hobza, P.; et al. Defining the hydrogen bond: An account (IUPAC Technical Report). *Pure Appl. Chem.* **2011**, *83*, 1619–1636. [CrossRef]
30. Fornaro, T.; Burini, D.; Biczysko, M.; Barone, V. Hydrogen-bonding effects on infrared spectra from anharmonic computations: Uracil-water complexes and uracil dimers. *J. Phys. Chem. A* **2015**, *119*, 4224–4236. [CrossRef] [PubMed]
31. Ananthanarayanan, V.S.; Cameron, T.S. Proline-containing β-turns. *Int. J. Pept. Protein Res.* **1988**, *31*, 399–411. [CrossRef]
32. Vincenzi, M.; Mercurio, F.A.; Leone, M. NMR Spectroscopy in the Conformational Analysis of Peptides: An Overview. *Curr. Med. Chem.* **2021**, *28*, 2729–2782. [CrossRef] [PubMed]
33. Wagner, G.; Pardi, A.; Wuethrich, K. Hydrogen bond length and proton NMR chemical shifts in proteins. *J. Am. Chem. Soc.* **1983**, *105*, 5948–5949. [CrossRef]
34. Liu, J.-Y.; Sun, X.-Y.; Tang, Q.; Song, J.-J.; Li, X.-Q.; Gong, B.; Liu, R.; Lu, Z.-L. An unnatural tripeptide structure containing intramolecular double H-bond mimics a turn-hairpin conformation. *Org. Biomol. Chem.* **2021**, *19*, 4359–4363. [CrossRef]
35. Llinás, M.M.; Klein, M.P. Solution conformation of the ferrichromes. VI. Charge relay at the peptide bond. Proton magnetic resonance study of solvation effects on the amide electron density distribution. *J. Am. Chem. Soc.* **1975**, *97*, 4731–4737. [CrossRef]
36. Stevens, E.S.; Sugawara, N.; Bonora, G.M.; Toniolo, C. Conformational analysis of linear peptides. 3. Temperature dependence of NH chemical shifts in chloroform. *J. Am. Chem. Soc.* **1980**, *102*, 7048–7050. [CrossRef]
37. Kessler, H. Conformation and Biological Activity of cyclic Peptides. *Angew. Chem. Int. Ed. Engl.* **1982**, *21*, 512–523. [CrossRef]
38. Iqbal, M.; Balaram, P. Aggregation of apolar peptides in organic solvents. Concentration dependence of 1H-nmr parameters for peptide NH groups in 310 helical decapeptide fragment of suzukacillin. *Biopolymers* **1982**, *21*, 1427–1433. [CrossRef]

39. Vijayakumar, E.K.S.; Balaram, P. Stereochemistry of a-aminoisobutyric acid peptides in solution. Helical conformations of protected decapeptides with repeating Aib-L-Ala and Aib-L-Val sequences. *Biopolymers* **1983**, *22*, 2133–2140. [CrossRef] [PubMed]
40. Andersen, N.H.; Neidigh, J.W.; Harris, S.M.; Lee, G.M.; Liu, Z.; Tong, H. Extracting Information from the Temperature Gradients of Polypeptide NH Chemical Shifts. 1. The Importance of Conformational Averaging. *J. Am. Chem. Soc.* **1997**, *119*, 8547–8561. [CrossRef]
41. Baxter, N.J.; Williamson, M.P. Temperature dependence of ^{1}H chemical shifts in proteins. *J. Biomol. NMR* **1997**, *9*, 359–369. [CrossRef]
42. Cierpicki, T.; Otlewski, J. Amide proton temperature coefficients as hydrogen bond indicators in proteins. *J. Biomol. NMR* **2001**, *21*, 249–261. [CrossRef] [PubMed]
43. Cierpicki, T.; Zhukov, I.; Byrd, R.A.; Otlewski, J. Hydrogen bonds in human ubiquitin reflected in temperature coefficients of amide protons. *J. Magn. Reson.* **2002**, *157*, 178–180. [CrossRef] [PubMed]
44. Sarkar, S.K.; Young, P.E.; Sullivan, C.E.; Torchia, D.A. Detection of *cis* and *trans* X-Pro peptide bonds in proteins by ^{13}C NMR: Application to collagen. *Proc. Natl. Acad. Sci. USA* **1984**, *81*, 4800–4803. [CrossRef]
45. O'Neal, K.D.; Chari, M.V.; Mcdonald, C.H.; Cook, R.G.; Yu-Lee, L.Y.; Morrisett, J.D.; Shearer, W.T. Multiple *cis-trans* conformers of the prolactin receptor proline-rich motif (PRM) peptide detected by reverse-phase HPLC, CD and NMR spectroscopy. *Biochem. J.* **1996**, *315*, 833–844. [CrossRef]
46. Bandara, H.M.; Friss, T.R.; Enriquez, M.M.; Isley, W.; Incarvito, C.; Frank, H.A.; Gascon, J.; Burdette, S.C. Proof for the concerted inversion mechanism in the trans->cis isomerization of azobenzene using hydrogen bonding to induce isomer locking. *J. Org. Chem.* **2010**, *75*, 4817–4827. [CrossRef]
47. Siebler, C.; Maryasin, B.; Kümin, M.; Erdmann, R.S.; Rigling, C.; Grünenfelder, C.; Ochsenfeld, C.; Wennemers, H. Importance of Dipole Moments and Ambient Polarity for the Conformation of Xaa-Pro Moieties–A Combined Experimental and Theoretical Study. *Chem. Sci.* **2015**, *6*, 6725–6730. [CrossRef] [PubMed]
48. Pollastrini, M.; Lipparini, F.; Pasquinelli, L.; Balzano, F.; Barretta, G.U.; Pescitelli, G.; Angelici, G. A Proline Mimetic for the Design of New Stable Secondary Structures: Solvent-Dependent Amide Bond Isomerization of (S)-Indoline-2-carboxylic Acid Derivatives. *J. Org. Chem.* **2021**, *86*, 7946–7954. [CrossRef]
49. Deetz, M.J.; Fahey, J.E.; Smith, B.D. NMR studies of hydrogen bonding interactions with secondary amide and urea groups. *J. Phys. Org. Chem.* **2001**, *14*, 463–467. [CrossRef]
50. Pignataro, M.F.; Herrera, M.G.; Dodero, V.I. Evaluation of Peptide/Protein Self-Assembly and Aggregation by Spectroscopic Methods. *Molecules* **2020**, *25*, 4854. [CrossRef]
51. Rogers, D.M.; Jasim, S.B.; Dyer, N.T.; Auvray, F.; Réfrégiers, M.; Hirst, J.D. Electronic Circular Dichroism Spectroscopy of Proteins. *Chemistry* **2019**, *5*, 2751–2774. [CrossRef]
52. Franken, N.A.P.; Stap, J.; Rodermond, H.; Haveman, J.; Van Bree, C. Clonogenic assay of cells in vitro. *Nat. Protoc.* **2006**, *1*, 2315–2319. [CrossRef]
53. Li, L.; Thomas, R.M.M.; Suzuki, H.; de Brabander, J.K.; Wang, X.; Harran, P.G. A small molecule Smac mimic potentiates TRAIL-and TNFalpha-mediated cell death. *Science* **2004**, *305*, 1471–1474. [CrossRef]
54. Mabonga, L.; Kappo, A.P. Peptidomimetics: A Synthetic Tool for Inhibiting Protein–Protein Interactions in Cancer. *Int. J. Pept. Res. Ther.* **2020**, *26*, 225–241. [CrossRef]
55. Schrödinger, E. *Maestro, Version 9.7*; Schrödinger: New York, NY, USA, 2014.
56. Schrödinger, E. MacroModel, version 10.3. Schrödinger: New York, NY, USA, 2014.
57. Mohamadi, F.; Richards, N.G.J.; Guida, W.C.; Liskamp, R.; Lipton, M.; Caufield, C.; Chang, G.; Hendrickson, T.; Still, W.C. Macromodel—An integrated software system for modeling organic and bioorganic molecules using molecular mechanics. *J. Comput. Chem.* **1990**, *11*, 440–467. [CrossRef]
58. Frisch, M.J.; Trucks, G.W.; Schlegel, H.B.; Scuseria, G.E.; Robb, M.A.; Cheeseman, J.R.; Scalmani, G.; Barone, V.; Petersson, G.A.; Nakatsuji, H.; et al. *Gaussian 16, Revision, C.01*; Gaussian, Inc.: Wallingford, CT, USA, 2016.
59. Marenich, A.V.; Cramer, C.J.; Truhlar, D.G. Universal solvation model based on solute electron density and a continuum model of the solvent defined by the bulk dielectric constant and atomic surface tensions. *J. Phys. Chem. B* **2009**, *113*, 6378–6396. [CrossRef]
60. Keith, T.A. *AIMAll, Version 19.02.13*; TK Gristmill Software: Overland Park, KS, USA, 2017.
61. Koch, U.; Popelier, P.L.A.J. Characterization of C-H-O Hydrogen Bonds on the Basis of the Charge Density. *J. Phys. Chem.* **1995**, *99*, 9747–9754. [CrossRef]
62. Popelier, P.L.A.J. Characterization of a dihydrogen bond on the basis of the electron density. *J. Phys. Chem. A* **1998**, *102*, 1873–1878. [CrossRef]
63. Rigaku, P.R.O. *CrysAlis, Version: 1.171.39.46*; Rigaku Oxford Diffraction Ltd.: Yarnton, UK, 2018.
64. Sheldrick, G.M. SHELXT-integrated space-group and crystal-structure determination. *Acta Crystallogr. A Found. Adv.* **2015**, *71*, 3–8. [CrossRef] [PubMed]
65. Sheldrick, G.M. Crystal structure refinement with SHELXL. *Acta Crystallogr. C* **2015**, *C71*, 3–8. [CrossRef]
66. Spek, A.L. CheckCIF validation ALERTS: What they mean and how to respond. *Acta Crystallogr. E76* **2020**, 1–11. [CrossRef] [PubMed]
67. Farrugia, L.J. ORTEP-3 for Windows-a version of ORTEP-III with a Graphical User Interface (GUI). *J. Appl. Cryst.* **1997**, *30*, 565. [CrossRef]

68. Macrae, C.F.; Sovago, I.; Cottrell, S.J.; Galek, P.T.A.; McCabe, P.; Pidcock, E.; Platings, M.; Shields, G.P.; Stevens, J.S.; Towler, M.; et al. Mercury 4.0: From visualization to analysis, design and prediction. *J. Appl. Crystallogr.* **2020**, *53*, 226–235. [CrossRef]
69. Kraljić, K.; Brkan, V.; Škevin, D.; Gaurina Srček, V.; Radošević, K. Canolol Dimer, a Biologically Active Phenolic Compound of Edible Rapeseed Oil. *Lipids* **2019**, *54*, 189–200. [CrossRef]
70. Radošević, K.; Rudjer, N.; Slivac, I.; Mihajlović, M.; Dumić, J.; Kniewald, Z.; Gaurina Srček, V. Cytotoxic and Apoptotic Effects of 17α-Ethynylestradiol and Diethylstilbestrol on CHO−K1 Cells. *Food Technol. Biotechnol.* **2011**, *49*, 447–452.

International Journal of
Molecular Sciences

Article

Pharmacologically Targeting the Fibroblast Growth Factor 14 Interaction Site on the Voltage-Gated Na^+ Channel 1.6 Enables Isoform-Selective Modulation

Nolan M. Dvorak , Cynthia M. Tapia, Aditya K. Singh, Timothy J. Baumgartner, Pingyuan Wang, Haiying Chen, Paul A. Wadsworth , Jia Zhou and Fernanda Laezza *

Department of Pharmacology and Toxicology, University of Texas Medical Branch, Galveston, TX 75901, USA; nmdvorak@utmb.edu (N.M.D.); cmtapia@utmb.edu (C.M.T.); adsingh@utmb.edu (A.K.S.); tjbaumga@utmb.edu (T.J.B.); wangpingyuan@ouc.edu.cn (P.W.); haichen@utmb.edu (H.C.); pawadswo@utmb.edu (P.A.W.); jizhou@utmb.edu (J.Z.)
* Correspondence: felaezza@utmb.edu; Tel.: +1-409-772-9672

Abstract: Voltage-gated Na^+ (Na_V) channels are the primary molecular determinant of the action potential. Among the nine isoforms of the Na_V channel α subunit that have been described (Na_V1.1-Na_V1.9), Na_V1.1, Na_V1.2, and Na_V1.6 are the primary isoforms expressed in the central nervous system (CNS). Crucially, these three CNS Na_V channel isoforms display differential expression across neuronal cell types and diverge with respect to their subcellular distributions. Considering these differences in terms of their localization, the CNS Na_V channel isoforms could represent promising targets for the development of targeted neuromodulators. However, current therapeutics that target Na_V channels lack selectivity, which results in deleterious side effects due to modulation of off-target Na_V channel isoforms. Among the structural components of the Na_V channel α subunit that could be pharmacologically targeted to achieve isoform selectivity, the C-terminal domains (CTD) of Na_V channels represent promising candidates on account of displaying appreciable amino acid sequence divergence that enables functionally unique protein–protein interactions (PPIs) with Na_V channel auxiliary proteins. In medium spiny neurons (MSNs) of the nucleus accumbens (NAc), a critical brain region of the mesocorticolimbic circuit, the PPI between the CTD of the Na_V1.6 channel and its auxiliary protein fibroblast growth factor 14 (FGF14) is central to the generation of electrical outputs, underscoring its potential value as a site for targeted neuromodulation. Focusing on this PPI, we previously developed a peptidomimetic derived from residues of FGF14 that have an interaction site on the CTD of the Na_V1.6 channel. In this work, we show that whereas the compound displays dose-dependent effects on the activity of Na_V1.6 channels in heterologous cells, the compound does not affect Na_V1.1 or Na_V1.2 channels at comparable concentrations. In addition, we show that the compound correspondingly modulates the action potential discharge and the transient Na+ of MSNs of the NAc. Overall, these results demonstrate that pharmacologically targeting the FGF14 interaction site on the CTD of the Na_V1.6 channel is a strategy to achieve isoform-selective modulation, and, more broadly, that sites on the CTDs of Na_V channels interacted with by auxiliary proteins could represent candidates for the development of targeted therapeutics.

Keywords: peptidomimetics; protein–protein interactions (PPIs); voltage-gated Na^+ (Na_V) channels; fibroblast growth factor 14 (FGF14); medium spiny neurons (MSNs); nucleus accumbens (NAc); neurotherapeutics

Citation: Dvorak, N.M.; Tapia, C.M.; Singh, A.K.; Baumgartner, T.J.; Wang, P.; Chen, H.; Wadsworth, P.A.; Zhou, J.; Laezza, F. Pharmacologically Targeting the Fibroblast Growth Factor 14 Interaction Site on the Voltage-Gated Na⁺ Channel 1.6 Enables Isoform-Selective Modulation. *Int. J. Mol. Sci.* **2021**, *22*, 13541. https://doi.org/10.3390/ijms222413541

Academic Editors: Menotti Ruvo and Nunzianna Doti

Received: 21 November 2021
Accepted: 15 December 2021
Published: 17 December 2021

1. Introduction

In excitable cells, voltage-gated Na^+ (Na_V) channels enable the initiation and propagation of the action potential [1,2]. Among the nine isoforms of the Na_V channel α subunit (Na_V1.1-Na_V1.9) that have been described, Na_V1.1, Na_V1.2, and Na_V1.6 are the primary isoforms expressed in the central nervous system (CNS) [1]. In addition to displaying unique electrophysiological profiles, $Na_v1.1$, $Na_v1.2$, and $Na_v1.6$ channels vary with respect to their distributions across neuronal cell types [3–7] and their subcellular distributions [8,9]. Given this heterogeneity of localization, isoform-selective targeting of one of the isoforms could enable targeted neuromodulatory effects. Unfortunately, current therapeutics that target Na_v channels lack isoform selectivity due to targeting structural regions of the α subunit that are highly conserved across the nine isoforms, which resultantly confers such drugs with deleterious side-effects due to modulation of off-target Na_v channel isoforms [10,11]. As such, the identification of less highly conserved structural regions that are amenable to pharmacological modulation is a necessary prerequisite to fully actualize the potential of Na_v channels as targets for neurologic and neuropsychiatric disorders [11].

On account of displaying appreciable amino acid sequence divergence, the C-terminal domains (CTDs) of Na_v channels could represent promising sites to pharmacologically target to achieve isoform selective modulation [11–16]. In particular, targeting sites on CTDs interacted with by Na_v channel auxiliary proteins could represent a novel strategy to achieve isoform-selective modulation given that these are sites that enable endogenously specific intermolecular regulation [11–13,17–23]. In the brain, the Na_v channel auxiliary protein fibroblast growth factor 14 (FGF14) interacts with the CTDs of the $Na_v1.1$, $Na_v1.2$, and $Na_v1.6$ channels, and its two splice variants, FGF14-1a and FGF14-1b, differentially regulate the gating and trafficking of the three CNS Nav channel isoforms [12,15,18,24]. Given these functionally unique protein–protein interactions (PPIs) between FGF14 splice variants and CNS Na_v channel isoforms, in tandem with FGF14 being an important regulator of neuronal activity and behavior [25–34], the interaction sites of FGF14 on the CTDs of these Na_v channel isoforms could potentially be pharmacologically targeted to develop novel neuromodulators.

Focusing on the FGF14 interaction site on the CTD of the $Na_v1.6$ channel on account of the FGF14:$Na_v1.6$ complex being central to the generation of electrical outputs of medium spiny neurons (MSNs) of the nucleus accumbens (NAc) [24], which is a critical brain region that regulates reward-related behavior [35], we previously developed a homology model of the PPI interface to guide drug discovery efforts [15]. These investigations identified three clusters of amino acids of FGF14 with interaction sites on the CTD of the $Na_v1.6$ channel, namely the Phe-Leu-Pro-Lys (FLPK) and Pro-Leu-Glu-Val (PLEV) motifs on the β12 sheet and the Tyr-Tyr-Val (YYV) motif on the β8/9 loop [15]. In subsequent works, these amino acid sequences were used as scaffolds for the development of peptidomimetics targeting the $Na_v1.6$ channel macromolecular complex [24,36–38]. Pertinent to the present investigation, we previously presented PW201, also referred to as compound 12, which is derived from the YYV peptide [37]. In our previous study [37], we found that PW201 modulated FGF14:$Na_v1.6$ complex assembly, bound appreciably to the CTD of the $Na_v1.6$ channel, decreased the $Na_v1.6$ channel-mediated transient Na^+ current (I_{Na}) in heterologous cells, and had predicted interactions with FGF14′s interaction site on the CTD of $Na_v1.6$. In the present investigation, we expand upon these findings and show that whereas PW201 modulates $Na_v1.6$ channel-mediated I_{Na} in heterologous cells in a dose-dependent manner, the compound displays no effects on $Na_v1.1$ channel- or $Na_v1.2$ channel-mediated I_{Na} at comparable concentrations. Additionally, we show that in MSNs of the NAc, PW201 correspondingly modulates I_{Na} and action potential discharge. Overall, these results demonstrate that pharmacologically targeting the FGF14 interaction site on the CTD of the $Na_v1.6$ channel enables isoform-selective modulation of the $Na_v1.6$ channel and resultantly alters MSN activity, which could collectively represent promising features for the development of future neuromodulators.

2. Results

2.1. PW201 Has Predicted Interactions with the $FGF14^{YYV}$ Interaction Site on the CTD of the $Na_v1.6$ Channel

In our previous study [15], we developed a homology model of the PPI interface between FGF14 and the CTD of the $Na_v1.6$ channel using the previously published crystal structure of the CTD of the $Na_v1.5$ channel in complex with calmodulin and FGF13 as a template [14]. Through assessment of the homology model, in tandem with biochemical and functional validation modules, we identified the Try158-Tyr159-Val160 motif on the β8/9 loop of FGF14 ($FGF14^{YYV}$) as a "hot segment" [39] at the FGF14:$Na_v1.6$ PPI interface [15]. Crucially, these three residues of FGF14 have predicted interaction sites on the CTD of the $Na_v1.6$ channel. Based upon $FGF14^{YYV}$ having this predicted interaction site on the CTD of the $Na_v1.6$ channel, we first sought to investigate if PW201, which is derived from the YYV motif of FGF14, similarly engaged with residues of the CTD of the $Na_v1.6$ channel.

To this end, we employed molecular modeling and docked PW201 with our previously reported homology model of the CTD of $Na_v1.6$ [15] (Figure 1A–C). Consistent with PW201's derivation from the YYV motif of the β8/9 loop of FGF14, the docking study of the compound showed that PW201 docks well with the $Na_v1.6$ CTD at the same site where the β8/9 loop of FGF14 interacts. In particular, PW201 forms H-bonds with Asp1833, Met1832, and Arg1891. In addition, the fluorenylmethoxycarbonyl (Fmoc) protecting group added to the N-terminus of the YYV scaffold to improve the compound's drug-like properties interacts with Arg1866, a residue of the CTD of the $Na_v1.6$ channel involved in an intramolecular salt bridge with Asp1846. Collectively considered, these molecular modeling studies provide insights into the putative binding mode of PW201 with the CTD of the $Na_v1.6$ channel, which has important implications for understanding its mechanism of action.

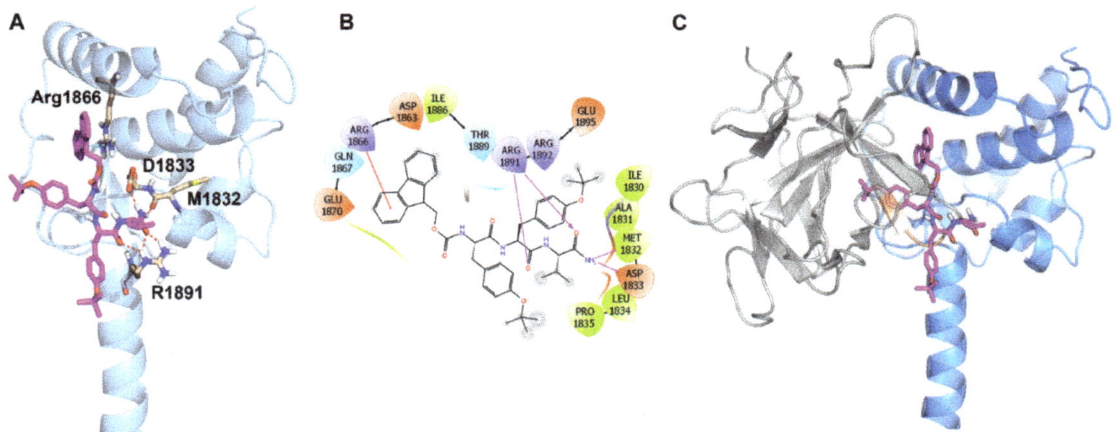

Figure 1. Molecular modeling of PW201 in complex with the homology model of the CTD of the $Na_v1.6$ channel (modified from Dvorak et al., 2020. *Molecules*. PMID: 32722255). (**A**) Ribbon representation of PW201 (magenta) docked with the homology model of the CTD of $Na_v1.6$ (blue). Key interaction residues are shown as sticks, H-bonds are shown as red dashed lines, and Pi–cation interactions are shown as cyan dashed lines. (**B**) Interaction diagram of PW201's predicted binding site with the homology model of the CTD of $Na_v1.6$. H-bonds are shown as purple lines and Pi–cation interactions are shown as red lines. (**C**) Docked pose of PW201 with the homology model of the CTD of the $Na_v1.6$ channel overlaid with FGF14. The $Na_v1.6$ CTD is shown as blue ribbon and FGF14 as gray ribbon. Ligand is shown as a magenta stick and the YYV motif of the FGF14 β8/9 loop is highlighted in orange.

2.2. PW201 Dose-Dependently Suppresses $Na_v1.6$ Channel-Mediated I_{Na} in Heterologous Cells

In our previous study [37], we showed that 20 μM PW201 suppressed $Na_v1.6$-mediated I_{Na} in heterologous cells, which is an effect similar to that observed due to co-expression of FGF14 with the $Na_v1.6$ channel in heterologous cells [12,15,24,38,40,41]. This effect is consistent with the compound's previously shown direct binding to the CTD of the $Na_v1.6$ channel [37]. Additionally, we previously showed that whereas PW201 modulated the peak I_{Na} density mediated by $Na_v1.6$ channels in heterologous cells, the compound did not affect the voltage dependences of activation or steady-state inactivation [37]. Given these previously shown electrophysiological changes conferred by the compound on $Na_v1.6$-mediated I_{Na}, we sought to assess the dose-dependency of PW201's effects on $Na_v1.6$-mediated peak I_{Na} density in HEK293 cells expressing the $Na_v1.6$ channel (HEK-$Na_v1.6$). To do so, HEK-$Na_v1.6$ cells were incubated for 30 min with either vehicle (0.1% DMSO; n = 6 cells) or one of seven concentrations of PW201 (range: 1–500 μM; n = 4–6 cells per concentration). After incubation, whole-cell patch-clamp electrophysiology was employed, and cells were recorded from using the voltage-clamp protocol shown in Figure 2A. Recordings performed in HEK-$Na_v1.6$ cells treated with 0.1% DMSO elicited an average peak current density of −65.6 ± 4.7 pA/pF (n = 6 cells). The peak current of each recording was then divided by this average and reported as the percent peak current density (Figure 2B,C).

Table 1. Summary of the effects of 15 μM PW201 on $Na_v1.1$-, $Na_v1.2$-, and $Na_v1.6$-mediated currents in heterologous cells [a].

Na_v Isoform	Condition	Peak Current Density (pA/pF) [b]	Tau of Fast Inactivation (ms) [c]	$V_{1/2}$ of Activation (mV) [d]	$V_{1/2}$ of Steady-State Inactivation (mV) [e]
$Na_v1.1$	DMSO	−139.0 ± 4.8 (8)	1.1 ± 0.1 (8)	−25.0 ± 1.4 (8)	−52.2 ± 1.9 (8)
	PW201	−132.3 ± 3.8 (9)	1.0 ± 0.1 (9)	−25.3 ± 1.5 (9)	−50.7 ± 1.0 (7)
$Na_v1.2$	DMSO	−113.4 ± 10.8 (8)	1.0 ± 0.1 (8)	−25.2 ± 2.5 (8)	−54.9 ± 2.3 (6)
	PW201	−107.8 ± 11.9 (7)	1.2 ± 0.1 (7)	−22.1 ± 1.2 (7)	−55.8 ± 3.7 (5)
$Na_v1.6$	DMSO	−65.6 ± 4.7 (6)	1.2 ± 0.1 (6)	−23.1 ± 0.7 (6)	−59.7 ± 0.3 (6)
	PW201	−43.4 ± 2.4 (6) **	1.2 ± 0.1 (6)	−21.8 ± 1.3 (6)	−61.0 ± 1.3 (6)

[a] Summary of the electrophysiological evaluation of 15 μM PW201 in HEK-$Na_v1.1$, HEK-$Na_v1.2$, and HEK-$Na_v1.6$ cells. Results are expressed as mean ± SEM. The number of independent experiments is shown in parentheses. A Student's t-test comparing cells treated with 0.1% DMSO and 15 μM PW201 was used to determine statistical significance. **, $p < 0.01$. [b] Peak current density, which described the number of channels in a conductive (open) state, is a measure of the maximum influx of I_{Na} (pA) into the cell normalized to membrane capacitance (pF) to control for variable cell sizes. [c] Tau of fast inactivation measures the decay phase of I_{Na} to characterize the time required for channels to transition from the conductive (open) state to a nonconductive state resulting from fast inactivation. [d] $V_{1/2}$ of activation is a measure of the voltage at which half of the available channels transition from the closed to the conductive (open) state. [e] $V_{1/2}$ of steady-state inactivation is a measure of the voltage at which half of channels are available to transition into the conductive (open) state, while the other half are non-conductive due to steady-state (closed-state) inactivation.

In Figure 2B, the half-maximal inhibitory concentration (IC_{50}) of PW201 in terms of suppressing $Na_v1.6$-mediated peak I_{Na} density was determined to be 15.1 μM. This finding is consistent with the bar graph representation of the data in Figure 2C, which shows that statistically significant inhibitory effects on $Na_v1.6$-mediated peak I_{Na} density are observed at 15 μM but not at single-digit micromolar concentrations. With escalating concentrations, a plateau effect appears to be reached at 50 μM, as the currents from cells treated with 100 and 500 μM PW201 are similarly suppressed. Overall, these dose-dependency studies support the findings of our previous investigation [37] and demonstrate that through direct binding to the CTD of the $Na_v1.6$ channel, PW201 is able to dose-dependently affect the peak transient I_{Na} density in heterologous cells in a fashion similar to that which is observed due to co-expression of FGF14 with the $Na_v1.6$ channel in heterologous cells.

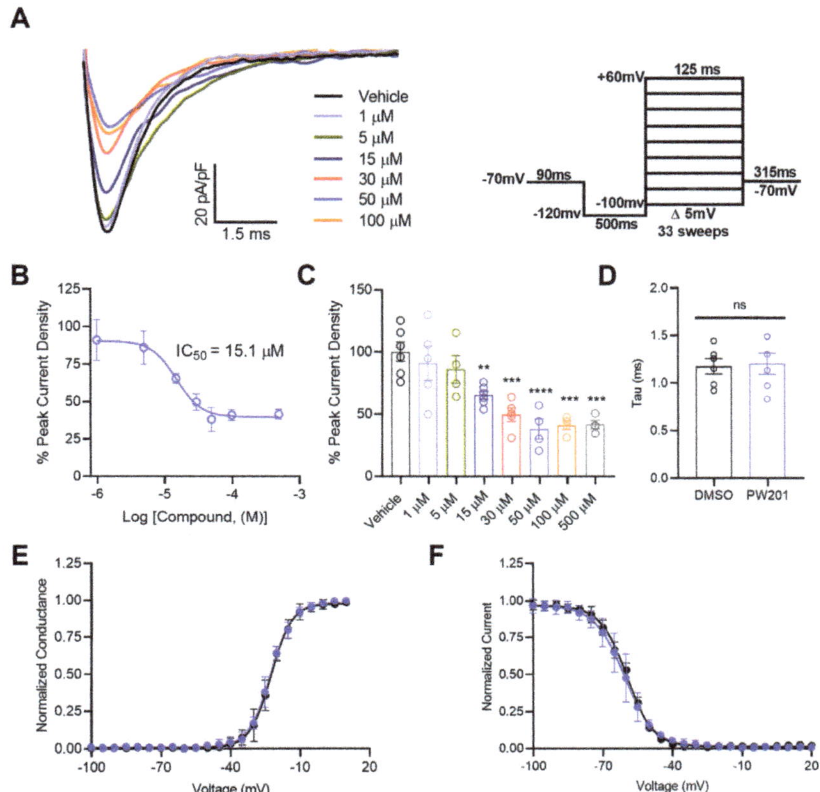

Figure 2. Dose-dependent effects of PW201 on peak I_{Na} densities elicited by HEK-Na$_V$1.6 cells. (**A**) Representative traces of peak transient currents recorded from HEK-Na$_V$1.6 cells treated with 0.1% DMSO (vehicle) or one of seven concentrations of PW201 (range: 1–500 μM). The trace of the peak current from cells treated with 500 μM is not shown to avoid overlap of traces and clarity of representation. Cells were recorded from using the voltage-clamp protocol shown in the inset with a P4 leak cancellation protocol. (**B**) Percentage peak current density plotted as a function of the log concentration of the compound to characterize the dose-dependency of the effects of PW201 on this electrophysiological parameter. Percent peak current was calculated by dividing the peak current density of each recording by the average peak current density of cells treated with 0.1% DMSO. Non-linear regression curve-fitting was performed using GraphPad Prism 8. (**C**) Bar graph representation of individual replicates from dose–response analyses shown in (**B**). (**D**) Comparison of tau of fast inactivation of I_{Na} between HEK-Na$_V$1.6 cells treated with DMSO and 15 μM PW201. (**E**) Normalized conductance plotted as a function of the voltage to characterize the effects of DMSO (black) and 15 μM PW201 (blue) on the voltage dependence of activation of I_{Na} elicited by HEK-Na$_V$1.6 cells. Plotted data were fitted with the Boltzmann equation to determine $V_{1/2}$ of activation (see Table 1). (**F**) Normalized current plotted as a function of the voltage to characterize the effects of DMSO (black) and 15 μM PW201 (blue) on the voltage dependence of steady-state inactivation of I_{Na} elicited by HEK-Na$_V$1.6 cells. Plotted data were fitted with the Boltzmann equation to determine $V_{1/2}$ of steady-state inactivation (see Table 1). Data shown are mean ± SEM. In (**C**,**D**), circles represent individual replicates. In (**C**), significance was assessed using a one-way ANOVA with post hoc Dunnett's multiple comparisons test. ns, not significant; **, $p < 0.01$; ***, $p < 0.001$; ****, $p < 0.0001$. In (**D**–**F**), significance was assessed using an unpaired *t*-test comparing cells treated with 0.1% DMSO or 15 μM (see Table 1).

Based on the dose-dependent effects of PW201 on peak I_{Na} density observed in Figure 2A–C, we elected to further test the effects of 15 μM PW201 on other electrophysiological parameters, a concentration selected on the basis of it being near the calculated IC_{50} value in Figure 2B. Consistent with our previous investigation, where 20 μM PW201 exerted no effects on the tau of fast inactivation, voltage dependence of activation, or voltage dependence of steady-state inactivation of I_{Na} elicited by HEK-Na$_V$1.6 cells [37], treatment of HEK-$Na_v1.6$ cells with 15 μM PW201 similarly did not affect these parameters (Figure 2D–F). As it pertains to the former, HEK-$Na_v1.6$ cells treated with 0.1% DMSO displayed a tau of fast inactivation value of 1.17 ± 0.08 ms, which was not significantly different than the tau of fast inactivation value of HEK-$Na_v1.6$ cells treated with 15 μM PW201 (1.21 ± 0.11 ms; $n = 6$ cells per group; $p = 0.83$; Figure 2D). As it relates to the voltage dependence of activation, HEK-$Na_v1.6$ cells treated with 0.1% DMSO or 15 μM PW201 displayed $V_{1/2}$ of activation values of -23.1 ± 0.72 mV or -21.8 ± 1.3, respectively ($n = 6$ per group; $p = 0.41$; Figure 2E). Lastly, as it relates to the voltage dependence of steady-state inactivation, HEK-$Na_v1.6$ cells treated with 0.1% DMSO or 15 μM PW201 displayed $V_{1/2}$ of steady-state inactivation values of -59.7 ± 0.31 mV or -61.0 ± 1.3 mV, respectively ($n = 6$ per group; $p = 0.35$; Figure 2F). Collectively considered, the results of Figure 2 demonstrate that PW201 confers dose-dependent effects on the I_{Na} amplitude, and that the effects of 15 μM PW201 on the I_{Na} amplitude are not accompanied by changes in the kinetics or voltage dependences of activation or inactivation of $Na_v1.6$ channels.

2.3. Profiling the Selectivity of PW201 for the $Na_v1.6$ Channel

Having shown previously [37] and demonstrated the dose-dependency (Figure 2) of the effects of PW201 on the transient I_{Na} of $Na_v1.6$ channels in heterologous cells, we next sought to characterize if these effects of PW201 were selective among CNS Na_v channel isoforms. To do so, HEK293 cells stably expressing either $Na_v1.1$ (HEK-$Na_v1.1$) [24,38,42] or $Na_v1.2$ (HEK-$Na_v1.2$) [24,38,43] channels were incubated for 30 min with either 0.1% DMSO or 15 μM PW201, a concentration of the ligand selected on the basis of its IC_{50} value for $Na_v1.6$ determined in Figure 2B. After incubation, the effects of vehicle and PW201 treatment on $Na_v1.1$ and $Na_v1.2$ channels were assessed using whole-cell voltage-clamp recordings (Figure 3).

As mentioned above, co-expression of FGF14 with the $Na_v1.6$ channel in heterologous cells results in a suppression of $Na_v1.6$-mediated I_{Na} [12,15,24,38,40,41], similar to the suppression of peak transient I_{Na} conferred by treatment of HEK-$Na_v1.6$ cells with PW201. Notably, co-expression of FGF14 with the $Na_v1.1$ channel [18] and the $Na_v1.2$ channel [12] in heterologous cells has similarly been shown to suppress $Na_v1.1$- and $Na_v1.2$-mediated peak transient I_{Na}. Despite these conserved modulatory effects of co-expression of FGF14 with $Na_v1.1$, $Na_v1.2$, and $Na_v1.6$ channels on peak I_{Na} density in heterologous cells, treatment of only HEK-$Na_v1.6$ cells with PW201 results in a suppression of peak I_{Na} density, whereas this parameter is unaffected in HEK-$Na_v1.1$ (Figure 3A–C) and HEK-$Na_v1.2$ (Figure 3I–K) cells treated with 15 μM PW201. Lending further credence to PW201′s isoform-selective effects on the $Na_v1.6$ channel, PW201 exerted no modulatory effects on tau of fast inactivation, the voltage dependence of activation, or the voltage dependence of steady-state inactivation of $Na_v1.1$ (Figure 3D–H) or $Na_v1.2$ (Figure L–P) channels stably expressed in heterologous systems.

2.4. PW201 Potentiates the Excitability of MSNs of the NAc through Na_v Channel Modulation

Having shown that PW201 modulates $Na_v1.6$-mediated I_{Na}, but not $Na_v1.1$- or $Na_v1.2$-mediated I_{Na}, in heterologous cells (Figures 2 and 3, respectively), we next sought to characterize how these collective modulatory effects might alter the activity of MSNs of the NAc. MSNs represent a promising cellular target to be affected by such a ligand as previous studies have shown that FGF14 and $Na_v1.6$ channels are enriched in these cells [24]. To test the effects of PW201 on intact MSNs of the NAc, acute brain slice preparations containing the NAc were incubated with either 0.01% DMSO or 15 μM PW201 for 30 min, after which either whole-cell current-clamp (Figure 4A–E) or whole-cell voltage-clamp (Figure 4F–H) electrophysiological recordings were performed.

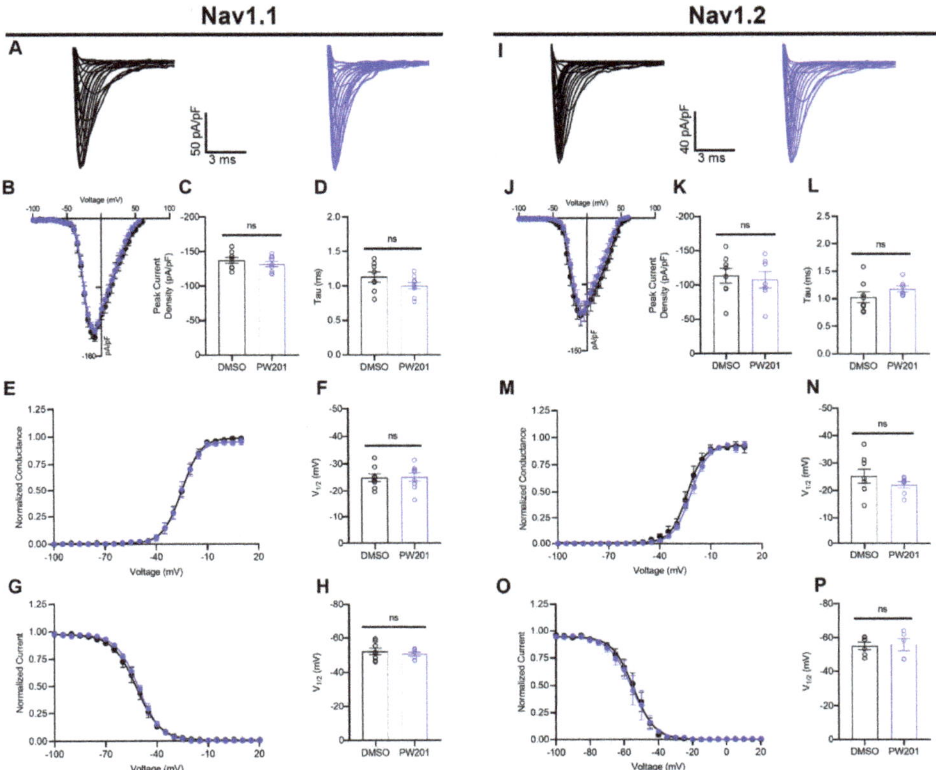

Figure 3. Electrophysiological evaluation of the effects of PW201 on $Na_V1.1$ and $Na_V1.2$ channels. (**A**,**I**) Representative traces of transient Na^+ currents elicited by the indicated cell type treated with either 0.1% DMSO (black) or 15 μM PW201 (blue). (**B**,**J**) Current–voltage relationships for cells of the indicated type treated with either 0.1% DMSO (black) or 15 μM PW201 (blue). (**C**,**K**) Comparison of the peak current density for the experimental groups described in (**B**) and (**J**), respectively. (**D**,**L**) Comparison of tau of fast inactivation of HEK-$Na_V1.1$ and HEK-$Na_V1.2$ cells, respectively, treated with either 0.1% DMSO (black) or 15 μM PW201 (blue). (**E**,**M**) Normalized conductance plotted as a function of the voltage for HEK-$Na_V1.1$ and HEK-$Na_V1.2$ cells, respectively, that were treated with 0.1% DMSO (black) or 15 μM PW201 (blue) to characterize the effects of vehicle and compound treatment on the voltage dependencies of activation of $Na_V1.1$ and $Na_V1.2$ channels. (**F**,**N**) Comparison of $V_{1/2}$ of activation of transient Na^+ currents elicited by HEK-$Na_V1.1$ and HEK-$Na_V1.2$ cells, respectively, that were treated with 0.1% DMSO or 15 μM PW201. (**G**,**O**) Normalized current plotted as a function of the voltage for HEK-$Na_V1.1$ and HEK-$Na_V1.2$ cells, respectively, that were treated with 0.1% DMSO (black) or 15 μM PW201 (blue) to characterize the effects of vehicle and compound treatment on the voltage dependencies of steady-state inactivation of $Na_V1.1$ and $Na_V1.2$ channels. (**H**,**P**) Comparison of $V_{1/2}$ of steady-state inactivation of transient Na^+ currents elicited by HEK-$Na_V1.1$ and HEK-$Na_V1.2$ cells, respectively, that were treated with 0.1% DMSO (black) or 15 μM PW201 (blue). Data are mean ± SEM. In bar graphs, circles represent individual replicates. Significance was assessed using an unpaired *t*-test comparing cells treated with 0.1% DMSO and 15 μM PW201. ns, not significant. A table summary of results is shown in Table 1.

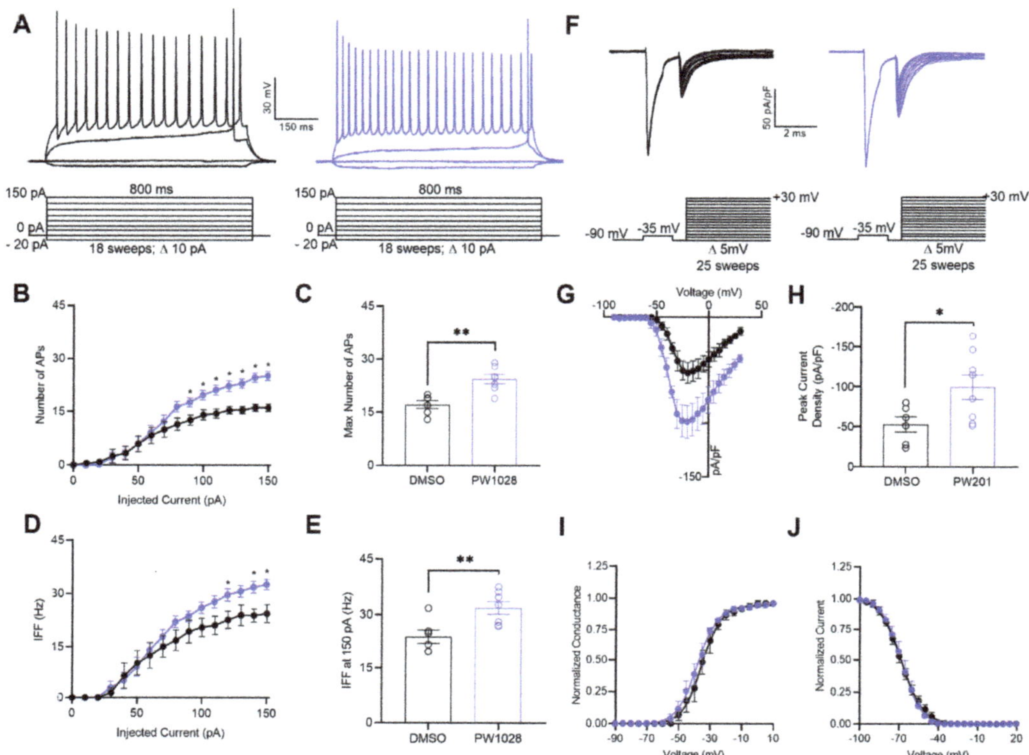

Figure 4. PW201 potentiates the intrinsic excitability and I_{Na} of MSNs in the NAc. (**A**) Representative traces of evoked action potentials from MSNs treated with 0.01% DMSO (black) or 15 µM PW201 (blue) in response to increasing current injections (schematic of the current-clamp protocol is shown below representative traces). (**B**) Average number of evoked action potentials at each current step from MSNs treated with 0.01% DMSO (black) or 15 µM PW201 (blue). (**C**) Comparison of the max number of evoked action potentials between MSNs treated with 0.01% DMSO or 15 µM PW201. (**D**) Average instantaneous firing frequencies (IFFs) at each current step from MSNs treated with 0.01% DMSO (black) or 15 µM PW201 (blue). (**E**) Comparison of IFFs of MSNs at the 150 pA current step treated with 0.01% DMSO or 15 µM PW201. (**F**) Representative traces of transient I_{Na} of MSNs of the NAc treated with either 0.01% DMSO (black) or 15 µM PW201 (blue) in response to the voltage-clamp protocol shown below the traces. (**G**) Current–voltage relationship for experimental groups descried in (**F**). (**H**) Bar graph derived from (**G**) comparing the peak I_{Na} density of MSNs treated with 0.01% DMSO or 15 µM PW201. (**I**) Normalized conductance plotted as a function of the voltage to characterize the effects of 0.01% DMSO (black) or 15 µM PW201 (blue) on the voltage dependence of activation of I_{Na} of MSNs. (**J**) Normalized current plotted as a function of the voltage to characterize the effects of 0.01% DMSO (black) or 15 µM PW201 (blue) on the voltage dependence of steady-state inactivation of I_{Na} of MSNs. Data are mean ± SEM. In bar graphs, circles represent individual replicates. Significance was assessed using an unpaired *t*-test comparing MSNs treated with 0.1% DMSO and 15 µM PW201. *, $p < 0.05$; **, $p < 0.01$. A table summary of the current-clamp results is shown in Table 2.

Table 2. Effects of PW201 on passive and electrical properties of MSNs of the NAc [a].

Treatment	Max Number of APs	IFF at 150 pA (Hz)	RMP (mV)	I_{thr} (pA)	V_{thr} (mV)	Max Rise (mV/ms)	Max Decay (mV/ms)	R_{in} (MΩ)	Tau (ms)	C_m (pF)
DMSO	17.2 ± 1.1 (6)	24.5 ± 2.6 (6)	−71.7 ± 3.8 (6)	40.0 ± 7.3 (6)	−43.1 ± 1.4 (6)	233.9 ± 33.7 (6)	−60.6 ± 4.9 (6)	199.6 ± 13.0 (6)	25.6 ± 4.6 (6)	125.2 ± 14.3 (6)
PW201	24.4 ± 1.3 (7) **	32.7 ± 1.4 (7) **	−71.1 ± 1.6 (7)	41.4 ± 6.3 (7)	−42.9 ± 1.5 (7)	227.4 ± 25.1 (7)	−58.4 ± 2.8 (7)	168.4 ± 13.5 (7)	15.6 ± 3.2 (7)	91.6 ± 14.1 (7)

[a] Summary of the ex vivo electrophysiological evaluation of PW201 in MSNs of the NAc. Results are expressed as mean ± SEM. The number of independent experiments is shown in parentheses. A Student's *t*-test comparing MSNs treated with 0.1% DMSO and 15 µM PW201 was used to determine statistical significance. **, $p < 0.01$.

In current-clamp recordings, treatment of MSNs of the NAc with PW201 resulted in a potentiation of their action potential discharge (Figure 4A–C; Table 2). In addition to increasing the maximal firing of MSNs, PW201 also increased the instantaneous firing frequency of these neurons (Figure 4D,E). These effects of PW201 on action potential discharge, coupled with a lack of effects on passive electrical properties, such as resting membrane potential and input resistance (Table 2), suggest that the ligand's modulatory effects on neuronal excitability likely arise due to changes in Na_v channel activity. Such a hypothesis is further supported by the effects of PW201 on $Na_v1.6$ channel conductance in heterologous cells (Figure 2), as well as by previous investigations demonstrating changes in the maximal firing and instantaneous firing frequency of neurons due to changes in Nav channel conductance [44–47].

As an additional test to ensure that the effects on the action potential discharge of MSNs of the NAc conferred by PW201 were mediated by changes in Na_v channel activity, whole-cell voltage-clamp recordings of I_{Na} were performed in intact MSNs in the acute brain slice preparation using the voltage-clamp protocol shown in Figure 4F. To circumvent space clamp issues that preclude recording of fast gating I_{Na} in brain slices using conventional voltage-clamp protocols, we employed a two-pulse step protocol described by Milescu et al. [48] and employed by others [49,50]. This protocol uses a depolarizing pre-pulse step to inactivate Na_v channels distant from the recording electrode that is followed shortly afterward with a second step to record Na_v channels close to the recording pipette. Using this protocol, we reliably resolved well-clamped I_{Na} of intact MSNs in the acute brain slice preparation. MSNs treated with 0.01% DMSO displayed an average peak I_{Na} density of -53.0 ± 9.6 pA/pF ($n = 6$), whereas MSNs treated with 15 μM PW201 displayed a significantly increased peak I_{Na} density of -100.2 ± 15.2 pA/pF ($n = 8$; $p < 0.05$; Figure 4F–H). This effect provides strong evidence that the compound's potentiation of action potential discharge of MSNs of the NAc is mediated by changes in the activity of their constituent Na_v channels.

In addition to assessing the effects of PW201 on the amplitude of I_{Na} of MSNs, the effects of the compound on the voltage dependence of activation (Figure 4I) and the voltage dependence of steady-state inactivation (Figure 4J) of I_{Na} of MSNs were also investigated. Consistent with the results shown in Figure 2 demonstrating that 15 μM PW201 modulates the amplitude of $Na_v1.6$-mediated I_{Na} in heterologous cells without affecting the voltage dependences of activation (Figure 2E) or steady-state inactivation (Figure 2F), treatment of acute brain slice preparations containing the NAc with 15 μM PW201 affected neither the voltage dependence of activation (Figure 4I) nor the voltage dependence of steady-state inactivation (Figure 4J) of the I_{Na} of MSNs compared to treatment with 0.01% DMSO. Specifically, MSNs treated with 0.01% DMSO displayed an I_{Na} with a $V_{1/2}$ of activation value of -35.4 ± 2.3 mV ($n = 6$), which was not significantly different than MSNs treated with 15 μM PW201 (-38.7 ± 2.0 mV; $n = 8$; $p = 0.3037$; Figure 4I). As it relates to inactivation, the $V_{1/2}$ of steady-state inactivation of I_{Na} of MSNs treated with 0.01% DMSO was -69.6 ± 1.5 mV ($n = 6$), which was not significantly different from the $V_{1/2}$ of steady-state inactivation value observed for MSNs treated with 15 μM PW201 (-68.0 ± 2.5 mV; $n = 8$; $p = 0.6224$; Figure 4J). Overall, the results of these recordings performed in MSNs, coupled with the recordings performed in HEK-$Na_v1.6$ cells, demonstrate that PW201 affects the I_{Na} amplitude without affecting the voltage dependences of activation or steady-state inactivation of I_{Na}.

3. Discussion

PPIs between the pore-forming α subunit of Na_v channels and auxiliary proteins regulate channel gating and trafficking [12,17,18,21,22,24,51,52]. Translationally, perturbation of these PPIs gives rise to neural circuity aberrations that are associated with neurologic and neuropsychiatric disorders [33,34], underscoring their role as critical sites for neuromodulation. Despite representing novel pharmacological targets for neuromodulation, such PPIs have historically proven difficult to appreciably modulate using conventional small molecules [53–56]. As this challenge largely arises from the large size of PPI interfaces making it difficult to identify druggable motifs that could confer functionally relevant modulation of the intermolecular interaction, efforts to map PPI interfaces using chemical probes, such as those employed in the present investigation, are a necessary pre-requisite for the development of small molecule modulators of PPIs.

In our previous work [37], we showed that PW201 modulated FGF14:Na_V1.6 complex assembly, displayed direct binding to the CTD of the Na_V1.6 channel, modulated Na_V1.6-mediated I_{Na} in heterologous cells, and docked well with residues that are similarly interacted with by the β8/9 loop of FGF14. In the present work, we expanded upon those findings and showed that whereas PW201 modulated Na_V1.6-mediated I_{Na} in heterologous cells in a dose-dependent manner with an IC_{50} of 15 μM (Figure 2), the ligand displayed no effects on Na_V1.1 or Na_V1.2 channels in heterologous cells when similarly tested at 15 μM (Figure 3). These findings could suggest that the ligand binding site of PW201 on the CTD of the Na_V1.6 channel is not conserved among the Na_V1.1 or Na_V1.2 channels; however, extensive structural and biophysical studies are required to unequivocally substantiate such a hypothesis. Nevertheless, these findings, coupled with the molecular modeling of PW201 shown in Figure 1, could help guide future rational design efforts seeking to develop isoform-selective small molecule modulators of the Na_V1.6 channel.

In addition to demonstrating isoform-selective effects of PW201 on Na_V1.6 channels in heterologous cells, we also assessed the effects of PW201 on the I_{Na} and intrinsic excitability of MSNs of the NAc. MSNs represent the principal cell type of the NAc [57,58], are highly vulnerable to neurodegeneration [59], and are enriched with FGF14 and the Na_V1.6 channel [24]. In current-clamp and voltage-clamp recordings, PW201 was shown to potentiate the action potential discharge (Figure 4A–E) and increase the I_{Na} amplitude of MSNs of the NAc (Figure 4F–H), respectively. Importantly, changes in Na_V channel conductance, such as those conferred by PW201, have previously been shown to increase neuronal excitability and confer changes in the instantaneous firing frequencies of neurons [44–47]. Coupled with the findings observed in heterologous cells, these results demonstrate that pharmacological manipulation of the Na_V1.6 channel achieved through targeting its PPI site with FGF14 can alter the activity of cells in clinically relevant brain regions, underscoring the potential translational value of the target for neurologic and neuropsychiatric diseases.

One seemingly paradoxical effect of PW201 is that whereas the ligand suppresses Na_V1.6-mediated I_{Na} in heterologous cells, the compound increases the I_{Na} of MSNs in the acute brain slice preparation. However, the opposite effects of FGF14, the protein from which PW201 is derived, in heterologous cells versus neurons is widely recognized [12,17,24]. Specifically, co-expression of FGF14 with the Na_V1.6 channel in heterologous systems has previously been shown to suppress Na_V1.6-mediated I_{Na} [12,15,24,38,40,41,60,61], whereas over-expression of FGF14 in neurons has been shown to increase I_{Na} [17]. As such, these opposite effects observed for PW201 in heterologous cells versus in neurons are unsurprising and provide supporting evidence for the compound functioning as a partial pharmacological mimic of FGF14.

Although PW201 is anticipated to not be blood–brain barrier permeable due to its high molecular weight and total polar surface area, the findings of the present investigation will inform rational design efforts to develop isoform-selective small molecule modulators of the Na_V1.6 channel macromolecular complex. Such neuromodulators that exert their effects through targeting of PPI interfaces within the CNS will represent an entirely novel class of neurotherapeutics and will demonstrate that PPIs represent hundreds of viable and unexplored targets for CNS drug development.

4. Materials and Methods

4.1. Molecular Docking

The molecular docking study was performed using Schrödinger Small-Molecule Drug Discovery Suite (Schrödinger, LLC, New York, NY, USA). The FGF14:Na_V1.6 homology model was built using the FGF13:Na_V1.5:CaM ternary complex crystal structure (PDB code:

4DCK) as a template [14]. The FGF14:$Na_V1.6$ CTD homology model was prepared with Schrödinger Protein Preparation Wizard using default settings. The SiteMap (Schrödinger, LLC) calculation was performed, and a potential binding site was identified on the PPI interface of FGF14 and the CTD of the $Na_V1.6$ channel. The docking was performed on the CTD of $Na_V1.6$ after removing the FGF14 chain structure. The grid center was chosen on the $Na_V1.6$ CTD at the previously identified binding site with a grid box sized in 24 Å covering the PPI surface on the $Na_V1.6$ CTD. The 3D structure of PW201 was created using Schrödinger Maestro and a low-energy conformation was generated using LigPrep. Docking was then employed with Glide using the SP precision. Docked poses were incorporated into Schrödinger Maestro for a ligand–receptor interactions visualization. The top docked pose of PW201 was superimposed with the FGF14:$Na_V1.6$ CTD complex homology model for an overlay analysis.

4.2. Chemicals

The synthetic route, as well as the chemical properties, of PW201 were previously described [37]. Lyophilized PW201 powder (purity > 95%) was reconstituted in 100% dimethyl sulfoxide (DMSO; Sigma-Aldrich, St. Louis, MO, USA) to achieve stock concentrations of 50 mM, which were frozen and stored at −20 °C until being thawed and further diluted for experimental purposes.

4.3. Cell Culture

HEK293 cells were maintained in a 1:1 mixture of Dulbecco's Modified Eagle Medium (DMEM) with 1 g/L glucose and F-12 (Invitrogen, Carlsbad, CA, USA) that was further supplemented with 10% fetal bovine serum, 100 units/mL of penicillin, and 100 μg/mL streptomycin (Invitrogen). Cells were maintained at 37 °C. The HEK293 cells stably expressing $hNa_V1.1$ [42], $hNa_V1.2$ [43], and $hNa_V1.6$ [15,24,40,41] channels have previously been described. These cells were maintained according to general cell culture protocols, with the caveat that 500 μg/mL of G418 (Invitrogen) was used to maintain stable $hNa_V1.2$ and $hNa_V1.6$ expression and 80 μg/mL of G418 was used to maintain stable expression of $hNa_V1.1$.

4.4. Animals

C57/BL6J mice were purchased from Jackson Laboratory (Bar Harbor, ME, USA). Mice were housed in the University of Texas Medical Branch vivarium, which operates in compliance with the United States Department of Agriculture Animal Welfare Act, the NIH Guide for the Care and Use of Laboratory Animals, the American Association for Laboratory Animal Science, and Institutional Animal Care and Use Committee approved protocols.

4.5. Electrophysiology

4.5.1. General

Borosilicate glass pipettes (Harvard Apparatus, Holliston, MA, USA) with resistance of 1.5–3 MΩ were fabricated using a PC-100 vertical Micropipette Puller (Narishige International Inc., East Meadow, NY, USA). Recordings were obtained using an Axopatch 200B amplifier (Molecular Devices, Sunnyvale, CA, USA). Membrane capacitance and series resistance were estimated using the dial settings on the amplifier, and capacitive transients and series resistances were compensated by 70–80%. Data acquisition and filtering occurred at 20 and 5 kHz, respectively, before digitization and storage. Clampex 9 software (Molecular Devices) was used to set experimental parameters, and electrophysiological equipment was interfaced to this software using a Digidata 1200 analog–digital interface (Molecular Devices). Analysis of electrophysiological data was performed using Clampfit 11 software (Molecular Devices) and GraphPad Prism 8 software (La Jolla, CA, USA). Results were expressed as mean ± standard error of the mean (SEM). Except where otherwise noted, statistical significance was determined using a Student's *t*-test comparing cells treated with vehicle (DMSO) or PW201, with $p < 0.05$ being considered statistically significant.

4.5.2. Whole-Cell Voltage-Clamp Recordings

Whole-cell voltage-clamp recordings in heterologous cell systems were performed as previously described [37,38]. Briefly, cells cultured as described in Section 4.3 were dissociated using TrypLE (Gibco, Waltham, MA, USA) and re-plated onto glass cover slips. After allowing cells at least 2–3 h to adhere, cover slips were transferred to a recording chamber. The recording chamber was filled with an extracellular recording solution comprised of the following salts: 140 mM NaCl; 3 mM KCl; 1 mM $MgCl_2$; 1 mM $CaCl_2$; 10 mM HEPES; and 10 mM glucose (final pH = 7.3; all salts purchased from Sigma-Aldrich, St. Louis, MO, USA). For control recordings, DMSO was added to the extracellular solution to reach a final concentration of 0.1%. For recordings to characterize the effects of PW201, the compound was added to the extracellular solution to reach the desired final concentration. Cover slips were incubated for 30 min in either vehicle only or PW201 containing extracellular solutions prior to the start of recordings. For voltage-clamp recordings, recording pipettes were filled with an intracellular solution comprised of the following salts: 130 mM CH_3O_3SCs; 1 mM EGTA; 10 mM NaCl; and 10 mM HEPES (pH = 7.3; all salts purchased from Sigma-Aldrich). After GΩ seal formation and entry into the whole-cell configuration, two voltage-clamp protocols were employed. The current-voltage (IV) protocol entailed voltage steps from −100 to +60 mV from a holding potential of −70 mV. The voltage dependence of steady-state inactivation protocol entailed a paired-pulse protocol during which, from the holding potential, cells were stepped to varying test potentials between −100 mV and +20 mV prior to a test pulse to −20 mV.

4.5.3. Voltage-Clamp Data Analysis

Current densities were obtained by dividing the Na^+ current (I_{Na}) amplitude by the membrane capacitance (C_m). Current–voltage relationships were then assessed by plotting the current density as a function of the applied voltage. Tau of fast inactivation was calculated by fitting the decay phase of currents at the −10 mV voltage step with a one-term exponential function. To assess the voltage dependence of activation, conductance (G_{Na}) was first calculated using the following equation:

$$G_{Na} = \frac{I_{Na}}{(V_m - E_{rev})}$$

where I_{Na} is the current amplitude at voltage V_m, and E_{rev} is the Na^+ reversal potential. Activation curves were then generated by plotting normalized G_{Na} as a function of the test potential. Data were then fitted with the Boltzmann equation to determine $V_{1/2}$ of activation using the following equation:

$$\frac{G_{Na}}{G_{Na,\ max}} = 1 + e^{V_a - E_m/k}$$

where $G_{Na,max}$ is the maximum conductance, V_a is the membrane potential of half-maximal activation, E_m is the membrane voltage, and k is the slope factor. For steady-state inactivation, the normalized current amplitude ($I_{Na}/I_{Na,max}$) at the test potential was plotted as a function of the pre-pulse potential (V_m) and fitted using the Boltzmann equation:

$$\frac{I_{Na}}{I_{Na,max}} = \frac{1}{1 + e^{V_h - E_m/k}}$$

where V_h is the potential of half-maximal inactivation, E_m is the membrane voltage, and k is the slope factor.

4.5.4. Acute Brain Slice Preparation

Whole-cell current-clamp and whole-cell voltage-clamp recordings were performed in acutely pre-prepared coronal brain slices containing the NAc from mice described in Section 4.4 that were 33–50 days old. For brain slice preparation, mice were anesthetized using isoflurane (Baxter, Deerfield, IL, USA) and quickly decapitated. After decapitation, brains were dissected and 300 μM coronal slices containing the NAc were prepared with a vibratome (Leica Biosystems, Buffalo Grove, IL, USA) in a continuously oxygenated (mixture of 95% O_2/5% CO_2) and chilled tris-based artificial cerebrospinal fluid (aCSF) containing the following salts: 72 mM Tris-HCL; 18 mM Tris-Base; 1.2 mM NaH_2PO_4; 2.5 mM KCl; 20 mM HEPES; 20 mM sucrose; 25 mM $NaHCO_3$; 25 mM glucose; 10 mM $MgSO_4$; 3 mM Na pyruvate; 5 mM Na ascorbate; and 0.5 mM $CaCl_2$ (pH = 7.4 and osmolarity = 300–310 mOsm; all salts purchased from Sigma-Aldrich). Prepared slices were first transferred to a continuously oxygenated and 31 °C recovery chamber containing fresh tris-based aCSF for 15 min. After 15 min, slices were transferred to a continuously oxygenated and 31 °C chamber containing standard aCSF, which was comprised of the following salts: 123.9 mM NaCl; 3.1 mM KCl; 10 mM glucose; 1 mM $MgCl_2$; 2 mM $CaCl_2$; 24 mM $NaHCO_3$; and 1.16 mM NaH_2PO_4 (pH = 7.4 and osmolarity = 300–310 mOsm; all salts were purchased from Sigma-Aldrich). After at least 30 min of recovery in standard aCSF, slices were incubated for 30 min in a chamber containing continuously oxygenated and 31 °C standard aCSF treated with either 0.01% DMSO or 15 μM PW201 before recording.

4.5.5. Whole-Cell Current-Clamp Recordings

After incubating for 30 min in either 0.01% DMSO or 15 μM PW201, slices were transferred to a recording chamber perfused with continuously oxygenated and heated standard aCSF. Somatic recordings of MSNs were then performed using electrodes filled with an internal solution comprised of the following salts: 145 mM K-gluconate; 2 mM $MgCl_2$; 0.1 mM EGTA; 2.5 mM Na_2ATP; 0.25 mM Na_2GTP; 5 mM phosphocreatine; and 10 mM HEPES (pH = 7.2 and osmolarity = 290 mOsm; all salts were purchased from Sigma-Aldrich). After GΩ formation and entry into the whole-cell configuration, the amplifier was switched to $I = 0$ mode for approximately 1 min to determine the resting membrane potential before switching to current-clamp mode to assess intrinsic excitability. During this 1 min interval in $I = 0$ mode, the following cocktail of synaptic blockers was perfused to halt changes in excitability driven by synaptic activity: 20 μM bicuculine; 20 μM NBQX; and 100 μM AP5 (synaptic blockers purchased from Tocris, Bristol, UK). To assess intrinsic excitability, evoked APs were measured in response to a range of current injections from −20 to +150 pA. Current steps were 800 ms in duration, and the change in the injected current between steps was 10 pA.

4.5.6. Current-Clamp Data Analysis

The maximum number of APs was determined by quantifying the maximum number of APs an MSN fired at any current step during the evoked protocol. The average instantaneous firing frequency was determined by calculating the mean value of the instantaneous firing frequency between APs at a given current step. The current threshold (I_{thr}) was defined as the current step at which at least one AP was evoked. Voltage threshold (V_{thr}) was defined as the voltage at which the first-order derivative of the rising phase of the AP exceeded 10 mV/ms [62]. The maximum rise and maximum decay of APs were defined as the maximal derivative value (dV/dt) of the depolarizing and repolarizing phases of the AP, respectively [63].

4.5.7. Ex Vivo Whole-Cell Voltage-Clamp Recordings of I_{Na}

The extracellular solution used for current-clamp recordings was also used to record I_{Na} of MSNs ex vivo, with the caveat that the superfusing solution was supplemented with 120 μM $CdCl_2$ (Sigma-Aldrich) to block Ca^{2+} currents. The intracellular solution to record I_{Na} of MSN ex vivo contained the following salts (in mM): 100 mM Cs-gluconate (Hello Bio

Inc., Princeton, NJ, USA); 10 mM tetraethylammonium chloride; 5 mM 4-aminopyridine; 10 mM EGTA; 1 mM $CaCl_2$; 10 mM HEPES; 4 mM Mg-ATP; 0.3 mM Na_3-GTP; 4 mM Na_2-phosphocreatine; and 4 mM NaCl (pH = 7.4 and osmolarity = 285 $\pm$ 5 mOsm/L; CsOH used to adjust pH and osmolarity; all salts except Cs-gluconate purchased from Sigma-Aldrich). After GΩ formation and entry into the whole-cell configuration, the same cocktail of synaptic blockers as used for the current-clamp recordings was perfused to block synaptic currents. Transient I_{Na} was elicited using the voltage-clamp protocol shown in Figure 4F and as described elsewhere [48–50]. I_{Na} density was calculated by normalizing the I_{Na} response by C_m.

Author Contributions: Conceptualization, N.M.D. and F.L.; methodology, N.M.D., C.M.T. and A.K.S.; validation, N.M.D., C.M.T., A.K.S. and F.L.; formal analysis, N.M.D.; investigation, N.M.D., C.M.T., A.K.S., T.J.B., P.W., H.C. and P.A.W.; resources, J.Z. and F.L.; data curation, N.M.D.; writing—original draft preparation, N.M.D.; writing—review and editing, N.M.D. and F.L.; visualization, N.M.D.; supervision, J.Z. and F.L.; project administration, J.Z. and F.L.; funding acquisition, J.Z. and F.L. All authors have read and agreed to the published version of the manuscript.

Funding: This work was supported by the National Institutes of Health (NIH) Grants R01 MH095995 (F.L.), R01 MH111107 (F.L. and J.Z.), P30 DA028821 (J.Z.), John D. Stobo, M.D., Distinguished Chair Endowment Fund (J.Z.), John Sealy Memorial Endowment Fund (F.L.), UTMB Technology Commercialization Program (F.L. and J.Z.), the Houston Area Molecular Biophysics Program Grant No. T32 GM008280 (N.M.D), Training Program funded by The National Institute of Aging (NIH Grant # T32AG067952-01; T.J.B.), and the National Institute of Environmental Health Sciences T32 ES007254 (C.M.T.).

Institutional Review Board Statement: The study was conducted according to the guidelines of the United States Department of Agriculture Animal Welfare Act, the NIH Guide for the Care and Use of Laboratory Animals, the American Association for Laboratory Animal Science, and UTMB Institutional Animal Care and Use Committee approved protocols (protocol code: 0904029D; approved 01/2019).

Informed Consent Statement: Not applicable.

Data Availability Statement: Data included in this study are available upon request from the corresponding author.

Acknowledgments: We acknowledge the Sealy Center for Structural Biology and Molecular Biology at the University of Texas Medical Branch at Galveston for providing research resources.

Conflicts of Interest: The corresponding author, F.L., is the founder and president of IonTx Inc., a start-up company focusing on developing regulators of voltage-gated Na^+ channels. However, this activity does not represent a conflict with the present study.

References

1. Catterall, W.A. Forty Years of Sodium Channels: Structure, Function, Pharmacology, and Epilepsy. *Neurochem. Res.* **2017**, *42*, 2495–2504. [CrossRef]
2. Bean, B.P. The Action Potential in Mammalian Central Neurons. *Nat. Rev. Neurosci.* **2007**, *8*, 451–465. [CrossRef]
3. Mechaly, I.; Scamps, F.; Chabbert, C.; Sans, A.; Valmier, J. Molecular Diversity of Voltage-Gated Sodium Channel Alpha Subunits Expressed in Neuronal and Non-Neuronal Excitable Cells. *Neuroscience* **2005**, *130*, 389–396. [CrossRef]
4. Ogiwara, I.; Miyamoto, H.; Morita, N.; Atapour, N.; Mazaki, E.; Inoue, I.; Takeuchi, T.; Itohara, S.; Yanagawa, Y.; Obata, K.; et al. Nav1.1 Localizes to Axons of Parvalbumin-Positive Inhibitory Interneurons: A Circuit Basis for Epileptic Seizures in Mice Carrying an Scn1a Gene Mutation. *J. Neurosci.* **2007**, *27*, 5903–5914. [CrossRef]
5. Wang, W.; Takashima, S.; Segawa, Y.; Itoh, M.; Shi, X.; Hwang, S.-K.; Nabeshima, K.; Takeshita, M.; Hirose, S. The Developmental Changes of Na(v)1.1 and Na(v)1.2 Expression in the Human Hippocampus and Temporal Lobe. *Brain Res.* **2011**, *1389*, 61–70. [CrossRef]
6. Tian, C.; Wang, K.; Ke, W.; Guo, H.; Shu, Y. Molecular Identity of Axonal Sodium Channels in Human Cortical Pyramidal Cells. *Front. Cell Neurosci.* **2014**, *8*, 297. [CrossRef]
7. Lorincz, A.; Nusser, Z. Cell-Type-Dependent Molecular Composition of the Axon Initial Segment. *J. Neurosci.* **2008**, *28*, 14329–14340. [CrossRef]
8. Spratt, P.W.E.; Alexander, R.P.D.; Ben-Shalom, R.; Sahagun, A.; Kyoung, H.; Keeshen, C.M.; Sanders, S.J.; Bender, K.J. Paradoxical Hyperexcitability from NaV1.2 Sodium Channel Loss in Neocortical Pyramidal Cells. *Cell Rep.* **2021**, *36*. [CrossRef]

9. Duflocq, A.; Le Bras, B.; Bullier, E.; Couraud, F.; Davenne, M. Nav1.1 Is Predominantly Expressed in Nodes of Ranvier and Axon Initial Segments. *Mol. Cell. Neurosci.* **2008**, *39*, 180–192. [CrossRef]
10. Catterall, W.A.; Swanson, T.M. Structural Basis for Pharmacology of Voltage-Gated Sodium and Calcium Channels. *Mol. Pharm.* **2015**, *88*, 141–150. [CrossRef]
11. Dvorak, N.M.; Wadsworth, P.A.; Wang, P.; Zhou, J.; Laezza, F. Development of Allosteric Modulators of Voltage-Gated Na(+) Channels: A Novel Approach for an Old Target. *Curr. Top. Med. Chem.* **2021**, *21*, 841–848. [CrossRef] [PubMed]
12. Laezza, F.; Lampert, A.; Kozel, M.A.; Gerber, B.R.; Rush, A.M.; Nerbonne, J.M.; Waxman, S.G.; Dib-Hajj, S.D.; Ornitz, D.M. FGF14 N-Terminal Splice Variants Differentially Modulate Nav1.2 and Nav1.6-Encoded Sodium Channels. *Mol. Cell. Neurosci.* **2009**, *42*, 90–101. [CrossRef] [PubMed]
13. Wang, C.; Wang, C.; Hoch, E.G.; Pitt, G.S. Identification of Novel Interaction Sites That Determine Specificity between Fibroblast Growth Factor Homologous Factors and Voltage-Gated Sodium Channels. *J. Biol. Chem.* **2011**, *286*, 24253–24263. [CrossRef] [PubMed]
14. Wang, C.; Chung, B.C.; Yan, H.; Lee, S.-Y.; Pitt, G.S. Crystal Structure of the Ternary Complex of a NaV C-Terminal Domain, a Fibroblast Growth Factor Homologous Factor, and Calmodulin. *Structure* **2012**, *20*, 1167–1176. [CrossRef]
15. Ali, S.R.; Singh, A.K.; Laezza, F. Identification of Amino Acid Residues in Fibroblast Growth Factor 14 (FGF14) Required for Structure-Function Interactions with Voltage-Gated Sodium Channel Nav1.6. *J. Biol. Chem.* **2016**, *291*, 11268–11284. [CrossRef]
16. Gardill, B.R.; Rivera-Acevedo, R.E.; Tung, C.-C.; Van Petegem, F. Crystal Structures of $Ca^{(2+)}$-Calmodulin Bound to Na(V) C-Terminal Regions Suggest Role for EF-Hand Domain in Binding and Inactivation. *Proc. Natl. Acad. Sci. USA* **2019**, *116*, 10763–10772. [CrossRef] [PubMed]
17. Laezza, F.; Gerber, B.R.; Lou, J.-Y.; Kozel, M.A.; Hartman, H.; Craig, A.M.; Ornitz, D.M.; Nerbonne, J.M. The FGF14(F145S) Mutation Disrupts the Interaction of FGF14 with Voltage-Gated Na+ Channels and Impairs Neuronal Excitability. *J. Neurosci.* **2007**, *27*, 12033–12044. [CrossRef]
18. Lou, J.-Y.; Laezza, F.; Gerber, B.R.; Xiao, M.; Yamada, K.A.; Hartmann, H.; Craig, A.M.; Nerbonne, J.M.; Ornitz, D.M. Fibroblast Growth Factor 14 Is an Intracellular Modulator of Voltage-Gated Sodium Channels. *J. Physiol.* **2005**, *569*, 179–193. [CrossRef]
19. Pitt, G.S.; Lee, S.-Y. Current View on Regulation of Voltage-Gated Sodium Channels by Calcium and Auxiliary Proteins. *Protein Sci.* **2016**, *25*, 1573–1584. [CrossRef]
20. Tseng, T.-T.; McMahon, A.M.; Johnson, V.T.; Mangubat, E.Z.; Zahm, R.J.; Pacold, M.E.; Jakobsson, E. Sodium Channel Auxiliary Subunits. *J. Mol. Microbiol. Biotechnol.* **2007**, *12*, 249–262. [CrossRef]
21. Goetz, R.; Dover, K.; Laezza, F.; Shtraizent, N.; Huang, X.; Tchetchik, D.; Eliseenkova, A.V.; Xu, C.-F.; Neubert, T.A.; Ornitz, D.M.; et al. Crystal Structure of a Fibroblast Growth Factor Homologous Factor (FHF) Defines a Conserved Surface on FHFs for Binding and Modulation of Voltage-Gated Sodium Channels. *J. Biol. Chem.* **2009**, *284*, 17883–17896. [CrossRef] [PubMed]
22. Goldfarb, M.; Schoorlemmer, J.; Williams, A.; Diwakar, S.; Wang, Q.; Huang, X.; Giza, J.; Tchetchik, D.; Kelley, K.; Vega, A.; et al. Fibroblast Growth Factor Homologous Factors Control Neuronal Excitability through Modulation of Voltage-Gated Sodium Channels. *Neuron* **2007**, *55*, 449–463. [CrossRef] [PubMed]
23. Effraim, P.R.; Huang, J.; Lampert, A.; Stamboulian, S.; Zhao, P.; Black, J.A.; Dib-Hajj, S.D.; Waxman, S.G. Fibroblast Growth Factor Homologous Factor 2 (FGF-13) Associates with Nav1.7 in DRG Neurons and Alters Its Current Properties in an Isoform-Dependent Manner. *Neurobiol. Pain* **2019**, *6*, 100029. [CrossRef]
24. Ali, S.R.; Liu, Z.; Nenov, M.N.; Folorunso, O.; Singh, A.; Scala, F.; Chen, H.; James, T.F.; Alshammari, M.; Panova-Elektronova, N.I.; et al. Functional Modulation of Voltage-Gated Sodium Channels by a FGF14-Based Peptidomimetic. *ACS Chem. Neurosci.* **2018**, *9*, 976–987. [CrossRef]
25. Hoxha, E.; Marcinnò, A.; Montarolo, F.; Masante, L.; Balbo, I.; Ravera, F.; Laezza, F.; Tempia, F. Emerging Roles of Fgf14 in Behavioral Control. *Behav. Brain Res.* **2019**, *356*, 257–265. [CrossRef]
26. Wang, Q.; Bardgett, M.E.; Wong, M.; Wozniak, D.F.; Lou, J.; McNeil, B.D.; Chen, C.; Nardi, A.; Reid, D.C.; Yamada, K.; et al. Ataxia and Paroxysmal Dyskinesia in Mice Lacking Axonally Transported FGF14. *Neuron* **2002**, *35*, 25–38. [CrossRef]
27. Wozniak, D.F.; Xiao, M.; Xu, L.; Yamada, K.A.; Ornitz, D.M. Impaired Spatial Learning and Defective Theta Burst Induced LTP in Mice Lacking Fibroblast Growth Factor 14. *Neurobiol. Dis.* **2007**, *26*, 14–26. [CrossRef] [PubMed]
28. Alshammari, T.K.; Alshammari, M.A.; Nenov, M.N.; Hoxha, E.; Cambiaghi, M.; Marcinno, A.; James, T.F.; Singh, P.; Labate, D.; Li, J.; et al. Genetic Deletion of Fibroblast Growth Factor 14 Recapitulates Phenotypic Alterations Underlying Cognitive Impairment Associated with Schizophrenia. *Transl. Psychiatry* **2016**, *6*, e806. [CrossRef]
29. Choquet, K.; La Piana, R.; Brais, B. A Novel Frameshift Mutation in FGF14 Causes an Autosomal Dominant Episodic Ataxia. *Neurogenetics* **2015**, *16*, 233–236. [CrossRef] [PubMed]
30. Miura, S.; Kosaka, K.; Fujioka, R.; Uchiyama, Y.; Shimojo, T.; Morikawa, T.; Irie, A.; Taniwaki, T.; Shibata, H. Spinocerebellar Ataxia 27 with a Novel Nonsense Variant (Lys177X) in FGF14. *Eur. J. Med. Genet.* **2019**, *62*, 172–176. [CrossRef] [PubMed]
31. Misceo, D.; Fannemel, M.; Barøy, T.; Roberto, R.; Tvedt, B.; Jaeger, T.; Bryn, V.; Strømme, P.; Frengen, E. SCA27 Caused by a Chromosome Translocation: Further Delineation of the Phenotype. *Neurogenetics* **2009**, *10*, 371–374. [CrossRef] [PubMed]
32. Brusse, E.; de Koning, I.; Maat-Kievit, A.; Oostra, B.A.; Heutink, P.; van Swieten, J.C. Spinocerebellar Ataxia Associated with a Mutation in the Fibroblast Growth Factor 14 Gene (SCA27): A New Phenotype. *Mov. Disord.* **2006**, *21*, 396–401. [CrossRef]
33. Di Re, J.; Wadsworth, P.A.; Laezza, F. Intracellular Fibroblast Growth Factor 14: Emerging Risk Factor for Brain Disorders. *Front. Cell. Neurosci.* **2017**, *11*, 103. [CrossRef]

34. Paucar, M.; Lundin, J.; Alshammari, T.; Bergendal, Å.; Lindefeldt, M.; Alshammari, M.; Solders, G.; Di Re, J.; Savitcheva, I.; Granberg, T.; et al. Broader Phenotypic Traits and Widespread Brain Hypometabolism in Spinocerebellar Ataxia 27. *J. Intern. Med.* **2020**, *288*, 103–115. [CrossRef] [PubMed]
35. Stanton, C.H.; Holmes, A.J.; Chang, S.W.C.; Joormann, J. From Stress to Anhedonia: Molecular Processes through Functional Circuits. *Trends Neurosci.* **2019**, *42*, 23–42. [CrossRef]
36. Liu, Z.; Wadsworth, P.; Singh, A.K.; Chen, H.; Wang, P.; Folorunso, O.; Scaduto, P.; Ali, S.R.; Laezza, F.; Zhou, J. Identification of Peptidomimetics as Novel Chemical Probes Modulating Fibroblast Growth Factor 14 (FGF14) and Voltage-Gated Sodium Channel 1.6 (Nav1.6) Protein-Protein Interactions. *Bioorg. Med. Chem. Lett.* **2019**, *29*, 413–419. [CrossRef]
37. Dvorak, N.M.; Wadsworth, P.A.; Wang, P.; Chen, H.; Zhou, J.; Laezza, F. Bidirectional Modulation of the Voltage-Gated Sodium (Nav1.6) Channel by Rationally Designed Peptidomimetics. *Molecules* **2020**, *25*, 3365. [CrossRef] [PubMed]
38. Wang, P.; Wadsworth, P.A.; Dvorak, N.M.; Singh, A.K.; Chen, H.; Liu, Z.; Zhou, R.; Holthauzen, L.M.F.; Zhou, J.; Laezza, F. Design, Synthesis, and Pharmacological Evaluation of Analogues Derived from the PLEV Tetrapeptide as Protein–Protein Interaction Modulators of Voltage-Gated Sodium Channel 1.6. *J. Med. Chem.* **2020**, *63*, 11522–11547. [CrossRef] [PubMed]
39. London, N.; Raveh, B.; Schueler-Furman, O. Druggable Protein-Protein Interactions—From Hot Spots to Hot Segments. *Curr. Opin. Chem. Biol.* **2013**, *17*, 952–959. [CrossRef] [PubMed]
40. Singh, A.K.; Wadsworth, P.A.; Tapia, C.M.; Aceto, G.; Ali, S.R.; Chen, H.; D'Ascenzo, M.; Zhou, J.; Laezza, F. Mapping of the FGF14:Nav1.6 Complex Interface Reveals FLPK as a Functionally Active Peptide Modulating Excitability. *Physiol. Rep.* **2020**, *8*, e14505. [CrossRef] [PubMed]
41. Wadsworth, P.A.; Singh, A.K.; Nguyen, N.; Dvorak, N.M.; Tapia, C.M.; Russell, W.K.; Stephan, C.; Laezza, F. JAK2 Regulates Nav1.6 Channel Function via FGF14(Y158) Phosphorylation. *Biochim. Biophys. Acta Mol. Cell Res.* **2020**, *1867*, 118786. [CrossRef] [PubMed]
42. James, T.F.; Nenov, M.N.; Tapia, C.M.; Lecchi, M.; Koshy, S.; Green, T.A.; Laezza, F. Consequences of Acute Na(v)1.1 Exposure to Deltamethrin. *Neurotoxicology* **2017**, *60*, 150–160. [CrossRef] [PubMed]
43. James, T.F.; Nenov, M.N.; Wildburger, N.C.; Lichti, C.F.; Luisi, J.; Vergara, F.; Panova-Electronova, N.I.; Nilsson, C.L.; Rudra, J.S.; Green, T.A.; et al. The Nav1.2 Channel Is Regulated by GSK3. *Biochim. Biophys. Acta* **2015**, *1850*, 832–844. [CrossRef]
44. Cantrell, A.R.; Catterall, W.A. Neuromodulation of Na+ Channels: An Unexpected Form of Cellular Platicity. *Nat. Rev. Neurosci.* **2001**, *2*, 397–407. [CrossRef] [PubMed]
45. Hodgkin, A. The Optimum Density of Sodium Channels in an Unmyelinated Nerve. *Philos. Trans. R. Soc. Lond. B Biol. Sci.* **1975**, *270*, 297–300. [CrossRef] [PubMed]
46. Schiffmann, S.N.; Lledo, P.M.; Vincent, J.D. Dopamine D1 Receptor Modulates the Voltage-Gated Sodium Current in Rat Striatal Neurones through a Protein Kinase A. *J. Physiol.* **1995**, *483 Pt 1*, 95–107. [CrossRef]
47. Moore, A.R.; Zhou, W.-L.; Potapenko, E.S.; Kim, E.-J.; Antic, S.D. Brief Dopaminergic Stimulations Produce Transient Physiological Changes in Prefrontal Pyramidal Neurons. *Brain Res.* **2011**, *1370*, 1–15. [CrossRef] [PubMed]
48. Milescu, L.S.; Bean, B.P.; Smith, J.C. Isolation of Somatic Na+ Currents by Selective Inactivation of Axonal Channels with a Voltage Prepulse. *J. Neurosci.* **2010**, *30*, 7740–7748. [CrossRef] [PubMed]
49. Pablo, J.L.; Wang, C.; Presby, M.M.; Pitt, G.S. Polarized Localization of Voltage-Gated Na^+ Channels Is Regulated by Concerted FGF13 and FGF14 Action. *Proc. Natl. Acad. Sci. USA* **2016**, *113*, E2665. [CrossRef]
50. Alexander, R.P.D.; Mitry, J.; Sareen, V.; Khadra, A.; Bowie, D. Cerebellar Stellate Cell Excitability Is Coordinated by Shifts in the Gating Behavior of Voltage-Gated Na(+) and A-Type K(+) Channels. *eNeuro* **2019**, *5*, 6. [CrossRef] [PubMed]
51. White, H.V.; Brown, S.T.; Bozza, T.C.; Raman, I.M. Effects of FGF14 and NaVβ4 Deletion on Transient and Resurgent Na Current in Cerebellar Purkinje Neurons. *J. Gen. Physiol.* **2019**, *151*, 1300–1318. [CrossRef]
52. Gade, A.R.; Marx, S.O.; Pitt, G.S. An Interaction between the III-IV Linker and CTD in NaV1.5 Confers Regulation of Inactivation by CaM and FHF. *J. Gen. Physiol.* **2020**, *152*. [CrossRef] [PubMed]
53. Wells, J.A.; McClendon, C.L. Reaching for High-Hanging Fruit in Drug Discovery at Protein-Protein Interfaces. *Nature* **2007**, *450*, 1001–1009. [CrossRef] [PubMed]
54. Arkin, M.R.; Wells, J.A. Small-Molecule Inhibitors of Protein-Protein Interactions: Progressing towards the Dream. *Nat. Rev. Drug Discov.* **2004**, *3*, 301–317. [CrossRef] [PubMed]
55. Whitty, A.; Kumaravel, G. Between a Rock and a Hard Place? *Nat. Chem. Biol.* **2006**, *2*, 112–118. [CrossRef]
56. Pelay-Gimeno, M.; Glas, A.; Koch, O.; Grossmann, T.N. Structure-Based Design of Inhibitors of Protein-Protein Interactions: Mimicking Peptide Binding Epitopes. *Angew. Chem. Int. Ed. Engl.* **2015**, *54*, 8896–8927. [CrossRef]
57. Kemp, J.M.; Powell, T.P. The Structure of the Caudate Nucleus of the Cat: Light and Electron Microscopy. *Philos. Trans. R. Soc. Lond. B Biol. Sci.* **1971**, *262*, 383–401. [CrossRef]
58. Tapia, C.M.; Folorunso, O.; Singh, A.K.; McDonough, K.; Laezza, F. Effects of Deltamethrin Acute Exposure on Nav1.6 Channels and Medium Spiny Neurons of the Nucleus Accumbens. *Toxicology* **2020**, *440*, 152488. [CrossRef]
59. Scala, F.; Nenov, M.N.; Crofton, E.J.; Singh, A.K.; Folorunso, O.; Zhang, Y.; Chesson, B.C.; Wildburger, N.C.; James, T.F.; Alshammari, M.A.; et al. Environmental Enrichment and Social Isolation Mediate Neuroplasticity of Medium Spiny Neurons through the GSK3 Pathway. *Cell Rep.* **2018**, *23*, 555–567. [CrossRef] [PubMed]

60. Singh, A.K.; Dvorak, N.M.; Tapia, C.M.; Mosebarger, A.; Ali, S.R.; Bullock, Z.; Chen, H.; Zhou, J.; Laezza, F. Differential Modulation of the Voltage-Gated Na^+ Channel 1.6 by Peptides Derived From Fibroblast Growth Factor 14. *Front. Mol. Biosci.* **2021**, *8*, 860. [CrossRef]
61. Dvorak, N.M.; Tapia, C.M.; Baumgartner, T.J.; Singh, J.; Laezza, F.; Singh, A.K. Pharmacological Inhibition of Wee1 Kinase Selectively Modulates the Voltage-Gated Na^+ Channel 1.2 Macromolecular Complex. *Cells* **2021**, *10*, 3103. [CrossRef]
62. Nenov, M.N.; Tempia, F.; Denner, L.; Dineley, K.T.; Laezza, F. Impaired Firing Properties of Dentate Granule Neurons in an Alzheimer's Disease Animal Model Are Rescued by PPARγ Agonism. *J. Neurophysiol.* **2015**, *113*, 1712–1726. [CrossRef]
63. Crofton, E.J.; Nenov, M.N.; Zhang, Y.; Scala, F.; Page, S.A.; McCue, D.L.; Li, D.; Hommel, J.D.; Laezza, F.; Green, T.A. Glycogen Synthase Kinase 3 Beta Alters Anxiety-, Depression-, and Addiction-Related Behaviors and Neuronal Activity in the Nucleus Accumbens Shell. *Neuropharmacology* **2017**, *117*, 49–60. [CrossRef]

International Journal of
Molecular Sciences

Communication

Relevance of AIF/CypA Lethal Pathway in SH-SY5Y Cells Treated with Staurosporine

Mariarosaria Conte [1], Rosanna Palumbo [2], Alessandra Monti [2], Elisabetta Fontana [1], Angela Nebbioso [1], Menotti Ruvo [2], Lucia Altucci [1,3] and Nunzianna Doti [2,*]

1 Department of Precision Medicine, University of Campania 'Luigi Vanvitelli', Via L. De Crecchio 7, 80138 Naples, Italy; mariarosaria.conte@unicampania.it (M.C.); elisabetta.fontana@unicampania.it (E.F.); angela.nebbioso@unicampania.it (A.N.); lucia.altucci@unicampania.it (L.A.)
2 Institute of Biostructures and Bioimaging (IBB), National Research Council (CNR), Via Mezzocannone 16, 80134 Napoli, Italy; rosanna.palumbo@cnr.it (R.P.); alessandra.monti@ibb.cnr.it (A.M.); menotti.ruvo@unina.it (M.R.)
3 Biogem, Institute of Molecular Biology and Genetics, Via Camporeale Area P.I.P., 83031 Ariano Irpino, Italy
* Correspondence: nunzianna.doti@cnr.it

Abstract: The AIF/CypA complex exerts a lethal activity in several rodent models of acute brain injury. Upon formation, it translocates into the nucleus of cells receiving apoptotic stimuli, inducing chromatin condensation, DNA fragmentation, and cell death by a caspase-independent mechanism. Inhibition of this complex in a model of glutamate-induced cell death in HT-22 neuronal cells by an AIF peptide (AIF(370-394)) mimicking the binding site on CypA, restores cell survival and prevents brain injury in neonatal mice undergoing hypoxia-ischemia without apparent toxicity. Here, we explore the effects of the peptide on SH-SY5Y neuroblastoma cells stimulated with staurosporine (STS), a cellular model widely used to study Parkinson's disease (PD). This will pave the way to understanding the role of the complex and the potential therapeutic efficacy of inhibitors in PD. We find that AIF(370-394) confers resistance to STS-induced apoptosis in SH-SY5Y cells similar to that observed with CypA silencing and that the peptide works on the AIF/CypA translocation pathway and not on caspases activation. These findings suggest that the AIF/CypA complex is a promising target for developing novel therapeutic strategies against PD.

Keywords: cyclophilin A (CypA); apoptosis-inducing factor (AIF); human neuroblastoma SH-SY5Y cells; staurosporine-mediated cell death; AIF(370-394) peptide; caspase-3; PARP

Citation: Conte, M.; Palumbo, R.; Monti, A.; Fontana, E.; Nebbioso, A.; Ruvo, M.; Altucci, L.; Doti, N. Relevance of AIF/CypA Lethal Pathway in SH-SY5Y Cells Treated with Staurosporine. *Int. J. Mol. Sci.* **2022**, 23, 265. https://doi.org/10.3390/ijms23010265

Academic Editor: Takuya Watanabe

Received: 2 December 2021
Accepted: 26 December 2021
Published: 27 December 2021

1. Introduction

Parkinson's disease (PD) is a devastating neurodegenerative disorder for which only symptomatic treatments are available. Developing effective therapies against PD is thereby a major need, and advancements in the knowledge of molecular and cellular mechanisms underlying its pathogenesis and/or progression are crucial. However, as human dopaminergic neurons, primary cells from PD patients are difficult to obtain and maintain. Therefore, studies on PD are almost exclusively performed with established neuronal cell models, including the undifferentiated neuroblastoma SH-SY5Y cell line [1]. To induce cellular stress, SH-SY5Y cells can be treated with staurosporine (STS), a protein kinase inhibitor, which provokes cell death through both caspase-dependent and independent pathways [2–4]. Indeed, SH-SY5Y cells treated with high concentrations of STS (over 0.5 μM) do not die following a characteristic necrotic phenotype but rather due to oxidative damage. Consistent with this idea, in the presence of high concentrations of STS, caspase inhibition by z-VAD-fmk, a broad-spectrum caspase inhibitor, reduces the apoptotic phenotype but does not inhibit cell death, which instead appears to be due to oxidative damage [5]. Specifically, high concentrations of STS have been shown to increase caspase-3 activity, Poli ADP-ribosio polimerasi (PARP) proteolysis, and morphological changes indicative of apoptosis, within a few hours of treatment [6–8]. It has been also demonstrated that STS treatment provides the nuclear translocation of apoptosis-inducing factor (AIF) from the mitochondria to the nucleus, where it exerts a proapoptotic activity [7,9,10].

AIF is a mitochondria-associated flavin-binding protein implicated in electron transport chain functions and reactive oxygen species (ROS) regulation [11–13]. However, it is also an important cell death effector in many cellular stress paradigms [14–16]. Upon several apoptotic stimuli, which induce outer mitochondrial membrane permeabilization, AIF is released from mitochondria as a truncated form of about ~57 kDa (AIF(Δ1-121), hereafter tAIF), translocating to the nucleus where induces chromatin condensation, DNA degradation, and cell death, through a caspase-independent mechanism [9,15]. Inhibition or down-regulation of AIF provides neuroprotection in vitro and in a variety of different rodent models of acute brain injury induced by cerebral hypoxia/ischemia (HI), arrest-induced brain damage, epileptic seizures, or even brain trauma [17–21]. Moreover, accumulating evidence also suggests that AIF-induced neuronal cell death can be involved in the progression of neurodegenerative diseases such as PD. In agreement with that, a massive nuclear translocation of tAIF has been observed in the ventral mesencephalon of autopsy samples of patients with PD [22]. In addition, its expression changes in the peripheral blood mononuclear cells of these patients [23].

In different cell and rodent models of acute brain injury, the lethal role of AIF is linked to its interaction with cyclophilin A (CypA) [24–26]. CypA is a ubiquitously expressed protein belonging to the immunophilin family with a peptidyl-prolyl *cis-trans* isomerase activity [27]. Current studies in animal models and humans have provided evidence of the critical role of CypA in several human diseases [27]. In neurons, CypA has a pro-apoptotic activity following its association with tAIF, because the complex promotes AIF nuclear translocation and/or DNAse activity [21,22]. Gene silencing of CypA indeed provides a significant neuroprotection effect by preventing the nuclear translocation of tAIF [24,25].

We have previously reported an AIF-based CypA-binding peptide named AIF(370-394) able to inhibit the interaction between the two proteins with an IC50 in the low micromolar range [25]. This molecule has been used in several in vitro models to evaluate the role of the AIF/CypA complex in different paradigms of cell death and also as a template for the design of new selective peptidomimetic inhibitors of the complex [28–31]. AIF(370-394) selectively inhibits the AIF/CypA complex formation, suppresses the glutamate-induced cell death in neuronal cells, and prevents brain injury in neonatal mice following HI [26]. More recently, AIF(370-394) has been used to demonstrate the crucial role of the AIF/CypA complex on myocyte death in arrhythmogenic cardiomyopathy, significantly expanding to other diseases the potential impact of targeting this complex for therapeutic approaches [32].

In this scenario, using AIF(370-394) as a prototypical inhibitor, we have investigated the possible crosstalk between the AIF/CypA complex and STS-evoked cell death in SH-SY5Y.

MTT and flow cytometry assays have been used to assess cell viability and apoptosis, whereas the associated molecular mechanism has been assessed by Western blotting (WB) analysis. Moreover, the efficiency and final outcome of using the AIF blocking peptide have been compared to the silencing of CypA. We find that CypA selective targeting confers significant resistance to STS-induced apoptosis in SH-SY5Y cells and that this effect is related to the blocking of CypA/AIF nuclear translocation without affecting caspases activation. The results provide evidence that the AIF/CypA complex is a promising target for the development of combined therapeutic strategies for the treatment of PD.

2. Results

2.1. *Down-Regulation of CypA Protects SH-SY5Y Cells from Death Induced by STS*

In order to evaluate the effects of AIF(370-394) on STS-treated SH-SY5Y cells, we first assessed the effects of the down-regulation of endogenous CypA in the cells. CypA was highly expressed in SH-SY5Y and its expression levels increased upon treatment with 10 μM STS, as shown by WB assays of lysates of cells exposed to the drug for 3 h (Figure 1A).

The relative densitometric analysis of bands was shown in Supplementary Figure S1A. SH-SY5Y cells were next transiently transfected with a siRNA directed against CypA (siRNACypA) or with an unrelated silencer (siRNACtrl) used as control. As shown in Figure 1B, transfection of the siRNACypA in SH-SY5Y cells provided a decrease of about 60% of CypA levels compared with control groups already after 24 h, as detected by densitometric analysis of WB bands (Supplementary Figure S1B).

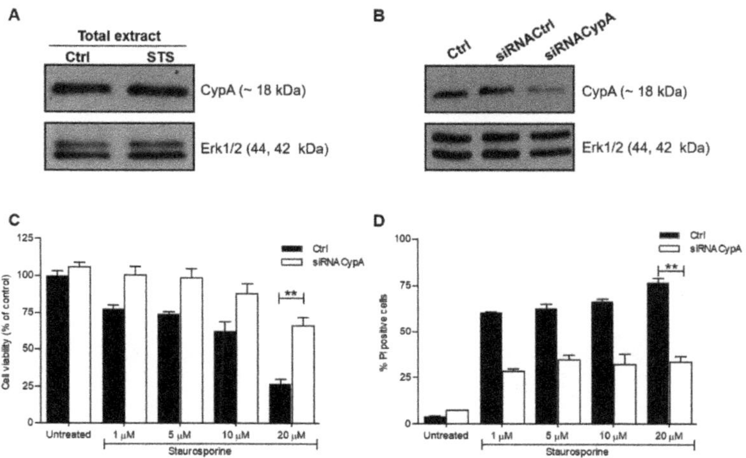

Figure 1. CypA-silencing inhibited STS-induced cell death in SH-SY5Y cells. (**A**) WB evaluation of the expression level of CypA in SH-SY5Y cells untreated and treated for 3 h with 10 μM of STS. (**B**) WB evaluation of CypA expression level after transfection of CypA small interfering RNA (siRNACypA). Erk1/2 proteins were used as a loading control. (**C**) MTT viability assay of SH-SY5Y transfected with siRNACtrl (Ctrl) and siRNACypA exposed to STS for 3 h at the indicated concentrations, (n = 8, ** $p < 0.01$). (**D**) Quantification of flow cytometry results of PI stained SH-SY5Y cells, transfected with siRNACtrl (Ctrl) or with the siRNACypA and treated with STS for 3 h at 10 μM, (n = 8, ** $p < 0.01$).

Cell viability 24 h after transfection of the siRNAs was assessed by MTT assays following treatment with different doses of STS (from 1 to 20 μM) for 3 h. In line with previous results [33], STS dose-dependently reduced cell viability reaching a 75% decrease at 20 μM (Figure 1C). Noteworthy, the downregulation of CypA promoted cell proliferation and at the highest concentration of STS (20 μM), cell viability was about 2.8 fold higher (from 25 to 72%, absolute change of about 47%) compared to SH-SY5Y cells treated only with STS (** $p < 0.01$). FACS analyses of apoptotic cells stained with PI were also performed, showing that STS treatment led to a significant percentage of PI positive cells (~60%) already at 1 μM with an increase up to 75% at 20 μM (Figure 1D). Consistent with MTT data, the strong pro-apoptotic effect of STS was neutralized by the downregulation of CypA, which lead to a significant reduction of PI positive cells (from >75% to about 30%) in all conditions tested (Figure 1D). Overall, the results show that CypA is implicated in STS-induced apoptosis in SH-SY5Y cells.

To better illustrate the pro-apoptotic effect of CypA in this experimental paradigm, we also performed experiments in SH-SY5Y cells over-expressing CypA. Cells were transfected with a plasmid coding for CypA fused with the GFP (green fluorescent protein). The efficiency of transfection was assessed by cell sorting monitoring GFP fluorescence. 24 h after the transfection about 75% of cells over-expressed the protein (Figure 2A). Notably, even if the overexpression of CypA induced a negligible cytotoxic effect on SH-SY5Y cells (Supplementary Figure S2), their treatment with STS at 1, 5, 10 and 20 μM for 3 h produced a significant increase of PI positive cells compared to cells not overexpressing CypA (Figure 2B).

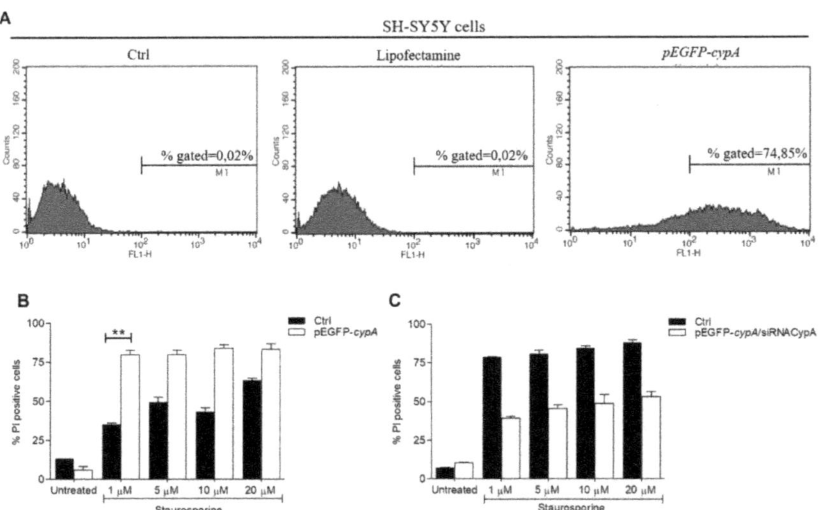

Figure 2. CypA overexpression increases STS-induced cell death in SH-SY5Y cells. (**A**) Assessment of pEGFP-*cypa* plasmid transfection efficiency after 24 h using flow cytometry monitoring the GFP fluorescence. (**B**) Quantification of flow cytometry results of PI stained SH-SY5Y cells, transfected or not (Ctrl) with pEGFP plasmid coding CypA (pEGFP-*cypa*) and treated with STS for 3 h at the indicated concentrations (n = 8, ** $p < 0.01$). (**C**) Quantification of flow cytometry results of PI stained SH-SY5Y cells, co-transfected with pEGFP-*cypa*/siRNACypA or transfected only with pEGFP-*cypA* (Ctrl) and treated with STS for 3 h at the indicated concentrations.

Finally, we co-transfected the cells with the plasmid coding for GFP-CypA and with the siRNACypA. The presence of siRNACypA induced a reduction of CypA expression levels of about 20% at 24 h and the effect increased at 48 and 72 h (Supplementary Figure S3). In line with previous results, the downregulation of CypA at 24 h reduced the percentage of PI stained cells after treatment with STS at all concentrations tested (Figure 2C). Altogether, the results show again that CypA plays a pro-apoptotic role in the cell death of SH-SY5Y induced by STS.

2.2. STS-Induced Cell Death Is Counteracted in AIF(370-394)-Treated Cells

Once assessed that CypA plays a role in the neuronal cell loss caused by STS, we used AIF(370-394) to inhibit the formation of the CypA/AIF complex and to evaluate its effect on cell viability compared with that observed following CypA silencing. AIF(370-394) conjugated with a TAT sequence (hereafter AIF(370-394)) was transfected in SH-SY5Y cells with a protocol previously optimized (see Materials and Methods for details).

The amount of peptide transfected into the cells was determined by FACS analysis using the FITC conjugated peptide at 3 different doses (25, 50, and 100 µM). The average transfection efficiency was about 27, 63, and 75% at 25, 50, and 100 µM, respectively at 24 h (Figure 3A). The effects of peptide transfection on cell viability was explored through FACS analysis by staining the apoptotic cells with the PI dye. Results show that the transfection of the peptide, at all concentrations tested at 24 h, provides no or negligible effects on cell viability compared to untreated cells. Indeed, in all cases, only about 1% of cells were positive to PI staining, just like untreated cells used as control (Figure 3B). Similar analyses performed at 72 h after transfection show that the peptide is not toxic in the concentration range tested up to 72 h (Supplementary Figure S4). On the basis of this evidence, the concentration of the peptide was maintained at 50 µM in all subsequent experiments. MTT experiments were thus performed on cells in the presence of STS at concentrations between 1.0 µM and 20 µM and with AIF(370-394) at 50 µM. Data show that the peptide provided a strong protective effect against cell death induced by the drug at all concentrations tested (Figure 2C). Importantly, up to 10 μM STS cell vitality was fully restored in the presence of peptide. Using STS at 20 μM, more than 75% of cells survived when exposed to AIF(370-394) (Figure 2C). These findings show that treating the cells with the peptide, the pro-apoptotic action of STS is significantly suppressed, and this effect is greater than that observed following silencing of CypA.

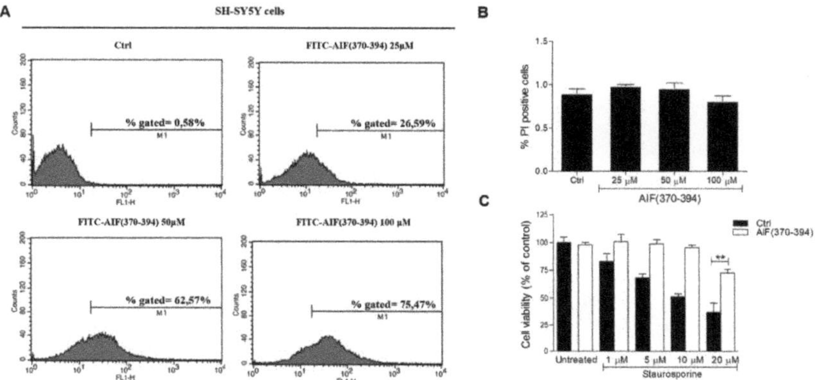

Figure 3. Transfection of the peptide AIF(370-394) protects SH-SY5Y cells from death induced by STS. (**A**) Assessment of AIF(370-394) transfection efficiency after 24 h using flow cytometry and the FITC-conjugated TAT peptide at 25, 50, and 100 μM. (**B**) Evaluation of the cytotoxic effects of FITC-AIF(370-394) by FACS analysis; apoptotic cells were stained with PI. (**C**) MTT viability assay of SH-SY5Y cells transfected with AIF(370-394) exposed to STS for 3 h at the indicated concentrations, (n = 8, ** $p < 0.01$).

2.3. *AIF(370-394) Influences the AIF/CypA Nuclear Translocation Induced by STS, without Affecting Caspase-3 Activation and PARP*

To investigate the mechanism underlying the protective effects of AIF(370-394) on STS-treated SH-SY5Y cells, we evaluated whether the peptide influenced the subcellular localization of CypA and AIF. SH-SY5Y cells were then transfected with the TAT-conjugated peptide and treated with STS. Peptide-treated cells not exposed to STS were used as controls. Nuclear and cytosolic fractions were extracted and AIF and CypA were detected by WB. As shown in Figure 4, CypA and AIF were stained in both the cytosol and the nucleus of untreated SH-SY5Y cells while, as observed in other cell lines [12,25,26,32], STS treatment induced the translocation of both proteins into the nucleus as a consequence of the kinase inhibitor-induced oxidative stress. Indeed, an increase of the AIF and CypA levels was observed in the nucleus upon STS treatment compared to control cells.

Importantly, in the presence of AIF(370-394), a reduced amount of both proteins was revealed in the nucleus of cells treated with STS as compared to control cells, leading to a significant accumulation of AIF in the cytosol (Figure 4A,B).

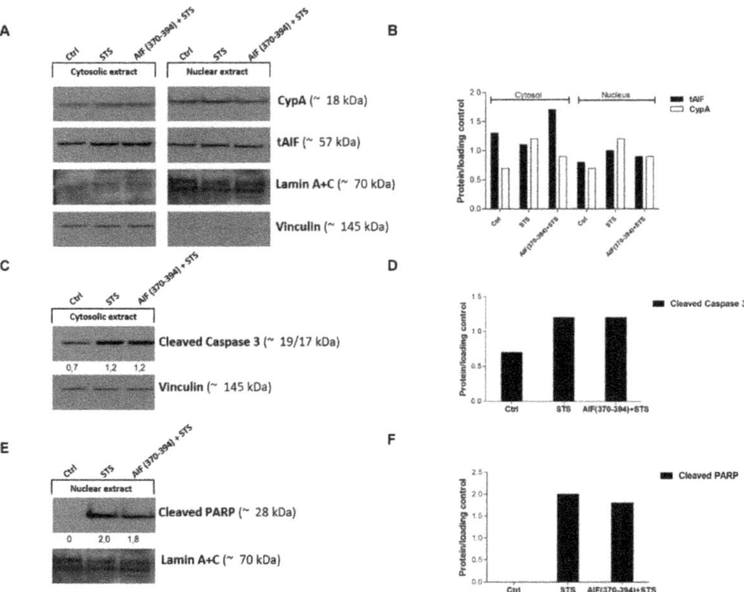

Figure 4. AIF(370-394) blocks AIF and CypA nuclear translocation induced by STS. (**A**) Representative immunoblots of CypA and tAIF and relative densitometric bar graph of proteins (**B**) in the cytosolic and nuclear fractions of SH-SY5Y cells untreated or treated with 5 μM STS for 3 h. (**C**) Representative immunoblot for the detection in the cytosolic extract of cleaved caspase 3 and (**D**) cleaved PARP in the nuclear extract and relative densitometric bar graphs of proteins (**E**). Densitometric analyses were performed using vinculin and lamin A/C as markers of cytosolic and nuclear proteins, respectively.

To further elucidate the mechanism underpinning the AIF(370-394) protective effect on the STS-treated SH-SY5Y cells, we also inspected the amounts of cleaved caspase-3 and PARP in both the cytosolic and nuclear extracts of cells treated and untreated with 10 μM STS. According to previous reports [5,34], in STS-treated cells, we detected a significant increase of activated caspase-3 and cleaved PARP, detected as the p24 subunit at ~24 kDa, in the cytosol and nucleus, respectively, compared to control cells (Figure 4C–F). The delivery of AIF(370-394) to STS-treated cells did not alter the levels of cleaved target proteins compared to cells treated with STS alone, indicating that the presence of the peptide did not influence the cell death mediated by caspase activation. Cleaved PARP was unexpectedly not stained in the STS-untreated cells. We hypothesize that at the time point evaluated, the level of cleaved PARP is still too low and is therefore not detected.

3. Discussion

Several reports have shown that the nuclear translocation of the AIF/CypA complex is associated with cell death in a variety of different cellular and rodent models of acute brain injury induced by oxidative stress, cerebral hypoxia/ischemia (HI), and even brain trauma. Following HI insults, the complex translocates to cell nuclei where induces chromatin condensation, DNA degradation, and cell death through a caspase-independent mechanism [25,26]. Recently, we have shown that this phenomenon is not restricted to neuronal tissues, but also occurs in in vitro and in vivo models of arrhythmogenic cardiomyopathy [32]. It is noteworthy that blocking the AIF/CypA complex and its nuclear translocation through CypA antisense oligonucleotides and/or the delivery of the inhibitory peptide AIF(370-394) protects against cell death induced by high doses of glutamate in HT22 hippocampal cells, prevents brain injury in neonatal mice undergoing HI, and averts myocyte death during myocardial dysfunction [25,26,32]. Despite the increasing evidence on the crucial role of the AIF/CypA complex in neurological diseases and the importance of its targeting for therapeutic purposes, the role of the AIF/CypA complex in PD has not so far been investigated.

In this framework, we explored the role of the AIF/CypA complex in SH-SY5Y treated with STS, which is a well-known model to study PD in cells [1]. In these cells, it has been demonstrated that, as a consequence of oxidative stress, high doses of STS induce the mitochondrial release of AIF in the cytoplasm and in the nucleus, where it is involved in cell death pathways [7,9,10]. Here, we have demonstrated that, in this experimental paradigm, AIF translocation to the nucleus and the subsequent effects on cell viability requires its association with CypA and that inhibiting the formation of this complex is a way to prevent cell damages induced by oxidative stress.

We found that CypA is highly expressed in this cell line and that treatment with STS increases its levels in both the cytoplasm and the nucleus, very likely due to an inflammatory response, as reported in previous reports [27]. In the same experimental conditions, AIF is also detected in the nucleus, suggesting that the complex AIF/CypA contributes to decreasing cell viability. The dependence of the death mechanism from the AIF/CypA complex formation is strongly supported by the observation that blocking CypA expression with the corresponding antisense significantly neutralizes STS-mediated cell killing and by the protective effects provided by AIF(370-394), which reportedly blocks the nuclear translocation of the complex. Moreover, the overexpression of CypA in SH-SY5Y significantly amplifies the lethal effect of STS while its downregulation strongly counteracts it. Altogether, these data demonstrate the implication of the AIF/CypA complex in the STS-driven mechanism of death of SH-SY5Y cells and that its targeting provides strong neuroprotection.

We have previously demonstrated that blocking the AIF/CypA nuclear translocation with AIF(370-394) not only suppresses apoptosis of HT-22 neuronal cells after glutamate-mediated oxidative stress but also preserves mitochondrial bioenergetics, suggesting alternative pathways of action of the peptide, upstream of lethal nuclear translocation of AIF/CypA [25]. However, this evidence was not confirmed in the model of neonatal mice brain injury after HI [26]. To assess this aspect in SH-SY5Y cells and to determine the impact of the peptide on caspases activation, we also evaluated the activation of caspase-3 upon STS treatment. In SH-SY5Y, STS promotes the permeabilization of the outer mitochondrial membrane [35], inducing the release of several proteins from mitochondria to the cytosol. Cytochrome C is one of the first proteins to translocate into the cytosol, where it activates by an allosteric mechanism, the apoptosis-protease activating factor 1 (APAF-1), which is in turn required for the proteolytic maturation of caspases, including the caspase-3-like proteases [36]. Preliminary experiments show that, as previously reported [36], in SH-SY5Y STS causes a significant increase of the cleaved/activated caspase-3 compared to untreated cells, but interestingly, the presence of the peptide does not appear to affect caspase-3 activation. To further investigate this observation, we have also examined the effect of STS and STS/peptide treatments on PARP cleavage, which is a major hallmark of caspase-3 activation. Consistent with the increase of caspase-3 cleavage, STS induces PARP inactivation [6–8], but, in line with previous results, AIF(370-394) does not affect the processing of these proteins in this cell model. These results globally show that inhibiting with AIF(370-394) the AIF/CypA complex in a cellular model of PD prevents cell death and that the mechanism is independent of the caspase pathways and does not affect mitochondrial bioenergetics. However, future experiments are needed to analyze in more detail the pro-apoptotic mechanism mediated by AIF/CypA complex in this experimental paradigm.

The development of effective therapies for PD is extremely challenging because of the limited understanding of the mechanisms of neurodegeneration in PD and the high heterogeneity of the pathology. Peptides are crucial tools for PD research studies and drug discovery [37]. Today several natural and synthetic peptides are used for the treatment of PD, such as Glucagon-like peptide-1 (GLP-1)-based receptor agonists [38] and NAPVSIPQ (NAP) [39]. We thus feel that AIF(370-394) may play a key role for understanding further the mechanisms underlying the disease onset and the potential pathways to target for its treatment, especially those associated with damage of the mitochondrial functions. The successful application of this synthetic peptide highlights the role of the AIF/CypA complex in the pathophysiological mechanisms leading to PD and suggests that it is a promising target for developing first-in-class therapeutics to treat this currently incurable disease.

4. Materials and Methods

4.1. Materials

Protected amino acids, coupling agents (HATU, Oxyma), and Fmoc-Rink Amide AM resin used for peptide synthesis were purchased from IRIS Biotech GmbH (Marktrewitz, DE). Solvents, including acetonitrile (CH3CN) and dimethylformamide (DMF) were purchased from Carlo Erba reagents (Milan, Italy). Other products such as trifluoroacetic acid (TFA), sym-collidine, diisopropylethylamine (DIPEA), piperidine, were from Sigma-Aldrich (Milan, Italy). HPLC analyses for peptides characterization were performed on an Alliance HT WATERS 2795 system, equipped with a PDA WATERS detector 2996, whereas preparative purifications were carried out on a WATERS 2545 preparative system (Waters, Milan, Italy) fitted out with a WATERS 2489 UV/Visible detector.

4.2. siRNA

RNA interference experiments were performed as previously described in the literature [25].

4.3. Peptide Synthesis and Characterization

AIF(370-394) conjugated at the N-terminus with the cell-penetrating TAT peptide (sequence: GRKKRRQRRRβAFC), which was introduced to favor membrane crossing [40], was assembled on solid phase (Rink-Amide MBHA resin) using a standard protocol for Fmoc chemistry with Oxyma-DIC and HATU-collidine as coupling reagents, as previously reported [41,42]. Peptide purity and identity were confirmed by liquid chromatography–mass spectrometry analysis (LC–MS), as reported in the literature [41,42]. The TAT-conjugated peptide is here called AIF(370-394) for simplicity. A TAT-AIF peptide variant N-terminally modified with fluorescein-5-isothiocyanate (FITC) was also similarly prepared and utilized to assess cell penetration by FACS analyses.

4.4. Neuroblastoma SH-SY5Y Cell Culture and Transfection

SH-SY5Y neuroblastoma cells were obtained from the American Type Culture Collection (ATCC) and cultured in high-glucose DMEM supplemented with 10% FBS (Hyclone), penicillin (100 mg/mL), streptomycin (100 mg/mL), and amphotericin B (250 mg/mL) (Merk Life Science S.r.l. Via Monte Rosa, 93 20149 Milan, Italy). The cells were incubated at 37 °C at a fixed concentration of CO_2 (5%), and culture medium was changed every 2–3 days.

Transient transfection with siRNA and pEGFP-*cypA* plasmid was performed using Lipofectamine-2000 (Invitrogen) following the manufacturer's procedure. Peptide transient transfection was performed using the AIF(370-394) peptide conjugated at N-terminus with the TAT peptide [26,32]. Briefly, the peptide was incubated with cells for 4 h in DMEM without serum. After the incubation, culture medium was changed with DMEM supplemented with 10% FBS (Hyclone), penicillin (100 mg/mL), streptomycin (100 mg/mL), and amphotericin B (250 mg/mL) (Sigma, UK).

4.5. Cell Treatment

Cells were treated with STS at the indicated times and concentrations. STS was dissolved in DMSO and added to the culture medium to obtain the final concentration indicated. Negative control cells were treated with an equal volume of DMSO (<0.1% *v*/*v*).

4.6. Cell Viability Assay (MTT)

The effect of STS on cell viability was determined by the MTT assay (MTT: 3-(4,5-dimethyl thiazol-2yl)-2, 5-diphenyl tetrazolium bromide) [43]. Cells were seeded in a 96-well flat-bottom plate at a density of 6×10^3 cells/well for 24 h at 37 °C in a CO_2 incubator. After 24 h incubation, the culture medium was replaced with a fresh medium, therefore treated with STS. Subsequently, 10 μL of MTT working solution (5 mg/mL in phosphate buffer solution) were added to each well and the plate was incubated for 4 h at 37 °C in a CO_2 incubator. The medium was then aspirated, and the formed formazan crystals were solubilized by adding 50 μL of DMSO. Absorbance intensity was measured using an Infinite M200 plate reader (TECAN) at 570 nm. Experiments were performed in triplicate and values are expressed as mean ± SD.

4.7. Cell Death Assay by Propidium Iodide (PI)

Cells were plated (2×10^5 cells/mL) and grown for 24 h. Cells were then transfected with the peptide and treated with the STS at the indicated concentration and time. Cell samples were left untreated and used as negative controls. Finally, cells were recovered and incubated with PI buffer containing 0.2 μg/mL of PI in PBS and analyzed by (FACS) calibur flow cytometer using Cell Quest software (Becton Dickinson, BD Biosciences, Drive Franklin Lakes, NJ 07417-1880 USA).

4.8. Subcellular Fractionation and Western Blot Analysis

For WB analysis, SH-SY5Y cells were lysed in 50 μL RIPA buffer supplemented with protease inhibitor cocktail and phenylmethylsulphonylfluoride (PMFS) (all from Sigma-Aldrich, Milano, Italy). After centrifugation at 13,000× *g* for 30 min at 4 °C the supernatant was stored at −80 °C until further use. For cytosolic extract preparation, cells were washed in cold PBS, centrifuged at 6000 rpm for 5 min at 4 °C and resuspended in the Cytoplasmic Extract (CE) buffer (10 mM HEPES pH 7.9, 10 mM KCl, 0.1 mM EDTA, 0.3% NP-40 supplemented with protease and phosphatase inhibitors cocktail) on ice for 5 min. After centrifugation at 3000 rpm for 5 min at 4 °C, the supernatant (cytosolic extract) was harvested. The pellet was washed twice with CE buffer without NP-40, centrifuged at 3000 rpm for 5 min at 4 °C, and incubated with an equal volume of Nuclear Extract (NE) buffer (20 mM HEPES, 0.4 M NaCl, 1 mM EDTA, 25% glycerol supplemented with protease inhibitors cocktail) for 10 min. After centrifugation at 14,000 rpm for 5 min at 4 °C the supernatant (nuclear extract) was harvested. The protein concentration was determined by the Bradford assay method. About 30 μg of proteins were separated on 4–12% pre-cast gel (Bolt Bis-Tris Plus, Thermo Fischer, Milano, Italy) followed by transfer to a PVDF membrane. After blocking with 5% skim milk powered in TRIS-buffered saline (TBS)-Tween for 1 h, the membranes were incubated overnight at 4 °C with specific primary antibodies: anti-CypA (GTX 104698, GeneTex, 2456 Alton Pkwy Irvine, CA 92606, USA), anti-Lamin-A/C (GTX 101127, GeneTex, 2456 Alton Pkwy Irvine, CA 92606, USA), anti-AIF (sc 9416, Santa Cruz Biotechnology, Inc. Bergheimer Str. 89-2, 69115 Heidelberg, Germany), anti-caspase 3 (ab32351-Abcam, Prodotti Gianni S.p.A., Via Quintiliano,30, 20138 Milan, Italy), anti-PARP (ab6079-Abcam, Prodotti Gianni S.p.A., Via Quintiliano,30, 20138 Milan, Italy), anti-Vinculin (orb 76294, Biorbyt Ltd., 5 Orwell Furlong Cowley Road Cambridge Cambridgeshire CB4 0WY, UK), and anti-Erk1/2 (sc-514302, Santa Cruz Biotechnology, Inc. Bergheimer Str. 89-2, 69115 Heidelberg, Germany). PVDF membranes were then exposed to the appropriate HRP-conjugated secondary antibody and immunocomplexes were visualized with the ECL detection system (Santa Cruz, CA, USA) and subsequently exposed to film. Relative band intensities were quantified by densitometric analysis with Image J software (NIH, Bethesda, MA, USA).

4.9. Statistical Data Analysis

Data were presented as the mean ± SD of biological replicates. Differences in the mean between different groups were calculated using analysis of variance (ANOVA) plus Student's t-test. p-values of less than 0.05 were recognized as significant.

Supplementary Materials: The following supporting information can be downloaded at: https://www.mdpi.com/article/10.3390/ijms23010265/s1.

Author Contributions: Conceptualization, M.C., R.P., L.A. and N.D.; methodology, M.C., A.M., E.F., R.P. and N.D.; validation, M.C., R.P., A.M. and N.D.; analysis, M.C., R.P., A.N. and N.D.; data curation, M.C., R.P., M.R., L.A. and N.D.; writing—original draft preparation, M.R., L.A. and N.D.; writing—review and editing, M.C., R.P., M.R., L.A. and N.D. All authors have read and agreed to the published version of the manuscript.

Funding: This research was funded by the Campania Regional Government Technology Platform Lotta alle Patologie Oncologiche: iCURE (B21C17000030007), Campania Regional Government FASE2: IDEAL (B63D18000560007), MIUR, Proof of Concept POC01_00043, Programma V: ALERE 2020—Progetto competitivo "NETWINS"—D.R. no. 138 of 17/02/2020. N.D., A.M., and M.R. acknowledge the support from Regione Campania for the projects: (i) "Fighting Cancer resistance: Multidisciplinary integrated Platform for a technologically Innovative Approach to Oncotherapies (Campania Oncotherapies)"; (ii) "Development of novel therapeutic approaches for treatment of resistant neoplastic diseases (SATIN)".

Institutional Review Board Statement: Not applicable.

Informed Consent Statement: Not applicable.

Data Availability Statement: All data are provided as figures and tables and included in this paper.

Conflicts of Interest: The authors declare no conflict of interest.

References

1. Xicoy, H.; Wieringa, B.; Martens, G.J. The SH-SY5Y cell line in Parkinson's disease research: A systematic review. *Mol. Neurodegener.* **2017**, *12*, 10. [CrossRef] [PubMed]
2. Belmokhtar, C.A.; Hillion, J.; Segal-Bendirdjian, E. Staurosporine induces apoptosis through both caspase-dependent and caspase-independent mechanisms. *Oncogene* **2001**, *20*, 3354–3362. [CrossRef] [PubMed]
3. Boix, J.; Llecha, N.; Yuste, V.J.; Comella, J.X. Characterization of the cell death process induced by staurosporine in human neuroblastoma cell lines. *Neuropharmacology* **1997**, *36*, 811–821. [CrossRef]
4. Lopez, E.; Ferrer, I. Staurosporine- and H-7-induced cell death in SH-SY5Y neuroblastoma cells is associated with caspase-2 and caspase-3 activation, but not with activation of the FAS/FAS-L-caspase-8 signaling pathway. *Brain Res. Mol. Brain Res.* **2000**, *85*, 61–67. [CrossRef]
5. Yuste, V.J.; Sanchez-Lopez, I.; Sole, C.; Encinas, M.; Bayascas, J.R.; Boix, J.; Comella, J.X. The prevention of the staurosporine-induced apoptosis by Bcl-X(L), but not by Bcl-2 or caspase inhibitors, allows the extensive differentiation of human neuroblastoma cells. *J. Neurochem.* **2002**, *80*, 126–139. [CrossRef] [PubMed]
6. Chakravarthy, B.R.; Walker, T.; Rasquinha, I.; Hill, I.E.; MacManus, J.P. Activation of DNA-dependent protein kinase may play a role in apoptosis of human neuroblastoma cells. *J. Neurochem.* **1999**, *72*, 933–942. [CrossRef]
7. Jantas, D.; Pytel, M.; Mozrzymas, J.W.; Leskiewicz, M.; Regulska, M.; Antkiewicz-Michaluk, L.; Lason, W. The attenuating effect of memantine on staurosporine-, salsolinol- and doxorubicin-induced apoptosis in human neuroblastoma SH-SY5Y cells. *Neurochem. Int.* **2008**, *52*, 864–877. [CrossRef]
8. Posmantur, R.; McGinnis, K.; Nadimpalli, R.; Gilbertsen, R.B.; Wang, K.K. Characterization of CPP32-like protease activity following apoptotic challenge in SH-SY5Y neuroblastoma cells. *J. Neurochem.* **1997**, *68*, 2328–2337. [CrossRef]
9. Daugas, E.; Susin, S.A.; Zamzami, N.; Ferri, K.F.; Irinopoulou, T.; Larochette, N.; Prevost, M.C.; Leber, B.; Andrews, D.; Penninger, J.; et al. Mitochondrio-nuclear translocation of AIF in apoptosis and necrosis. *FASEB J.* **2000**, *14*, 729–739. [CrossRef]
10. Susin, S.A.; Lorenzo, H.K.; Zamzami, N.; Marzo, I.; Snow, B.E.; Brothers, G.M.; Mangion, J.; Jacotot, E.; Costantini, P.; Loeffler, M.; et al. Molecular characterization of mitochondrial apoptosis-inducing factor. *Nature* **1999**, *397*, 441–446. [CrossRef]
11. Cheung, E.C.; Joza, N.; Steenaart, N.A.; McClellan, K.A.; Neuspiel, M.; McNamara, S.; MacLaurin, J.G.; Rippstein, P.; Park, D.S.; Shore, G.C.; et al. Dissociating the dual roles of apoptosis-inducing factor in maintaining mitochondrial structure and apoptosis. *EMBO J.* **2006**, *25*, 4061–4073. [CrossRef]
12. Hangen, E.; Feraud, O.; Lachkar, S.; Mou, H.; Doti, N.; Fimia, G.M.; Lam, N.V.; Zhu, C.; Godin, I.; Muller, K.; et al. Interaction between AIF and CHCHD4 Regulates Respiratory Chain Biogenesis. *Mol. Cell* **2015**, *58*, 1001–1014. [CrossRef]

13. Sevrioukova, I.F. Apoptosis-inducing factor: Structure, function, and redox regulation. *Antioxid. Redox Signal.* **2011**, *14*, 2545–2579. [CrossRef] [PubMed]
14. Cao, G.; Clark, R.S.; Pei, W.; Yin, W.; Zhang, F.; Sun, F.Y.; Graham, S.H.; Chen, J. Translocation of apoptosis-inducing factor in vulnerable neurons after transient cerebral ischemia and in neuronal cultures after oxygen-glucose deprivation. *J. Cereb. Blood Flow. Metab.* **2003**, *23*, 1137–1150. [CrossRef] [PubMed]
15. Otera, H.; Ohsakaya, S.; Nagaura, Z.; Ishihara, N.; Mihara, K. Export of mitochondrial AIF in response to proapoptotic stimuli depends on processing at the intermembrane space. *EMBO J.* **2005**, *24*, 1375–1386. [CrossRef]
16. Zhang, X.; Chen, J.; Graham, S.H.; Du, L.; Kochanek, P.M.; Draviam, R.; Guo, F.; Nathaniel, P.D.; Szabo, C.; Watkins, S.C.; et al. Intranuclear localization of apoptosis-inducing factor (AIF) and large scale DNA fragmentation after traumatic brain injury in rats and in neuronal cultures exposed to peroxynitrite. *J. Neurochem.* **2002**, *82*, 181–191. [CrossRef] [PubMed]
17. Klein, J.A.; Longo-Guess, C.M.; Rossmann, M.P.; Seburn, K.L.; Hurd, R.E.; Frankel, W.N.; Bronson, R.T.; Ackerman, S.L. The harlequin mouse mutation downregulates apoptosis-inducing factor. *Nature* **2002**, *419*, 367–374. [CrossRef]
18. Piao, C.S.; Loane, D.J.; Stoica, B.A.; Li, S.; Hanscom, M.; Cabatbat, R.; Blomgren, K.; Faden, A.I. Combined inhibition of cell death induced by apoptosis inducing factor and caspases provides additive neuroprotection in experimental traumatic brain injury. *Neurobiol. Dis.* **2012**, *46*, 745–758. [CrossRef]
19. Ravagnan, L.; Gurbuxani, S.; Susin, S.A.; Maisse, C.; Daugas, E.; Zamzami, N.; Mak, T.; Jaattela, M.; Penninger, J.M.; Garrido, C.; et al. Heat-shock protein 70 antagonizes apoptosis-inducing factor. *Nat. Cell Biol.* **2001**, *3*, 839–843. [CrossRef]
20. Susin, S.A.; Daugas, E.; Ravagnan, L.; Samejima, K.; Zamzami, N.; Loeffler, M.; Costantini, P.; Ferri, K.F.; Irinopoulou, T.; Prevost, M.C.; et al. Two distinct pathways leading to nuclear apoptosis. *J. Exp. Med.* **2000**, *192*, 571–580. [CrossRef]
21. Susin, S.A.; Zamzami, N.; Castedo, M.; Hirsch, T.; Marchetti, P.; Macho, A.; Daugas, E.; Geuskens, M.; Kroemer, G. Bcl-2 inhibits the mitochondrial release of an apoptogenic protease. *J. Exp. Med.* **1996**, *184*, 1331–1341. [CrossRef]
22. Burguillos, M.A.; Hajji, N.; Englund, E.; Persson, A.; Cenci, A.M.; Machado, A.; Cano, J.; Joseph, B.; Venero, J.L. Apoptosis-inducing factor mediates dopaminergic cell death in response to LPS-induced inflammatory stimulus: Evidence in Parkinson's disease patients. *Neurobiol. Dis.* **2011**, *41*, 177–188. [CrossRef] [PubMed]
23. Yalcinkaya, N.; Haytural, H.; Bilgic, B.; Ozdemir, O.; Hanagasi, H.; Kucukali, C.I.; Ozbek, Z.; Akcan, U.; Idrisoglu, H.A.; Gurvit, H.; et al. Expression changes of genes associated with apoptosis and survival processes in Parkinson's disease. *Neurosci. Lett.* **2016**, *615*, 72–77. [CrossRef] [PubMed]
24. Artus, C.; Boujrad, H.; Bouharrour, A.; Brunelle, M.N.; Hoos, S.; Yuste, V.J.; Lenormand, P.; Rousselle, J.C.; Namane, A.; England, P.; et al. AIF promotes chromatinolysis and caspase-independent programmed necrosis by interacting with histone H2AX. *EMBO J.* **2010**, *29*, 1585–1599. [CrossRef]
25. Doti, N.; Reuther, C.; Scognamiglio, P.L.; Dolga, A.M.; Plesnila, N.; Ruvo, M.; Culmsee, C. Inhibition of the AIF/CypA complex protects against intrinsic death pathways induced by oxidative stress. *Cell Death Dis.* **2014**, *5*, e993. [CrossRef]
26. Rodriguez, J.; Xie, C.; Li, T.; Sun, Y.; Wang, Y.; Xu, Y.; Li, K.; Zhang, S.; Zhou, K.; Wang, Y.; et al. Inhibiting the interaction between apoptosis-inducing factor and cyclophilin A prevents brain injury in neonatal mice after hypoxia-ischemia. *Neuropharmacology* **2020**, *171*, 108088. [CrossRef]
27. Nigro, P.; Pompilio, G.; Capogrossi, M.C. Cyclophilin A: A key player for human disease. *Cell Death Dis.* **2013**, *4*, e888. [CrossRef] [PubMed]
28. Farina, B.; Di Sorbo, G.; Chambery, A.; Caporale, A.; Leoni, G.; Russo, R.; Mascanzoni, F.; Raimondo, D.; Fattorusso, R.; Ruvo, M.; et al. Structural and biochemical insights of CypA and AIF interaction. *Sci. Rep.* **2017**, *7*, 1138. [CrossRef]
29. Farina, B.; Sturlese, M.; Mascanzoni, F.; Caporale, A.; Monti, A.; Di Sorbo, G.; Fattorusso, R.; Ruvo, M.; Doti, N. Binding mode of AIF(370-394) peptide to CypA: Insights from NMR, label-free and molecular docking studies. *Biochem. J.* **2018**, *475*, 2377–2393. [CrossRef]
30. Monti, A.; Sturlese, M.; Caporale, A.; Roger, J.A.; Mascanzoni, F.; Ruvo, M.; Doti, N. Design, synthesis, structural analysis and biochemical studies of stapled AIF(370–394) analogues as ligand of CypA. *Biochim. Biophys Acta Gen. Subj.* **2020**, *1864*, 129717. [CrossRef]
31. Russo, L.; Mascanzoni, F.; Farina, B.; Dolga, A.M.; Monti, A.; Caporale, A.; Culmsee, C.; Fattorusso, R.; Ruvo, M.; Doti, N. Design, Optimization, and Structural Characterization of an Apoptosis-Inducing Factor Peptide Targeting Human Cyclophilin A to Inhibit Apoptosis Inducing Factor-Mediated Cell Death. *J. Med. Chem.* **2021**, *64*, 11445–11459. [CrossRef]
32. Chelko, S.P.; Keceli, G.; Carpi, A.; Doti, N.; Agrimi, J.; Asimaki, A.; Beti, C.B.; Miyamoto, M.; Amat-Codina, N.; Bedja, D.; et al. Exercise triggers CAPN1-mediated AIF truncation, inducing myocyte cell death in arrhythmogenic cardiomyopathy. *Sci. Transl. Med.* **2021**, *13*, eabf0891. [CrossRef]
33. Curci, A.; Mele, A.; Camerino, G.M.; Dinardo, M.M.; Tricarico, D. The large conductance $Ca^{(2+)}$ -activated $K^{(+)}$ (BKCa) channel regulates cell proliferation in SH-SY5Y neuroblastoma cells by activating the staurosporine-sensitive protein kinases. *Front. Physiol.* **2014**, *5*, 476. [CrossRef]
34. Mashimo, M.; Onishi, M.; Uno, A.; Tanimichi, A.; Nobeyama, A.; Mori, M.; Yamada, S.; Negi, S.; Bu, X.; Kato, J.; et al. The 89-kDa PARP1 cleavage fragment serves as a cytoplasmic PAR carrier to induce AIF-mediated apoptosis. *J. Biol. Chem.* **2021**, *296*, 100046. [CrossRef]
35. Mookherjee, P.; Quintanilla, R.; Roh, M.S.; Zmijewska, A.A.; Jope, R.S.; Johnson, G.V. Mitochondrial-targeted active Akt protects SH-SY5Y neuroblastoma cells from staurosporine-induced apoptotic cell death. *J. Cell Biochem.* **2007**, *102*, 196–210. [CrossRef]

36. Kim, H.E.; Du, F.; Fang, M.; Wang, X. Formation of apoptosome is initiated by cytochrome c-induced dATP hydrolysis and subsequent nucleotide exchange on Apaf-1. *Proc. Natl. Acad. Sci. USA* **2005**, *102*, 17545–17550. [CrossRef]
37. Baig, M.H.; Ahmad, K.; Saeed, M.; Alharbi, A.M.; Barreto, G.E.; Ashraf, G.M.; Choi, I. Peptide based therapeutics and their use for the treatment of neurodegenerative and other diseases. *Biomed. Pharmacother.* **2018**, *103*, 574–581. [CrossRef] [PubMed]
38. Sisson, E.M. Liraglutide: Clinical pharmacology and considerations for therapy. *Pharmacotherapy* **2011**, *31*, 896–911. [CrossRef] [PubMed]
39. Tiwari, S.K.; Chaturvedi, R.K. Peptide therapeutics in neurodegenerative disorders. *Curr. Med. Chem.* **2014**, *21*, 2610–2631. [CrossRef] [PubMed]
40. Patel, S.G.; Sayers, E.J.; He, L.; Narayan, R.; Williams, T.L.; Mills, E.M.; Allemann, R.K.; Luk, L.Y.P.; Jones, A.T.; Tsai, Y.H. Cell-penetrating peptide sequence and modification dependent uptake and subcellular distribution of green florescent protein in different cell lines. *Sci. Rep.* **2019**, *9*, 6298. [CrossRef]
41. Caporale, A.; Doti, N.; Monti, A.; Sandomenico, A.; Ruvo, M. Automatic procedures for the synthesis of difficult peptides using oxyma as activating reagent: A comparative study on the use of bases and on different deprotection and agitation conditions. *Peptides* **2018**, *102*, 38–46. [CrossRef] [PubMed]
42. Caporale, A.; Doti, N.; Sandomenico, A.; Ruvo, M. Evaluation of combined use of Oxyma and HATU in aggregating peptide sequences. *J. Pept. Sci.* **2017**, *23*, 272–281. [CrossRef] [PubMed]
43. Hansen, M.B.; Nielsen, S.E.; Berg, K. Re-examination and further development of a precise and rapid dye method for measuring cell growth/cell kill. *J. Immunol. Methods* **1989**, *119*, 203–210. [CrossRef]

International Journal of
Molecular Sciences

MDPI

Review

Multitalented Synthetic Antimicrobial Peptides and Their Antibacterial, Antifungal and Antiviral Mechanisms

Tania Vanzolini [1,†], Michela Bruschi [1,*,†], Andrea C. Rinaldi [2], Mauro Magnani [1] and Alessandra Fraternale [1]

1 Department of Biomolecular Sciences, University of Urbino Carlo Bo, 61029 Urbino, PU, Italy; t.vanzolini@campus.uniurb.it (T.V.); mauro.magnani@uniurb.it (M.M.); alessandra.fraternale@uniurb.it (A.F.)
2 Department of Biomedical Sciences, University of Cagliari, 09042 Monserrato, CA, Italy; rinaldi@unica.it
* Correspondence: michela.bruschi@uniurb.it
† Contributed equally.

Abstract: Despite the great strides in healthcare during the last century, some challenges still remained unanswered. The development of multi-drug resistant bacteria, the alarming growth of fungal infections, the emerging/re-emerging of viral diseases are yet a worldwide threat. Since the discovery of natural antimicrobial peptides able to broadly hit several pathogens, peptide-based therapeutics have been under the lenses of the researchers. This review aims to focus on synthetic peptides and elucidate their multifaceted mechanisms of action as antiviral, antibacterial and antifungal agents. Antimicrobial peptides generally affect highly preserved structures, e.g., the phospholipid membrane via pore formation or other constitutive targets like peptidoglycans in Gram-negative and Gram-positive bacteria, and glucan in the fungal cell wall. Additionally, some peptides are particularly active on biofilm destabilizing the microbial communities. They can also act intracellularly, e.g., on protein biosynthesis or DNA replication. Their intracellular properties are extended upon viral infection since peptides can influence several steps along the virus life cycle starting from viral receptor-cell interaction to the budding. Besides their mode of action, improvements in manufacturing to increase their half-life and performances are also taken into consideration together with advantages and impairments in the clinical usage. Thus far, the progress of new synthetic peptide-based approaches is making them a promising tool to counteract emerging infections.

Keywords: antimicrobial peptides; antifungal; antibacterial; antiviral; peptide-based therapies; synthetic peptides

Citation: Vanzolini, T.; Bruschi, M.; Rinaldi, A.C.; Magnani, M.; Fraternale, A. Multitalented Synthetic Antimicrobial Peptides and Their Antibacterial, Antifungal and Antiviral Mechanisms. *Int. J. Mol. Sci.* **2022**, *23*, 545. https://doi.org/10.3390/ijms23010545

Academic Editors: Nunzianna Doti and Menotti Ruvo

Received: 11 November 2021
Accepted: 30 December 2021
Published: 4 January 2022

Publisher's Note: MDPI stays neutral with regard to jurisdictional claims in published maps and institutional affiliations.

1. Introduction

When Fleming in 1922 discovered the first natural antibiotic, the lysozyme, [1] able to "lyse" bacterial cells and in 1928, Penicillin, from the fungus *Penicillium notatum*, able to inhibit bacterial growth, [2] the dawn of the antibiotic age started. Later on, in 1939, René Dubos isolated an antibacterial agent from *Bacillus brevi*, called gramicidin [3]. Gramicidin demonstrated its broad-spectrum activity against Gram-positive and Gram-negative bacteria becoming the first antibiotic commercially manufactured and sold up to this day [4]. Since the discovery of human defensins, histatins and cathelicidins, antimicrobial peptides (AMPs) have been studied, sequenced, and synthesized in laboratory in order to be used in the clinic for the treatment of several bacterial, fungal and viral infections. Besides representing the first defense of the innate immune system against pathogens, [5] they also have immunomodulatory effects working as mediators of the infection-associated inflammation, recruiting, and enhancing the activity of leukocytes and the release of cytokines but also contributing to the infection control and resolution [6,7].

Besides humans, natural AMPs have been found in different kingdoms (animals, plants, bacteria, fungi but also archaea and protists) and registered in the AMP database (https://aps.unmc.edu/, accessed on 3rd November 2021) [8,9]. Briefly, all AMPs share common features, such as a sequence composed of less than 100 amino acids (aa), [10] with the majority having between 10 and 60 aa [11]. Even if some anionic AMPs, rich in glutamic and aspartic acids, are negatively charged [12], almost all antimicrobial peptides have a net positive charge for the presence of a high number of lysine, arginine and histidine (protonated in acidic conditions) [13]. Finally, another common feature is represented by the hydrophobicity conferred by hydrophobic aa that often overcomes 50% of the total amino acid sequence [14]. The high lipophilicity is useful especially for the penetration in the biological membranes but considering the net charge, overall, AMPs are amphipathic molecules. The classifications are based on their structure or the presence/absence of recognizable motifs. AMPs could be α-helix, β-sheet, linearly extended, both α-helix and β-sheet, cyclic and with complex structure or, seen from a different perspective, tryptophan- and arginine-rich, histidine-rich, proline-rich and glycine-rich [15,16].

In the last decades, the increasing resistance to antibiotic treatments, i.e., Methicillin, Vancomycin-resistant *Staphilococus aureus* and the rise of species with intrinsic multi-drug resistance, such as *Candida auris*, highlights the need for the development of new agents [17–19]. It has been estimated that nowadays in the US every year 2.8 million people are infected by antibiotic-resistant microorganisms with a death rate of 35,000 people [20] and just in recent years the world was affected by a new pandemic virus (SARS-CoV-2) with 236 million cases and 5.9 million deaths up-to-date [21].

Studies on the AMPs synthetic analogs provided a new tool to understand the different and unique modes of actions against diverse microorganisms. Thus, this review will focus on the improvements of their properties with respect to their natural counterpart, their activity on bacterial and fungal conserved structures, i.e., membranes and cell walls, as well as on biofilm formation, their antiviral properties and execution dynamics.

The latest studies in vivo and in vitro will be discussed, with highlights on the successful therapeutic application despite drawbacks like toxicity and immunogenicity.

2. Synthetic Antimicrobial Peptides

Natural antimicrobial peptides have been always present during the evolutionary process [22], however, many natural AMPs showed host toxicity, rapid degradation by proteases, instability due to pH changes, loss of activity in presence of serum and high salt concentrations, lack of suitable delivery systems able to limit the drawbacks, and high costs of production [23–25]. Moreover, their complex design, low antimicrobial activity and pharmacokinetics led many laboratories to improve their structure and amino acid sequence to enhance their therapeutic properties [26]. Despite the multiple obstacles in the clinical application, synthetic peptides were developed to overcome the difficulties linked to the natural peptides while mimicking their pharmacological qualities [27].

The approaches commonly used for the development of non-natural AMPs are (1) the site-directed mutations characterized by the addition, the deletion or the substitution of aa, (2) the de novo design which doesn't use any template sequence, (3) the template-based design that uses fragments of the parental compound as starting point for the construction of new AMPs (in this case, antibodies seem to be a big source of patterns, especially those which recognize and bind components of the cell membrane and wall), and lastly (4) the self-assembly-based design that exploits the formation of simple nanostructures like dimers, or more complex as micelles, vesicles and nanotubes [11].

Semi-synthetic AMPs maintained the active sites of the natural source, but chemical changes were brought in order to reach the optimal properties whereas synthetic AMPs are obtained from chemical synthesis with frequent usage of the solid phase. This technique is based on the addition of one aa at a time, thus favoring the investigation of the role of each amino acid in the sequence [28].

Apart from the solid-phase method, synthetic AMPs can also derive from the catalytic ring-opening polymerization (ROP) of α-amino acid *N*-carboxyanhydride (NCA), an exquisite tool for the fabrication of long polypeptides with low polydispersity but variable chemical composition and topology [29]. Chemical synthesis represents a great step forward in peptide production with higher efficiency, reliability, and speed, especially when compared to the AMPs produced through the technology of the recombinant DNA followed by bacterial expression and purification.

The advances in the AMPs synthesis are the result of several studies about machine learning and algorithms able to predict or identify potential sequences based on the physicochemical and structural properties and on the quantitative structure-activity relationship (QSAR) of AMPs and targets already present in databases followed by high-throughput screenings [30]. Therefore, several strategies were tested to achieve a superior half-life e.g., the usage of D-amino acids [31], peptide cyclization [32], unnatural amino acids [33]. With peptidases able to recognize mainly L-amino acids sequences, stereogenic D-variants of amphipathic peptides could be resistant to proteolysis [34], as well as peptides with uncommon amino acids, i.e., ω and β-amino-acids [35,36]. Protection from cleavage could be also conferred by modifying or protecting vulnerable peptide bonds so that they cannot be easily accessed [37]. In some cases, such modifications could be applied just to the *N*- and C-terminus i.e., *C*-amidation or *N*-acetylation [38].

Similarly, PEGylation, the covalent attachment of polyethylene glycol (PEG) chains to lysine or to the N-terminus [39], could also be applied to mask other residues like arginine [40]. On the other hand, lipidation, consisting in the attachment of one or more fatty acid chains to a lysine residue or to the amine of the N-terminus, [41] could improve AMPs properties by enhancing their interaction with the membranes. Introduction of sulfonamide groups has been also investigated to exploit their bio-active properties, enhance their proteolytic stability and hydrogen bonding ability [42].

Another approach to improve the half-life of peptides in vivo is to synthesize them as dendrimers around a residue or a linear polymer core [43]. These multiple antigen peptides (MAP) developed by Tam and colleagues [44] are mainly constituted by a lysine core to which peptide chains are attached [45]. The number of bi-, tri-, tetra and more sequence patterns define the multivalency of those peptides and confers an increased cationic charge as well as hydrophobic groups. The steric hindrance given by the bulk, firstly, limit the access to the proteolytic site [46,47] and, secondly, seems to improve their activity by increasing the local concentration of peptide units with membranolytic activity [48]. Peptide structure is a pivotal point for the interaction with the membranes: the cationic charge allows the initial binding to a negatively charged layer; afterwards, while amphipathicity is necessary for membrane perturbation and peptide uptake, the hydrophobic groups are responsible for the carving [49]. Studies on the mechanism of action would divide the AMPs in two categories: *membrane disruptive* [50,51] and *non-membrane disruptive* (activity on other targets) (Figure 1) [52,53].

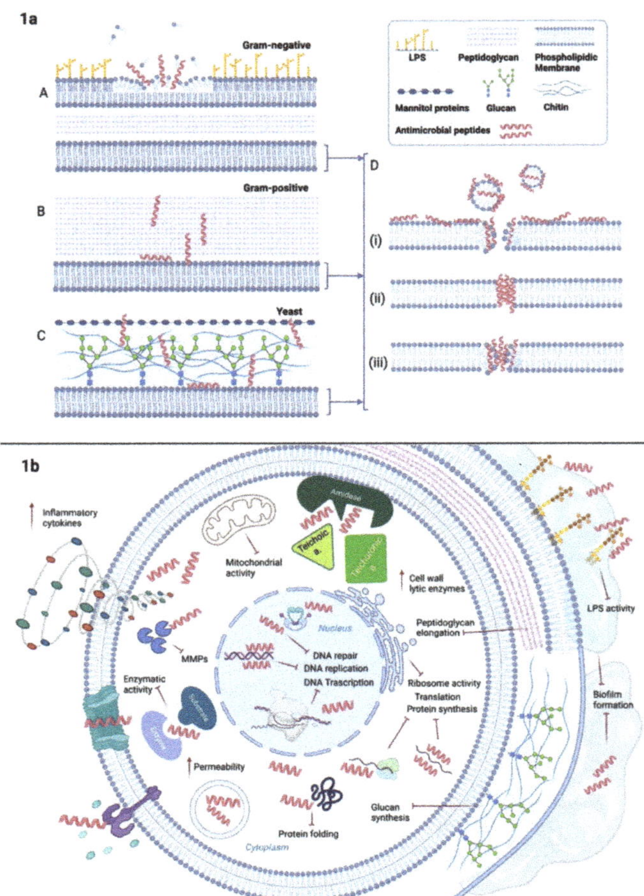

Figure 1. AMPs broad-spectrum antimicrobial activity. (**a**) Primarily, AMPs'action is based on their action on cytoplasmic membranes, i.e., perturbation or disruption. However, in presence of Gram-negative bacteria (**A**) AMPs have to firstly cross the outer phospholipidic membrane and secondly traverse the peptidoglycan layer before reaching the inner membrane. In Gram-positive bacteria (**B**) they navigate through the thick cell wall of peptidoglycan and in fungi (**C**), they encounter mannitol proteins, glucans and chitin prior to access to the cytoplasmic membrane. Once reached the phospholipidic bilayer, they induce perturbation via pore formation following either (**D**) (i) *carpet-like*, (ii) *barrel-stave* or (iii) or *toroidal pore* model depending on the peptide composition. (**b**) Besides pore formation, some AMPs bind some components and receptors on the extracellular side of the membrane, i.e., Toll-like receptors; others manage to enter the cytosol through direct penetration in vesicles or channels thus destabilizing the permeability and activating the inflammatory cytokines cascade. Intracellularly, they could also interfere with DNA or RNA leading to degradation and cell death. They may also affect mitochondrial activity or protein synthesis by targeting ribosome subunits or protein folding. In the case of bacterial cell wall, they can prevent elongation of peptidoglycan chains or hinder teichoic and teichuronic binding acids to amidases. Cell wall components inhibition will promote cell autolysis. In the extracellular space, AMPs can sequestrate LPS reducing the impact of endotoxins on the host's immune response. In fungal cells, AMPs can intervene on glucan synthesis thus blocking the building pieces of their wall. Further inhibitory action on biofilm matrix impairs the *quorum sensing* and improves the susceptibility of the single pathogens in both bacterial and fungal communities.

3. Antibacterial Peptides and Their Mechanism of Action

Many factors can influence membrane perturbation and disruption by AMPs, i.e., amino acids sequence, the lipid composition of the membrane, peptide concentration as well as differences in membrane composition between eukaryotic and bacterial cells allow the AMPs to distinguish a microbial target from the host. Bacterial membranes are negatively charged due to the presence of anionic phospholipids groups, e.g., phosphatidylglycerol, phosphatidylserine, while eukaryotic cells possess groups with a neutral charge, e.g., phosphatidylcholine and phosphatidylethanolamine [54]. Moreover, the presence of cholesterol, a common feature in eukaryotic cells, is able to interact with AMPs either neutralizing or reducing their activity or stabilizing the phospholipid bilayer [55].

In Gram-positive bacteria, AMPs have to cross first the cell wall composed of crosslinked peptidoglycan with lipoteichoic acid prior to reaching the membrane whereas in Gram-negative they face a coat of lipopolysaccharide (LPS) followed by a phospholipidic outer membrane and a less cross-linked peptidoglycan layer [56]. Electrostatic interactions between the cationic peptide and the negatively charged components, e.g., lipopolysaccharide in Gram-negative and teichoic acid in Gram-positive, are the first steps to contribute to bacterial membrane affinity [57]. However, while AMPs seem to traverse the peptidoglycan layer with ease and access to the cytoplasmic membrane of the Gram-positive, they need to disrupt or perturb both outer and cytoplasmic membrane in Gram-negatives. Impedance in crossing or permeabilization results in loss of antimicrobial activity (Figure 1a (A,B)) [58].

In order to explain the perturbation of the phospholipidic membranes operated by the AMPs, three main models have been proposed: *carpet-like*, *barrel-stave* and *toroidal pore* (Figure 1a (D)). Generally, when the ratio of peptide/lipids is low, AMPs interact with the phospholipidic layer of the membrane in a parallel manner, defined as *carpet-like* model, and interaction among the peptides or penetration in the hydrophobic core of the bilayer are not taking place [59]. Membrane integrity is disrupted and micelles are formed as in a detergent-like process [60]. With increasing AMPs ratio, they move to a perpendicular orientation until reaching such a concentration that they can cross the membrane forming pores (1:50–1:500 and more) [61,62]. A minimum length of ~22 amino acid for α-helix peptides is required to span the phospholipid layer, while β-sheet structures necessitate a minimum of 8 [63].

In the *barrel-stave*, interaction among peptides is a prerequisite as they mimic a transmembrane pore, whereas, in the case of the *toroidal* model, peptides are loosely arranged [64,65]. Despite the perturbation of the membrane seems to vary depending on the peptides, actually, the mechanisms of action are not completely well-defined and they are partially overlapping [66]. Moreover, all these models are based on the membrane perturbation but, then, the killing effect is not always enough to provide antimicrobial activity [67].

Besides membrane disruption, recent studies showed how peptides could act on other targets as well (Figure 1b) [68]. Some AMPs have shown their efficacy by binding some components and receptors on the extracellular side of the membrane and wall, thus destabilizing the permeability and/or activating intracellular signaling pathways that have, as a response, the inhibition or the activation of several functions. An interesting example is represented by the binding of Toll-like receptors and the consequential amplification of the inflammatory response via NFkB cascade followed by activation of the immune system towards microbiological pathogens [69,70]. Other antimicrobial peptides manage to enter the cytosol through direct penetration, endocytosis (both micropinocytosis and receptor-mediated) [71], or the exploit of delivery systems [72]. There, they can affect different enzymes and intermediates involved in vital processes.

The inhibitors of the nucleic acid biosynthesis seem to have a high binding affinity for both DNA and RNA because they share with nucleic acid-binding enzymes or substrates, homologous fragments of their sequences; an interesting example is represented by DNA-binding protein histone H2A [73]. Other mechanisms use the inhibition of the enzymes involved in the DNA/RNA biosynthesis, like DNA topoisomerase I preventing DNA relaxation [74], RNA polymerase blocking the transcription [75] and gyrase impairing the supercoiling of DNA. [76] As a result, DNA/RNA degradation is induced and consequentially also cell death. There are several inhibitors of protein biosynthesis which alter the transcription and the translation but also the correct folding and the degradation of the protein. Usually, the AMPs that act on the protein biosynthesis target the ribosome subunits [77] but some others can interfere with the incorporation of histidine, uridine and thymidine [78,79], the amino acid synthesis pathways [80], the release factors on the ribosome [81], the regulation of sigma factors [82], the nucleotide and coenzyme transport [80] and the degradation of DNA-replication-associated proteins [83]. Some peptides influence protein folding, in particular, DnaK, the major Hsp70 of the chaperone pathway in *Escherichia coli*, which has been seen as an optimal target to prevent the refolding of misfolded proteins [84]. Another approach is linked to the inhibition of matrix metalloproteases, essential enzymes in microbial cell growth and homeostasis, i.e., serine protease, trypsin-like protease, elastase and chymotrypsins [85–87]. There are also inhibitors of cell division that block DNA replication or the mechanisms essential for the repair of DNA damages, then resulting in the block of the cell cycle, in the impairment of the chromosome separation, in the failure of septation, in the alteration of mitochondrial activity and in a substantial change in the cell morphology with clearly visible blebbing and elongation towards a filamentous shape [88,89].

Cell wall synthesis is another suitable target. Some AMPs act on lipid II by sequestrating it from the functional site [90,91] or by binding D-Ala-D-Ala residues of its precursor preventing the addition of *N*-acetylglucosamine and *N*-acetylmuramic acid in the structure, hence the peptidoglycan elongation [92]. Other peptides have shown antimicrobial activity by activating cell wall-associated lytic enzymes, for example, some AMPs binding teichoic and teichuronic acids which otherwise are linked to amidases. The release of amidase stimulates premature autolysin activity and, consequently, cell lysis.

Moreover, lipopolysaccharides (LPS) are components of the membrane as well but, when released, are also well-known endotoxins able to raise an excessive and harmful pro-inflammatory response. AMPs that bind and neutralize LPS avoid the excessive stimulation of the immune system favoring a correct and balanced infection resolution [93].

Recently, the AMPs inhibitory activity on biofilm has been reported. Biofilm, consisting of an extracellular matrix of mainly polysaccharides, provides virulence, persistence and drug resistance to the microbial community [94,95]. Anti-biofilm mechanisms, similar to the membrane-targeting ones, are also very diverse and sequence dependent. A database of biofilm-active peptides can be found online [http://www.baamps.it/, accessed on 10 November 2021]. AMPs could prevent biofilm formation by affecting cell attachment, or could act on preformed biofilm by disrupting the *quorum-sensing*, dispersing the cells within it, or affecting the expression of the related genes [96,97]. Destabilization of matrix architecture impairing secretion or interaction between the matrix polymers has been also hypothesized [98]. Another target is the stress-responder guanosine pentaphosphate [(p)ppGpp] a major player for biofilm growth and environmental stress resistance [99]. Weakening of the biofilm increases the susceptibility of the pathogen to the AMPs or to the conventional antibiotics, therefore, even a synergistic action could be appealing for clinical purposes [100].

4. Antifungal Peptides

The concern generated by bacterial infections goes hand in hand with that of fungal infections especially considering both the frequency and the rapidity their resistance develops and spreads and the poor arsenal of available antifungal drugs. Fungal infections become extremely threatening especially for certain categories represented by patients with a compromised immune system due to pathological conditions, such as HIV/AIDS or autoimmune diseases and to therapeutic outcomes like chemotherapy and organ transplantation [101]. Among the fungal species *Candida albicans*, *Aspergillus fumigatus*, *Cryptococcus neoformans* and *Pneumocystis jirovecii* are the main ones responsible for the majority of severe mycoses [102] with 90% of reported deaths [103]. Among the emerging and reemerging species, such as *Histoplasma capsulatum* and *Fusarium* spp., of note is *Candida auris* which is considered by the Centre for Disease Control and Prevention (CDC) as an urgent global threat for its multi-drug resistance [18,104].

The latest reports highlight also the need for efficient treatments that nowadays are based only on three major classes of antifungal drugs: azoles, echinocandins and polyenes. Of these classes, echinocandins originated from non-ribosomal AMPs synthetically optimized [101]. Fungi are eukaryotic organisms; hence, they share with mammalian cells high similarities making it difficult to identify suitable targets while minimizing the risk of adverse effects. Although toxicity is an important issue, synthetic modifications of AMPs structures have extremely improved safety leaving just a few exceptions mainly represented by erythrocyte hemolysis and nucleic acid damages [105–107]. As previously seen for antibacterial AMPs, peptides with antifungal activity may present improved affinity towards phospholipids of the fungal membrane (phosphatidylserine and phosphatidylethanolamine) suggesting a distinctive relation between structure and activity (Figure 1a (C)) [108].

The three models used to describe the pore generated in bacterial membranes (*carpet-like*, *barrel-stave* and *toroidal*) are applicable also for AMPs acting on fungal membranes. Interestingly amphotericin B, the major representative of the polyene class of antifungal drugs, behaves as *barrel-stave*-pore forming peptide [101]. Evidence has demonstrated the existence of AMPs acting on the fungal membrane and on its components without having always clear information about their mechanism of action. Often these peptides affect the permeability of the membrane leading to ROS accumulation, oxidative stress damages, ATP release and the activation of stress-response pathways as HOG and MAPK cascade [109–111]. On the other hand, just a few AMPs have been revealed to interact with membrane components like glucosylceramides and β-1,3-glucans or with enzymes involved in the production of membrane components as the inositol phosphoryl ceramide synthase which is essential for the sphingolipid biosynthesis [112–114]. Membrane-active peptides have good potential and a broad-spectrum that sometimes includes both bacteria and fungi, nevertheless, as some AMPs with exclusive antifungal properties exist, it is also the case of antimicrobial peptides active against the cell wall (Figure 1b). The cell wall is an external structure proper of fungi unique in its composition since rich in glucans, chitin and mannan. The development of cell wall-active-AMPs grants high levels of safety with no or minor toxicity for mammalian cells. Most of the AMPs interfere with the synthesis of the wall components, such as β-1,3-glucan synthase fundamental enzyme for the production of β-1,3-glucans hence for the maintenance of the structural integrity (echinocandins exert this mechanism of action) and chitin synthase essential for chitin production [106]. Mannan and its glyco—and proteo-conjugates are deeply involved in fungal virulence, biofilm formation and adhesion to both biotic and abiotic surfaces included. Mannan-binding peptides form ternary complexes with calcium able to disrupt the fungal structural integrity [115]. Other AMPs that have been investigated have identified in nucleic acids their targets, in particular, several peptides bind and intercalate the DNA or inhibit the enzymes involved in its synthesis and repair [74]. In certain cases, some antifungal AMPs altered consistently the cell morphology and the organelle functions (in particular mitochondria, nucleus and vacuole) and interact with intracellular proteins [116–119]. In addition to these modes of action, it is important to mention the innovative use in the fungal world of the cations hijacking strategy using an Aluminum and/or Iron chelator translocatable inside the fungal cell through the siderophore iron transporter 1 (Sit1) [120].

Worthy of remark is the antibiofilm activity of some antifungal peptides. Biofilm is a virulence factor that, similarly to bacteria, a community of fungal cells adopts to evade the immune system. Moreover, it provides protection from antifungal drugs since the extracellular matrix works as a penetration-delayer factor. The colonization of both biotic and abiotic surfaces followed by biofilm formation represents a great risk especially in nosocomial settings where the use of invasive devices is a normal practice. Biofilm is associated with high morbidity and mortality rates and the development of AMPs with antibiofilm potential is urgently needed. Several antifungal peptides have been widely characterized and, among their abilities, they managed to both inhibit the biofilm formation and eradicate mature biofilm [121–125]. A negative point is the lack of precise information about the mechanism that sometimes could be considered as a downstream consequence attributable to the modes of action just described.

5. Antiviral Peptides

Viruses represent a major cause of human disease, and the emergence of viral drug resistance and epidemics induce to search for new antivirals. Natural AMPs are an interesting source of innovative antiviral agents, but more interestingly, antiviral peptides (AVPs) can be designed and optimized to block critical steps of the viral life cycle (Figure 2) [126]. In 2014, Kumar et al. described the AVP targeting about 60 medically significant viruses [127]. Usually, AVPs exhibit antiviral effects by inhibiting the virus directly, but their inhibition sites and the mechanism of action vary within the viral replication cycle.

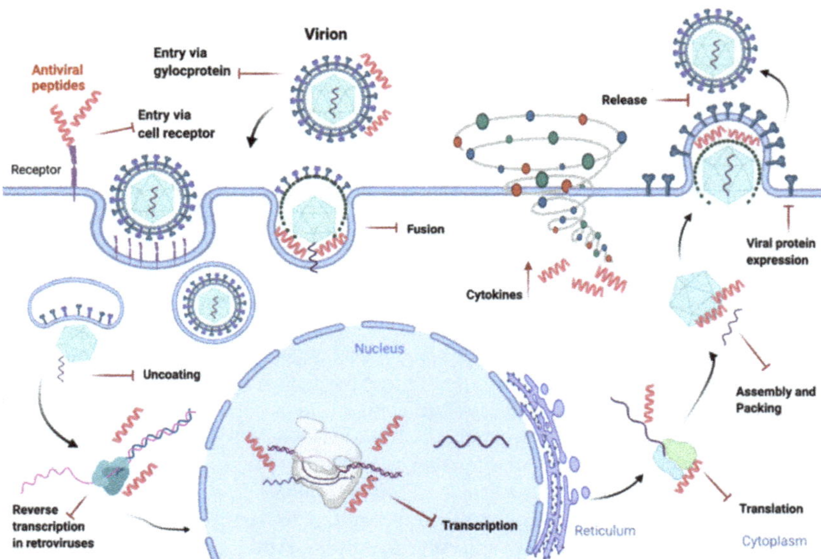

Figure 2. AVPs targets in viral life cycle. Depending on the type of virus and on the mode of action of the peptides, AVPs can block viral entry by binding with specific cellular receptors or interaction with viral glycoproteins, which are involved in both entry and fusion process. They may also hinder the fusion via physicochemical interaction with hydrophobic membrane–protein interfaces. AVPs can act intracellularly as well by direct influence of viral nucleic acid synthesis or blocking viral protein expression. Others modulate the antiviral immune system of the host cell by up-regulating expression of interferons and cytokines.

Most viral pathogens are present in the Emerging Infectious Diseases/Pathogens list of the US National Institute of Allergy and Infectious Diseases (NIAID), such as Smallpox virus, viral hemorrhagic fever viruses (arenaviruses, bunyaviruses, flaviviruses and filoviruses), and coronaviruses are membrane-enveloped viruses. Virus and host cell membrane fusion is necessary for virus entry and biophysical as well as biochemical features of the membrane fusion process can be common among enveloped viruses. Targeting these conserved characteristics that are necessary for membrane fusion, is emerging as a new tool for the development of broad-spectrum antivirals [128].

Viral entry, which is the earliest phase of infection in the viral life cycle, is the favored target for AVPs. Most AVPs block viral entry by one of the next mechanisms: (1) interaction with heparan sulfate, (2) blocking of cell-to-cell spread, (3) interaction with specific cellular receptors, (4) interaction with viral glycoproteins, (5) membrane or viral envelope interaction [129].

Viral surface glycoproteins are involved in both the entry and penetration process and undergo conformational changes because of the interactions with the receptor proteins. Most AVP inhibit enveloped viruses' entry by physico-chemical interaction with hydrophobic membrane–protein interfaces [130]. A few examples of peptide entry inhibitors are reported. Enfuvirtide is a peptide entry inhibitor for HIV that acts by blocking the HR1 domain of the viral envelope glycoprotein 41; it was approved by the US Food and Drug Administration (FDA) and the European Medicines Agency (EMA) for human use in 2003 [131]. Another class of HIV entry inhibitors, termed anchor inhibitors, target the fusion peptide [132].

The mimetic peptide, DN59, which consists of the amino acids corresponding to the amphipathic stem region of the dengue virus envelope glycoprotein was shown to interfere with the normal infective process [133]. Peptides homologous to the surface glycoproteins of HSV-1 and HSV-2 envelopes were demonstrated to be active against the herpes virus [134]. Peptide entry inhibitors were also used against other viruses, such as cytomegaloviruses, influenza virus and coronaviruses [130]. ACE2-derived peptides were already used to contrast SARS-CoV infection [135], and the approaches used to synthesize peptides against coronaviruses in the past may be re-considered to design new peptides for inhibition of SARS-CoV-2 infection on the documented evidence of efficacy against SARS-CoV, MERS-CoV, SARS-related CoVs. For example, among these peptides, which had already been used against SARS-CoV-1, 15 were selected against the receptor-binding domain (RBD) of the spike protein of SARS-CoV-2 potentially able to inhibit the entry of SARS-CoV-2. Moreover, peptides targeting domains in the S protein other than the RBD may also interfere with viral entry [136]. The approaches followed for the development of peptides targeting SARS-CoV-2 entry have been recently summarized by Schütz and colleagues [136].

AVPs with potential anti-SARS-CoV-2 activities could target the host as well. The mouse β-defensins-4 derived P9, thanks to its polycationic property, prevents endosomal acidification necessary for viral-host endosomal membrane fusion and consequent viral uncoating and RNA release, resulting in inhibition of the virus [137].

More recently, a dual-functional cross-linking peptide 8P9R has been demonstrated to inhibit both the endocytic pathway and the TMPRSS2-mediated pathway of SARS-CoV-2 hypothesizing its employment in effective cocktail therapy with repurposed drugs [138]. Otherwise, another target may be the ACE2 receptors instead of the viral S1 subunit [139].

Distinctively, other AVPs have been designed to modulate intracellular targets [129]. It is known that antimicrobial host defense peptides, such as PR39 and LL-37 can cross lipid membranes, while others are found as precursors inside host cell vacuoles. Cellular internalization of these peptides can stimulate gene/protein expression by blocking viral protein expression, influencing viral nucleic acid synthesis, or stimulating host cell antiviral defenses [128,140]. while others modulate the antiviral immune system of the host cell by up-regulating the expression of interferons and cytokines [141]. For example, rhesus theta-defensin 1 (RTD-1) is a cyclic antimicrobial peptide first identified in rhesus macaque leukocytes, that was demonstrated to alter pulmonary infection outcome induced by SARS-CoV in mice by potentiating cytokine responses [142,143].

Therefore, AVPs can be designed and optimized through a deep knowledge of the structures of viral proteins and cellular targets. Host cell factors proteins or pathways required by numerous viruses to complete their replication cycle are attractive targets for broad-spectrum antivirals, included AVPs. Indeed, this strategy would offer a versatile solution that could work against many viruses, including the emerging ones, offering a low possibility of inducing drug resistance. However, the major concerns duly noted are the cellular proteins function in the complex network of interactions as well as cytotoxicity.

For this reason, many peptides are designed, as described previously, to act extracellularly, i.e., to target early steps of viral replication, such as viral envelope glycoprotein activation, receptor attachment, or fusion.

Although smaller than standard AVPs, reduced glutathione (GSH) deserves to be cited as an effective antiviral against different viruses. GSH is a tripeptide, present in all mammalian cells, constituted of the amino acid L-glutamate, L-cysteine, and glycine. Its synthesis is catalyzed sequentially by γ-glutamylcysteine and GSH synthetase. Inside the cells, 98% of glutathione is found in reduced form, and only 2% is oxidized (GSSG) or joined with other molecules [144]. Glutathione (GSH) has a key role in cellular physiology and metabolism [145]. Furthermore, in the last years, an imbalance in the GSH/GSSG ratio has been described in several pathologies including viral infections [146]. It has been widely demonstrated that intracellular redox status alterations, associated with depletion of GSH are essential for the completion of the viral cycle. However, the mechanisms by which viruses induce a decrease in intracellular GSH content are different and not completely clear. Accordingly, GSH has been proposed as a potent antiviral acting with different mechanisms depending on the type of virus. Recently, the role of GSH in determining individual responsiveness to COVID-19 infection and the possibility of using GSH for the treatment and prevention of COVID-19 illness has been also described [147].

Unfortunately, GSH has a short half-life in blood plasma and hardly crosses the cell membrane; for this reason, design strategies have emerged in the development of GSH derivatives with improved permeability or small molecules able to release intracellularly precursors for GSH synthesis [148]. Many papers have reported the efficacy of GSH and pro-GSH molecules in inhibiting replication of several viruses and many reviews have summarized the results achieved over the years [146,149–151].

In conclusion, AVPs, due to their ability to target various aspects of the viral lifecycle, their low molecular weight and low toxicity, can be considered a potential resource to combat emerging and re-emerging viral pathogens for which drug-resistance was developed or specific therapies do not exist. Especially, in the light of recent fast-replicating viruses with high rate of mutation frequency, novel candidates with multiple mechanisms of action or synergistic effects are indeed highly desirable [152].

6. AMPs—Goods vs. Bads, and the Long Way towards Clinical Application

There are obvious, multiple advantages of AMPs over classical antibiotics. As previously describes, AMPs are easy to synthesize, thanks to recent advances in automated protein synthesis, or can alternatively be produced in large quantities in heterologous expression systems, either in microbial cells or in plants [153]. In addition, AMPs are largely prone to chemical modification, aimed at overcoming inherent problems, such as susceptibility to enzymatic degradation, chemical/physical instability and toxicity to host cells, thus optimizing molecules' features and smoothing their pathway towards the clinics [154]. Broad-spectrum activity and rapid killing are other much-appreciated characteristics. Finally, AMPs are increasingly seen as a promising therapeutic alternative for treating biofilm-associated infections, one of the major threats in the field of bacterial infections [155]. Similar to the fungal biofilm structure, bacteria as well usually acquire significant resistance against conventional antibiotics and the immune system defenses, thanks to the features of the biofilm itself, including the matrix of extracellular polymeric substances produced by the same microorganisms. Of the several molecules that have been already studied for their antibiofilm activity, dendrimeric AMPs seem particularly promising, in particular by displaying the property to inhibit biofilm formation in host-mimicking conditions [156,157].

A suitable instance of both the limitations to therapeutic use inherent to the nature itself of AMPs and the ways to overcome these is offered by the recent study of Wang Manchuriga and colleagues on temporins [158]. As many natural AMPs isolated from the skin of anuran amphibians (frogs and toads), temporins display a potent antimicrobial activity but this quality is often thwarted by elevated cytotoxicity, in particular against erythrocytes [159]. Working on temporin-GHa from *Hylarana guentheri*, Manchuriga and colleagues designed several analogs of the naturally-occurring sequence, modifying the type, position and number of charged residues. Some of the derived peptides displayed a significant reduction of hemolytic activity with respect to parent peptide while retaining potent antibacterial activity, but it was not possible to reduce cytotoxicity to zero without compromising antibacterial activity, confirming that a delicate balance of charge and other physico-chemical parameters (e.g., amphipathic and extension of hydrophobic surfaces) is necessary to obtain a plausible therapeutic lead [158].

Other key criteria of AMPs that should always be studied in detail when considering these molecules for use in clinical settings are immunogenicity and pharmacodynamics/pharmacodynamics properties. Proline-rich AMPs (PrAMPs) are a class of membrane-permeable AMPs that have been identified more than 20 years ago in mammals and insects; they have an intracellular mode of action, inhibiting protein synthesis leading to a bactericidal outcome [160]. Apidaecin Api88 (18 aa) and oncocin Onc72 (19 aa)—PrAMPs based on natural peptides isolated from milkweed bug *Oncopeltus fasciatus*—were shown to be nonimmunogenic in mice, unless conjugated to protein carriers, a fact attributed to the small size of these molecules [161]. A pharmacokinetics analysis showed that Onc72 reached several organs within 10 min and that the peptide's concentrations in blood were well above the minimal inhibitory concentrations for gram-negative key pathogens like *K. pneumoniae* [161]. More recently, the long-lasting post-antibiotic effect (PAE)—an important criterion of antimicrobial pharmacodynamics indicating the persistent growth of bacteria briefly exposed to antibiotics independently of host defense mechanisms—of several PrAMPs was tested, revealing prolonged PAEs against several strains of *E. coli*, *P. aeruginosa* and *K. pneumoniae* for all tested peptides but especially Api88, Api137, Bac7(1–60) and A3-APO [162]. "The PAEs presented here provide an additional hypothesis besides immunomodulatory effects that can explain the good in vivo efficacies of PrAMPs", notwithstanding the fast clearance rate measured for some of these peptides, authors discussed [162], "This again highlights that MIC values determined for AMPs in vitro cannot be simply used to predict in vivo efficacies, as often assumed in the literature. Instead, MIC values should be seen as one important criterion among other parameters to be considered," authors appropriately remarked [162].

One of the aspects that are often quoted in support of the (potential) use of AMPs in clinical practice is their low tendency to evoke antibiotic resistance. This tenet stems from the fact that AMPs generally (but not always, as specified above) hit the lipid component of the plasma membrane, a cellular component that is believed *per se* to be not easily modifiable in its basic physicochemical features by microbial targets. Although the slower emergence of resistance to AMPs with respect to conventional antibiotics is a reality, however, experience and much work have clearly shown that the reassuring thought that the complex phenomenon of resistance would not eventually thwart AMPs' value, is somewhat naïve and misleading. In fact, the long coevolution of microorganisms and AMPs has spurred the development of several resistance mechanisms. These include sequestration by bacterial enzymes, proteolytic degradation of peptides, efflux pumps to remove AMPs from the periplasmic space, alteration of components of bacterial surface to reduce surface attachment and permeability, down-regulation by immunomodulation [163–166].

The concept of coevolution and its effect on the rise of bacterial resistance to AMPs' action is well explained by the example of *Helicobacter pylori*. Sabine Nuding and colleagues tested the pattern of induction of gastric antimicrobial peptides by *H. pylori* as well as its susceptibility to the same peptides [167]. Researchers found that the induction of antimicrobial peptides, such as the inducible defensin HBD2 in the gastric mucosa by *H. pylori*, did not enhance the killing capacity against *H. pylori* itself. On the other hand, the expression levels of the constitutive defensin HBD1, inducible HBD3 and LL37, remained unchanged. Tested *H. Pylori* strains proved resistant to HBD1, but susceptible to the killing activities of HBD3 and LL37. "The combination of selective defensin induction and resistance to others may enable *Helicobacter* to colonize the gastric mucus layer where it can adhere to epithelial cells and induce inflammatory as well as malignant processes," concluded the authors, that remarked the need for further studies aimed at understanding the mechanisms regarding *H. pylori* selective antimicrobial resistance [168].

Despite the limitations briefly outlined above, that have hampered their development in the classical drug discovery pipeline, AMPs are attracting continuous and ever-increasing interest as new antimicrobials agents. Out of some ~3000 molecules that have been isolated from different sources, just a handful have been the object of preclinical studies and further proceeded to clinical trials [166]. A recent analysis of AMPs patents from 2015 through 2020 has confirmed a long-standing trend, i.e., the fact that AMPs earmarked for clinical development are in vast majority analogs or derivatives of natural peptides, obtained through a template-based strategy aimed at enhancing the activity and stability of natural AMPs while reducing their toxicity [168].

Currently, just three AMPs have been approved by the U.S. Food and Drug Administration (FDA) for therapeutic use, i.e., gramicidin, colistin and daptomycin. Gramicidin has a long history. First isolated from *Bacillus brevis* over 70 years ago, gramicidin is active against a range of Gram-positive and Gram-negative bacteria, although its severe toxicity for human erythrocytes has a limited clinical indication to topical applications [169]. Polymyxin and colistin, which are cationic peptides in use for decades, have regained interest lately, due to their strong activity against multi-drug resistant Gram-negative pathogens. Their ability to bind the lipid A component of LPS makes them precious, the last resource weapons to fight septic shock, notwithstanding their known nephrotoxicity. Resistance has emerged, however, and is spreading at an alarming pace, putting the effectiveness of these valuable therapeutics at risk [170,171]. Last but not least, daptomycin. This membrane-active cyclic lipopeptide has received the green light from the FDA in 2003 to treat Gram-positive infections. It is believed that its mechanism of action differs from that of other AMPs since daptomycin causes bacterial membrane depolarization rather than membrane disruption and pore formation [172]. In recent years, resistance in *Staphylococcus aureus* has been more and more frequently reported, and the search for substitutes that might prolong the clinical use of this important antibiotic is actively underway [173].

The concern caused by AMPs resistance is clearly transmitted by a very recent clinical trial aimed at evaluating the efficacy of oral colistin-neomycin in preventing multidrug-resistant *Enterobacterales* (MDR-E) infections in solid organ transplant recipients. In the trial's frame, a 14-day regimen of oral colistin and neomycin did not reduce MDR-E infections, and four liver-recipients developed colistin resistance [174]. A study of the molecular mechanisms of colistin resistance in environmental isolates of *Acinetobacter baumannii*, recovered from hospital wastewater and wastewater treatment plant, has shown that all isolates had increased levels of eptA mRNA and decreased levels of lpxA and lpxD mRNA; the eptA gene, in particular, could indicate its main role in colistin resistance through lipid A modification [175]. Authors hypothesized that when untreated hospital wastewater is released into the urban sewage, it might contain colistin-susceptible *A. baumannii*, and that resistance might emerge in wastewater itself following exposure to pollutants, such as cationic surfactants, and subsequently spread in the environment [175]. Looking at the bright side, things can always improve. Recent work has shown that kynomycin, a new daptomycin analog, was endowed with enhanced activity against both methicillin-resistant *S. aureus* and vancomycin-resistant *Enterococcus*, with improved pharmacokinetics and lower cytotoxicity than daptomycin [176]. Freshly acquired data suggest that physicochemical features like Ca^{2+} binding and Ca^{2+}-mediated oligomerization could explain kynomycin's enhanced antibacterial activity [177].

Even a hasty glance at the AMPs pipeline conveys the level of difficulty at bringing these molecules to the market, either for topical or systemic treatment [166,178]. After many failures, however, a couple of promising candidates loom on the horizon, at least for some therapeutic indications. Polyphor is developing the synthetic lipopeptide murepavadin, a member of a novel class of antibiotics that combine high-affinity binding to both LPS and outer membrane proteins, resulting in high specificity towards Gram-negative bacteria and effective bactericidal activity. Murepavadin, in particular, targets the lipopolysaccharide transport protein D (LptD), an outer membrane protein on *Pseudomonas aeruginosa*, leading to cell death. Phase 3 clinical trials investigating the safety and efficacy of intravenous murepavadin have been prematurely stopped due to a rise of creatinine concentration in the serum of patients treated with the AMP, indicating renal failures [179,180]. Despite these disappointing results, Polyphor plans to continue the development of inhaled murepavadin to treat chronic *P. aeruginosa* infections associated with cystic fibrosis. Exeporfinium chloride (XF-73), a derivative of AMP concept containing two cationic ammoniums and one porphyrin core, is currently the main protagonist of the anti-infectives program at Destiny Pharma [181]. XF-73 is a membrane-active antibiotic, particularly potent against Gram-positive bacteria, including MRSA. A phase 2 trial of XF-73 for the prevention of post-surgical staphylococcal nasal infections is ongoing. An in vitro study of bacterial resistance that compared XF-73 to standard antibiotics currently in use did not demonstrate the emergence of any resistance to XF-73 even after 55 repeat exposures [182].

7. Conclusions

The challenging research for new antimicrobial entities is still ongoing but not without difficulties. New species of bacteria, fungi and viruses are emerging, and the most alarming fact is their intrinsic and sometimes multi-drug resistance to first-line drugs. These aspects together with the fast and global spread of resistance through horizontal transfer represent a serious threat for global health. An innovative approach involves the use of compounds inspired by nature and subsequently optimized to reach suitable features, i.e., low toxicity and strong activity. The result of this process is represented by synthetic peptides. Their broad mechanisms of action and the unlikely resistance that they generate, are important advantages and perhaps the key point for a shift towards new antimicrobial synthetic peptides-based treatments for the near future.

Author Contributions: T.V., M.B., A.C.R. and A.F. carried out the literature study and drafted the manuscript. M.M. edited and reviewed. All authors have read and agreed to the published version of the manuscript.

Funding: This work was supported by the University of Urbino Carlo Bo (A.F. and T.V. grants).

Institutional Review Board Statement: Not applicable.

Informed Consent Statement: Not applicable.

Data Availability Statement: Data are contained within the article.

Acknowledgments: Cartoons of Figure 1a,b were created using BioRender.com. Figure 2 was adapted from "Generic Viral Life Cycle", by BioRender.com (2021). Retrieved from biorender.com/biorender-templates.

Conflicts of Interest: The authors declare no conflict of interest.

References

1. Fleming, A.; Allison, V.D. Observations on a Bacteriolytic Substance ("Lysozyme") Found in Secretions and Tissues. *Br. J. Exp. Pathol.* **1922**, *3*, 252–260.
2. Garrod, L.P. Alexander Fleming. A Dedication on the 50th Anniversary of the Discovery of Penicillin. *Br. J. Exp. Pathol.* **1979**, *60*, 1–2. [PubMed]
3. Dubos, R.J. Studies on a Bactericidal Agent Extracted from a Soil Bacillus: I. Preparation of the Agent. Its Activity in Vitro. *J. Exp. Med.* **1939**, *70*, 1–10. [CrossRef] [PubMed]
4. Nakatsuji, T.; Gallo, R.L. Antimicrobial Peptides: Old Molecules with New Ideas. *J. Investig. Dermatol.* **2012**, *132*, 887–895. [CrossRef] [PubMed]
5. De Smet, K.; Contreras, R. Human Antimicrobial Peptides: Defensins, Cathelicidins and Histatins. *Biotechnol. Lett.* **2005**, *27*, 1337–1347. [CrossRef]
6. Diamond, G.; Beckloff, N.; Weinberg, A.; Kisich, K.O. The Roles of Antimicrobial Peptides in Innate Host Defense. *Curr. Pharm. Des.* **2009**, *15*, 2377–2392. [CrossRef]
7. Van der Does, A.M.; Hiemstra, P.S.; Mookherjee, N. Antimicrobial Host Defence Peptides: Immunomodulatory Functions and Translational Prospects. In *Antimicrobial Peptides: Basics for Clinical Application*; Matsuzaki, K., Ed.; Advances in Experimental Medicine and Biology; Springer: Singapore, 2019; pp. 149–171. ISBN 9789811335884.

8. Boparai, J.K.; Sharma, P.K. Mini Review on Antimicrobial Peptides, Sources, Mechanism and Recent Applications. *Protein Pept. Lett.* **2020**, *27*, 4–16. [CrossRef]
9. Geitani, R.; Moubareck, C.A.; Xu, Z.; Karam Sarkis, D.; Touqui, L. Expression and Roles of Antimicrobial Peptides in Innate Defense of Airway Mucosa: Potential Implication in Cystic Fibrosis. *Front. Immunol.* **2020**, *11*, 1198. [CrossRef]
10. Wang, G. *Antimicrobial Peptides: Discovery, Design and Novel Therapeutic Strategies*, 2nd ed.; CABI: Wallingford, UK, 2017; ISBN 978-1-78639-039-4.
11. Huan, Y.; Kong, Q.; Mou, H.; Yi, H. Antimicrobial Peptides: Classification, Design, Application and Research Progress in Multiple Fields. *Front. Microbiol.* **2020**, *11*, 2559. [CrossRef]
12. Harris, F.; Dennison, S.R.; Phoenix, D.A. Anionic Antimicrobial Peptides from Eukaryotic Organisms. *Curr. Protein Pept. Sci.* **2009**, *10*, 585–606. [CrossRef]
13. López Cascales, J.J.; Zenak, S.; García de la Torre, J.; Lezama, O.G.; Garro, A.; Enriz, R.D. Small Cationic Peptides: Influence of Charge on Their Antimicrobial Activity. *ACS Omega* **2018**, *3*, 5390–5398. [CrossRef]
14. Hancock, R.E.W.; Scott, M.G. The Role of Antimicrobial Peptides in Animal Defenses. *Proc. Natl. Acad. Sci. USA* **2000**, *97*, 8856–8861. [CrossRef]
15. Mishra, A.K.; Choi, J.; Moon, E.; Baek, K.-H. Tryptophan-Rich and Proline-Rich Antimicrobial Peptides. *Molecules* **2018**, *23*, 815. [CrossRef]
16. The Vast Structural Diversity of Antimicrobial Peptides | Elsevier Enhanced Reader. Available online: https://reader.elsevier.com/reader/sd/pii/S0165614719300896?token=AC26A2A3A45C3A5FDA55579937D526F726B0848E72EAAF15FBC5BC9E4C015C21A91A1FDE71DE83D781B6A91FB50B08CE&originRegion=eu-west-1&originCreation=20211028065806 (accessed on 28 October 2021).
17. Friedman, D.Z.P.; Schwartz, I.S. Emerging Fungal Infections: New Patients, New Patterns, and New Pathogens. *J. Fungi* **2019**, *5*, 67. [CrossRef]
18. Lockhart, S.R.; Guarner, J. Emerging and Reemerging Fungal Infections. *Semin. Diagn. Pathol.* **2019**, *36*, 177–181. [CrossRef]
19. Lima, P.G.; Oliveira, J.T.A.; Amaral, J.L.; Freitas, C.D.T.; Souza, P.F.N. Synthetic Antimicrobial Peptides: Characteristics, Design, and Potential as Alternative Molecules to Overcome Microbial Resistance. *Life Sci.* **2021**, *278*, 119647. [CrossRef]
20. CDC Antibiotic-Resistant Germs: New Threats. Available online: https://www.cdc.gov/drugresistance/biggest-threats.html (accessed on 27 October 2021).
21. WHO Coronavirus (COVID-19) Dashboard. Available online: https://covid19.who.int (accessed on 11 October 2021).
22. Hancock, R.E. Cationic Peptides: Effectors in Innate Immunity and Novel Antimicrobials. *Lancet Infect. Dis.* **2001**, *1*, 156–164. [CrossRef]
23. Sarkar, T.; Chetia, M.; Chatterjee, S. Antimicrobial Peptides and Proteins: From Nature's Reservoir to the Laboratory and Beyond. *Front. Chem.* **2021**, *9*, 691532. [CrossRef]
24. Giuliani, A.; Rinaldi, A. Beyond Natural Antimicrobial Peptides: Multimeric Peptides and Other Peptidomimetic Approaches. *Cell. Mol. Life Sci.* **2011**, *68*, 717. [CrossRef]
25. Nordström, R.; Malmsten, M. Delivery Systems for Antimicrobial Peptides. *Adv. Colloid Interface Sci.* **2017**, *242*, 17–34. [CrossRef]
26. Marr, A.K.; Gooderham, W.J.; Hancock, R.E. Antibacterial Peptides for Therapeutic Use: Obstacles and Realistic Outlook. *Curr. Opin. Pharmacol.* **2006**, *6*, 468–472. [CrossRef]
27. Lei, J.; Sun, L.; Huang, S.; Zhu, C.; Li, P.; He, J.; Mackey, V.; Coy, D.H.; He, Q. The Antimicrobial Peptides and Their Potential Clinical Applications. *Am. J. Transl. Res.* **2019**, *11*, 3919–3931.
28. Jaradat, D.M.M. Thirteen Decades of Peptide Synthesis: Key Developments in Solid Phase Peptide Synthesis and Amide Bond Formation Utilized in Peptide Ligation. *Amino Acids* **2018**, *50*, 39–68. [CrossRef]
29. González-Henríquez, C.M.; Sarabia-Vallejos, M.A.; Rodríguez-Hernández, J. Strategies to Fabricate Polypeptide-Based Structures via Ring-Opening Polymerization of N-Carboxyanhydrides. *Polymers* **2017**, *9*, 551. [CrossRef]
30. Cardoso, M.H.; Orozco, R.Q.; Rezende, S.B.; Rodrigues, G.; Oshiro, K.G.N.; Cândido, E.S.; Franco, O.L. Computer-Aided Design of Antimicrobial Peptides: Are We Generating Effective Drug Candidates? *Front. Microbiol.* **2020**, *10*, 3097. [CrossRef]
31. Mohamed, M.F.; Abdelkhalek, A.; Seleem, M.N. Evaluation of Short Synthetic Antimicrobial Peptides for Treatment of Drug-Resistant and Intracellular Staphylococcus Aureus. *Sci. Rep.* **2016**, *6*, 29707. [CrossRef]
32. Goodman, M.; Zapf, C.; Rew, Y. New Reagents, Reactions, and Peptidomimetics for Drug Design. *Biopolymers* **2001**, *60*, 229–245. [CrossRef]
33. Chou, K.C. Prediction of Protein Cellular Attributes Using Pseudo-Amino Acid Composition. *Proteins* **2001**, *43*, 246–255. [CrossRef] [PubMed]
34. Adessi, C.; Soto, C. Converting a Peptide into a Drug: Strategies to Improve Stability and Bioavailability. *Curr. Med. Chem.* **2002**, *9*, 963–978. [CrossRef] [PubMed]
35. Banerjee, A.; Pramanik, A.; Bhattacharjya, S.; Balaram, P. Omega Amino Acids in Peptide Design: Incorporation into Helices. *Biopolymers* **1996**, *39*, 769–777. [CrossRef]
36. Godballe, T.; Nilsson, L.L.; Petersen, P.D.; Jenssen, H. Antimicrobial β-Peptides and α-Peptoids. *Chem. Biol. Drug Des.* **2011**, *77*, 107–116. [CrossRef]
37. Evans, B.J.; King, A.T.; Katsifis, A.; Matesic, L.; Jamie, J.F. Methods to Enhance the Metabolic Stability of Peptide-Based PET Radiopharmaceuticals. *Molecules* **2020**, *25*, 2314. [CrossRef]

38. Gan, B.H.; Gaynord, J.; Rowe, S.M.; Deingruber, T.; Spring, D.R. The Multifaceted Nature of Antimicrobial Peptides: Current Synthetic Chemistry Approaches and Future Directions. *Chem. Soc. Rev.* **2021**, *50*, 7820–7880. [CrossRef]
39. Falciani, C.; Lozzi, L.; Scali, S.; Brunetti, J.; Bracci, L.; Pini, A. Site-Specific Pegylation of an Antimicrobial Peptide Increases Resistance to Pseudomonas Aeruginosa Elastase. *Amino Acids* **2014**, *46*, 1403–1407. [CrossRef]
40. Gong, Y.; Andina, D.; Nahar, S.; Leroux, J.-C.; Gauthier, M.A. Releasable and Traceless PEGylation of Arginine-Rich Antimicrobial Peptides. *Chem. Sci.* **2017**, *8*, 4082–4086. [CrossRef]
41. Rounds, T.; Straus, S.K. Lipidation of Antimicrobial Peptides as a Design Strategy for Future Alternatives to Antibiotics. *Int. J. Mol. Sci.* **2020**, *21*, 9692. [CrossRef]
42. Tang, J.; Chen, H.; He, Y.; Sheng, W.; Bai, Q.; Wang, H. Peptide-Guided Functionalization and Macrocyclization of Bioactive Peptidosulfonamides by Pd(II)-Catalyzed Late-Stage C-H Activation. *Nat. Commun.* **2018**, *9*, 3383. [CrossRef]
43. Abbasi, E.; Aval, S.F.; Akbarzadeh, A.; Milani, M.; Nasrabadi, H.T.; Joo, S.W.; Hanifehpour, Y.; Nejati-Koshki, K.; Pashaei-Asl, R. Dendrimers: Synthesis, Applications, and Properties. *Nanoscale Res. Lett.* **2014**, *9*, 247. [CrossRef]
44. Tam, J.P. Synthetic Peptide Vaccine Design: Synthesis and Properties of a High-Density Multiple Antigenic Peptide System. *Proc. Natl. Acad. Sci. USA* **1988**, *85*, 5409–5413. [CrossRef]
45. Tam, J.P.; Spetzler, J.C. Synthesis and Application of Peptide Dendrimers as Protein Mimetics. *Curr. Protoc. Immunol.* **2001**, *9*, 96. [CrossRef]
46. Bracci, L.; Falciani, C.; Lelli, B.; Lozzi, L.; Runci, Y.; Pini, A.; De Montis, M.G.; Tagliamonte, A.; Neri, P. Synthetic Peptides in the Form of Dendrimers Become Resistant to Protease Activity. *J. Biol. Chem.* **2003**, *278*, 46590–46595. [CrossRef]
47. Falciani, C.; Lozzi, L.; Pini, A.; Corti, F.; Fabbrini, M.; Bernini, A.; Lelli, B.; Niccolai, N.; Bracci, L. Molecular Basis of Branched Peptides Resistance to Enzyme Proteolysis. *Chem. Biol. Drug Des.* **2007**, *69*, 216–221. [CrossRef] [PubMed]
48. Wadhwani, P.; Reichert, J.; Bürck, J.; Ulrich, A.S. Antimicrobial and Cell-Penetrating Peptides Induce Lipid Vesicle Fusion by Folding and Aggregation. *Eur. Biophys. J.* **2012**, *41*, 177–187. [CrossRef] [PubMed]
49. Dong, N.; Chou, S.; Li, J.; Xue, C.; Li, X.; Cheng, B.; Shan, A.; Xu, L. Short Symmetric-End Antimicrobial Peptides Centered on β-Turn Amino Acids Unit Improve Selectivity and Stability. *Front. Microbiol.* **2018**, *9*, 2832. [CrossRef] [PubMed]
50. Benfield, A.H.; Henriques, S.T. Mode-of-Action of Antimicrobial Peptides: Membrane Disruption vs. Intracellular Mechanisms. *Front. Med. Technol.* **2020**, *2*, 20. [CrossRef]
51. Mwangi, J.; Hao, X.; Lai, R.; Zhang, Z.-Y. Antimicrobial Peptides: New Hope in the War against Multidrug Resistance. *Zool. Res.* **2019**, *40*, 488–505. [CrossRef]
52. Scocchi, M.; Mardirossian, M.; Runti, G.; Benincasa, M. Non-Membrane Permeabilizing Modes of Action of Antimicrobial Peptides on Bacteria. *Curr. Top. Med. Chem.* **2016**, *16*, 76–88. [CrossRef]
53. Gabriel, G.J.; Madkour, A.E.; Dabkowski, J.M.; Nelson, C.F.; Nüsslein, K.; Tew, G.N. Synthetic Mimic of Antimicrobial Peptide with Nonmembrane-Disrupting Antibacterial Properties. *Biomacromolecules* **2008**, *9*, 2980–2983. [CrossRef]
54. Yeaman, M.R.; Yount, N.Y. Mechanisms of Antimicrobial Peptide Action and Resistance. *Pharmacol. Rev.* **2003**, *55*, 27–55. [CrossRef]
55. McHenry, A.J.; Sciacca, M.F.M.; Brender, J.R.; Ramamoorthy, A. Does Cholesterol Suppress the Antimicrobial Peptide Induced Disruption of Lipid Raft Containing Membranes? *Biochim. Biophys. Acta* **2012**, *1818*, 3019–3024. [CrossRef]
56. Silhavy, T.J.; Kahne, D.; Walker, S. The Bacterial Cell Envelope. *Cold Spring Harb. Perspect. Biol.* **2010**, *2*, a000414. [CrossRef]
57. Malanovic, N.; Lohner, K. Antimicrobial Peptides Targeting Gram-Positive Bacteria. *Pharmaceuticals* **2016**, *9*, 59. [CrossRef]
58. Li, J.; Koh, J.-J.; Liu, S.; Lakshminarayanan, R.; Verma, C.S.; Beuerman, R.W. Membrane Active Antimicrobial Peptides: Translating Mechanistic Insights to Design. *Front. Neurosci.* **2017**, *11*, 73. [CrossRef]
59. Wimley, W.C. Describing the Mechanism of Antimicrobial Peptide Action with the Interfacial Activity Model. *ACS Chem. Biol.* **2010**, *5*, 905–917. [CrossRef]
60. Brogden, K.A. Antimicrobial Peptides: Pore Formers or Metabolic Inhibitors in Bacteria? *Nat. Rev. Microbiol.* **2005**, *3*, 238–250. [CrossRef]
61. Manzini, M.C.; Perez, K.R.; Riske, K.A.; Bozelli, J.C.; Santos, T.L.; da Silva, M.A.; Saraiva, G.K.V.; Politi, M.J.; Valente, A.P.; Almeida, F.C.L.; et al. Peptide:Lipid Ratio and Membrane Surface Charge Determine the Mechanism of Action of the Antimicrobial Peptide BP100. Conformational and Functional Studies. *Biochim. Biophys. Acta* **2014**, *1838*, 1985–1999. [CrossRef]
62. Yang, L.; Harroun, T.A.; Weiss, T.M.; Ding, L.; Huang, H.W. Barrel-Stave Model or Toroidal Model? A Case Study on Melittin Pores. *Biophys. J.* **2001**, *81*, 1475–1485. [CrossRef]
63. Kumar, S.; Singh, D.; Kumari, P.; Malik, R.S.; Poonam, P.K.; Parang, K.; Tiwari, R.K. PEGylation and Cell-Penetrating Peptides: Glimpse into the Past and Prospects in the Future. *Curr. Top. Med. Chem.* **2020**, *20*, 337–348. [CrossRef]
64. Sengupta, D.; Leontiadou, H.; Mark, A.E.; Marrink, S.-J. Toroidal Pores Formed by Antimicrobial Peptides Show Significant Disorder. *Biochim. Biophys. Acta* **2008**, *1778*, 2308–2317. [CrossRef]
65. Tuerkova, A.; Kabelka, I.; Králová, T.; Sukeník, L.; Pokorná, Š.; Hof, M.; Vácha, R. Effect of Helical Kink in Antimicrobial Peptides on Membrane Pore Formation. *eLife* **2020**, *9*, e47946. [CrossRef]
66. Clark, S.; Jowitt, T.A.; Harris, L.K.; Knight, C.G.; Dobson, C.B. The Lexicon of Antimicrobial Peptides: A Complete Set of Arginine and Tryptophan Sequences. *Commun. Biol.* **2021**, *4*, 605. [CrossRef] [PubMed]

67. Moravej, H.; Moravej, Z.; Yazdanparast, M.; Heiat, M.; Mirhosseini, A.; Moosazadeh Moghaddam, M.; Mirnejad, R. Antimicrobial Peptides: Features, Action, and Their Resistance Mechanisms in Bacteria. *Microb. Drug Resist.* **2018**, *24*, 747–767. [CrossRef] [PubMed]
68. Le, C.-F.; Fang, C.-M.; Sekaran, S.D. Intracellular Targeting Mechanisms by Antimicrobial Peptides. *Antimicrob. Agents Chemother.* **2017**, *61*, e02340–16. [CrossRef] [PubMed]
69. Lee, E.Y.; Lee, M.W.; Wong, G.C.L. Modulation of Toll-like Receptor Signaling by Antimicrobial Peptides. *Semin. Cell Dev. Biol.* **2019**, *88*, 173–184. [CrossRef] [PubMed]
70. Ryu, M.; Park, J.; Yeom, J.-H.; Joo, M.; Lee, K. Rediscovery of Antimicrobial Peptides as Therapeutic Agents. *J. Microbiol.* **2021**, *59*, 113–123. [CrossRef]
71. Madani, F.; Lindberg, S.; Langel, Ü.; Futaki, S.; Gräslund, A. Mechanisms of Cellular Uptake of Cell-Penetrating Peptides. *J. Biophys.* **2011**, *2011*, 414729. [CrossRef]
72. Tang, Z.; Ma, Q.; Chen, X.; Chen, T.; Ying, Y.; Xi, X.; Wang, L.; Ma, C.; Shaw, C.; Zhou, M. Recent Advances and Challenges in Nanodelivery Systems for Antimicrobial Peptides (AMPs). *Antibiotics* **2021**, *10*, 990. [CrossRef]
73. Birkemo, G.A.; Lüders, T.; Andersen, Ø.; Nes, I.F.; Nissen-Meyer, J. Hipposin, a Histone-Derived Antimicrobial Peptide in Atlantic Halibut (*Hippoglossus Hippoglossus* L.). *Biochim. Biophys. Acta* **2003**, *1646*, 207–215. [CrossRef]
74. Marchand, C.; Krajewski, K.; Lee, H.-F.; Antony, S.; Johnson, A.A.; Amin, R.; Roller, P.; Kvaratskhelia, M.; Pommier, Y. Covalent Binding of the Natural Antimicrobial Peptide Indolicidin to DNA Abasic Sites. *Nucleic Acids Res.* **2006**, *34*, 5157–5165. [CrossRef]
75. Braffman, N.R.; Piscotta, F.J.; Hauver, J.; Campbell, E.A.; Link, A.J.; Darst, S.A. Structural Mechanism of Transcription Inhibition by Lasso Peptides Microcin J25 and Capistruin. *Proc. Natl. Acad. Sci. USA* **2019**, *116*, 1273–1278. [CrossRef]
76. Heddle, J.G.; Blance, S.J.; Zamble, D.B.; Hollfelder, F.; Miller, D.A.; Wentzell, L.M.; Walsh, C.T.; Maxwell, A. The Antibiotic Microcin B17 Is a DNA Gyrase Poison: Characterisation of the Mode of Inhibition11Edited by J. Karn. *J. Mol. Biol.* **2001**, *307*, 1223–1234. [CrossRef]
77. Polikanov, Y.S.; Aleksashin, N.A.; Beckert, B.; Wilson, D.N. The Mechanisms of Action of Ribosome-Targeting Peptide Antibiotics. *Front. Mol. Biosci.* **2018**, *5*, 48. [CrossRef]
78. Friedrich, C.L.; Rozek, A.; Patrzykat, A.; Hancock, R.E.W. Structure and Mechanism of Action of an Indolicidin Peptide Derivative with Improved Activity against Gram-Positive Bacteria*. *J. Biol. Chem.* **2001**, *276*, 24015–24022. [CrossRef]
79. Patrzykat, A.; Friedrich, C.L.; Zhang, L.; Mendoza, V.; Hancock, R.E.W. Sublethal Concentrations of Pleurocidin-Derived Antimicrobial Peptides Inhibit Macromolecular Synthesis in Escherichia Coli. *Antimicrob. Agents Chemother.* **2002**, *46*, 605–614. [CrossRef]
80. Ho, Y.-H.; Shah, P.; Chen, Y.-W.; Chen, C.-S. Systematic Analysis of Intracellular-Targeting Antimicrobial Peptides, Bactenecin 7, Hybrid of Pleurocidin and Dermaseptin, Proline–Arginine-Rich Peptide, and Lactoferricin B, by Using Escherichia Coli Proteome Microarrays*. *Mol. Cell. Proteom.* **2016**, *15*, 1837–1847. [CrossRef]
81. Florin, T.; Maracci, C.; Graf, M.; Karki, P.; Klepacki, D.; Berninghausen, O.; Beckmann, R.; Vázquez-Laslop, N.; Wilson, D.N.; Rodnina, M.V.; et al. An Antimicrobial Peptide That Inhibits Translation by Trapping Release Factors on the Ribosome. *Nat. Struct. Mol. Biol.* **2017**, *24*, 752–757. [CrossRef]
82. El-Mowafi, S.A.; Sineva, E.; Alumasa, J.N.; Nicoloff, H.; Tomsho, J.W.; Ades, S.E.; Keiler, K.C. Identification of Inhibitors of a Bacterial Sigma Factor Using a New High-Throughput Screening Assay. *Antimicrob. Agents Chemother.* **2015**, *59*, 193–205. [CrossRef]
83. Van Eijk, E.; Wittekoek, B.; Kuijper, E.J.; Smits, W.K. DNA Replication Proteins as Potential Targets for Antimicrobials in Drug-Resistant Bacterial Pathogens. *J. Antimicrob. Chemother.* **2017**, *72*, 1275–1284. [CrossRef]
84. Knappe, D.; Goldbach, T.; Hatfield, M.P.D.; Palermo, N.Y.; Weinert, S.; Sträter, N.; Hoffmann, R.; Lovas, S. Proline-Rich Antimicrobial Peptides Optimized for Binding to Escherichia Coli Chaperone DnaK. *Protein Pept. Lett.* **2016**, *23*, 1061–1071. [CrossRef]
85. Bhowmick, M.; Tokmina-Roszyk, D.; Onwuha-Ekpete, L.; Harmon, K.; Robichaud, T.; Fuerst, R.; Stawikowska, R.; Steffensen, B.; Roush, W.; Wong, H.R.; et al. Second Generation Triple-Helical Peptide Inhibitors of Matrix Metalloproteinases. *J. Med. Chem.* **2017**, *60*, 3814–3827. [CrossRef]
86. Tay, C.X.; Quah, S.Y.; Lui, J.N.; Yu, V.S.H.; Tan, K.S. Matrix Metalloproteinase Inhibitor as an Antimicrobial Agent to Eradicate Enterococcus Faecalis Biofilm. *J. Endod.* **2015**, *41*, 858–863. [CrossRef]
87. Rahman, F.; Nguyen, T.-M.; Adekoya, O.A.; Campestre, C.; Tortorella, P.; Sylte, I.; Winberg, J.-O. Inhibition of Bacterial and Human Zinc-Metalloproteases by Bisphosphonate- and Catechol-Containing Compounds. *J. Enzyme Inhib. Med. Chem.* **2021**, *36*, 819–830. [CrossRef]
88. Salomón, R.A.; Farías, R.N. Microcin 25, a Novel Antimicrobial Peptide Produced by Escherichia Coli. *J. Bacteriol.* **1992**, *174*, 7428–7435. [CrossRef]
89. Chileveru, H.R.; Lim, S.A.; Chairatana, P.; Wommack, A.J.; Chiang, I.-L.; Nolan, E.M. Visualizing Attack of Escherichia Coli by the Antimicrobial Peptide Human Defensin 5. *Biochemistry* **2015**, *54*, 1767–1777. [CrossRef]
90. Scherer, K.M.; Spille, J.-H.; Sahl, H.-G.; Grein, F.; Kubitscheck, U. The Lantibiotic Nisin Induces Lipid II Aggregation, Causing Membrane Instability and Vesicle Budding. *Biophys. J.* **2015**, *108*, 1114–1124. [CrossRef]
91. Wiedemann, I.; Böttiger, T.; Bonelli, R.R.; Schneider, T.; Sahl, H.-G.; Martínez, B. Lipid II-Based Antimicrobial Activity of the Lantibiotic Plantaricin C. *Appl. Environ. Microbiol.* **2006**, *72*, 2809–2814. [CrossRef]

92. Brötz, H.; Bierbaum, G.; Reynolds, P.E.; Sahl, H.-G. The Lantibiotic Mersacidin Inhibits Peptidoglycan Biosynthesis at the Level of Transglycosylation. *Eur. J. Biochem.* **1997**, *246*, 193–199. [CrossRef]
93. Bruschi, M.; Pirri, G.; Giuliani, A.; Nicoletto, S.F.; Baster, I.; Scorciapino, M.A.; Casu, M.; Rinaldi, A.C. Synthesis, Characterization, Antimicrobial Activity and LPS-Interaction Properties of SB041, a Novel Dendrimeric Peptide with Antimicrobial Properties. *Peptides* **2010**, *31*, 1459–1467. [CrossRef]
94. Costa-Orlandi, C.B.; Sardi, J.C.O.; Pitangui, N.S.; de Oliveira, H.C.; Scorzoni, L.; Galeane, M.C.; Medina-Alarcón, K.P.; Melo, W.C.M.A.; Marcelino, M.Y.; Braz, J.D.; et al. Fungal Biofilms and Polymicrobial Diseases. *J. Fungi* **2017**, *3*, 22. [CrossRef]
95. Das, B.; Bhadra, R.K. (P)PpGpp Metabolism and Antimicrobial Resistance in Bacterial Pathogens. *Front. Microbiol.* **2020**, *11*, 2415. [CrossRef]
96. Luo, Y.; Song, Y. Mechanism of Antimicrobial Peptides: Antimicrobial, Anti-Inflammatory and Antibiofilm Activities. *Int. J. Mol. Sci.* **2021**, *22*, 11401. [CrossRef] [PubMed]
97. Hancock, R.E.W.; Alford, M.A.; Haney, E.F. Antibiofilm Activity of Host Defence Peptides: Complexity Provides Opportunities. *Nat. Rev. Microbiol.* **2021**, *19*, 786–797. [CrossRef] [PubMed]
98. Dostert, M.; Belanger, C.R.; Hancock, R.E.W. Design and Assessment of Anti-Biofilm Peptides: Steps toward Clinical Application. *J. Innate Immun.* **2019**, *11*, 193–204. [CrossRef]
99. Luong, H.X.; Thanh, T.T.; Tran, T.H. Antimicrobial Peptides—Advances in Development of Therapeutic Applications. *Life Sci.* **2020**, *260*, 118407. [CrossRef] [PubMed]
100. Dostert, M.; Trimble, M.; Hancock, R.E. Antibiofilm Peptides: Overcoming Biofilm-Related Treatment Failure. *RSC Adv.* **2021**, *11*, 2718–2728. [CrossRef]
101. Buda De Cesare, G.; Cristy, S.A.; Garsin, D.A.; Lorenz, M.C. Antimicrobial Peptides: A New Frontier in Antifungal Therapy. *mBio* **2020**, *11*, e02123–20. [CrossRef]
102. Firacative, C. Invasive Fungal Disease in Humans: Are We Aware of the Real Impact? *Mem. Inst. Oswaldo Cruz* **2020**, *115*, e200430. [CrossRef]
103. Brown, G.D.; Denning, D.W.; Gow, N.A.R.; Levitz, S.M.; Netea, M.G.; White, T.C. Hidden Killers: Human Fungal Infections. *Sci. Transl. Med.* **2012**, *4*, 165rv13. [CrossRef]
104. Candida Auris | Candida Auris | Fungal Diseases | CDC. Available online: https://www.cdc.gov/fungal/candida-auris/index.html (accessed on 31 October 2021).
105. Aranda, F.J.; Teruel, J.A.; Ortiz, A. Further Aspects on the Hemolytic Activity of the Antibiotic Lipopeptide Iturin A. *Biochim. Biophys. Acta* **2005**, *1713*, 51–56. [CrossRef]
106. Matejuk, A.; Leng, Q.; Begum, M.D.; Woodle, M.C.; Scaria, P.; Chou, S.-T.; Mixson, A.J. Peptide-Based Antifungal Therapies against Emerging Infections. *Drugs Future* **2010**, *35*, 197. [CrossRef]
107. Yamamoto, K.N.; Hirota, K.; Kono, K.; Takeda, S.; Sakamuru, S.; Xia, M.; Huang, R.; Austin, C.P.; Witt, K.L.; Tice, R.R. Characterization of Environmental Chemicals with Potential for DNA Damage Using Isogenic DNA Repair-Deficient Chicken DT40 Cell Lines. *Environ. Mol. Mutagenesis* **2011**, *52*, 547–561. [CrossRef]
108. Li, T.; Li, L.; Du, F.; Sun, L.; Shi, J.; Long, M.; Chen, Z. Activity and Mechanism of Action of Antifungal Peptides from Microorganisms: A Review. *Molecules* **2021**, *26*, 3438. [CrossRef]
109. Hayes, B.M.E.; Bleackley, M.R.; Wiltshire, J.L.; Anderson, M.A.; Traven, A.; van der Weerden, N.L. Identification and Mechanism of Action of the Plant Defensin NaD1 as a New Member of the Antifungal Drug Arsenal against *Candida albicans*. *Antimicrob. Agents Chemother.* **2013**, *57*, 3667–3675. [CrossRef]
110. Han, Q.; Wu, F.; Wang, X.; Qi, H.; Shi, L.; Ren, A.; Liu, Q.; Zhao, M.; Tang, C. The Bacterial Lipopeptide Iturins Induce Verticillium Dahliae Cell Death by Affecting Fungal Signalling Pathways and Mediate Plant Defence Responses Involved in Pathogen-Associated Molecular Pattern-Triggered Immunity. *Environ. Microbiol.* **2015**, *17*, 1166–1188. [CrossRef]
111. Edgerton, M.; Koshlukova, S.E.; Araujo, M.W.B.; Patel, R.C.; Dong, J.; Bruenn, J.A. Salivary Histatin 5 and Human Neutrophil Defensin 1 Kill Candida Albicans via Shared Pathways. *Antimicrob. Agents Chemother.* **2000**, *44*, 3310–3316. [CrossRef]
112. Trudel, J.; Grenier, J.; Potvin, C.; Asselin, A. Several Thaumatin-Like Proteins Bind to β-1,3-Glucans. *Plant Physiol.* **1998**, *118*, 1431–1438. [CrossRef]
113. Thevissen, K.; Warnecke, D.C.; François, I.E.J.A.; Leipelt, M.; Heinz, E.; Ott, C.; Zähringer, U.; Thomma, B.P.H.J.; Ferket, K.K.A.; Cammue, B.P.A. Defensins from Insects and Plants Interact with Fungal Glucosylceramides. *J. Biol. Chem.* **2004**, *279*, 3900–3905. [CrossRef]
114. Aeed, P.A.; Young, C.L.; Nagiec, M.M.; Elhammer, A.P. Inhibition of Inositol Phosphorylceramide Synthase by the Cyclic Peptide Aureobasidin A. *Antimicrob. Agents Chemother.* **2009**, *53*, 496–504. [CrossRef]
115. Walsh, T.J.; Giri, N. Pradimicins: A Novel Class of Broad-Spectrum Antifungal Compounds. *Eur. J. Clin. Microbiol. Infect. Dis.* **1997**, *16*, 93–97. [CrossRef]
116. Liu, Y.; Lu, J.; Sun, J.; Zhu, X.; Zhou, L.; Lu, Z.; Lu, Y. C16-Fengycin A Affect the Growth of *Candida albicans* by Destroying Its Cell Wall and Accumulating Reactive Oxygen Species. *Appl. Microbiol. Biotechnol.* **2019**, *103*, 8963–8975. [CrossRef]
117. Guzmán-de-Peña, D.L.; Correa-González, A.M.; Valdés-Santiago, L.; León-Ramírez, C.G.; Valdés-Rodríguez, S. In Vitro Effect of Recombinant Amaranth Cystatin (AhCPI) on Spore Germination, Mycelial Growth, Stress Response and Cellular Integrity of *Aspergillus niger* and *Aspergillus parasiticus*. *Mycology* **2015**, *6*, 168–175. [CrossRef]

118. Li, X.S.; Reddy, M.S.; Baev, D.; Edgerton, M. Candida Albicans Ssa1/2p Is the Cell Envelope Binding Protein for Human Salivary Histatin 5. *J. Biol. Chem.* **2003**, *278*, 28553–28561. [CrossRef] [PubMed]
119. Puri, S.; Edgerton, M. How Does It Kill?: Understanding the Candidacidal Mechanism of Salivary Histatin 5. *Eukaryot. Cell* **2014**, *13*, 958–964. [CrossRef] [PubMed]
120. Nakamura, I.; Ohsumi, K.; Takeda, S.; Katsumata, K.; Matsumoto, S.; Akamatsu, S.; Mitori, H.; Nakai, T. ASP2397 Is a Novel Natural Compound That Exhibits Rapid and Potent Fungicidal Activity against Aspergillus Species through a Specific Transporter. *Antimicrob. Agents Chemother.* **2019**, *63*, e02689-18. [CrossRef] [PubMed]
121. Yang, Y.; Chen, F.; Chen, H.-Y.; Peng, H.; Hao, H.; Wang, K.-J. A Novel Antimicrobial Peptide Scyreprocin from Mud Crab Scylla Paramamosain Showing Potent Antifungal and Anti-Biofilm Activity. *Front. Microbiol.* **2020**, *11*, 1589. [CrossRef]
122. Hacioglu, M.; Oyardi, O.; Bozkurt-Guzel, C.; Savage, P.B. Antibiofilm Activities of Ceragenins and Antimicrobial Peptides against Fungal-Bacterial Mono and Multispecies Biofilms. *J. Antibiot.* **2020**, *73*, 455–462. [CrossRef]
123. Singulani, J.d.L.; Oliveira, L.T.; Ramos, M.D.; Fregonezi, N.F.; Gomes, P.C.; Galeane, M.C.; Palma, M.S.; Fusco Almeida, A.M.; Mendes Giannini, M.J.S. The Antimicrobial Peptide MK58911-NH2 Acts on Planktonic, Biofilm, and Intramacrophage Cells of Cryptococcus Neoformans. *Antimicrob. Agents Chemother.* **2021**, *65*, e00904-20. [CrossRef]
124. Graham, C.E.; Cruz, M.R.; Garsin, D.A.; Lorenz, M.C. Enterococcus Faecalis Bacteriocin EntV Inhibits Hyphal Morphogenesis, Biofilm Formation, and Virulence of Candida Albicans. *Proc. Natl. Acad. Sci. USA* **2017**, *114*, 4507–4512. [CrossRef]
125. Roscetto, E.; Contursi, P.; Vollaro, A.; Fusco, S.; Notomista, E.; Catania, M.R. Antifungal and Anti-Biofilm Activity of the First Cryptic Antimicrobial Peptide from an Archaeal Protein against Candida Spp. Clinical Isolates. *Sci. Rep.* **2018**, *8*, 17570. [CrossRef]
126. Vilas Boas, L.C.P.; Campos, M.L.; Berlanda, R.L.A.; de Carvalho Neves, N.; Franco, O.L. Antiviral Peptides as Promising Therapeutic Drugs. *Cell. Mol. Life Sci.* **2019**, *76*, 3525–3542. [CrossRef]
127. Qureshi, A.; Thakur, N.; Tandon, H.; Kumar, M. AVPdb: A Database of Experimentally Validated Antiviral Peptides Targeting Medically Important Viruses. *Nucleic Acids Res.* **2014**, *42*, D1147–D1153. [CrossRef]
128. Vigant, F.; Santos, N.C.; Lee, B. Broad-Spectrum Antivirals against Viral Fusion. *Nat. Rev. Microbiol.* **2015**, *13*, 426–437. [CrossRef]
129. Jenssen, H.; Hamill, P.; Hancock, R.E.W. Peptide Antimicrobial Agents. *Clin. Microbiol. Rev.* **2006**, *19*, 491–511. [CrossRef]
130. Badani, H.; Garry, R.F.; Wimley, W.C. Peptide Entry Inhibitors of Enveloped Viruses: The Importance of Interfacial Hydrophobicity. *Biochim. Biophys. Acta* **2014**, *1838*, 2180–2197. [CrossRef]
131. LaBonte, J.; Lebbos, J.; Kirkpatrick, P. Enfuvirtide. *Nat. Rev. Drug Discov.* **2003**, *2*, 345–346. [CrossRef]
132. Eggink, D.; Bontjer, I.; de Taeye, S.W.; Langedijk, J.P.M.; Berkhout, B.; Sanders, R.W. HIV-1 Anchor Inhibitors and Membrane Fusion Inhibitors Target Distinct but Overlapping Steps in Virus Entry. *J. Biol. Chem.* **2019**, *294*, 5736–5746. [CrossRef]
133. Lok, S.-M.; Costin, J.M.; Hrobowski, Y.M.; Hoffmann, A.R.; Rowe, D.K.; Kukkaro, P.; Holdaway, H.; Chipman, P.; Fontaine, K.A.; Holbrook, M.R.; et al. Release of Dengue Virus Genome Induced by a Peptide Inhibitor. *PLoS ONE* **2012**, *7*, e50995. [CrossRef]
134. Cetina-Corona, A.; López-Sánchez, U.; Salinas-Trujano, J.; Méndez-Tenorio, A.; Barrón, B.L.; Torres-Flores, J. Peptides Derived from Glycoproteins H and B of Herpes Simplex Virus Type 1 and Herpes Simplex Virus Type 2 Are Capable of Blocking Herpetic Infection in Vitro. *Intervirology* **2016**, *59*, 235–242. [CrossRef]
135. Han, D.P.; Penn-Nicholson, A.; Cho, M.W. Identification of Critical Determinants on ACE2 for SARS-CoV Entry and Development of a Potent Entry Inhibitor. *Virology* **2006**, *350*, 15–25. [CrossRef]
136. Schütz, D.; Ruiz-Blanco, Y.B.; Münch, J.; Kirchhoff, F.; Sanchez-Garcia, E.; Müller, J.A. Peptide and Peptide-Based Inhibitors of SARS-CoV-2 Entry. *Adv. Drug Deliv. Rev.* **2020**, *167*, 47–65. [CrossRef]
137. Zhao, H.; To, K.K.W.; Sze, K.-H.; Yung, T.T.-M.; Bian, M.; Lam, H.; Yeung, M.L.; Li, C.; Chu, H.; Yuen, K.-Y. A Broad-Spectrum Virus- and Host-Targeting Peptide against Respiratory Viruses Including Influenza Virus and SARS-CoV-2. *Nat. Commun.* **2020**, *11*, 4252. [CrossRef]
138. Zhao, H.; To, K.K.W.; Lam, H.; Zhou, X.; Chan, J.F.-W.; Peng, Z.; Lee, A.C.Y.; Cai, J.; Chan, W.-M.; Ip, J.D.; et al. Cross-Linking Peptide and Repurposed Drugs Inhibit Both Entry Pathways of SARS-CoV-2. *Nat. Commun.* **2021**, *12*, 1517. [CrossRef]
139. Wang, C.; Wang, S.; Li, D.; Zhao, X.; Han, S.; Wang, T.; Zhao, G.; Chen, Y.; Chen, F.; Zhao, J.; et al. Lectin-like Intestinal Defensin Inhibits 2019-NCoV Spike Binding to ACE2. *BioRxiv* **2020**. [CrossRef]
140. Hsu, C.-H.; Chen, C.; Jou, M.-L.; Lee, A.Y.-L.; Lin, Y.-C.; Yu, Y.-P.; Huang, W.-T.; Wu, S.-H. Structural and DNA-Binding Studies on the Bovine Antimicrobial Peptide, Indolicidin: Evidence for Multiple Conformations Involved in Binding to Membranes and DNA. *Nucleic Acids Res.* **2005**, *33*, 4053–4064. [CrossRef]
141. Scott, M.G.; Davidson, D.J.; Gold, M.R.; Bowdish, D.; Hancock, R.E.W. The Human Antimicrobial Peptide LL-37 Is a Multifunctional Modulator of Innate Immune Responses. *J. Immunol.* **2002**, *169*, 3883–3891. [CrossRef]
142. Wohlford-Lenane, C.L.; Meyerholz, D.K.; Perlman, S.; Zhou, H.; Tran, D.; Selsted, M.E.; McCray, P.B. Rhesus Theta-Defensin Prevents Death in a Mouse Model of Severe Acute Respiratory Syndrome Coronavirus Pulmonary Disease. *J. Virol.* **2009**, *83*, 11385–11390. [CrossRef]
143. Mahendran, A.S.K.; Lim, Y.S.; Fang, C.-M.; Loh, H.-S.; Le, C.F. The Potential of Antiviral Peptides as COVID-19 Therapeutics. *Front. Pharmacol.* **2020**, *11*, 1475. [CrossRef] [PubMed]
144. Wang, W.; Ballatori, N. Endogenous Glutathione Conjugates: Occurrence and Biological Functions. *Pharmacol. Rev.* **1998**, *50*, 335–356. [PubMed]
145. Lu, S.C. Glutathione Synthesis. *Biochim. Biophys. Acta* **2013**, *1830*, 3143–3153. [CrossRef] [PubMed]

146. Fraternale, A.; Paoletti, M.F.; Casabianca, A.; Nencioni, L.; Garaci, E.; Palamara, A.T.; Magnani, M. GSH and Analogs in Antiviral Therapy. *Mol. Asp. Med.* **2009**, *30*, 99–110. [CrossRef] [PubMed]
147. Polonikov, A. Endogenous Deficiency of Glutathione as the Most Likely Cause of Serious Manifestations and Death in COVID-19 Patients. *ACS Infect. Dis.* **2020**, *6*, 1558–1562. [CrossRef]
148. Cacciatore, I.; Cornacchia, C.; Pinnen, F.; Mollica, A.; Di Stefano, A. Prodrug Approach for Increasing Cellular Glutathione Levels. *Molecules* **2010**, *15*, 1242–1264. [CrossRef]
149. Ghezzi, P. Redox Regulation of Immunity and the Role of Small Molecular Weight Thiols. *Redox Biol.* **2021**, *44*, 102001. [CrossRef]
150. Fraternale, A.; Zara, C.; De Angelis, M.; Nencioni, L.; Palamara, A.T.; Retini, M.; Di Mambro, T.; Magnani, M.; Crinelli, R. Intracellular Redox-Modulated Pathways as Targets for Effective Approaches in the Treatment of Viral Infection. *Int. J. Mol. Sci.* **2021**, *22*, 3603. [CrossRef]
151. Checconi, P.; De Angelis, M.; Marcocci, M.E.; Fraternale, A.; Magnani, M.; Palamara, A.T.; Nencioni, L. Redox-Modulating Agents in the Treatment of Viral Infections. *Int. J. Mol. Sci.* **2020**, *21*, 4084. [CrossRef]
152. Tonk, M.; Růžek, D.; Vilcinskas, A. Compelling Evidence for the Activity of Antiviral Peptides against SARS-CoV-2. *Viruses* **2021**, *13*, 912. [CrossRef]
153. Shanmugaraj, B.; Bulaon, C.J.I.; Malla, A.; Phoolcharoen, W. Biotechnological Insights on the Expression and Production of Antimicrobial Peptides in Plants. *Molecules* **2021**, *26*, 4032. [CrossRef]
154. Rezende, S.B.; Oshiro, K.G.N.; Júnior, N.G.O.; Franco, O.L.; Cardoso, M.H. Advances on Chemically Modified Antimicrobial Peptides for Generating Peptide Antibiotics. *Chem. Commun.* **2021**, *57*, 3793. [CrossRef]
155. Batoni, G.; Maisetta, G.; Esin, S. Antimicrobial Peptides and Their Interaction with Biofilms of Medically Relevant Bacteria. *Biochim. Biophys. Acta* **2016**, *1858*, 1044–1060. [CrossRef]
156. Grassi, L.; Pompilio, A.; Kaya, E.; Rinaldi, A.C.; Sanjust, E.; Maisetta, G.; Crabbé, A.; Di Bonaventura, G.; Batoni, G.; Esin, S. The Anti-Microbial Peptide (Lin-SB056-1)2-K Reduces Pro-Inflammatory Cytokine Release through Interaction with Pseudomonas Aeruginosa Lipopolysaccharide. *Antibiotics* **2020**, *9*, 585. [CrossRef]
157. Grassi, L.; Batoni, G.; Ostyn, L.; Rigole, P.; Van den Bossche, S.; Rinaldi, A.C.; Maisetta, G.; Esin, S.; Coenye, T.; Crabbé, A. The Antimicrobial Peptide Lin-SB056-1 and Its Dendrimeric Derivative Prevent Pseudomonas Aeruginosa Biofilm Formation in Physiologically Relevant Models of Chronic Infections. *Front. Microbiol.* **2019**, *10*, 198. [CrossRef] [PubMed]
158. Wei, H.; Xie, Z.; Tan, X.; Guo, R.; Song, Y.; Xie, X.; Wang, R.; Li, L.; Wang, M.; Zhang, Y. Temporin-Like Peptides Show Antimicrobial and Anti-Biofilm Activities against Streptococcus Mutans with Reduced Hemolysis. *Molecules* **2020**, *25*, 5724. [CrossRef] [PubMed]
159. Rinaldi, A.C.; Conlon, J.M. The Temporins. In *Hanbook of Biologically Active Peptides*; Kastin, A.J., Ed.; Elsevier: San Diego, CA, USA, 2013; pp. 400–406.
160. Welch, N.G.; Li, W.; Hossain, M.A.; Separovic, F.; O'Brien-Simpson, N.M.; Wade, J.D. (Re)Defining the Proline-Rich Antimicrobial Peptide Family and the Identification of Putative New Members. *Front. Chem.* **2020**, *8*, 607769. [CrossRef] [PubMed]
161. Holfeld, L.; Herth, N.; Singer, D.; Hoffmann, R.; Knappe, D. Immunogenicity and Pharmacokinetics of Short, Proline-Rich Antimicrobial Peptides. *Future Med. Chem.* **2015**, *7*, 1581–1596. [CrossRef]
162. Holfeld, L.; Knappe, D.; Hoffmann, R. Proline-Rich Antimicrobial Peptides Show a Long-Lasting Post-Antibiotic Effect on Enterobacteriaceae and Pseudomonas Aeruginosa. *J. Antimicrob. Chemother.* **2018**, *73*, 933–941. [CrossRef]
163. Nuri, R.; Shprung, T.; Shai, Y. Defensive Remodeling: How Bacterial Surface Properties and Biofilm Formation Promote Resistance to Antimicrobial Peptides. *Biochim. Biophys. Acta* **2015**, *1848*, 3089–3100. [CrossRef]
164. Band, V.I.; Weiss, D.S. Mechanisms of Antimicrobial Peptide Resistance in Gram-Negative Bacteria. *Antibiotics* **2015**, *4*, 18–41. [CrossRef]
165. Joo, H.-S.; Fu, C.-I.; Otto, M. Bacterial Strategies of Resistance to Antimicrobial Peptides. *Philos. Trans. R. Soc. Lond. B Biol. Sci.* **2016**, *371*, 20150292. [CrossRef]
166. Magana, M.; Pushpanathan, M.; Santos, A.L.; Leanse, L.; Fernandez, M.; Ioannidis, A.; Giulianotti, M.A.; Apidianakis, Y.; Bradfute, S.; Ferguson, A.L.; et al. The Value of Antimicrobial Peptides in the Age of Resistance. *Lancet Infect. Dis.* **2020**, *20*, e216–e230. [CrossRef]
167. Nuding, S.; Gersemann, M.; Hosaka, Y.; Konietzny, S.; Schaefer, C.; Beisner, J.; Schroeder, B.O.; Ostaff, M.J.; Saigenji, K.; Ott, G.; et al. Gastric Antimicrobial Peptides Fail to Eradicate Helicobacter Pylori Infection Due to Selective Induction and Resistance. *PLoS ONE* **2013**, *8*, e73867. [CrossRef]
168. Annunziato, G.; Costantino, G. Antimicrobial Peptides (AMPs): A Patent Review (2015–2020). *Expert Opin. Ther. Pat.* **2020**, *30*, 931–947. [CrossRef]
169. Guan, Q.; Huang, S.; Jin, Y.; Campagne, R.; Alezra, V.; Wan, Y. Recent Advances in the Exploration of Therapeutic Analogues of Gramicidin S, an Old but Still Potent Antimicrobial Peptide. *J. Med. Chem.* **2019**, *62*, 7603–7617. [CrossRef]
170. Rodríguez-Santiago, J.; Cornejo-Juárez, P.; Silva-Sánchez, J.; Garza-Ramos, U. Polymyxin Resistance in Enterobacterales: Overview and Epidemiology in the Americas. *Int. J. Antimicrob. Agents* **2021**, *16*, 106426. [CrossRef]
171. Gogry, F.A.; Siddiqui, M.T.; Sultan, I.; Haq, Q.M.R. Current Update on Intrinsic and Acquired Colistin Resistance Mechanisms in Bacteria. *Front. Med.* **2021**, *8*, 677720. [CrossRef]
172. Huang, H.W. DAPTOMYCIN, Its Membrane-Active Mechanism vs. That of Other Antimicrobial Peptides. *Biochim. Biophys. Acta Biomembr.* **2020**, *1862*, 183395. [CrossRef]

173. Liu, W.-T.; Chen, E.-Z.; Yang, L.; Peng, C.; Wang, Q.; Xu, Z.; Chen, D.-Q. Emerging Resistance Mechanisms for 4 Types of Common Anti-MRSA Antibiotics in Staphylococcus Aureus: A Comprehensive Review. *Microb. Pathog.* **2021**, *156*, 104915. [CrossRef]
174. Fariñas, M.C.; González-Rico, C.; Fernández-Martínez, M.; Fortún, J.; Escudero-Sanchez, R.; Moreno, A.; Bodro, M.; Muñoz, P.; Valerio, M.; Montejo, M.; et al. Oral Decontamination with Colistin plus Neomycin in Solid Organ Transplant Recipients Colonized by Multidrug-Resistant Enterobacterales: A Multicentre, Randomized, Controlled, Open-Label, Parallel-Group Clinical Trial. *Clin. Microbiol. Infect.* **2021**, *27*, 856–863. [CrossRef]
175. Jovcic, B.; Novovic, K.; Dekic, S.; Hrenovic, J. Colistin Resistance in Environmental Isolates of Acinetobacter Baumannii. *Microb. Drug Resist.* **2021**, *27*, 328–336. [CrossRef]
176. Chow, H.Y.; Po, K.H.L.; Jin, K.; Qiao, G.; Sun, Z.; Ma, W.; Ye, X.; Zhou, N.; Chen, S.; Li, X. Establishing the Structure-Activity Relationship of Daptomycin. *ACS Med. Chem. Lett.* **2020**, *11*, 1442–1449. [CrossRef]
177. Blasco, P.; Zhang, C.; Chow, H.Y.; Chen, G.; Wu, Y.; Li, X. An Atomic Perspective on Improving Daptomycin's Activity. *Biochim. Biophys. Acta Gen. Subj.* **2021**, *1865*, 129918. [CrossRef]
178. Jiang, Y.; Chen, Y.; Song, Z.; Tan, Z.; Cheng, J. Recent Advances in Design of Antimicrobial Peptides and Polypeptides toward Clinical Translation. *Adv. Drug. Deliv. Rev.* **2021**, *170*, 261–280. [CrossRef]
179. Provenzani, A.; Hospodar, A.R.; Meyer, A.L.; Leonardi Vinci, D.; Hwang, E.Y.; Butrus, C.M.; Polidori, P. Multidrug-Resistant Gram-Negative Organisms: A Review of Recently Approved Antibiotics and Novel Pipeline Agents. *Int. J. Clin. Pharm.* **2020**, *42*, 1016–1025. [CrossRef]
180. Jabbour, J.-F.; Sharara, S.L.; Kanj, S.S. Treatment of Multidrug-Resistant Gram-Negative Skin and Soft Tissue Infections. *Curr. Opin. Infect. Dis.* **2020**, *33*, 146–154. [CrossRef]
181. Farrell, D.J.; Robbins, M.; Rhys-Williams, W.; Love, W.G. In Vitro Activity of XF-73, a Novel Antibacterial Agent, against Antibiotic-Sensitive and -Resistant Gram-Positive and Gram-Negative Bacterial Species. *Int. J. Antimicrob. Agents* **2010**, *35*, 531–536. [CrossRef]
182. MacLean, R.C. Assessing the Potential for Staphylococcus Aureus to Evolve Resistance to XF-73. *Trends Microbiol.* **2020**, *28*, 432–435. [CrossRef]

MDPI
St. Alban-Anlage 66
4052 Basel
Switzerland
Tel. +41 61 683 77 34
Fax +41 61 302 89 18
www.mdpi.com

International Journal of Molecular Sciences Editorial Office
E-mail: ijms@mdpi.com
www.mdpi.com/journal/ijms

www.ingramcontent.com/pod-product-compliance
Lightning Source LLC
LaVergne TN
LVHW070149170726
843515LV00007B/1881
* 9 7 8 3 0 3 6 5 4 3 9 6 3 *